Particle Accelerator Physics II

Helmut Wiedemann

Particle
Accelerator Physics II

Nonlinear and Higher-Order Beam Dynamics

With 118 Figures

 Springer

Professor Dr. Helmut Wiedemann

Applied Physics Department and Stanford Synchrotron Radiation Laboratory
Stanford University, Stanford, CA 94309-0210, USA

ISBN 3-540-57564-2 Springer-Verlag Berlin Heidelberg New York
ISBN 0-387-57564-2 Springer-Verlag New York Berlin Heidelberg

Library of Congress Cataloging-in-Publication Data. Wiedemann, Helmut, 1938- Particle accelerator physics II: nonlinear and higher-order beam dynamics / Helmut Wiedemann. p. cm. Includes bibliographical references and indexes. ISBN 3-540-57564-2 (Berlin: acid-free paper). – ISBN 0-387-57564-2 (New York: acid-free paper) 1. Beam dynamics. 2. Particle accelerators–Design and construction. I. Title. QC793.3.B4W55 1995 539.7'3– dc20 94-35447

This work is subject to copyright. All rights are reserved, whether the whole or part of the material is concerned, specifically the rights of translation, reprinting, reuse of illustrations, recitation, broadcasting, reproduction on microfilm or in any other way, and storage in data banks. Duplication of this publication or parts thereof is permitted only under the provisions of the German Copyright Law of September 9, 1965, in its current version, and permission for use must always be obtained from Springer-Verlag. Violations are liable for prosecution under the German Copyright Law.

© Springer-Verlag Berlin Heidelberg 1995
Printed in Germany

The use of general descriptive names, registered names, trademarks, etc. in this publication does not imply , even in the absence of a specific statement, that such names are exempt from the relevant protective laws and regulations and therefore free for general use.

Typesetting: Camera-ready copy from the author using a Springer T$_{\mathrm{E}}$X macro package
SPIN 10070489   54/3144 - 5 4 3 2 1 0 - Printed on acid-free paper

Preface

Q C
793
.3
B4
W55
1995
PHYS

This text is a continuation of the first volume of "Particle Accelerator Physics" on "Basic Principles and Linear Beam Dynamics". While the first volume was written as an introductory overview into beam dynamics, it does not include more detailed discussion of nonlinear and higher-order beam dynamics or the full theory of synchrotron radiation from relativistic electron beams. Both issues are, however, of fundamental importance for the design of modern particle accelerators.

In this volume, beam dynamics is formulated within the realm of Hamiltonian dynamics, leading to the description of multiparticle beam dynamics with the Vlasov equation and including statistical processes with the Fokker Planck equation. Higher-order perturbations and aberrations are discussed in detail, including Hamiltonian resonance theory and higher-order beam dynamics. The discussion of linear beam dynamics in Vol. I is completed here with the derivation of the general equation of motion, including kinematic terms and coupled motion. To build on the theory of longitudinal motion in Vol. I, the interaction of a particle beam with the rf system, including beam loading, higher-order phase focusing, and the combination of acceleration and transverse focusing, is discussed. The emission of synchrotron radiation greatly affects the beam quality of electron or positron beams and we therefore derive the detailed theory of synchrotron radiation, including spatial and spectral distribution as well as properties of polarization. The results of this derivation are then applied to insertion devices such as undulator and wiggler magnets. Beam stability in linear and circular accelerators is compromized by the interaction of the electrical charge in the beam with its environment, leading to instabilities. Theoretical models of such instabilities are discussed and scaling laws for the onset and rise time of instabilities are derived.

Although this text builds upon Vol. I, it relates to it only as a reference for basic issues of accelerator physics, which could be obtained as well elsewhere. This volume is aimed specifically at those students, engineers, and scientists who desire to aqcuire a deeper knowledge of particle beam dynamics in accelerators. To facilitate the use of this text as a reference, many of the more important results are emphasized by a frame for quick detection. Consistent with Vol. I we use the cgs system of units. However, for the convenience of the reader used to the system of international units, conversion factors have been added whenever such conversion is necessary, e.g. whenever electrical or magnetic units are used. These conversion factors are enclosed in square brackets like $[\sqrt{4\pi\epsilon_o}]$ and should be ignored by those who use formulas in the cgs system. The conversion factors are easy to identify since they include only the constants $c, \pi, \epsilon_o, \mu_o$ and should therefore not be mixed up with other factors in square brackets. For the convenience of the reader, the sources of these conversion factors are compiled in the Appendix together with other useful tools.

I would like to thank Joanne Kwong, who typed the initial draft of this text and introduced me to the intricacies of TEX typesetting, and to my students who guided me through numerous inquisitive questions. Partial support by the Division of Basic Energy Sciences in the Department of Energy through the Stanford Synchrotron Radiation Laboratory in preparing this text is gratefully acknowledged. Special thanks to Dr. C. Maldonado for painstakingly reading the manuscript and to the editorial staff of Springer Verlag for support during the preparation of this text.

Palo Alto, California   *Helmut Wiedemann*
March 1994

Contents

Contents to Volume I

1. Hamiltonian Formulation of Beam Dynamics

Particles in electromagnetic fields behave like oscillators and are conservative systems as long as we ignore statistical effects like the emission of quantized photons in the form of synchrotron radiation. Even in such cases particles can be treated as conservative systems on average. In particular, particle motion in beam transport systems and circular accelerators is oscillatory and can be described in terms of perturbed oscillators. The *Hamiltonian formalism* provides a powerful tool to analyze particle motion and define conditions of stability and onset of instability. In this chapter we will derive the *Lagrangian* and *Hamiltonian* formalism with special consideration to particle beam dynamics.

1.1 Hamiltonian Formalism

Like any other mechanical system, particle beam dynamics in the presence of external electromagnetic fields can be described and studied very generally through the *Hamiltonian formalism*. The motion of particles in beam transport systems, expressed in normalized coordinates, is that of a harmonic oscillator and deviations caused by nonlinear restoring forces appear as perturbations of the harmonic oscillation. Such systems have been studied extensively in the past and powerful mathematical tools have been developed to describe the dynamics of harmonic oscillators under the influence of perturbations. Of special importance is the Hamiltonian formalism which we will apply to the dynamics of charged particles. Although this theory is well documented in many text books, for example in [1.1, 2], we will rederive the Hamiltonian theory with special attention to the application for charged particle dynamics.

1.1.1 Lagrange Equations

Based on a general principle of mechanics, a function exists for any mechanical system, called the *action*, which assumes a minimum value for real motion. In beam dynamics the mechanical systems are charged particles. The statement of the principle then is that the action for the particle to move from point P_o at time t_o to point P_1 at time t_1 is a minimum for the real trajectory. This *variational principle* is expressed by

$$\delta \int_{t_o}^{t_1} L \, dt = 0 \tag{1.1}$$

which is also called the *Hamiltonian variational principle* and the function L is called the *Lagrange function*. For a system with n degrees of freedom the Lagrange function depends on the generalized coordinates q_i, velocities $\dot{q}_i$ and time t

$$L = L(q_i, \dot{q}_i, t), \tag{1.2}$$

where $i = 1, 2, 3, \dots n$. The *time variable* need not be the real time but is the independent variable used to describe the system. In accelerator physics this can be the time but also the position along the beam transport line expressed in units of length or betatron oscillation phase. It is part of the beauty of the Hamiltonian formalism that it allows the use of the most convenient variables for any particular system.

We perform now the variation (1.1) by integrating the *Lagrange function* along different but close paths. In doing so we assume that both trajectories start and end at the same point as indicated in Fig. 1.1.

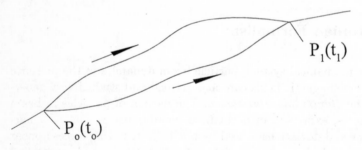

Fig. 1.1. Variation of trajectories

We also assume that the Lagrange function and its derivatives are continuous functions and that it does not explicitly depend on the time $\partial L/\partial t = 0$. In this case the variation and the integral can be exchanged in (1.1) and we get for the integrand.

$$\delta L = \sum_i \frac{\partial L}{\partial q_i} \delta q_i + \sum_i \frac{\partial L}{\partial \dot{q}_i} \delta \dot{q}_i. \tag{1.3}$$

The integral of the second term can be modified by partial integration

$$\int_{t_o}^{t_1} \frac{\partial L}{\partial \dot{q}_i} \delta \dot{q}_i = \int_{t_o}^{t_1} \frac{\partial L}{\partial \dot{q}_i} \frac{d}{dt} \delta q_i = \frac{\partial L}{\partial \dot{q}_i} \delta q_i \Big|_{t_o}^{t_1} - \int_{t_o}^{t_1} \frac{d}{dt} \frac{\partial L}{\partial \dot{q}_i} \delta q_i \, dt. \tag{1.4}$$

The first term on the right-hand side vanishes, since the variations δq_i vanish at the start and end point of the integration path $\delta q_i(t_o) = \delta q_i(t_1) = 0$. With (1.1,2) we then get

$$\delta \int L \, \mathrm{d}t \ = \ \int \sum_i \left(\frac{\partial L}{\partial q_i} - \frac{\mathrm{d}}{\mathrm{d}t} \frac{\partial L}{\partial \dot{q}} \right) \delta q_i \, \mathrm{d}t \ = 0 \,. \tag{1.5}$$

This integral must vanish for any arbitrary path and therefore, the integrand must vanish individually for each coordinate q_i resulting in the well-known *Lagrange equations*

$$\frac{\partial L}{\partial q_i} - \frac{\mathrm{d}}{\mathrm{d}t} \frac{\partial L}{\partial \dot{q}} \ = \ 0 \,. \tag{1.6}$$

The Lagrange function cannot be derived from more basic principles but must be formulated as a creative act to describe physics by a mathematical formula. For further general discussions on this point and actual formulation of Lagrange functions the reader is referred to relevant textbooks.

Following the discussion by *Landau* and *Lifshitz* [1.3] we can in particular formulate the Lagrange function for a charged particle in an electromagnetic field. The action is composed of two parts, the pure mechanical motion of a particle and the interaction with the field. In a field free environment the only physical parameter to describe a particle is its mass m and we set for the action

$$- \int mc^2 \, \mathrm{d}t^* \ = \ - \int mc^2 \sqrt{1 - \beta^2} \, \mathrm{d}t \,, \tag{1.7}$$

where t^* is the time in the particle system, $c^2 \beta^2 = \dot{\mathbf{q}}^2$ and the constant factor c for the speed of light is added to make the expression dimensionally convenient. The interaction with the electromagnetic field is effected only by the charge of a particle and the strength of the magnetic field but not by the mass and we therefore express the action in *four vector notation*

$$- \frac{e}{c} \int A_i \, \mathrm{d}q^i \ = \ e \int \left(\frac{[c]}{c} \mathbf{A} \, \mathrm{d}\mathbf{q} - \phi \, \mathrm{d}t \right) \,, \tag{1.8}$$

where $\mathbf{A}$ is the magnetic *vector potential* and ϕ the *electrical potential*. The total action then can be expressed by

$$\int (-mc^2 \sqrt{1 - \beta^2} + [c] \frac{e}{c} \mathbf{A} \, \dot{\mathbf{q}} - e \, \phi) \, \mathrm{d}t \,. \tag{1.9}$$

The *Lagrange function* for a charged particle in an electromagnetic field is identical to the integrand

$$L \ = \ -mc^2 \sqrt{1 - \beta^2} + [c] \frac{e}{c} \mathbf{A} \, \dot{\mathbf{q}} - e \, \phi \,. \tag{1.10}$$

From the derivative $\partial L/\partial \dot{q}_i$ we get for a particle in field free space with $c\partial \beta/\partial \dot{q}_i = \partial/\partial \dot{q}_i \sqrt{\sum_j \dot{q}_j^2}$

$$\frac{\partial L}{\partial \dot{q}_i} = \frac{m \dot{q}_i}{\sqrt{1 - \beta^2}} \,. \tag{1.11}$$

The significance of this equation is that the right-hand side agrees with the momentum of a particle in field free space. Generalizing (1.11) the components of the *canonical momentum* p_i of a mechanical system are defined by

$$p_i = \frac{\partial L}{\partial \dot{q}_i} \,. \tag{1.12}$$

For a particle moving in an electromagnetic field with the vector potential **A** the *canonical momentum* is therefore, in vector form

$$\boxed{\mathbf{p} = \frac{m \dot{\mathbf{q}}}{\sqrt{1 - \beta^2}} + [c] \frac{e}{c} \mathbf{A} = m\gamma \dot{\mathbf{q}} + [c] \frac{e}{c} \mathbf{A} \,.} \tag{1.13}$$

In the presence of an electromagnetic field the canonical momentum differs from the mechanical momentum by an additive field term. In most cases of transverse beam dynamics we use transverse magnetic fields and the additional term vanishes since purely transverse magnetic fields are derived from a vector potential which has only a longitudinal component. The definition of the transverse momentum is therefore not affected by the magnetic field. However, we must be careful when longitudinal fields are present as is the case in a solenoid magnet.

1.1.2 Hamiltonian Equations

While the Lagrange function depends on the coordinates q_i and velocities $\dot{q}_i$ it becomes desirable to describe a mechanical system by the coordinates and momenta. This can be achieved by a *Legendre transformation* derived from the *generating function*

$$H(q_i, p_i) = \sum \dot{q}_i \, p_i - L(q_i, \dot{q}_i) \,. \tag{1.14}$$

The function $H(q_i, p_i)$ is called the *Hamiltonian function* or short the *Hamiltonian* of the system. For systems that do not explicitly depend on the time, as we assume here, the Hamiltonian is a constant of motion. To show this we calculate the variation of the Lagrange function with respect to the time

$$\frac{\mathrm{d}L}{\mathrm{d}t} = \sum_i \frac{\partial L}{\partial q_i} \dot{q}_i + \sum_i \frac{\partial L}{\partial \dot{q}_i} \ddot{q}_i \,. \tag{1.15}$$

Using the Lagrange equations (1.6) we set $\frac{\partial L}{\partial q_i} = \frac{d}{dt} \frac{\partial L}{\partial \dot{q}_i}$ and get with (1.12)

$$\frac{d}{dt} \left(\sum_i \dot{q}_i p_i - L \right) = \frac{d}{dt} H(q_i, p_i) = 0 \tag{1.16}$$

expressing the constancy of the Hamiltonian.

From (1.14) as well as from the properties of the Lagrange function we can derive general equations of motion for the system. To do this we calculate the variation of the Hamiltonian δH from (1.14) and get with (1.16)

$$dH = \sum_i (\dot{q}_i \, dp_i + p_i \, d\dot{q}_i) - \sum_i \left(\frac{\partial L}{\partial q_i} dq_i + \frac{\partial L}{\partial \dot{q}_i} d\dot{q}_i \right)$$

which becomes after some manipulations using the Lagrange equation and the definition of the momentum (1.12)

$$dH = \sum_i (\dot{q}_i \, dp_i - \dot{p}_i \, dq_i). \tag{1.17}$$

From this equation we obtain directly the *Hamiltonian equations* of motion for a mechanical system

$$\boxed{\begin{aligned} \frac{\partial H}{\partial q_i} &= -\dot{p}_i \,, \\ \frac{\partial H}{\partial p_i} &= +\dot{q}_i \,. \end{aligned}} \tag{1.18}$$

The coordinates (q_i, p_i) as defined above are called *canonical variables* or *canonical coordinates*. In general any pair of variables (q, p), for which a function $H(q, p)$ exists with the property (1.18), are canonical variables.

The Hamiltonian for a particle in an electromagnetic field can now be derived by introducing the generalized momenta $\partial L / \partial \dot{q}_i$ from (1.12,13) into (1.14) and we get with the expression (1.10) for the Lagrange function the Hamiltonian

$$H(q_i, p_i) = \sum_i \frac{m \dot{q}_i^2}{\sqrt{1 - \beta^2}} + mc^2 \sqrt{1 - \beta^2} + e \phi. \tag{1.19}$$

With $c^2 \beta^2 = \sum_i \dot{q}_i^2$ we get

$$(H - e\phi)^2 = \frac{m^2 c^4}{1 - \beta^2}. \tag{1.20}$$

and the Hamiltonian for a particle in an electromagnetic field becomes finally with (1.13)

$$\boxed{(H - e\phi)^2 \ = \ m^2 c^4 + (c\mathbf{p} - [c]\,e\,\mathbf{A})^2\,.}$$ (1.21)

Since the fields are functions of the coordinates only, we have finally expressed the Hamiltonian in terms of coordinates and momenta alone.

With the knowledge of the electromagnetic field $(\phi, \mathbf{A})$ we may now express the Lagrange function (1.10), derive the generalized momenta (1.13), and the equations of motion (1.18) for a charged particle in an electromagnetic field. The problem solving power of the Hamiltonian formalism, however, is much greater than what has become obvious so far. Without solving any differential equation it will become possible to extract important information about the motion of a particle in the presence of electromagnetic fields, its invariants and possible instabilities and resonances.

1.1.3 Canonical Transformations

The solution of the equations of motion become greatly simplified in such cases, where the Hamiltonian does not depend on one or more of the coordinates or momenta. In this case one or more of the Hamiltonian equations (1.18) are zero and the corresponding *conjugate variables* are *constants of motion*. Of particular interest for particle dynamics or harmonic oscillators are the cases where the Hamiltonian does not depend on say the coordinate q_i but only on the momenta p_i. In this case we have

$$H = H(q_1, \ \dots \ q_{i-1}, q_{i+1} \ \cdots \ , p_1, p_2 \ \dots, p_i, \dots)$$

and the first Hamiltonian equation becomes

$$\frac{\partial H}{\partial q_i} \ = \ -\dot{p}_i \ = \ 0 \qquad \text{or} \qquad p_i = \text{const}\,.$$ (1.22)

Coordinates q_i which do not appear in the Hamiltonian are called *cyclic coordinates* and their conjugate momenta are constants of motion. From the second Hamiltonian equation we get

$$\frac{\partial H}{\partial p_i} \ = \ \dot{q}_i \ = \ \text{const}\,,$$ (1.23)

since $p_i = \text{const}$, which can be integrated immediately for

$$q_i(t) \ = \ \omega_i t + c_i\,,$$ (1.24)

where $\omega_i = \dot{q}_i$ is a constant defined by (1.23) and c_i is the integration constant. It is obvious that the complexity of a mechanical system can be greatly reduced if by a proper choice of canonical variables some or all dependence of the Hamiltonian on space coordinates can be eliminated. We will derive the formalism that allows the transformation of canonical coordinates into new ones, where some of them might be cyclic.

For mechanical systems which allow in principle a formulation in terms of cyclic variables, we need to derive rules to transform one set of variables to another set, while preserving their property of being conjugate variables appropriate to formulate the Hamiltonian for the system. In other words, the coordinate transformation must preserve the variational principle (1.1). Such transformations are called *canonical transformations*.

The transformation from the coordinates (q_i, p_i, t) to the new coordinates $(\bar{q}_i, \bar{p}_i, t)$ can be expressed by the $2N$ equations

$$
\begin{aligned}
\bar{q}_k &= f_k(q_i, p_i, t)\,, \\
\bar{p}_k &= g_k(q_i, p_i, t)\,,
\end{aligned}
\tag{1.25}
$$

where N is the number of linearly independent coordinates $k = 1, 2 \ldots N$, and the new coordinates $\bar{q}_k$, $\bar{p}_k$ are functions of the coordinates q_i, p_i with $i = 1, 2, \ldots N$. In both cases the variational principle must hold and we have with (1.14)

$$
\begin{aligned}
\delta \int \left(\sum_k \dot{q}_k\, p_k - H \right) \mathrm{d}t &= 0\,, \\
\delta \int \left(\sum_k \dot{\bar{q}}_k\, \bar{p}_k - \bar{H} \right) \mathrm{d}t &= 0\,.
\end{aligned}
\tag{1.26}
$$

The new Hamiltonian $\bar{H}$ need not be the same as the old Hamiltonian H nor need both integrants be the same. Both integrants can differ, however, only by a total time derivative of an otherwise arbitrary function G

$$
\sum_k \dot{q}_k\, p_k - H = \sum_k \dot{\bar{q}}_k \bar{p}_k - \bar{H} + \frac{\mathrm{d}G}{\mathrm{d}t}\,.
\tag{1.27}
$$

After integration $\int \frac{\mathrm{d}G}{\mathrm{d}t}\,\mathrm{d}t$ becomes a constant and the variation of the integral obviously vanishes under the variational principle (Fig. 1.1). The arbitrary function G is called the *generating function* and may depend on some or all of the old and new variables

$$
G = G(q_k, \bar{q}_k, p_k, \bar{p}_k, t)\,.
\tag{1.28}
$$

The generating functions are functions of only $2N$ variables. Of the $4N$ variables only $2N$ are independent because of the $2N$ transformation equations (1.25). We may now choose any two of four variables to be independent keeping only in mind that one must be an old and one a new variable. Depending on our choice for the independent variables the generating function may have one of four forms

$$
\begin{aligned}
G_1 &= G_1(q, \bar{q}, t)\,, & G_3 &= G_3(p, \bar{q}, t)\,, \\
G_2 &= G_2(q, \bar{p}, t)\,, & G_4 &= G_4(p, \bar{p}, t)\,,
\end{aligned}
\tag{1.29}
$$

where we have set $q = (q_1, q_2, \ldots q_N)$ etc. We take, for example, the generating function G_1, insert the total time derivative in (1.27) and get after some sorting

$$\sum_k \dot{q}_k \left(p_k - \frac{\partial G_1}{\partial q_k} \right) - \sum_k \dot{\bar{q}}_k \left(\bar{p}_k + \frac{\partial G_1}{\partial \bar{q}_k} \right) - \left(H - \bar{H} + \frac{\partial G_1}{\partial t} \right) = 0 . \quad (1.30)$$

Both old and new variables are independent and the expressions in the brackets must therefore vanish separately leading to the defining equations

$$p_k = \frac{\partial G_1}{\partial q_k} ,$$

$$\bar{p}_k = -\frac{\partial G_1}{\partial \bar{q}_k} , \qquad (1.31)$$

$$H = \bar{H} - \frac{\partial G_1}{\partial t} .$$

Variables for which (1.31) hold are called *canonical variables* and the transformations (1.25) are called canonical. Generating functions to obtain other pairings of new and old canonical variables can be obtained from G_1 by *Legendre transformations* of the form

$$G_2(q, \bar{p}, t) = G_1(q, \bar{q}, t) + q \bar{p} . \qquad (1.32)$$

Similar to the derivation above, other conjugate coordinates $(p, \bar{q})$ can be found from equations analogous to (1.32).

Equations (1.31) can be expressed in a general form for all four different types of generating functions. We set $x_k = q_k$ or $x_k = p_k$ and for the new variables we define $\bar{x}_k = \bar{q}_k$ or $\bar{x}_k = \bar{p}_k$. Further we define y_k and $\bar{y}_k$ as the conjugate variables of x_k and $\bar{x}_k$ respectively, and obtain from the general generating equation $G = G(x_k, \bar{x}_k, t)$

$$\boxed{\begin{aligned} y_k &= \pm \frac{\partial}{\partial x_k} G(x_k, \bar{x}_k, t) , \\ \bar{y}_k &= \mp \frac{\partial}{\partial \bar{x}_k} G(x_k, \bar{x}_k, t) , \\ H &= \bar{H} - \frac{\partial}{\partial t} G(x_k, \bar{x}_k, t) . \end{aligned}} \qquad (1.33)$$

The upper signs are to be used if the derivatives are taken with respect to coordinates and the lower signs if the derivatives are taken with respect to moments. It is not obvious which type of generating function should be used for a particular problem. However, the objective of canonical transformations is to express the problem at hand in as many cyclic variables as possible. Any form of generating function that achieves this goal is therefore appropriate.

To illustrate the use of generating functions for canonical transformation we will discuss a few very general examples. For an identity transformation we use a generating function of the form

$$G = q_1 \bar{p}_1 + q_2 \bar{p}_2 + \dots \qquad (1.34)$$

and get with (1.33) and $i = 1, 2, \dots N$ the identities

$$
\begin{aligned}
p_i &= \frac{\partial G}{\partial q_i} = \bar{p}_i \,, \\
\bar{q}_i &= \frac{\partial G}{\partial \bar{p}_i} = q_i \,.
\end{aligned}
\qquad (1.35)
$$

A transformation from rectangular (x, y, z) to cylindrical (r, φ, z) coordinates is determined by the generating function

$$G = -p_x\, r \cos\varphi - p_y \sin\varphi - p_z\, z \qquad (1.36)$$

and the transformation relations are

$$
\begin{aligned}
x &= -\frac{\partial G}{\partial p_x} = r \cos\varphi \,, & p_r &= -\frac{\partial G}{\partial r} = +p_x \cos\varphi + p_y \sin\varphi \,, \\
y &= -\frac{\partial G}{\partial p_y} = r \sin\varphi \,, & p_\varphi &= -\frac{\partial G}{\partial \varphi} = -p_x r \sin\varphi + p_y r \cos\varphi \,, \\
z &= -\frac{\partial G}{\partial p_z} = z \,, & p_z &= -\frac{\partial G}{\partial z} = p_z \,.
\end{aligned}
\qquad (1.37)
$$

Similarly relations for the transformation from rectangular to polar coordinates are derived from the generating function

$$G = -p_x\, r \cos\varphi \sin\vartheta - p_y r \sin\varphi \sin\vartheta - p_z r \cos\vartheta \,. \qquad (1.38)$$

It is not always obvious if a coordinate transformation is canonical. To identify a canonical transformation we use *Poisson brackets* defined by

$$\left[f_k(q_i, p_j), g_k(q_i, p_j) \right] = \sum_i \left(\frac{\partial f_k}{\partial q_i} \frac{\partial g_k}{\partial p_i} - \frac{\partial f_k}{\partial p_i} \frac{\partial g_k}{\partial q_i} \right) \,. \qquad (1.39)$$

It can be shown [1.4] that the new variables $\bar{q}_k\, \bar{p}_k$ or (1.25) are canonical only if

$$[\bar{p}_i, \bar{p}_j] = 0 \qquad [\bar{q}_i, \bar{q}_j] = 0 \qquad [\bar{q}_i, \bar{p}_j] = \delta_{ij} \,, \qquad (1.40)$$

where δ_{ij} is the Kronecker symbol.

While the formalism for canonical transformation is straight-forward we do not get a hint as to the optimum set of variables for a particular mechanical system. In the next sections we will see, however, that specific transformations are applicable to a whole class of mechanical systems.

1.1.4 Action-Angle Variables

Particularly important for particle beam dynamics is the canonical transformation from cartesian to *action-angle variables* (ψ, J). This class of transformations is best suited for *harmonic oscillators* like charged particles under the influence of restoring forces. We assume the equations of motion to be expressed in *normalized coordinates* of particle beam dynamics, where we use the variables $(w, \frac{\mathrm{d}w}{\mathrm{d}\varphi} = \dot{w}, \varphi)$ with the independent variable φ instead of the time. The *generating function* for the transformation to action-angle variables is of the form G_1 in (1.29) and, using some convenient constant factors, can be written as

$$G = -\tfrac{1}{2}\nu_{\mathrm{o}}\, w^2 \tan(\psi - \vartheta)\,, \tag{1.41}$$

where the variables (ψ, J) are *action-angle variables* and ϑ an arbitrary phase. Applying (1.33) to the generating function (1.41) we get

$$\begin{aligned}
\frac{\partial G}{\partial w} &= \dot{w} = -\nu_{\mathrm{o}}\, w\, \tan(\psi - \vartheta)\,, \\
\frac{\partial G}{\partial \psi} &= -J = -\tfrac{1}{2}\frac{\nu_{\mathrm{o}}\, w^2}{\cos^2(\psi - \vartheta)}\,.
\end{aligned} \tag{1.42}$$

After some manipulation the transformation equations take the form

$$\boxed{\begin{aligned}
w &= \sqrt{\frac{2J}{\nu_{\mathrm{o}}}}\,\cos(\psi - \vartheta)\,, \\
\dot{w} &= -\sqrt{2\nu_{\mathrm{o}}\, J}\,\sin(\psi - \vartheta)\,.
\end{aligned}} \tag{1.43}$$

To determine whether the transformation to action-angle variables has led us to *cyclic variables* we will use the unperturbed Hamiltonian, while ignoring perturbations, and substitute the old variables by new ones through the transformations (1.43). The generating function (1.41) does not explicitly depend on the independent variable φ and the new Hamiltonian is therefore from [Ref. 1.5, Eq. (5.151)] given by

$$H = \nu_{\mathrm{o}}\, J\,. \tag{1.44}$$

The variable ψ is obviously cyclic and from $\partial H/\partial \psi = 0 = \dot{J}$ we find the first invariant or *constant of motion*

$$J = \mathrm{const}\,. \tag{1.45}$$

The second Hamiltonian equation

$$\frac{\partial H}{\partial J} = \dot{\psi} = \nu \tag{1.46}$$

returns the frequency of the oscillator which is a *constant of motion* since the action J is invariant. The frequency or tune

$$\nu = \nu_o = \text{const},\tag{1.47}$$

and the angle variable ψ is the betatron phase. Eliminating the betatron phase ψ from (1.43), we obtain an expression of the action in normalized coordinates

$$J = \tfrac{1}{2}\nu_o w^2 + \tfrac{1}{2}\frac{\dot{w}^2}{\nu_o} = \text{const}.\tag{1.48}$$

Both terms on the r.h.s. can be associated with the potential and kinetic energy of the oscillator, respectively, and the constancy of the action J is synonymous with the constancy of the total energy of the oscillator. Going further back from the normalized coordinates $(w, \dot{w}, \varphi)$ to regular cartesian coordinates (u, u', s) we finally get

$$\boxed{J = \tfrac{1}{2}\nu_o\left(\gamma u^2 + 2\alpha u u' + \beta u'^2\right) = \tfrac{1}{2}\nu_o\,\epsilon.}\tag{1.49}$$

The invariant J is, except for a constant factor, identical to the *Courant-Snyder invariant* and expresses the invariance of the particle motion in phase-space along an elliptical trajectory. In (ψ, J) phase-space the particle moves along a circle with radius J at a revolution frequency ν_o. The motion is uniform, periodic and stable. Including the independent variable φ to form a three-dimensional phase-space, we find a particle to spiral along the surface of a torus as shown in Fig. 1.2. The ensemble of all particles oscillating with the same amplitude J follow spirals occupying the full surface of the torus.

This result is not particularly interesting in itself, since it only corroborates what we have found earlier for harmonic oscillators with simpler mathematical tools. The circle in (ψ, J) - phase space, however, provides us with a reference against which to compare perturbed motions and derive stability criteria. Indeed, we will later use canonical transformations to eliminate well-known linear motions, like the circular motion in (ψ, J)-space to exhibit the effect of perturbations only. Including perturbations into the Hamiltonian, we apply (1.46) and are able to derive perturbed tunes and study resonance phenomena. Having defined canonical variables for the system, we also will be able to study the evolution of particle beams by applying Vlasov's equation in Chap. 8. The Fokker-Planck equation finally will allow us to determine beam parameters even in the presence of statistical events.

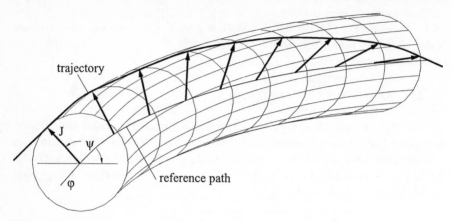

Fig. 1.2. Unperturbed particle trajectories in (ψ, J, φ) phase-space

1.2 Hamiltonian Resonance Theory

Particle resonances in circular accelerators have been discussed in [Ref. 1.5, Chap. 7] and occur as a result of perturbation terms involving particular Fourier harmonics. That approach is based on the common knowledge that periodic perturbations of a harmonic oscillator can cause a resonance when the frequency of the perturbation is equal to an eigenfrequency of the oscillator. However, we were not able to determine characteristic features of resonances. In the realm of Hamiltonian resonance theory we will be able to derive not only obvious resonant behaviour but also resonant dynamics which does not necessarily lead to a loss of the beam but to a significant change of beam parameters. We also will be able to determine the strength of resonances, effectiveness, escape mechanisms and more.

1.2.1 Nonlinear Hamiltonian

While simple Fourier expansions of perturbations around a circular accelerator allow us to derive the locations of lattice resonances in the tune diagram, we can obtain much deeper insight into the characteristics of resonances through the application of the Hamiltonian theory of linear and nonlinear oscillators. Soon after the discovery of strong focusing, particle dynamicists noticed the importance of perturbations with respect to beam stability and the possibility of beam instability even in the presence of very strong focusing.

Extensive simulations and development of theories were pursued in an effort to understand beam stability in high-energy proton synchrotrons then being designed at the *Brookhaven National Laboratory* and CERN. The first Hamiltonian theory of linear and non linear perturbations has been published by *Schoch* [1.6] which includes also references to early attempts to solve perturbation problems. A modern, consistent and complete theory of all resonances has been developed by *Guignard* [1.7]. In this text we will concentrate on main features of resonance theory and point the interested reader for more details to these references.

Multipole perturbations have been discussed in [Ref. 1.5, Sect. 7.4] where we specifically noted the appearance of resonance conditions. We will discuss in this chapter the characteristics of Hamiltonian resonance theory. To simplify the discussion we reproduce from [1.5] a table of perturbation terms which are the source of resonances up to third order, where the parameters n and r are the order of horizontal and vertical perturbations, respectively. The equation of motion under the influence of an nth-order perturbation is in normalized coordinates

$$\ddot{w} + \nu^2\, w \;=\; \bar{p}_n(\psi)\, w^{n-1}, \tag{1.50}$$

which can be also derived from the *nonlinear Hamiltonian*

$$H_w \;=\; \tfrac{1}{2}\,\dot{w}^2 + \tfrac{1}{2}\,\nu_o^2\, w^2 + p_n(\varphi)\left(\frac{\nu_o}{2}\right)^{n/2} w^n. \tag{1.51}$$

Here we expand the expression for the perturbation $\bar{p}_n(\varphi)$ from Table 1.1 by some constant factors like

$$p_n(\varphi) \;=\; -\bar{p}_n(\varphi)\,\frac{1}{n}\,\left(\frac{\nu_o}{2}\right)^{-n/2} \tag{1.52}$$

for future convenience. The use of the designation p_n or $\bar{p}_n$ for perturbations should not be confused with the earlier use of the same letter for the particle momentum.

Table 1.1. Multipole perturbation terms

order		$\bar{p}_{nx}(\varphi)\, w^{n-1}\, v^{r-1}$	$\bar{p}_{ny}(\varphi)\, v^{n-1}\, w^{r-1}$
n	r		
1	2	$-\nu_{xo}^2\, \beta_x^{3/2}\, \beta_y^{1/2}\, \underline{k}\, v$	
2	1		$-\nu_{yo}^2\, \beta_y^{3/2}\, \beta_x^{1/2}\, \underline{k}\, w$
3	1	$-\nu_{xo}^2\, \beta_x^{5/2}\, \tfrac{1}{2} m\, w^2$	
2	2		$-\nu_{yo}^2\, \beta_x^{1/2}\, \beta_y^2\, m w v$
1	3	$+\nu_{xo}^2\, \beta_x^{3/2}\, \beta_y\, \tfrac{1}{2} m\, v^2$	
4	1	$-\nu_{xo}^2\, \beta_x^3\, \tfrac{1}{6} r\, w^3$	
3	2		$+\nu_{yo}^2\, \beta_x\, \beta_y^2\, \tfrac{1}{2} r\, w^2\, v$
2	3	$+\nu_{xo}^2\, \beta_x^2\, \beta_y \tfrac{1}{2} r\, w\, v^2$	
1	4		$-\nu_{yo}^2\, \beta_y^3\, \tfrac{1}{6} r\, v^3$

To discuss resonance phenomena it is useful to perform a canonical transformation from the coordinates $(w, \dot{w})$ to *action-angle variables* (ψ, J) which can be derived from the generating function (1.41) and the new Hamiltonian expressed in action-angle variables is

$$H = \nu_{\rm o} J + p_n(\varphi)\, J^{n/2} \cos^n(\psi - \vartheta). \tag{1.53}$$

The action-angle variables take on the role of an "energy" and frequency of the oscillatory system which becomes evident from (1.48). Due to the phase dependent perturbation, the oscillation amplitude J is no more a constant of motion like in (1.45) and the circular motion of Fig. 1.2 becomes distorted as shown in Fig. 1.3 for a sextupolar perturbation. The *oscillator frequency* $\dot{\psi} = \partial \psi / \partial \varphi = \nu$ is similarly perturbed and can be derived from the other Hamiltonian equation of motion

$$\boxed{\frac{\partial H}{\partial J} = \dot{\psi} = \nu = \nu_{\rm o} + \tfrac{n}{2}\, p_n(\varphi)\, J^{n/2-1} \cos^n(\psi - \vartheta).} \tag{1.54}$$

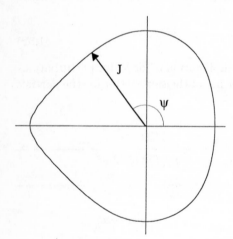

Fig. 1.3. Nonlinear perturbation of phase-space motion

Perturbation terms seem to modify the oscillator frequency $\nu_{\rm o}$ but because of the oscillatory trigonometric factor it is not obvious if there is a net shift or spread in the tune. We therefore expand the perturbation $p_n(\varphi)$ as well as the trigonometric factor $\cos^n \psi$ to determine its spectral content. The distribution of the multipole perturbations in a circular accelerator is periodic with a periodicity equal to the length of a superperiod or of the whole ring circumference and we are therefore able to expand the perturbation $p_n(\varphi)$ into a Fourier series

$$p_n(\varphi) = \sum_q p_{nq}\, {\rm e}^{-iqN\varphi}, \tag{1.55}$$

where N is the *superperiodicity* of the circular accelerator. We also expand the trigonometric factor in (1.54) into exponential functions, while dropping the arbitrary phase ϑ

$$\cos^n \psi = \sum_{|m| \leq n} c_{nm}\, e^{im\psi} \qquad (1.56)$$

and get

$$
\begin{aligned}
p_n(\varphi)\, \cos^n \psi &= \sum_q p_{nq}\, e^{-iqN\varphi} \sum_{|m| \leq n} c_{nm}\, e^{im\psi}\,, \\[4pt]
&= \sum_{\substack{q \\ |m| \leq n}} c_{nm}\, p_{nq}\, e^{i(m\psi - qN\varphi)}\,, \\[4pt]
&= c_{no}\, p_{no} + \sum_{\substack{q \geq 0 \\ 0 < m \leq n}} 2\, c_{nm}\, p_{nq}\, \cos(m\psi - qN\varphi)\,.
\end{aligned}
\qquad (1.57)
$$

In the last equation the perturbation $p_n(\varphi)$ is expanded about a symmetry point merely to simplify the expressions of resonant terms. For asymmetric lattices the derivation is similar but includes extra terms. We have also separated the nonoscillatory term $c_{no}\, p_{no}$ from the oscillating terms to distinguish between systematic frequency shifts and mere periodic variations of the tune. The Hamiltonian (1.53) now becomes with (1.57)

$$H = \nu_o\, J + c_{no}\, p_{no}\, J^{n/2} + J^{n/2} \sum_{\substack{q \geq 0 \\ 0 < m \leq n}} 2 c_{nm}\, p_{nq}\, \cos(m\psi - qN\varphi)\,. \qquad (1.58)$$

The third term on the r.h.s. consists mostly of fast oscillating terms which in this approximation do not lead to any specific consequences. For the moment we will ignore these terms and remember to come back later in this section.

The shift of the oscillator frequency due to the lowest-order perturbation becomes obvious and may be written as

$$\frac{\partial H}{\partial J} = \dot{\psi} = \nu = \nu_o + \tfrac{n}{2}\, c_{no}\, p_{no}\, J^{n/2 - 1} + \text{oscillatory terms}\,. \qquad (1.59)$$

Since $c_{no} \neq 0$ for even values of n only, we find earlier results in [Ref. 1.5, Chap. 7] confirmed, where we observed the appearance of amplitude-dependent tune shifts and tune spreads for even-order perturbations. Specifically we notice, that there is a coherent tune shift for all particles within a beam in case of a gradient field perturbation with $n = 2$ and a tune spread within a finite beam size for all other higher- and even-order multipole perturbations.

We should recapitulate at this point where we stand and what we have achieved. The canonical transformation of the normalized variables to action-angle variables has indeed eliminated the angle coordinate as long as we neglect oscillatory terms. The angle variable therefore is in this approximation a cyclic variable and the Hamiltonian formalism tells us that the conjugate variable, in this case the amplitude J is a constant of motion or an invariant. This is an important result which we obtained by simple application of the Hamiltonian formalism confirming our earlier expectation to isolate constants of motion. This has not been possible in a rigorous way since we had to obtain approximate invariants by neglecting summarily all oscillatory terms. In certain circumstances this approximation can lead to totally wrong results. To isolate these circumstances we pursue further canonical transformations to truly separate from the oscillating terms all nonoscillating terms of order $\frac{n}{2}$ while the rest of the oscillating terms are transformed to a higher order in the amplitude J.

1.2.2 Resonant Terms

Neglecting oscillating terms is justified only in such cases where these terms oscillate rapidly. Upon closer inspection of the arguments in the trigonometric functions we notice that for each value of m in (1.58) there exists a value q which causes the phase

$$m_r \psi_r \approx m\psi - qN\varphi \qquad (1.60)$$

to vary slowly possibly leading to a *resonance*. The condition for the occurrence of a resonance is $\psi_r \approx 0$ or with $\psi \approx \nu_o \varphi$

$$m_r \nu_o \approx rN, \qquad (1.61)$$

where we have set $q = r$ to identify the index for which the *resonance condition* (1.61) is met. The index m_r is the order of the resonance and can assume any integer value between unity and n. The order n of the multipole field determines the highest order of the resonance. Specifically we find that a sextupole can excite any resonance up to third order or a decapole any resonance up to fifth order. It is customary to ignore the lower-order resonances of a higher-order multipole assuming that these lower resonances are already excited by lower-order multipoles. We will therefore concentrate mainly on the nth-order resonance caused by nth-order multipoles.

The effects of resonances do not only appear when the resonance condition is exactly fulfilled for $\psi_r = 0$. Significant changes in the particle motion can be observed when the particle oscillation frequency approaches the resonance condition. We therefore keep all terms which vary slowly compared to the betatron frequency $\dot{\psi}$.

After isolating resonant terms we may now neglect all remaining fast oscillating terms with $m \neq m_r$. Later we will show that these terms can be transformed to higher order and are therefore of no consequence to the order of approximation of interest. Keeping only resonant terms defined by (1.61), we get from (1.58) the nth-order Hamiltonian in normalized coordinates

$$H = \nu_{\rm o} J + c_{no} \, p_{no} \, J^{n/2} + J^{n/2} \sum_{\substack{r \\ 0 < m_r \leq n}} 2 \, c_{nm_r} \, p_{nr} \, \cos(m_r \psi_{\rm r}) \,. \tag{1.62}$$

The value of m_r indicates the order of the resonance and we note that the maximum order of resonance driven by a multipole of order n is not higher than n. A dipole field therefore can drive only an integer resonance, a quadrupole field up to a half-integer resonance, a sextupole up to a third-order resonance, an octupole up to a quarter resonance and so forth. As we have noticed before, whenever we derive mathematical results we should keep in mind that such results are valid only within the approximation under consideration. It is, for example, known [1.8] that sextupoles can also drive quarter integer resonances through higher-order terms. In nonlinear particle beam dynamics any statement about stability or instability must be accompanied by a statement defining the order of approximation made to allow independent judgement for the validity of a result to a particular problem.

The interpretation of the Hamiltonian (1.62) becomes greatly simplified after another canonical transformation to eliminate the appearance of the independent variable φ. We thereby transform to a coordinate system that moves with the reference particle, thus eliminating the linear motion that we already know. This can be achieved by a canonical similarity transformation from the coordinates (J, ψ) to (J_1, ψ_1) which we derive from the generating function

$$G_1 = J_1 \left(\psi - \frac{rN\varphi}{m_r} \right) . \tag{1.63}$$

From this we get the relations between the old and new coordinates

$$\frac{\partial G_1}{\partial J_1} = \psi_1 = \psi - \frac{rN}{m_r} \varphi \tag{1.64}$$

and

$$\frac{\partial G_1}{\partial \psi} = J = J_1 \,. \tag{1.65}$$

The quantity ψ_1 now describes the phase deviation of a particle from that of the reference particle. Since the generating function depends on the independent variable φ we get for the new Hamiltonian $H_1 = H + \partial G_1/\partial \varphi$ or

$$H_1 = \left(\nu_{\rm o} - \frac{rN}{m_r} \right) J_1 + c_{no} p_{no} J_1^{n/2} + \tilde{p}_{nr} J_1^{n/2} \cos(n\psi_1) \,, \tag{1.66}$$

where we have retained for simplicity only the highest-order resonant term as discussed earlier and have set

$$\tilde{p}_{nq} = 2 \, c_{nm_r} \, p_{nq} \,. \tag{1.67}$$

With $\dot{\psi} = (d\psi/d\varphi) = \nu$ and (1.61) a resonance condition occurs whenever

$$\nu_{\mathrm{o}} \approx \frac{rN}{m_r} = \nu_r. \tag{1.68}$$

Setting $\Delta\nu_r = \nu_{\mathrm{o}} - \nu_r$ for the distance of the tune ν_{o} from the resonance tune ν_r, the Hamiltonian becomes with all perturbation terms

$$H_1 = \Delta\nu_r \, J_1 + \sum_n c_{n\mathrm{o}} \, p_{n\mathrm{o}} \, J_1^{n/2} + \sum_n \sum_{\substack{r \\ 0 < m_r \leq n}} \tilde{p}_{nr} \, J_1^{n/2} \cos(m_r \psi_1). \tag{1.69}$$

The coefficients $c_{n\mathrm{o}}$ are defined by (1.56) and the harmonic amplitudes of the perturbations are defined with (1.67) by the Fourier expansion (1.55). The resonance order r and integer m_r depend on the ring tune and are selected such that (1.68) is approximately met. A selection of most common multipole perturbations are compiled in Table 1.1 and picking an nth-order term we get from (1.52) the expression for $p_n(\varphi)$.

In the course of the mathematical derivation we started out in (1.51) with only one multipole perturbation of order n. For reasons of generality, however, all orders of perturbation n have been included again in (1.69). We will, however, not deal with the complexity of this multiresonance Hamiltonian nor do we need to in order to investigate the character of individual resonances. Whatever the number of actual resonances may be in a real system the effects are superpositions of individual resonances. We will therefore investigate in more detail single resonances and discuss superpositions of more than one resonance later in this chapter.

1.2.3 Resonance Patterns and Stop-Band Width

Equation (1.69) can be used to calculate the *stop-band width* of resonances and to explore *resonance patterns* which are a superposition of particle trajectories $H_1 = $ const in (ψ_1, J_1) phase-space. Depending on the nature of the problem under study, we may use selective terms from both sums in (1.69). Specifically to illustrate characteristic properties of resonances, we will use from the first sum the term $c_{4\mathrm{o}} \, p_{4\mathrm{o}}$ which originates from an octupole field. From the second sum we choose a single nth-order term driving the rth resonance and get the simplified Hamiltonian

$$H_1 = \Delta\nu_r \, J_1 + c_{4\mathrm{o}} \, p_{4\mathrm{o}} \, J_1^2 + \tilde{p}_{nr} \, J_1^{n/2} \cos(m_r \psi_1) = \mathrm{const}. \tag{1.70}$$

To further simplify the writing of equations and the discussion of results we divide (1.70) by $\tilde{p}_{nr} J_{\mathrm{o}}^{n/2}$, where the amplitude J_{o} is an arbitrary reference amplitude at the starting point $J_{\mathrm{o}} = J_1 \, (\varphi = 0)$. Defining an amplitude ratio or *beat factor*

$$R = \frac{J}{J_{\mathrm{o}}}, \tag{1.71}$$

and considering only resonances of order $m_r = n$ (1.70) becomes

$$\Delta R + \Omega R^2 + R^{n/2} \cos n\psi_1 = \text{const}, \tag{1.72}$$

where the *detuning* from the resonance is

$$\Delta = \frac{\Delta\nu_r}{\tilde{p}_{nr} J_o^{n/2-1}} \tag{1.73}$$

and the *tune-spread parameter*

$$\Omega = \frac{c_{4o} \, p_{4o}}{\tilde{p}_{nr} J_o^{n/2-2}}. \tag{1.74}$$

This expression has been derived first by *Schoch* [1.6] for particle beam dynamics. Because the ratio R describes the variation of the oscillation amplitude in units of the starting amplitude J_o we call the quantity R the *beat factor* of the oscillation.

Before we discuss *stop-bands* and *resonance patterns* we make some general observations concerning the stability of particles. The stability of particle motion in the vicinity of resonances depends strongly on the distance of the tune from the nearest nth-order resonance and on the tune-spread parameter Ω. When both parameters Δ and Ω vanish we have no stability for any finite oscillation amplitude, since (1.72) can be solved for all values of ψ_1 only if $R = 0$. For a finite tune-spread parameter $\Omega \neq 0$ while $\Delta = 0$ and (1.72) becomes $R^2 (\Omega + R^{n/2-2} \cos n\psi_1) = \text{const}$ and resonances of order $n > 4$ exhibit some range of stability for amplitudes $R^{n/2-2} < |\Omega|$. Oscillations in the vicinity of, for example, a quarter resonance are all stable for $|\Omega| > 1$ and all unstable for smaller values of the tune-spread parameter $|\Omega| < 1$. A finite tune-spread parameter Ω appears in this case due to an octupolar field and has a stabilizing effect at least for small amplitudes.

For very small oscillation amplitudes $R \to 0$ the oscillating term in (1.72) becomes negligible for $n > 4$ compared to the detuning term and the particle trajectory approaches the form of a circle with radius R. This well behaved character of particle motion at small amplitudes becomes distorted for resonances of order $n = 2$ and $n = 3$ in case of small detuning and a finite tune spread parameter. We consider $\Delta = 0$ and have

$$\Omega R^2 + R^{n/2} \cos n\psi = \text{const},$$

where $n = 2$ or $n = 3$. For very small amplitudes the quadratic term is negligible and the dominant oscillating term alone is unstable. The amplitude for a particle starting at $R \approx 0$ and $\psi_1 = 0$ grows to large amplitudes as ψ_1 increases, reaching values which make the quadratic tune-spread term dominant before the trigonometric term becomes negative. The resulting trajectory in phase space becomes a figure of eight for the half-integer resonance as shown in Fig. 1.4.

In the case of a third-order resonance small-amplitude oscillations behave similarly and follow the outline of a clover leave as shown in Fig. 1.5.

Half-integer stop-band: A more detailed discussion of (1.72) will reveal that instability due to resonances does not only happen exactly at resonant tunes. Particle oscillations become unstable within a finite vicinity of resonance lines in the resonance diagram and such areas of instability are known as *stop-bands*. The most simple case occurs for $\Omega = 0$ and a half-integer resonance, where $n = 2$ and

$$R\left(\Delta + \cos 2\psi_1\right) = \text{const}. \tag{1.75}$$

For this equation to be true for all values of the angle variable ψ_1 we require that the quantity in the brackets does not change sign while ψ_1 varies from 0 to 2π. This condition cannot be met if $|\Delta| \leq 1$. To quantify this we observe a particle starting with an amplitude $J = J_o$ at $\psi_1 = 0$ and (1.75) becomes

$$R\Delta + R\cos 2\psi_1 = \Delta + 1. \tag{1.76}$$

Now we calculate the variation of the oscillation amplitude R as the angle variable ψ_1 increases. The beat factor R reaches its maximum value at $2\psi_1 = \pi$ and is

$$R_{\max} = \frac{\Delta + 1}{\Delta - 1} > 0. \tag{1.77}$$

The variation of the amplitude R is finite as long as $\Delta > 1$. If $\Delta < 0$ we get a similar stability condition

$$R_{\max} = \frac{|\Delta| - 1}{|\Delta| + 1} > 0 \tag{1.78}$$

and stability occurs for $\Delta < -1$. The complete resonance *stability criterion* for the half-integer resonance is therefore

$$|\Delta| > 1. \tag{1.79}$$

Beam instability due to a *half-integer resonance*, where $n = 2$, occurs within a finite vicinity $\Delta\nu_r = \pm\widetilde{p}_{2r}$ as defined by (1.73) and the total *stop-band width* for a half-integer resonance becomes

$$\Delta\nu_{\text{stop}}^{(2)} = 2\,\widetilde{p}_{2r}. \tag{1.80}$$

The width of the stop-band increases with the strength of the perturbation but does not depend on the oscillation amplitude J_o. However, for higher-order resonances the stop band width does depend on the oscillation amplitudes as will be discussed later.

To observe the particle trajectories in phase space, we calculate the contour lines for (1.72) setting $n = 2$ and obtain patterns as shown in Fig. 1.4.

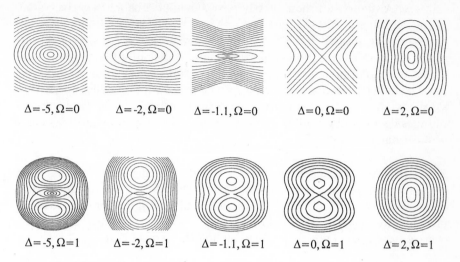

Fig. 1.4. (R,ψ_1) phase-space motion for a half-integer resonance

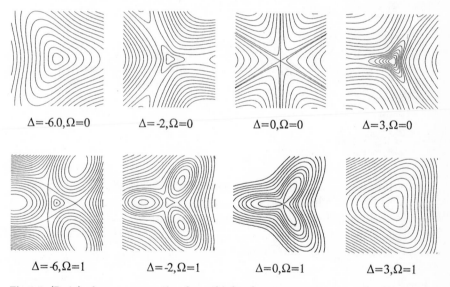

Fig. 1.5. (R,ψ_1) phase-space motion for a third-order resonance

Here the particle trajectories are plotted in (ψ, J) phase-space for a variety of *detuning parameters* Δ and *tune-spread parameters* Ω. Such diagrams are called *resonance patterns*. The first row of Fig. 1.4 shows particle trajectories for the case of a half-integer resonance with a vanishing tune-spread parameter $\Omega = 0$. As the detuning Δ is increased we observe a deformation of the particle trajectories but no appearance of a stable island as long as

$|\Delta| < 1$. Although we show mostly resonance patterns for negative values of the detuning $\Delta < 0$ the patterns look exactly the same for $\Delta > 0$ except that they are rotated by 90°. For $|\Delta| > 1$ the unstable trajectories part vertically from the origin and allow the appearance of a stable island that grows as the detuning grows. In the second row of resonance patterns, we have included a finite tune-spread parameter of $\Omega = 1$ which leads to a stabilization of all large amplitude particle trajectories. Only for small amplitudes do we still recognize the irregularity of a figure of eight trajectory as mentioned above.

Separatrices: The appearance of an island structure as noticeable from the resonance patterns is a common phenomenon and is due to tune-spread terms of even order like that of an octupole field. In Fig. 1.6 common features of resonance patterns are shown and we note specifically the existence of a central stable part and islands surrounding the central part.

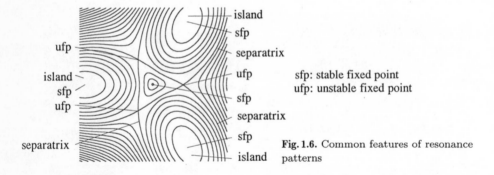

Fig. 1.6. Common features of resonance patterns

sfp: stable fixed point
ufp: unstable fixed point

The boundaries of the areas of stable motion towards the islands are called *separatrices*. These separatrices also separate the area of stable motion from that for unstable motion. The crossing points of these separatrices, as well as the center of the islands, are called *fixed points* of the dynamic system and are defined by the conditions

$$\frac{\partial H_1}{\partial \psi_1} = 0 \quad \text{and} \quad \frac{\partial H_1}{\partial J_1} = 0. \tag{1.81}$$

Application of these conditions to (1.70) defines the location of the fixed points and we find from the first equation (1.81) the azimuthal positions $\psi_1 = \psi_f$ of the fixed points from

$$\sin(m_r \, \psi_{1f}) = 0$$

or

$$m_r \, \psi_{1fk} = k \, \pi, \tag{1.82}$$

where k is an integer number in the range $0 \leq k \leq 2m_r$. From the second equation (1.81) we get an expression for the radial location of the fixed points J_{fk}

$$\Delta\nu_r + 2\,c_{4o}\,p_{4o}\,J_{fk} + \tfrac{n}{2}\,\widetilde{p}_{nr}\,J_{fk}^{n/2-1}\,\cos(k\pi) = 0 . \qquad (1.83)$$

There are in principle $2\,m_r$ separate fixed points in each resonance diagram as becomes evident from Figs. 1.4–10. Closer inspections shows that alternately every second fixed point is a *stable fixed point* or an *unstable fixed point*, respectively. The unstable fixed points coincide with the crossing points of separatrices and exist even in the absence of octupole terms. Stable fixed points define the center of stable *islands* and, except for the primary stable fixed point at the origin of the phase diagram, exist only in the presence of a tune spread caused by octupole like terms $c_{no}\,p_{no}\,J^{n/2}$ in (1.66) which contribute to beam stability. Trajectories that were unstable without the octupole term become closed trajectories within an island area centered at stable fixed points. This island structure is characteristic for resonances since the degree of symmetry is equal to the order of the resonance.

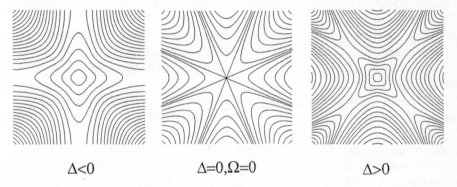

$$\Delta<0 \qquad\qquad \Delta=0,\Omega=0 \qquad\qquad \Delta>0$$

Fig. 1.7. Fourth-order resonance patterns

General stop-band width: From the discussion of the half-integer resonance, it became apparent that certain conditions must be met to obtain stability for particle motion. Specifically we expect instability in the vicinity of resonances and we will try to determine quantitatively the area of instability or stop-band width for general resonances. Similar to (1.76) we look for stable solutions from

$$R\,\Delta + R^{n/2}\,\cos n\psi_1 = \Delta \pm 1 \qquad (1.84)$$

which describes a particle starting with an amplitude $R = 1$. Equation (1.84) must be true along all points of the trajectory and for reasons of symmetry the particle oscillation amplitude approaches again the starting amplitude for $\psi_1 = 0$ as $\psi_1 \to 2\pi/n$. Solving for Δ we get real solutions for R only if

$$\Delta^+ \geq -\frac{R^{n/2} - 1}{R - 1} = -\frac{n}{2}, \tag{1.85}$$

where the index $^+$ indicates the sign to be used on the r.h.s. of (1.84). Similarly, following a particle starting with $R = 1$ at $\psi_1 = \pi/n$ to $\psi_1 = 3\pi/n$ we get the condition

$$\Delta^- \leq \frac{n}{2}. \tag{1.86}$$

The total nth-order *stop-band width* is therefore with (1.73)

$$\Delta\nu_{\text{stop}}^{(n)} = n \, |\widetilde{p}_{nr}| \, J_{\text{o}}^{n/2-1} \tag{1.87}$$

indicating that stable particle motion is possible only for tunes outside this stop-band. The stop-band width of nonlinear resonances $(n > 2)$ is strongly amplitude dependent and special effort must be exercised to minimize higher-order perturbations. Where higher-order magnetic fields cannot be eliminated it is prudent to minimize the value of the betatron functions at those locations. Such a case occurs, for example, in *colliding-beam storage rings*, where the strongly nonlinear field of one beam perturbs the trajectories of particles in the other beam. This effect is well known as the *beam-beam effect*.

Through a series of canonical transformations and redefinitions of parameters we seem to have parted significantly from convenient laboratory variables and parameters. We will therefore convert (1.87) back to variables we are accustomed to use. Equation (1.67) becomes $\widetilde{p}_{nr} = 2\,c_{nn}\,p_{nr}$ with $q = r$, and $m_r = n$, where $r \approx \frac{n}{N}\,\nu_{\text{o}}$. Tacitly, lower-order resonances $m_r < n$ have been ignored. From (1.55) we find the Fourier component

$$p_{nr} = \frac{1}{2\pi} \int_0^{2\pi} p_n(\varphi) \, e^{irN\varphi} \, d\varphi, \tag{1.88}$$

and from (1.56) we have $c_{nn} = 1/2$. The amplitude factor $J_{\text{o}}^{n/2-1}$ is replaced by (1.49) which becomes with (1.43) and $\psi_1 = 0$

$$J_{\text{o}} = \frac{\nu_{\text{o}}}{2} \, w_{\text{o}}^2 = \frac{\nu_{\text{o}}}{2} \, \frac{x_{\text{o}}^2}{\beta}. \tag{1.89}$$

Finally we recall the definition (1.52)

$$p_n(\varphi) = \frac{1}{n} \, \bar{p}_n(\varphi) \left(\frac{\nu_{\text{o}}}{2} \right)^{-n/2} \tag{1.90}$$

and get for the nth-order *stop-band width*

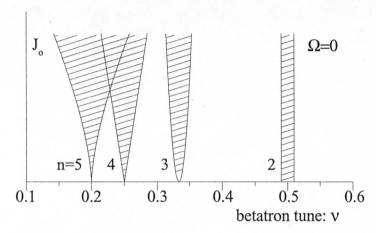

Fig. 1.8. Stop-band width as a function of the amplitude J_o for resonances of order $n = 2, 3, 4, 5$ and detuning parameter $\Omega = 0$

$$\Delta^{(n)}_{\text{stop}} = \frac{w_o^{n-2}}{\pi \nu_o} \left| \int_0^{2\pi} \bar{p}_n(\varphi) \, e^{irN\varphi} \, d\varphi \right| \tag{1.91}$$

where $\bar{p}_n$ is the nth-order perturbation from Tab. 1.1.

This result is general and includes our earlier finding for the half-integer resonance. For resonances of order $n > 2$ the stop-band width increases with amplitude limiting the stability of particle beams to the vicinity of the axis (Fig. 1.8). The introduction of sufficiently strong octupole terms can lead to a stabilization of resonances and we found, for example, that the quarter resonance is completely stabilized if $\Omega \geq 1$. For resonances of order $n > 4$ however, we find that the term $R^{n/2} \cos n\psi_1$ will become dominant for large values of the amplitude and resonance therefore cannot be avoided.

Figure 1.9 shows, for example, a stable area for small amplitudes at the fifth-order resonance, as we would expect, but at larger amplitudes the motion becomes unstable.

1.2.4 Third-Order Resonance

The third-order resonance plays a special role in accelerator physics and we will therefore discuss this resonance in more detail. The special role is generated by the need to use sextupoles for chromaticity correction. While such magnets are beneficial in one respect, they may introduce third-order resonances that need to be avoided or at least kept under control. Sometimes the properties of a third-order resonance are also used constructively to eject particles at the end of a synchrotron acceleration cycle slowly over many turns.

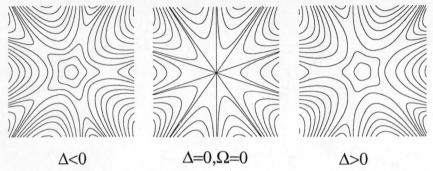

$\Delta<0$ $\Delta=0, \Omega=0$ $\Delta>0$

Fig. 1.9. Fifth-order resonance patterns

In the absence of octupole fields the Hamiltonian for the third-order resonance is from (1.70) for $n = 3$

$$H_1 = \Delta\nu_{1/3}\, J_1 + \widetilde{p}_{3r}\, J_1^{3/2}\, \cos(3\,\psi_1)\,. \tag{1.92}$$

We expand $\cos 3\psi_1 = \cos^3\psi_1 - 3\cos\psi_1\sin^2\psi_1$ and return to normalized coordinates

$$\begin{aligned} w &= \sqrt{\frac{2J_1}{\nu_{\mathrm o}}}\,\cos\psi_1\,, \\ \dot{w} &= \sqrt{2\nu_{\mathrm o}J_1}\,\sin\psi_1\,. \end{aligned} \tag{1.93}$$

In these coordinates the Hamiltonian reveals the boundaries of the stable region from the unstable resonant region. Introducing the normalized coordinates into (1.92) we get the Hamiltonian

$$H_1 = \Delta\nu_{1/3}\frac{\nu_{\mathrm o}}{2}\left(w^2 + \frac{\dot{w}^2}{\nu_{\mathrm o}^2}\right) + \widetilde{p}_{3r}\frac{\nu_{\mathrm o}^{3/2}}{2^{3/2}}\left(w^3 - 3w\frac{\dot{w}^2}{\nu_{\mathrm o}^2}\right)\,. \tag{1.94}$$

Dividing by $\widetilde{p}_{3r}(\nu_{\mathrm o}/2)^2$ and subtracting a constant term $\frac{1}{2}W_{\mathrm o}^3$, where

$$W_{\mathrm o} = \frac{4}{3}\frac{\Delta\nu_{1/3}}{\widetilde{p}_{3r}\,\nu_{\mathrm o}}\,, \tag{1.95}$$

the Hamiltonian assumes a convenient form to exhibit the boundaries between the stable and unstable area

$$\boxed{\begin{aligned} \widetilde{H}_1 &= \tfrac{3}{2}W_{\mathrm o}\left(w^2 + \frac{\dot{w}^2}{\nu_{\mathrm o}^2}\right) + \left(w^3 - 3w\frac{\dot{w}^2}{\nu_{\mathrm o}^2}\right) - \tfrac{1}{2}W_{\mathrm o}^3 \\ &= (w - \tfrac{1}{2}W_{\mathrm o})\,(w - \sqrt{3}\,\frac{\dot{w}}{\nu_{\mathrm o}} + W_{\mathrm o})\,(w + \sqrt{3}\,\frac{\dot{w}}{\nu_{\mathrm o}} + W_{\mathrm o})\,. \end{aligned}} \tag{1.96}$$

This Hamiltonian has three linear solutions for $\widetilde{H}_1 = 0$ defining the *separatrices*. The resonance plot for (1.96) is shown in Fig. 1.10 where

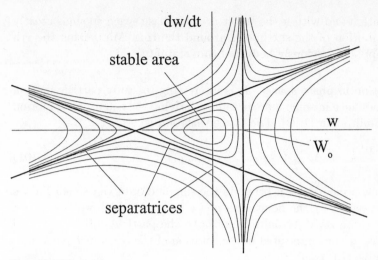

Fig. 1.10. Third-order resonance

we have assumed that W_o is positive. For a given distribution of the sex-tupoles $\widetilde{p}_{3r}$ the resonance pattern rotates by $180°$ while moving the tune from one side of the resonance to the other. Clearly, there is a stable central part bounded by separatrices. The area of the central part depends on the strength and distribution of the sextupole fields summarized by $\widetilde{p}_{3r}$ and the distance $\Delta\nu_{1/3}$ of the tune from the third-order resonance.

The *higher-order field perturbation* $\widetilde{p}_{3r}$ depends on the distribution of the sextupoles around the circular accelerator. In the horizontal plane

$$\bar{p}_{3x}(\varphi) = -\nu_{xo}^2\,\beta_x^{5/2}\,\tfrac{1}{2}\,m \tag{1.97}$$

or with (1.52)

$$p_3(\varphi) = \tfrac{1}{3}\,\sqrt{2\,\nu_{xo}}\,\beta_x^{5/2}\,m\,. \tag{1.98}$$

The Fourier components of this perturbation are given by

$$p_{3r} = \tfrac{1}{2\pi}\int\limits_o^{2\pi} p_3(\varphi)\,\mathrm{e}^{irN\varphi}\,\mathrm{d}\varphi \tag{1.99}$$

and the perturbation term (1.67) becomes finally with $m_r = 3$ and $c_{33} = 1/8$ from (1.56)

$$\widetilde{p}_{3r} = \frac{\sqrt{2\,\nu_{xo}}}{24\pi}\int\limits_o^{2\pi}\beta_x^{5/2}\,m\,\mathrm{e}^{irN\varphi}\,\mathrm{d}\varphi \tag{1.100}$$

where $\varphi = \int_o^s \frac{\mathrm{d}s}{\nu_{ox}\beta_x}$, $m = m(\varphi)$ is the sextupole distribution and $\beta_x = \beta_x(\varphi)$ the horizontal betatron function. From this expression, it becomes clear that the perturbation and with it the stable area in phase space depends greatly on the distribution of the sextupoles around the ring. Minimizing the rth Fourier component obviously benefits beam stability.

Particle motion in phase space: It is interesting to study particle motion close to a resonance in some more detail by deriving the equations of motion from the Hamiltonian (1.92). The phase variation is

$$\frac{\partial H_1}{\partial J_1} = \frac{\mathrm{d}\psi_1}{\mathrm{d}\varphi} = \Delta\nu_{1/3} + \tfrac{3}{2}\,\widetilde{p}_{3r}\,J_1^{1/2}\cos 3\psi_1\,. \tag{1.101}$$

Now we follow a particle as it orbits in the ring and observe its coordinates every time it passes by the point with phase φ_{o} or ψ_{o} which we assume for convenience to be zero. Actually we observe the particle only every third turn, since we are not interested in the rotation of the resonance pattern in phase space by 120° every turn.

For small amplitudes the first term is dominant and we note that the particles move in phase space clockwise or counter clockwise depending on $\Delta\nu_{1/3}$ being negative or positive, respectively. The motion becomes more complicated in the vicinity and outside the separatrices, where the second term is dominant. For a particle starting at $\psi_1 = 0$ the phase ψ_1 increases or decreases from turn to turn and asymptotically approaches $\psi_1 = \pm 30°$ depending on the perturbation $\widetilde{p}_{3r}$ being positive or negative respectively. The particles therefore move clockwise or counter clockwise and the direction of this motion is reversed, whenever we move into an adjacent area separated by separatrices because the trigonometric term has changed sign.

To determine exactly the position of a particle after $3q$ turns we have with $\psi(q) = 3q\,2\pi\nu_{\mathrm{o}}$

$$\psi_1(q) = 2\pi\,(3\nu_{\mathrm{o}} - rN)\,q\,. \tag{1.102}$$

With this phase expression we derive the associated amplitude J_{1q} from the Hamiltonian (1.92) and may plot the particle positions for successive triple turns $3q = 0, 3, 6, 9, \ldots$ in a figure similar to Fig. 1.10. The change in the oscillation amplitude is from the second Hamiltonian equation of motion is

$$\frac{\partial H_1}{\partial \psi_1} = -\frac{\mathrm{d}J_1}{\mathrm{d}\varphi} = -3\,\widetilde{p}_{3r}\,J^{3/2}\sin 3\psi_1 \tag{1.103}$$

and is very small in the vicinity of $\psi_1 \approx 0$ or even multiples of 30°. For ψ_1 equal to odd multiples of 30°, on the other hand, the oscillation amplitude changes rapidly as shown in Fig. 1.10.

1.3 Hamiltonian and Coupling

In practical beam transport systems particle motion is not completely contained in one or the other plane although special care is being taken to avoid *coupling* effects as much as possible. Coupling of the motion from one plane into the other plane can be generated through the insertion of actual rotated magnets or in a more subtle way by rotational misalignments of upright magnets. Since such misalignments are unavoidable, it is customary to place weak rotated quadrupoles in a transport system to provide the ability to counter what is known as *linear coupling* caused by unintentional magnet misalignments. Whatever the source of coupling, we consider such fields as small perturbations to the particle motion.

The Hamiltonian treatment of coupled motion follows that for motion in a single plane in the sense that we try to find cyclic variables while transforming away those parts of the motion which are well known. For a single particle normalized coordinates can be defined which eliminate the s-dependence of the unperturbed part of the equations of motion. Such transformations cannot be applied in the case of coupled motion since they involve the oscillation frequency or betatron phase function which is different for both planes.

1.3.1 Linearly Coupled Motion

We will derive some properties of coupled motion for the case of linear coupling introduced, for example, by a rotated quadrupole, while a more general derivation of all resonances can be found in [1.7]. The equations of linearly coupled motion are of the form

$$x'' + k\,x = -p(s)\,y\,,$$
$$y'' - k\,y = -p(s)\,x\,,$$

which can be derived from the Hamiltonian for linearly coupled motion

$$H \;=\; \tfrac{1}{2}\,x'^2 + \tfrac{1}{2}\,y'^2 + \tfrac{1}{2}\,k\,x^2 - \tfrac{1}{2}\,k\,y^2 + p(s)\,x\,y\,. \qquad (1.104)$$

This Hamiltonian is composed of an uncoupled Hamiltonian H_{o} and the perturbation Hamiltonian

$$H_1 \;=\; p(s)\,x\,y\,, \qquad (1.105)$$

describing the effect of coupling. The solutions for the uncoupled equations with integration constants c_u and ϕ are of the form

$$u(s) \;=\; c_u\,\sqrt{\beta_u}\,\cos[\psi_u(s) + \phi]\,,$$
$$u'(s) \;=\; -\frac{c_u}{\sqrt{\beta_u}}\,\{\alpha_u(s)\,\cos[\psi_u(s) + \phi] + \sin[\psi_u(s) + \phi]\}\,, \qquad (1.106)$$

and applying the method of variation of integration constants we try the ansatz

$$u(s) = \sqrt{2\,a(s)}\,\sqrt{\beta_u}\,\cos[\psi_u(s) + \phi(s)]\,, \qquad (1.107)$$

$$u'(s) = -\frac{\sqrt{2\,a(s)}}{\sqrt{\beta_u}}\,\{\alpha(s)\,\cos[\psi_u(s) + \phi(s)] + \sin[\psi_u(s) + \phi(s)]\}\,,$$

for the coupled motion, where $u = x$ or y and the new variables have been chosen such that the *Jacobian determinant* for the transformation $(u, u') \rightarrow (\phi, a)$ is equal to unity. To show that the new variables are canonical we use the Hamiltonian equations $\partial H/\partial u' = du/ds$ and $\partial H/\partial u = -du'/ds$ and get

$$\frac{\partial H}{\partial u'} = \frac{\partial H_o}{\partial u'} + \frac{\partial H_1}{\partial u'} = \frac{du}{ds} = \frac{\partial u}{\partial s} + \frac{\partial u}{\partial a} \cdot \frac{\partial a}{\partial s} + \frac{\partial u}{\partial \phi} \cdot \frac{\partial \phi}{\partial s} \qquad (1.108)$$

and a similar expression for the second Hamiltonian equation of motion

$$\frac{\partial H}{\partial u} = \frac{\partial H_o}{\partial u} + \frac{\partial H_1}{\partial u} = -\frac{du'}{ds} = -\frac{\partial u'}{\partial s} - \frac{\partial u'}{\partial a} \cdot \frac{\partial a}{\partial s} - \frac{\partial u'}{\partial \phi} \cdot \frac{\partial \phi}{\partial s}. \qquad (1.109)$$

For uncoupled oscillators we know that $a = \mathrm{const}$ and $\phi = \mathrm{const}$ and therefore $\partial u/\partial s = \partial H_o/\partial u'$ and $\partial u'/\partial s = -\partial H_o/\partial u$. With this we derive from (1.105,106) the equations

$$-\frac{\partial H_1}{\partial u}\frac{\partial u}{\partial \phi} - \frac{\partial H_1}{\partial u'}\frac{\partial u'}{\partial \phi} = -\frac{\partial H_1}{\partial \phi} = \frac{da}{ds}\,,$$
$$\frac{\partial H_1}{\partial u}\frac{\partial u}{\partial a} + \frac{\partial H_1}{\partial u'}\frac{\partial u'}{\partial a} = +\frac{\partial H_1}{\partial a} = \frac{d\phi}{ds}\,, \qquad (1.110)$$

demonstrating that the new variables (ϕ, a) are canonical variables and (1.107) are canonical transformations. Applying (1.107) to the perturbation Hamiltonian (1.105) we get with appropriate indices to distinguish between horizontal and vertical plane the perturbation Hamiltonian

$$H_1 = 2\,p(s)\,\sqrt{\beta_x\,\beta_y}\,\sqrt{a_x\,a_y}\,\cos(\psi_x + \phi_x)\,\cos(\psi_y + \phi_y)\,, \qquad (1.111)$$

where s is still the independent variable. The dynamics of linearly coupled motion becomes more evident after isolating the periodic terms in (1.111). For the trigonometric functions we set

$$\cos(\psi_u + \phi_u) = \tfrac{1}{2}\left(e^{i(\psi_u + \phi_u)} + e^{-i(\psi_u + \phi_u)}\right) \qquad (1.112)$$

and the Hamiltonian assumes the form

$$H_1 = \tfrac{1}{2}\,p(s)\,\sqrt{\beta_x\,\beta_y}\,\sqrt{a_x\,a_y}\sum_{l_x, l_y} e^{i\,[l_x\,(\psi_x + \phi_x) + l_y\,(\psi_y + \phi_y)]}\,, \qquad (1.113)$$

where the integers l_x and l_y are defined by

$$-1 \leq l_x, l_y \leq +1 . \tag{1.114}$$

Similar to the one-dimensional case we try to separate constant or slowly varying terms from the fast oscillating terms and expand the exponent in (1.113) like

$$\begin{aligned} l_x\psi_x + l_y\psi_y - l_x\nu_{ox}\varphi - l_y\nu_{oy}\varphi \\ + l_x\nu_{ox}\varphi + l_y\nu_{oy}\varphi + l_x\phi_x + l_y\phi_y , \end{aligned} \tag{1.115}$$

where ν_i are the tunes for the periodic lattice $\varphi = 2\pi\, s/L$ and L is the length of the lattice period. The first four terms in (1.115) are periodic with the period $\varphi(L) = 2\pi + \varphi(0)$. Inserting (1.115) into (1.113) we get with $\psi_i(L) = 2\pi\nu_{oi} + \psi_i(0)$

$$\bar{H}_1 = \tfrac{1}{2} \sum_{l_x, l_y} p(s)\sqrt{\beta_x\,\beta_y}\, \mathrm{e}^{\mathrm{i}\,(l_x\,\psi_x + l_y\,\psi_y - l_x\nu_{ox}\varphi - l_y\nu_{oy}\varphi)} \tag{1.116}$$

$$\times \sqrt{a_x\,a_y}\, \mathrm{e}^{\mathrm{i}\,(l_x\phi_x + l_y\phi_y + l_x\nu_{ox}\varphi + l_y\nu_{oy}\varphi)} .$$

In this form we recognize the periodic factor

$$A(\varphi) = p(s)\,\sqrt{\beta_x\,\beta_y}\, \mathrm{e}^{\mathrm{i}\,(l_x\,\psi_x + l_y\,\psi_y - l_x\nu_{ox}\varphi - l_y\nu_{oy}\varphi)} , \tag{1.117}$$

since betatron functions and perturbations $p(s) = \underline{k}(s)$ are periodic. After expanding (1.117) into a Fourier series

$$\frac{L}{2\pi}\, A(\varphi) = \sum_q \kappa_{q l_x l_y}\, \mathrm{e}^{-\mathrm{i} q N\varphi} \tag{1.118}$$

coupling coefficients can be defined by

$$\kappa_{q l_x l_y} = \frac{1}{2\pi} \int\limits_0^{2\pi} \frac{L}{2\pi}\, A(\varphi)\, \mathrm{e}^{\mathrm{i} q N\varphi}\, \mathrm{d}\varphi \tag{1.119}$$

$$= \frac{1}{2\pi} \int\limits_0^L \sqrt{\beta_x\beta_y}\,\underline{k}\, \mathrm{e}^{\mathrm{i}\,[l_x\,\psi_x + l_y\,\psi_y - (l_x\nu_{ox} + l_y\nu_{oy} - qN)\,2\pi\frac{s}{L}]}\, \mathrm{d}s .$$

Since $\kappa_{q,1,1} = \kappa_{q,-1,-1}$ and $\kappa_{q,1,-1} = \kappa_{q,-1,1}$ we have with $-1 \leq l \leq +1$

$$\boxed{\; \kappa_{ql} = \frac{1}{2\pi} \int\limits_0^L \sqrt{\beta_x\beta_y}\,\underline{k}\, \mathrm{e}^{\mathrm{i}\,[\psi_x + l\,\psi_y - (\nu_{ox} + l\,\nu_{oy} - qN)\,2\pi\frac{s}{L}]}\, \mathrm{d}s . \;} \tag{1.120}$$

The coupling coefficient is a complex quantity indicating that there are two orthogonal and independent contributions which require also two orthogonally independent corrections.

Now that the coupling coefficients are defined in a convenient form for numerical evaluation we replace the independent variable s by the angle variable $\varphi = \frac{2\pi}{L} s$ and obtain the new Hamiltonian $\widetilde{H}_1 = \frac{2\pi}{L} H_1$ or

$$\widetilde{H} = \sum_q \kappa_{ql} \sqrt{a_x a_y} \cos(\phi_x + l\,\phi_y + \Delta\,\varphi) , \tag{1.121}$$

where

$$\Delta = \nu_{ox} + l\,\nu_{oy} - Nq . \tag{1.122}$$

Most terms in (1.121) are fast oscillating and therefore cancel before any damage can be done to the particle stability. One term, however, is only slowly varying for $q = r$ defining the resonance condition for coupled motion

$$r N \approx \nu_{ox} + l\,\nu_{oy} . \tag{1.123}$$

In this resonant case the quantity Δ_r is the distance of the tunes from the coupling resonance as defined by (1.122) with $q = r$. Neglecting all fast oscillating terms we apply one more canonical transformation $(a_i, \phi_i) \rightarrow (\tilde{a}_i, \widetilde{\phi}_i)$ to eliminate the independent variable φ from the Hamiltonian. In essence we thereby use a coordinate system that follows with the unperturbed particle and exhibits only the deviations from the ideal motion. From the generating function

$$G = \tilde{a}_x \left(\phi_x + \tfrac{1}{2}\,\Delta_r \cdot \varphi\right) + \tilde{a}_y \left(\phi_y + l\,\tfrac{1}{2}\,\Delta_r \cdot \varphi\right) \tag{1.124}$$

we get for the new variables

$$
\begin{aligned}
\widetilde{\phi}_x &= \frac{\partial G}{\partial \tilde{a}_x} = \phi_x + \tfrac{1}{2}\Delta_r \cdot \varphi , & a_x &= \frac{\partial G}{\partial \phi_x} = \tilde{a}_x , \\
\widetilde{\phi}_y &= \frac{\partial G}{\partial \tilde{a}_y} = \phi_y + l\,\tfrac{1}{2}\Delta_r \cdot \varphi , & a_y &= \frac{\partial G}{\partial \phi_y} = \tilde{a}_y ,
\end{aligned}
\tag{1.125}
$$

and the new Hamiltonian for the rotating coordinate system is

$$\widetilde{H}_{\mathrm{r}} = \widetilde{H} + \frac{\partial G}{\partial \varphi} = \widetilde{H} + \tfrac{1}{2}\,\Delta_r\,a_x + l\,\tfrac{1}{2}\,\Delta_r\,a_y . \tag{1.126}$$

The resonant Hamiltonian becomes after this transformation

$$\widetilde{H}_{\mathrm{r}} = \tfrac{1}{2}\,\Delta_r\,(a_x + l\,a_y) + \kappa_{rl} \sqrt{a_x a_y} \cos(\widetilde{\phi}_x + l\,\widetilde{\phi}_y) , \tag{1.127}$$

and application of the Hamiltonian formalism gives the equations of motion

$$
\begin{aligned}
\frac{\partial a_x}{\partial \varphi} &= -\frac{\partial \widetilde{H}_{\mathrm{r}}}{\partial \widetilde{\phi}_x} = \kappa_{rl} \sqrt{a_x a_y} \sin(\widetilde{\phi}_x + l\,\widetilde{\phi}_y) , \\
\frac{\partial a_y}{\partial \varphi} &= -\frac{\partial \widetilde{H}_{\mathrm{r}}}{\partial \widetilde{\phi}_y} = l\,\kappa_{rl} \sqrt{a_x a_y} \sin(\widetilde{\phi}_x + l\,\widetilde{\phi}_y) ,
\end{aligned}
\tag{1.128}
$$

and

$$\frac{\partial \tilde{\phi}_x}{\partial \varphi} = \frac{\partial \tilde{H}_r}{\partial a_x} = \tfrac{1}{2}\Delta_r + \tfrac{1}{2}\kappa_{rl}\sqrt{\frac{a_y}{a_x}}\,\cos(\tilde{\phi}_x + l\,\tilde{\phi}_y)\,,$$

$$\frac{\partial \tilde{\phi}_y}{\partial \varphi} = \frac{\partial \tilde{H}_r}{\partial a_y} = l\,\tfrac{1}{2}\Delta_r + \tfrac{1}{2}\kappa_{rl}\sqrt{\frac{a_x}{a_y}}\,\cos(\tilde{\phi}_x + l\,\tilde{\phi}_y)\,. \qquad (1.129)$$

From these equations we can derive criteria for the stability or resonance condition of coupled systems. Depending on the value of l we distinguish a *sum resonance* if $l = +1$ or a *difference resonance* if $l = -1$.

Linear difference resonance: In case of a *difference resonance*, where $l = -1$, we add both equations (1.128) and get

$$\frac{\mathrm{d}}{\mathrm{d}\varphi}\,(a_x + a_y) = 0\,. \qquad (1.130)$$

The coupled motion is stable because the sum of both amplitudes does not change. Both amplitudes a_x and a_y will change such that one amplitude increases at the expense of the other amplitude but the sum of both will not change and therefore neither amplitude will grow indefinitely. Since a_x and a_y are proportional to the beam emittance, we note that the sum of the horizontal and vertical emittance stays constant as well,

$$\epsilon_x + \epsilon_y = \text{const}\,. \qquad (1.131)$$

The resonance condition (1.123) for a difference resonance becomes [1.9]

$$\nu_x - \nu_y = m_r N\,. \qquad (1.132)$$

Our discussion of linear coupling resonances reveals the feature that a difference resonances will cause an exchange of oscillation amplitudes between the horizontal and vertical plane but will not lead to beam instability. This result is important for lattice design. If emittance coupling is desired, one would choose tunes which closely meet the resonance condition. Conversely, when coupling is to be avoided or minimized, tunes are chosen at a safe distance from the coupling resonance.

There exists a finite *stop-band width* also for the coupling resonance just as for any other resonance and we have all the mathematical tools to calculate that width. Since the beam is not lost at a difference coupling resonance, we are also able to measure experimentally the stop-band width by moving the tunes through the resonance. The procedure becomes obvious after linearizing the equations of motion (1.125,126). Following a suggestion by *Guignard* [1.7], we define new variables similar in form to normalized coordinates

$$w = \sqrt{a_x}\,e^{\mathrm{i}\tilde{\phi}_x}\,,$$

$$v = \sqrt{a_y}\,e^{\mathrm{i}\tilde{\phi}_y}\,. \qquad (1.133)$$

Taking derivatives of (1.133) with respect to φ and using (1.128) and (1.129) we get after some manipulation the linear equations

$$\begin{aligned}
\frac{dw}{d\varphi} &= \frac{i}{2}\left(\kappa\, v + \Delta_r\, w\right) \\
\frac{dv}{d\varphi} &= \frac{i}{2}\left(\kappa\, w - \Delta_r\, v\right),
\end{aligned} \tag{1.134}$$

where we have set for simplicity $\kappa_{r,-1} = \kappa$.

These equations can be solved analytically and will provide further insight into the dynamics of coupled oscillations. We will look for characteristics of coupled motion which do not depend on initial conditions but are general for all particles. Expecting the solutions w and v to describe oscillations, we assume that the motion in both planes depends on the initial conditions w_o, v_o in both planes due to the effect of coupling. For simplicity, however, we study the dynamics of a particle which starts with a finite amplitudes $w_o \neq 0$ in the horizontal plane only and set $v_o = 0$. The ansatz for the oscillations be

$$\begin{aligned}
w(\varphi) &= +w_o\left(a\,e^{i\nu\varphi} + b\,e^{-i\nu\varphi}\right), \\
v(\varphi) &= +w_o\left(c\,e^{i\nu\varphi} + d\,e^{-i\nu\varphi}\right),
\end{aligned} \tag{1.135}$$

where we define an as yet undefined frequency ν. Inserting (1.135) into (1.134) the coefficients of the exponential functions vanish separately and we get from the coefficients of $e^{i\nu\varphi}$ the two equations

$$\begin{aligned}
2\nu a &= \kappa c + \Delta_r\, a, \\
2\nu c &= \kappa a - \Delta_r\, c,
\end{aligned} \tag{1.136}$$

from which we may eliminate the unknowns a and c to get the defining equation for the oscillation frequency

$$\nu = \tfrac{1}{2}\sqrt{\Delta_r^2 + \kappa^2}. \tag{1.137}$$

While determining the coefficients $a, b, c,$ and d we note that due to the initial conditions $a + b = 1$ and $c + d = 0$. Similar to (1.136) we derive another pair of equations from the coefficients of $e^{-i\nu\varphi}$

$$\begin{aligned}
2\nu b &= \kappa d - \Delta_r\, b, \\
2\nu d &= \kappa b + \Delta_r\, d,
\end{aligned} \tag{1.138}$$

which completes the set of four equations required to determine with (1.137) the four unknown coefficients

$$\begin{aligned}
a &= \frac{2\nu + \Delta_r}{4\nu}, & b &= \frac{2\nu - \Delta_r}{4\nu}, \\
c &= +\frac{\kappa}{4\nu}, & d &= -\frac{\kappa}{4\nu}.
\end{aligned} \tag{1.139}$$

With this, the solutions (1.135) become

$$w(\varphi) = w_o \cos\nu\varphi + i\,w_o \frac{\Delta_r}{2\nu}\sin\nu\varphi,$$
$$v(\varphi) = i\,w_o \frac{\kappa}{2\nu}\sin\nu\varphi,$$

(1.140)

and by multiplication with the complex conjugate and (1.134) we get expressions for the coupled beam emittances $\epsilon_u = 2a_u$

$$a_x = a_{xo}\frac{1}{4\nu^2}\left(\Delta_r^2 + \kappa^2\cos^2\nu\varphi\right),$$
$$a_y = a_{xo}\frac{\kappa^2}{4\nu^2}\sin^2\nu\varphi.$$

(1.141)

The ratio of maximum values for beam emittances in both planes under the influence of linear coupling is from (1.141)

$$\boxed{\frac{\epsilon_y}{\epsilon_x} = \frac{\kappa^2}{\Delta_r^2 + \kappa^2}.}$$

(1.142)

The emittance coupling increases with the strength of the coupling coefficient and is equal to unity at the coupling resonance or for very large values of κ. At the coupling resonance we observe complete exchange of the emittances at the frequency ν. If on the other hand, the tunes differ and $\Delta_r \neq 0$ there will always be a finite oscillation amplitude left in the horizontal plane because we started with a finite amplitude in this plane. A completely symmetric result would be obtained only for a particle starting with a finite vertical amplitude as well.

We may now collect all results and derive the particle motion as a function of time or φ. For example, the horizontal particle position is determined from (1.107), where we set $\sqrt{a_x} = w\,e^{-i\tilde\phi_x}$ and further replace w by (1.135). Here we are only interested in the oscillation frequencies of the particle motion and note that the oscillatory factor in (1.107) is $\mathrm{Re}\{e^{(\psi_x + \phi_x)}\}$. Together with other oscillatory quantities $e^{-i\tilde\phi_x}$ and w we get both in the horizontal and vertical plane terms with oscillatory factors

$$\mathrm{Re}\{e^{i(\psi_u + \phi_u - \tilde\phi_u \pm \nu\,\varphi)}\},$$

(1.143)

where the index $_u$ stands for either x or y. The phase $\psi_u = \nu_u\,\varphi$ and we have from (1.125) with $l = -1$ for the difference resonance $\tilde\phi_u = \phi_u \pm \tfrac{1}{2}\Delta_r\varphi$. These expressions used in (1.143) define two oscillation frequencies

$$\nu_{I,II} = \nu_{x,y} \mp \tfrac{1}{2}\Delta_r \pm \nu$$

(1.144)

or with (1.119,134)

$$\boxed{\nu_{I,II} = \nu_{x,y} \mp \tfrac{1}{2}\Delta_r \pm \tfrac{1}{2}\sqrt{\Delta_r^2 + \kappa^2}\,.}$$
(1.145)

We have again found the result that under coupling conditions the beta-tron oscillations assume two modes. In a real accelerator only these mode frequencies can be measured under conditions which occur close to the coupling resonance. For very weak coupling, $\kappa \approx 0$, the mode frequencies are approximately equal to the uncoupled frequencies $\nu_{x,y}$, respectively. Even for large coupling this equality is preserved as long as the tunes are far away from the coupling resonance or $\Delta_r \gg \kappa$.

The mode frequencies can be measured while adjusting quadrupoles such that the beam is moved through the coupling resonance. During this adjustment the detuning parameter Δ_r varies and changes sign as the coupling resonance is crossed. For example, if we vary the vertical tune across a coupling resonance from below, we note that the horizontal tune or ν_I does not change appreciably until the resonance is reached, because $-\Delta_r + \sqrt{\Delta_r^2 + \kappa^2} \approx 0$. Above the coupling resonance, however Δ_r has changed sign and ν_I increase with Δ_r. The opposite occurs with the vertical tune. Going through the coupling resonance the horizontal tune has been transformed into the vertical tune and vice versa without ever getting equal.

Actual tune measurements [1.10] are shown in Fig. 1.11 as a function of the excitation current of a vertically focusing quadrupole. The vertical tune change is proportional to the quadrupole current and so is the parameter Δ_r. While increasing the quadrupole current, the vertical tune is increased and the horizontal tune stays practically constant. We note that the tunes actually do not cross the linear coupling resonance during that procedure, rather the tune of one plane is gradually transformed into the tune of the other plane and vice versa. Both tunes never become equal and the closest distance is determined by the magnitude of the coupling coefficient κ.

The coupling coefficient may be nonzero for various reasons. In some cases coupling may be caused because special beam characteristics are desired. In most cases, however, coupling is not desired or planned for and a finite linear coupling of the beam emittances is the result of rotational misalignments of upright quadrupoles. Where this coupling is not desired and must be minimized, we may introduce a pair or two sets of rotated quadrupoles into the lattice to cancel the coupling due to misalignments. The coupling coefficient is defined in (1.120) in the form of a complex quantity. Both orthogonal components must therefore be compensated by two orthogonally located rotated quadrupoles and the proper adjustment of these quadrupoles can be determined by measuring the width of the linear coupling resonance.

Linear sum resonance: To complete the discussion, we will now set $l = +1$ and get from (1.123) the resonance condition for a *sum resonance*

$$\nu_x + \nu_y = m_r N\,.$$
(1.146)

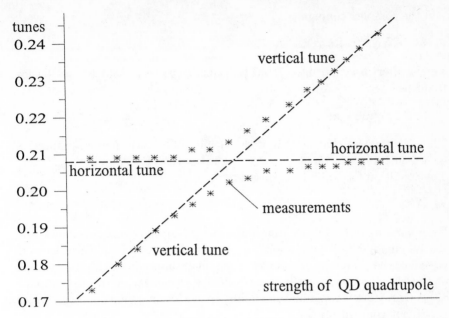

Fig. 1.11. Measurements of mode frequencies as a function of detuning for linearly coupled motion [1.10]

Taking the difference of both equations (1.128) we get

$$\frac{\mathrm{d}}{\mathrm{d}\varphi}\,(a_x - a_y) = 0\,,\tag{1.147}$$

which states only that the difference of the emittances remains constant. Coupled motion in the vicinity of a sum resonance is therefore unstable allowing both emittances to grow unlimited.

To solve the equations of motion (1.125,126) we try the ansatz

$$u = \sqrt{a_x}\,\mathrm{e}^{\mathrm{i}\Phi_x} + \mathrm{i}\,\sqrt{a_y}\,\mathrm{e}^{\mathrm{i}\Phi_y}\,.\tag{1.148}$$

From the derivative $\mathrm{d}u/\mathrm{d}\varphi$, we get with (1.125,126)

$$\frac{\mathrm{d}u}{\mathrm{d}\varphi} = \frac{\mathrm{i}}{2}\,\Delta_r\,w - \frac{\mathrm{i}}{2}\,\kappa\,w^*\tag{1.149}$$

and for the complex conjugate

$$\frac{\mathrm{d}u^*}{\mathrm{d}\varphi} = -\frac{\mathrm{i}}{2}\,\Delta_r\,w^* - \frac{\mathrm{i}}{2}\,\kappa\,w\,.\tag{1.150}$$

Solving these differential equations with the ansatz

$$u = a\,\mathrm{e}^{\mathrm{i}\nu\varphi} + b\,\mathrm{e}^{-\mathrm{i}\nu\varphi}$$

and the complex conjugate

$$u^* = a\,\mathrm{e}^{-\mathrm{i}\nu\varphi} + b\,\mathrm{e}^{\mathrm{i}\nu\varphi}$$

we get after insertion into (1.146,147) analogous to (1.136) the oscillation frequency

$$\nu = \tfrac{1}{2}\sqrt{\Delta_r^2 - \kappa^2}\,. \tag{1.151}$$

This result shows that motion in the vicinity of a linear sum resonance becomes unstable as soon as the detuning is less than the coupling coefficient. The condition for stability is therefore

$$\Delta_r > \kappa\,. \tag{1.152}$$

By a careful choice of the tune difference to avoid a sum resonance and careful alignment of quadrupoles, it is possible in real circular accelerators to reduce the coupling coefficient to very small values. Perfect compensation of the linear coupling coefficient eliminates the linear emittance coupling altogether. However, nonlinear coupling effects become then dominant which we cannot compensate for.

1.3.2 Higher-Order Coupling Resonances

So far all discussions on coupled motions and resonances have been based on linear coupling effects caused by rotated quadrupole fields. For higher-order coupling the mathematical treatment of the beam dynamics is similar although more elaborate. The general form of the nth-order *resonance condition* (1.123) is

$$l_x\nu_x + l_y\nu_y = m_r\,N \qquad \text{with} \qquad |l_x| + |l_y| \le n\,. \tag{1.153}$$

The factors l_x and l_y are integers and the sum $|l_x| + |l_y|$ is called the order of the resonance.

In most cases it is sufficient to choose a location in the resonance diagram which avoids such resonances since circular accelerators are generally designed for minimum coupling. In special cases, however, where strong sextupoles are used to correct chromaticities, coupling resonances can be excited in higher order. The difference resonance $2\nu_x - 2\nu_y$ for example has been observed at the 400 GeV proton synchrotron at the *Fermi National Laboratory*, FNAL [1.8]. Further information on higher-order coupling resonances can be obtained from [1.7], where all sum and difference resonances are discussed in great detail.

1.3.3 Multiple Resonances

We have only discussed isolated resonances. In general, however, nonlinear fields of different orders do exist, each contributing to the stop-band of resonances. A particularly strong source of nonlinearities occurs due to the *beam-beam effect* in colliding-beam facilities where strong and highly nonlinear fields generated by one beam cause significant perturbations to particles in the other beam. The resonance patterns from different resonances are superimposed creating new features of particle instability which were not present in any of the resonances while treated as isolated resonances. Of course, if one of these resonances is unstable for any oscillation amplitude the addition of other weaker resonances will not change this situation.

Combining the effects of several resonances should cause little change for small amplitude oscillations since the trajectory in phase space is close to a circle for resonances of any order provided there is stability at all. Most of the perturbations of resonance patterns will occur in the vicinity of the island structures. When island structures from different resonances start to overlap, *chaotic motion* can occur and may lead to *stochastic instability*. The onset of island overlap is often called the *Chirikov criterion* after *Chirikov* [1.11], who has studied extensively particle motion in such situations.

It is beyond the scope of this text to evaluate the mathematical criteria of multiresonance motion. For further insight and references the interested reader may consult articles in [1.12 – 15]. A general overview and extensive references can also be found in [1.16].

1.4 Symplectic Transformation

Charged particles in electromagnetic fields are conservative systems and their dynamics can be described in the realm of Hamiltonian formalism. This feature assures also for particle beams the validity of far reaching general conclusions for Hamiltonian systems. In particular, we note that transformations in beam dynamics are symplectic, area preserving transformations assuring the conservation of beam phase space. Since this property is of fundamental importance for particle beam dynamics, we recollect the features of symplectic transformations.

The properties of the Wronskian can be generalized from one degree of freedom to an arbitrary number of dimensions in phase space including coupling effects between different degrees of freedom. To do this it is useful to use canonical formulation for the dynamic system. In the realm of linear beam dynamics we need only to make use of the quadratic nature of the Hamiltonian in the canonical variables (q_i, p_i). Following *Courant* and *Snyder* [1.17] we express the coefficients of the quadratic Hamiltonian H by a symmetric matrix $\mathcal{H}$ and get

$$H = \tfrac{1}{2}\mathbf{u}^T \mathcal{H} \mathbf{u}, \tag{1.154}$$

where

$$\mathbf{u} = \begin{pmatrix} q_1 \\ p_1 \\ \vdots \\ q_n \\ p_n \end{pmatrix} \tag{1.155}$$

is the coordinate vector for n degrees of freedom and $\mathbf{u}^T$ is its transpose. With this the equations of motion are simply

$$\frac{d}{ds}\mathbf{u} = \mathcal{S}\mathcal{H}\mathbf{u}, \tag{1.156}$$

where $\mathcal{S}$ is a $2n \times 2n$ matrix composed of n submatrices $\begin{pmatrix} 0 & -1 \\ 1 & 0 \end{pmatrix}$ along the diagonal

$$\mathcal{S} = \begin{pmatrix} \begin{matrix} 0 & -1 \\ 1 & 0 \end{matrix} & & 0 \\ & \ddots & \\ 0 & & \begin{matrix} 0 & -1 \\ 1 & 0 \end{matrix} \end{pmatrix}. \tag{1.157}$$

This formalism can be used to derive invariants of motion for the dynamical system. If $\mathbf{u}_i$ and $\mathbf{u}_k$ are two linearly independent solutions of the equation of motion, we form the vector $\mathbf{u}_i^T \mathcal{S} \mathbf{u}_k$ and investigate its variation along the beam line

$$\frac{d}{ds}[\mathbf{u}_i^T \mathcal{S} \mathbf{u}_k] = \mathbf{u}_i'^T \mathcal{S} \mathbf{u}_k + \mathbf{u}_i^T \mathcal{S} \mathbf{u}_k'. \tag{1.158}$$

Using on the r.h.s. the relation $\mathbf{u}_i'^T = (\mathcal{S}\mathcal{H}\mathbf{u}_i)^T$ from (1.156) we get

$$(\mathcal{S}\mathcal{H}\mathbf{u}_i)^T \mathcal{S}\mathbf{u}_k = \mathbf{u}_i^T \mathcal{H}^T \mathcal{S}^T \mathcal{S}\mathbf{u}_k.$$

From (1.157) we note that $\mathcal{S}^T = -\mathcal{S}$ and $\mathcal{S}^2 = \mathcal{I}$ where $\mathcal{I}$ is the unit matrix. By virtue of symmetry of the quadratic form in the coordinates the transpose of the Hamiltonian is $\mathcal{H}^T = \mathcal{H}$. With this and a similar procedure for the second term on the r.h.s. of (1.158) we find

$$\frac{d}{ds}\mathbf{u}_i^T \mathcal{S}\mathbf{u}_k = 0. \tag{1.159}$$

Applying this to one degree of freedom only, (1.159) becomes after integration

$$\begin{pmatrix} u_2 & p_2 \end{pmatrix} \begin{pmatrix} 0 & -1 \\ 1 & 0 \end{pmatrix} \begin{pmatrix} u_1 \\ p_1 \end{pmatrix} = u_1 p_2 - u_2 p_1 = \text{const}, \tag{1.160}$$

which is identical to the constancy of the Wronskian. These general constants of motion for any pairs of linearly independent solutions u_i and u_k

$$\mathbf{u}_i^{\mathrm{T}} \mathbf{S} \mathbf{u}_k = \text{const} \tag{1.161}$$

are also called *Lagrange invariants*. We may equate expressions (1.161) for two locations s_o and s_1. Using the transformation $\mathbf{u}(s_1) = \mathcal{M}(s_1, s_0)\,\mathbf{u}(s_0)$ we find after some manipulation the *symplecticity condition* [1.18]

$$\mathcal{M}^{\mathrm{T}} \mathbf{S} \mathcal{M} = \mathbf{S}. \tag{1.162}$$

Any transformation matrix $\mathcal{M}$ which meets condition (1.162) is called a *symplectic transformation* matrix. We use this property to check the validity of a transformation matrix since all transformation matrices in beam dynamics which do not involve damping or quantum effects must be symplectic. This is particularly important in particle tracking programs where we may want to use some approximation. Deviations from symplecticity due to such approximations may lead to artifacts while tracking particles through many betatron or synchrotron oscillations.

Problems

Problem 1.1.   Prove that the Jacobian determinant for the transformation (1.107) is indeed equal to unity.

Problem 1.2.   Determine which of the following transformations are canonical and which are not:

a) $q_1 = x_1$ $p_1 = \dot{x}_1$

             $q_2 = x_2$ $p_2 = \dot{x}_2$

b) $q = r \cos \psi$ $p = r \sin \psi$

c) $q_1 = x_1$ $p_1 = \dot{x}_1 \pm \dot{x}_2$

             $q_2 = x_1 \pm x_2$ $p_2 = \dot{x}_2$

d) $q = q_\mathrm{o}\, e^\epsilon$ $p = p_\mathrm{o}\, e^\epsilon$

Prove the validity of formalism you use.

Problem 1.3.   Using expressions from this chapter only, show that the Hamiltonian transforms like $H_\varphi = \frac{\mathrm{d}t}{\mathrm{d}\varphi} H_t$ if the independent variable is changed from t to $\varphi(t)$.

Problem 1.4.   Plot a resonance diagram up to fourth order for the PEP lattice with tunes $\nu_x = 21.28$ and $\nu_y = 18.16$ and a superperiodicity of six or any other circular accelerator lattice with multiple superperiodicity. Choose the parameters of the diagram such that a wide resonance environment for the above tunes of at least ± 3 ($\pm$ half the number of superperiods) integers is covered.

Problem 1.5. Choose numerical values for parameters of a single multipole in the Hamiltonian (1.69) and plot a resonance diagram $H(J, \psi) = \text{const}$. Determine the stability limit for your choice of parameters. What would the tolerance on the multipole field perturbation be if you require a stability for an emittance as large as $\epsilon = 100$ mm-mrad?

Problem 1.6. Consider a simple FODO lattice of your choice forming a circular ring. Calculate the natural chromaticity (ignore focusing in bending magnets) and correct the chromaticities to zero by placing thin sextupoles in the center of the quadrupoles. Calculate and plot the horizontal third-order stop-band width as a function of the horizontal tune.

Problem 1.7. Take the lattice of problem 1.6 and adjust close to the third-order resonance so that the unstable fixed point on the symmetry axis are 5 cm from the beam center. Determine the equations for the separatrices. Choose a point P just outside the stable area and close to the crossing of two separatrices along the symmetry axis. Where in the diagram would a particle starting at P be after 3, 6, and 9 turns? At what amplitude could you place a 5 mm thin septum magnet to eject the beam from the accelerator?

Problem 1.8. In circular accelerators rotated quadrupoles may be inserted to compensate for coupling due to misalignments. Assume a statistical distribution of rotational quadrupole errors which need to be compensated by special rotated quadrupoles. How many such quadrupoles are required and and what criteria would you use for optimum placement in the ring?

Problem 1.9. Use the lattice from problem 1.6 and introduce a gaussian distribution of rotational quadrupole misalignments. Calculate and plot the coupling coefficient for the ring and the emittance ratio as a function of the rms misalignment. If the emittance coupling is to be held below 1% how well must the quadrupoles be aligned? Insert two rotated quadrupoles into the lattice such that they can be used to compensate the coupling due to misalignments. Calculate the required quadrupole strength.

Problem 1.10. Use the measurement in Fig. 1.11 and determine the the the coupling coefficient κ.

2. General Electromagnetic Fields

In this Chapter we will derive expressions for general field and alignment errors to be used in the Hamiltonian perturbation theory. Pure electromagnetic multipole fields have been derived in [2.1] from the *Laplace equation* while ignoring *kinematic terms* as well as *higher-order field perturbations*. In preparation for more sophisticated beam transport systems and accelerator designs aiming, for example, at ever smaller beam emittances it becomes imperative to consider higher-order perturbations to preserve desired beam characteristics. We will therefore derive the general electromagnetic fields from the curvilinear Laplace equation and formulate the general equations of motion in both planes with all higher-order geometric and chromatic perturbations up to third order. To meet also the increasing use of wiggler magnets and superconducting magnets, we discuss magnetic parameters of both magnet types.

2.1 General Transverse Magnetic-Field Expansion

The equations of motion in the presence of pure *multipole magnet fields* has been derived in [2.1]. Solving the *Laplace equation*, we made a restrictive ansatz which included only pure multipole components. We also neglected all geometric effects caused by the curvilinear coordinate system except for the linear focusing term $1/\rho_o^2$ occurring in sector magnets. These approximations eliminate many higher-order terms which may become of significance for particular beam transport designs. To obtain all field components allowed by the Laplace equation, a more general ansatz for the field expansion must be made.

The particular distribution of the magnetic field in a magnet is greatly determined by the technical design defining the exact location of boundary surfaces. To obtain, however, the composition of a general magnetic field we ignore at this point the boundary conditions and assume only that the magnets used in beam transport systems provide the desired field in a material free *aperture* for the passage of the particle beam. The general field is then determined by the solution of the Laplace equation [2.2, 3] for the magnetic *scalar potential*. Since we use a curvilinear coordinate system for beam dynamics, we use the same for the magnetic-field expansion and express the Laplace equation in these *curvilinear coordinates*

$$\Delta V = \frac{1}{H} \left[\frac{\partial}{\partial x} \left(H \frac{\partial V}{\partial x} \right) + \frac{\partial}{\partial y} \left(H \frac{\partial V}{\partial y} \right) + \frac{\partial}{\partial s} \left(\frac{1}{H} \frac{\partial V}{\partial s} \right) \right], \qquad (2.1)$$

where

$$H = 1 + \kappa_x x + \kappa_y y. \qquad (2.2)$$

We also assume without restricting full generality that the particle beam may be bend horizontally as well as vertically. For the general solution of the Laplace equation (2.1) we use an ansatz in the form of a power expansion

$$V(x, y, s) = -\frac{cp}{e} \sum_{p,q \geq 0} A_{pq}(s) \frac{x^p}{p!} \frac{y^q}{q!}. \qquad (2.3)$$

Terms with negative indices p and q are excluded to avoid nonphysical divergences of the potential at $x = 0$ or $y = 0$. We insert this ansatz into (2.1), collect all terms of equal powers in x and y and get

$$\sum_{p \geq 0} \sum_{q \geq 0} \{ F_{pq} \} \frac{x^p}{(p-2)!} \frac{y^q}{(q-2)!} \equiv 0, \qquad (2.4)$$

where $\{F_{pq}\}$ represents the collection of all coefficients for the term $x^p y^q$.

We have chosen here the *scalar potential* to derive the general magnetic fields. Of course those fields also could be derived from a *vector potential* **A**. For pure transverse fields, however, only the longitudinal component of the *vector potential* is nonzero, $\mathbf{A} = (0, 0, A_z)$ and the magnetic fields are just the derivatives of A_z with respect to the transverse coordinates, $\mathbf{B} = \partial A_z / \partial y, -\partial A_z / \partial x, 0$. The mathematical formalism would be identical to the one we follow to derive the general magnetic fields from the scalar potential.

For (2.4) to be true for all values of the coordinates x and y we require that every coefficient F_{pq} must vanish individually. Setting $F_{pq} = 0$ leads to the *recursion formula*

$$
\begin{aligned}
A_{p,q+2} + A_{p+2,q} = &\, -\kappa_x (3p+1) A_{p+1,q} - \kappa_y (3q+1) A_{p,q+1} \\
& - 3\kappa_y q\, A_{p+2,q-1} - 3\kappa_x p\, A_{p-1,q+2} \\
& - 2\kappa_x \kappa_y (3p+1) A_{p+1,q-1} - 2\kappa_x \kappa_y p (3q+1) A_{p-1,q+1} \\
& - 3\kappa_y^2 q (q-1) A_{p+2,q-2} - 3\kappa_x^2 p (p-1) A_{p-2,q+2} \\
& - \kappa_x^3 p (p^2 - 3p + 2) A_{p-3,q+2} - \kappa_y^3 q (q^2 - 3q + 2) A_{p+2,q-3} \\
& - \kappa_x \kappa_y^2 q (q - 1 + 3pq - 3p) A_{p+1,q-2} \\
& - \kappa_x^2 \kappa_y p (p - 1 + 3pq - 3q) A_{p-2,q+1} \\
& - \kappa_y q (3\kappa_x^2 p^2 - \kappa_x^2 p + \kappa_y^2 q^2 - 2\kappa_y^2 q + \kappa_y^2) A_{p,q-1} \\
& - \kappa_x p (3\kappa_y^2 q^2 - \kappa_y^2 q + \kappa_x^2 p^2 - 2\kappa_x^2 p + \kappa_x^2) A_{p-1,q} \\
& - (3p-1) p\kappa_x^2 A_{p,q} - (3q-1) q\kappa_y^2 A_{p,q} \\
& - A_{p,q}'' - \kappa_x p A_{p-1,q}'' - \kappa_y q A_{p,q-1}'' - \kappa_x' p A_{p-1,q}' - \kappa_y' q A_{p,q-1}'
\end{aligned}
\qquad (2.5)
$$

which allows us to determine all coefficients A_{pq}. The derivatives, indicated by a prime, are understood to be taken with respect to the independent variable s, like $A' = \mathrm{d}A/\mathrm{d}s$, etc. Although most beam transport lines include only horizontal bending magnets, we have chosen a fully symmetric field expansion to be completely general. Equation (2.5) is a recursion formula for the field coefficients A_{pq} and we have to develop a procedure to obtain all terms consistent with this expression.

The Laplace equation is of quadratic order and therefore we cannot derive coefficients of quadratic or lower order from the recursion formula. The lowest-order coefficient A_{00} represents a constant potential independent of the transverse coordinates x and y and since this term does not contribute to a transverse field component, we will ignore it in this section. However, where *longitudinal field components* become important we cannot continue to neglect this term. Such fields will be discussed in more detail in Chap. 3 and we set for simplicity therefore

$$A_{00} = 0 \,. \tag{2.6}$$

The terms linear in x or y are the curvatures in the horizontal and vertical plane

$$A_{10} = -\kappa_y \qquad \text{and} \qquad A_{01} = \kappa_x \,, \tag{2.7}$$

defined in [2.1] by $(\kappa_x, \kappa_y) = (-x'', -y'')$ or

$$\boxed{
\begin{aligned}
\kappa_x &= -x'' = +\frac{[c]\,e}{cp}\,B_y \quad \text{with} \quad \left|\frac{[c]\,e}{cp}\,B_y\right| = \frac{1}{\rho_x}\,, \\[2mm]
\kappa_y &= -y'' = -\frac{[c]\,e}{cp}\,B_x \quad \text{with} \quad \left|\frac{[c]\,e}{cp}\,B_x\right| = \frac{1}{\rho_y}\,.
\end{aligned}
}
\tag{2.8}$$

Finally, the quadratic term proportional to x and y is identical to the quadrupole strength parameter

$$A_{11} = k \,. \tag{2.9}$$

With these definitions of the linear coefficients we may start exploiting the recursion formula. All terms on the right-hand side of (2.5) are of lower order than the two terms on the left-hand side which are of order $N = p + q + 2$. The left-hand side therefore is composed of two contributions, one resulting from pure multipole fields of order N and the other from higher-order field terms of lower-order multipoles.

We identify and separate from all other terms the *pure multipole terms* of order N which do not depend on lower-order multipole terms by setting

$$A_{p,q+2,N} + A_{p+2,q,N} = 0 \tag{2.10}$$

and adding the index N to indicate that these terms are the pure Nth-order multipoles. Only the sum of two terms can be determined which means both terms have the same value but opposite signs. For $N = 2$ we have, for example, $A_{20} = -A_{02}$ and a comparison with the potentials of pure multipoles in Tab. 2.1, which is the reproduction of [Ref. 2.1, Tab. 4.2] shows that $A_{20} = -\underline{k}$ [1].

Table 2.1. Correspondence between the potential coefficients and multipole strength parameters[1]

				A_{00}						
			A_{10}		A_{01}					
		A_{20}		A_{11}		A_{02}				
	A_{30}		A_{21}		A_{12}		A_{03}			
A_{40}		A_{31}		A_{22}		A_{13}		A_{04}		
A_{50}	A_{41}		A_{32}		A_{23}		A_{14}		A_{05}	

$$\Updownarrow$$

				0						
			$-\kappa_y$		κ_x					
		$-\underline{k}$		k		$\underline{k}$				
	$-\underline{m}$		m		$\underline{m}$		$-m$			
$-\underline{r}$		r		$\underline{r}$		$-r$		$-\underline{r}$		
$-\underline{d}$	d		$\underline{d}$		$-d$		$-\underline{d}$		d	

In third order we get similarly $A_{30} = -A_{12} = -\underline{m}$ and $A_{21} = -A_{03} = m$. The fourth-order terms are determined by $A_{40} = -A_{22} = -\underline{r}$, $A_{22} = -A_{04} = \underline{r}$ and $A_{31} = -A_{13} = r$. By a systematic application of all allowed values for the indices a correspondence between the coefficients $A_{jk,N}$ and the multipole strength parameters defined in Tab. 2.1 can be established. All pure multipoles are determined by (2.10) alone.

Having identified the pure multipole components, we concentrate now on using the recursion formula for other terms which we chose so far to neglect. First we note that coefficients of the same order N on the left-hand side of (2.5) must be split into two parts to distinguish pure multipole components $A_{jk,N}$ of order N from the Nth-order terms A_{jk}^* of lower-order multipoles which we label by an asterisk $*$. Since we have already derived the pure multipole terms, we explore (2.5) for the A^* coefficients only

$$A_{p,q+2}^* + A_{p+2,q}^* = \text{r.h.s. of (2.5)}. \tag{2.11}$$

---

[1] Consistent with the definitions of magnet strengths in [2.1] the underlined quantities represent the magnet strengths of rotated multipole magnets.

For the predetermined coefficients A_{10}, A_{01} and A_{11} there are no corresponding terms A^* since that would require indices p and q to be negative. For $p = 0$ and $q = 0$ we have

$$A_{02}^* + A_{20}^* = -\kappa_x A_{10} - \kappa_y A_{01} = 0 \,. \tag{2.12}$$

This solution is equivalent to (2.10) and does not produce any new field terms. The next higher-order terms for $p = 0$ and $q = 1$ or for $p = 1$ and $q = 0$ are determined by the equations

$$\begin{aligned} A_{03}^* + A_{21}^* &= -\kappa_x k - \kappa_y \underline{k} - \kappa_x'' = C \,, \\ A_{12}^* + A_{30}^* &= -\kappa_y k + \kappa_x \underline{k} + \kappa_y'' = D \,, \end{aligned} \tag{2.13}$$

where we set in preparation for the following discussion the right-hand sides equal to the as yet undetermined quantities C and D. Since we have no lead how to separate the coefficients we set

$$\begin{aligned} A_{21}^* &= f\,C \,, & A_{03}^* &= (1-f)\,C \,, \\ A_{12}^* &= g\,D \,, & A_{30}^* &= (1-g)\,D \,, \end{aligned} \tag{2.14}$$

where $0 \le (f,g) \le 1$ and $f = g$. The indeterminate nature of this result is an indication that these terms may depend on the actual design of the magnets.

Trying to interpret the physical meaning of these terms, we assume a magnet with a pure vertical dipole field in the center of the magnet, $B_y(0,0,0) \ne 0$, but no horizontal or finite longitudinal field components, $B_x(0,0,0) = 0$ and $B_s(0,0,0) = 0$. Consistent with these assumptions the magnetic potential is

$$\begin{aligned} [c]\frac{e}{cp} V(x,y,s) = &-A_{01}y - \tfrac{1}{2}A_{21}^* x^2 y - \tfrac{1}{2}A_{12}^* xy^2 \\ &-\tfrac{1}{6}A_{30}^* x^3 - \tfrac{1}{6}A_{03}^* y^3 + \mathcal{O}(4) \,. \end{aligned} \tag{2.15}$$

From (2.13) we get $D \equiv 0$, $C = -\kappa_x''$ and with (2.14) $A_{12}^* = A_{30}^* = 0$. The magnetic-field potential reduces therefore to

$$\frac{[c]e}{cp} V(x,y,s) = -\kappa_x y + \tfrac{1}{2} f\kappa_x'' x^2 y + \tfrac{1}{6}(1-f)\kappa_x'' y^3 \tag{2.16}$$

and the magnetic-field components are

$$\begin{aligned} \frac{[c]e}{cp} B_x &= -f\kappa_x'' xy \,, \\ \frac{[c]e}{cp} B_y &= +\kappa_x - \tfrac{1}{2}f\kappa_x'' x^2 - \tfrac{1}{2}(1-f)\kappa_x'' y^2 \,. \end{aligned} \tag{2.17}$$

The physical origin of these terms becomes apparent if we investigate the two extreme cases for which $f = 0$ or $f = 1$ separately. The magnetic fields in these cases are for $f = 0$

$$\frac{[c]e}{cp} B_x = 0 \quad \text{and} \quad \frac{[c]e}{cp} B_y = \kappa_x - \frac{1}{2}\kappa_x'' y^2 \tag{2.18}$$

and for $f = 1$

$$\frac{[c]e}{cp} B_x = -\kappa_x'' xy \quad \text{and} \quad \frac{[c]e}{cp} B_y = \kappa_x - \frac{1}{2}\kappa_x'' x^2 , \tag{2.19}$$

where the curvatures are functions of s.

Both cases differ only in the κ_x''-terms describing the magnet fringe field. In the case of a straight bending magnet ($\kappa_x \neq 0$) with infinitely wide poles in the x-direction, horizontal field components B_x must vanish consistent with $f = 0$. The field configuration in the fringe field region is of the form shown in Fig. 2.1 and independent of x.

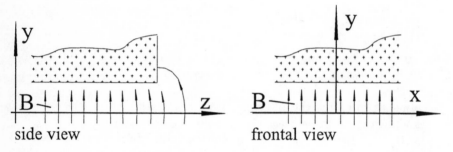

side view   frontal view

Fig. 2.1. Dipole end field configuration for $f = 0$

Conversely, the case $0 < f < 1$ describes the field pattern in the fringe field of a bending magnet with poles of finite width in which case finite horizontal field components B_x appear off the symmetry planes. The fringe fields not only bulge out of the magnet gap along s but also spread horizontally due to the finite pole width as shown in Fig. 2.2 thus creating a finite horizontal field component off the midplane. While it is possible to identify the origin of these field terms, we are not able to determine the exact value of the factor f in a general way but may apply three-dimensional magnet codes to determine the field configuration numerically. The factor f is different for each type of magnet depending on its actual mechanical dimensions.

Following general praxis in beam dynamics and magnet design, however, we ignore these effects of finite pole width, since they are generally being kept small by designing for wide pole pieces, and we set $f = g = 0$. In this approximation we get

$$A_{21}^* = A_{12}^* = 0 \tag{2.20}$$

and
$$A_{03}^* = -\kappa_x k - \kappa_y \underline{k} - \kappa_x'',$$
$$A_{30}^* = -\kappa_y k + \kappa_x \underline{k} + \kappa_y''. \tag{2.21}$$

Similar effects of a finite pole sizes appear for all multipole terms where $m \neq n$. As before, we set $f = 0$ for lack of accurate knowledge of the actual magnet design and assume that these terms are very small because the magnet pole width has been chosen broad enough to provide sufficient good field region for the particle beam. For the fourth-order terms we have therefore with $A_{22}^* \equiv 0$ and

$$A_{40}^* = \kappa_x \underline{m} - \kappa_y m - 4\kappa_x \kappa_y k + 4\kappa_x^2 \underline{k} + \underline{k}'' + 2\kappa_x \kappa_y'' + 2\kappa_x' \kappa_y',$$
$$A_{04}^* = \kappa_y m - \kappa_x \underline{m} - 4\kappa_x \kappa_y k - 4\kappa_y^2 \underline{k} - \underline{k}'' - 2\kappa_y \kappa_x'' - 2\kappa_y' \kappa_x'. \tag{2.22}$$

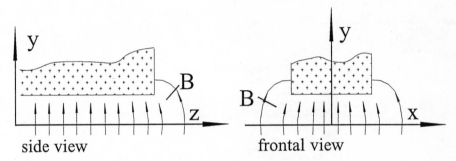

side view   frontal view

Fig. 2.2. Dipole end field configuration for $0 < f < 1$

In the case $p = q$ we expect $A_{ij} = A_{ji}$ from symmetry and get

$$2A_{13}^* = 2A_{31}^* = -\kappa_x m - \kappa_y \underline{m} + 2\kappa_x^2 k + 2\kappa_y^2 k - k''$$
$$+ 2\kappa_y \kappa_y'' - 2\kappa_x \kappa_x'' - \kappa_x'^2 + \kappa_y'^2. \tag{2.23}$$

With these terms we have finally determined all coefficients of the magnetic potential up to fourth order. Higher-order terms can be derived along similar arguments. Using these results the general *magnetic-field potential* up to fourth order is from (2.3)

$$-\frac{[c]e}{cp} V(x,y,s) = + A_{10}x + A_{01}y + \tfrac{1}{2}A_{20}x^2 + \tfrac{1}{2}A_{02}y^2 + A_{11}xy$$
$$+ \tfrac{1}{6}A_{30}x^3 + \tfrac{1}{2}A_{21}x^2 y + \tfrac{1}{2}A_{12}xy^2 + \tfrac{1}{6}A_{03}y^3$$
$$+ \tfrac{1}{6}A_{30}^* x^3 + \tfrac{1}{6}A_{03}^* y^3 \tag{2.24}$$
$$+ \tfrac{1}{24}A_{40}x^4 + \tfrac{1}{6}A_{31}x^3 y + \tfrac{1}{4}A_{22}x^2 y^2 + \tfrac{1}{6}A_{13}xy^3$$
$$+ \tfrac{1}{24}A_{04}y^4 + \tfrac{1}{24}A_{40}^* x^4 + \tfrac{1}{6}A_{31}^* xy(x^2 + y^2)$$
$$+ \tfrac{1}{24}A_{04}^* y^4 + \mathcal{O}(5).$$

From the magnetic potential we obtain the *magnetic field expansion* by differentiation with respect to x or y for B_x and B_y, respectively. Up to third order we obtain the transverse field components

$$
\begin{aligned}
\frac{[c]e}{cp} B_x = &-\kappa_y - \underline{k}x + ky \\
&- \tfrac{1}{2}\underline{m}(x^2-y^2) + mxy + \tfrac{1}{2}\left(-\kappa_y k + \kappa_x \underline{k} + \kappa_y''\right)x^2 \\
&- \tfrac{1}{6}\underline{r}(x^3 - 3xy^2) - \tfrac{1}{6}r(y^3 - 3x^2 y) \\
&- \tfrac{1}{12}\left(\kappa_x m + \kappa_y \underline{m} + 2\kappa_x^2 k + 2\kappa_y^2 k + k'' - \kappa_y \kappa_y'' \right. \\
&\left. \quad + \kappa_x \kappa_x'' + \kappa_x'^2 - \kappa_y'^2\right)(3x^2 y + y^3) \\
&+ \tfrac{1}{6}(\kappa_x \underline{m} - \kappa_y m - 4\kappa_x \kappa_y k + 4\kappa_x^2 \underline{k} \\
&\quad + \underline{k}'' + 2\kappa_x \kappa_y'' + 2\kappa_x' \kappa_y')x^3 + \mathcal{O}(4)
\end{aligned}
\tag{2.25}
$$

and

$$
\begin{aligned}
\frac{[c]e}{cp} B_y = &+\kappa_x + \underline{k}y + kx \\
&+ \tfrac{1}{2}m(x^2-y^2) + \underline{m}xy - \tfrac{1}{2}\left(\kappa_x k + \kappa_y \underline{k} + \kappa_x''\right)y^2 \\
&+ \tfrac{1}{6}r(x^3 - 3xy^2) - \tfrac{1}{6}\underline{r}(y^3 - 3x^2 y) \\
&- \tfrac{1}{12}\left(\kappa_x m + \kappa_y \underline{m} + 2\kappa_x^2 k + 2\kappa_y^2 k + k'' - \kappa_y \kappa_y'' \right. \\
&\left. \quad + \kappa_x \kappa_x'' + \kappa_x'^2 - \kappa_y'^2\right)(x^3 + 3xy^2) \\
&+ \tfrac{1}{6}(\kappa_y m - \kappa_x \underline{m} - 4\kappa_x \kappa_y k - 4\kappa_y^2 \underline{k} \\
&\quad - \underline{k}'' - 2\kappa_x'' \kappa_y - 2\kappa_x' \kappa_y')y^3 + \mathcal{O}(4).
\end{aligned}
\tag{2.26}
$$

The third component of the gradient in a curvilinear coordinate system is $\frac{[c]e}{cp} B_s = -\frac{1}{H}\frac{\partial V}{\partial s}$ and collecting all terms up to second order we get

$$
\begin{aligned}
\frac{[c]e}{cp} B_s = &+\kappa_x' y - \kappa_y' x + (\kappa_y \kappa_y' - \kappa_x \kappa_x' + k')xy \\
&+ (\kappa_x \kappa_y' - \tfrac{1}{2}\underline{k}')x^2 - (\kappa_x' \kappa_y - \tfrac{1}{2}\underline{k}')y^2 + \mathcal{O}(3).
\end{aligned}
\tag{2.27}
$$

While equations (2.25 – 27) describe the general fields in a magnet, in praxis, special care is taken to limit the number of fundamentally different field components present in any one magnet. In fact most magnet are designed as single multipoles like dipoles or quadrupoles or sextupoles etc. A beam transport system utilizing only such magnets is also called a *separated-function lattice* since bending and focusing is performed in different types of magnets.

A combination of bending and focusing, however, is being used for some special applications and in the magnet structure of most early synchrotrons. A transport system composed of such combined-field magnets is called a *combined-function lattice*. Sometimes even a sextupole term is incorporated in a magnet together with the dipole and quadrupole fields. Rotated magnets, like rotated sextupoles $\underline{m}$ and octupoles $\underline{r}$ are either not used or in the case of a rotated quadrupole the chosen strength is generally weak and its effect on the beam dynamics is treated by perturbation methods.

No mention has been made about *electric field patterns*. However, since the Laplace equation for *electrical fields* in material free areas is the same as for magnetic fields we conclude that the electrical potentials are expressed by (2.24) as well and the *electrical multipole field* components are also given by (2.25) to (2.27) after replacing the nomenclature for the magnetic field (B_x, B_y, B_s) by electric-field components (E_x, E_y, E_s).

2.2 Third-Order Differential Equation of Motion

Equations of motions have been derived in [2.1] for the transverse $(\mathbf{x}, \mathbf{s})$ and $(\mathbf{y}, \mathbf{s})$ planes up to second order which is adequate for most applications. For very special and sophisticated beam lines and accelerators, however, we need to derive the equations of motion in higher order of precision. A *curvilinear coordinate system* moving along the curved trajectory of the reference particle was used and we generalize this system to include curvatures in both transverse planes as shown in Fig. 2.3.

In this (x, y, s)-coordinate system a particle at the location σ and under the influence of the *Lorentz force* follows a path described by the vector $\mathbf{S}$ as shown in Fig. 2.3. The change in the momentum vector per unit time is due only to a change in the direction of the momentum while the value is unchanged in static magnetic fields. We set therefore $\mathbf{p} = p \, d\mathbf{S}/d\sigma$ where p is the value of the particle momentum and $d\mathbf{S}/d\sigma$ is the unit vector along the particle trajectory. With $\frac{d\mathbf{p}}{d\tau} = \frac{d\mathbf{p}}{d\sigma}\beta c$, where $\tau = \frac{\sigma}{\beta c}$ and $\mathbf{v}_\sigma = \frac{d\mathbf{S}}{d\tau} = \frac{d\mathbf{S}}{d\sigma}\beta c$, we finally obtain the differential equation describing the particle trajectory under the influence of a Lorentz force $\mathbf{F}_L$ from $\frac{d\mathbf{p}}{d\tau} = \mathbf{F}_L = \frac{[c]e}{c}[\mathbf{v}_\sigma \times \mathbf{B}]$

$$\frac{d^2\mathbf{S}}{d\sigma^2} = \frac{[c]\,e}{cp}\left[\frac{d\mathbf{S}}{d\sigma} \times \mathbf{B}\right]. \tag{2.28}$$

To evaluate (2.28) further we note that

$$\frac{d\mathbf{S}}{d\sigma} = \frac{d\mathbf{S}/ds}{d\sigma/ds} = \frac{\mathbf{S}'}{\sigma'} \tag{2.29}$$

and set

$$\frac{d^2\mathbf{S}}{d\sigma^2} = \frac{1}{\sigma'}\frac{d}{ds}\frac{\mathbf{S}'}{\sigma'}. \tag{2.30}$$

With this the general equation of motion is from (2.28)

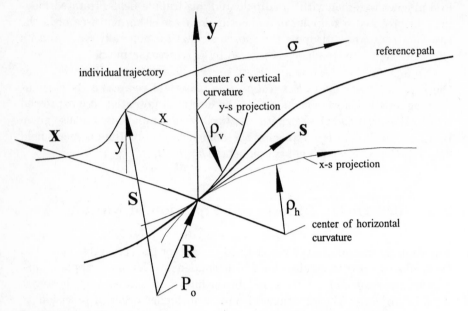

Fig. 2.3. Coordinate system

$$\frac{\mathrm{d}^2 \mathbf{S}}{\mathrm{d}s^2} - \frac{1}{2\sigma'^2}\frac{\mathrm{d}\mathbf{S}}{\mathrm{d}s}\frac{\mathrm{d}\sigma'^2}{\mathrm{d}s} = \frac{[c]\,e}{cp}\,\sigma'\left[\frac{\mathrm{d}\mathbf{S}}{\mathrm{d}s}\times\mathbf{B}\right]. \qquad (2.31)$$

In the remainder of this section, we will re-evaluate this equation in terms of more simplified parameters. From Fig. 2.3 we obtain

$$\mathbf{S} = x\,\mathbf{x}_\mathrm{o} + y\,\mathbf{y}_\mathrm{o} + \mathbf{R}, \qquad (2.32)$$

where the vectors $\mathbf{x}_\mathrm{o}$, $\mathbf{y}_\mathrm{o}$ and $\mathbf{s}_\mathrm{o}$ are the unit vectors defining the curvilinear coordinate system and variation of (2.32) yields

$$\mathrm{d}\mathbf{S} = x\,\mathrm{d}\mathbf{x}_\mathrm{o} + \mathrm{d}x\,\mathbf{x}_\mathrm{o} + y\,\mathrm{d}\mathbf{y}_\mathrm{o} + \mathrm{d}y\,\mathbf{y}_\mathrm{o} + \mathrm{d}\mathbf{R}. \qquad (2.33)$$

We also derive from Fig. 2.3 for small values of the deflection angles $(\delta_\mathrm{h}, \delta_\mathrm{v})$ the geometric relations

$$\begin{aligned}
\mathrm{d}\mathbf{x}_\mathrm{o} &= \mathbf{s}_\mathrm{o}\tan\delta_h \approx \kappa_x\,\mathbf{s}_\mathrm{o}\,\mathrm{d}s \\
\mathrm{d}\mathbf{y}_\mathrm{o} &= -\mathbf{s}_\mathrm{o}\tan\delta_v \approx \kappa_y\,\mathbf{s}_\mathrm{o}\,\mathrm{d}s, \\
\mathrm{d}\mathbf{s}_\mathrm{o} &= \kappa_x\,\mathbf{x}_\mathrm{o}\,\mathrm{d}s + \kappa_y\,\mathbf{y}_\mathrm{o}\,\mathrm{d}s, \\
\mathrm{d}\mathbf{R} &= \mathbf{s}_\mathrm{o}\,\mathrm{d}s.
\end{aligned} \qquad (2.34)$$

Inserting (2.34) into (2.33) we get

$$\frac{\mathrm{d}\mathbf{S}}{\mathrm{d}s} = x'\,\mathbf{x}_\mathrm{o} + y'\,\mathbf{y}_\mathrm{o} + (1 + \kappa_x x + \kappa_y y)\,\mathbf{s}_\mathrm{o} \qquad (2.35)$$

and after squaring with $\mathbf{S}'^2 = \sigma'^2$

$$\sigma'^2 = x'^2 + y'^2 + (1 + \kappa_x x + \kappa_y y)^2 \,. \tag{2.36}$$

To completely evaluate (2.31) the second derivative $\mathrm{d}^2 \mathbf{S}/\mathrm{d}s^2$ must be derived from (2.35) with (2.34)

$$\begin{aligned}
\frac{\mathrm{d}^2 \mathbf{S}}{\mathrm{d}s^2} = & + \mathbf{x}_\mathrm{o} \left(x'' + \kappa_x + x\kappa_x^2 + y\kappa_x\kappa_y \right) \\
& + \mathbf{y}_\mathrm{o} \left(y'' + \kappa_y + x\kappa_x\kappa_y + y\kappa_y^2 \right) \\
& + \mathbf{s}_\mathrm{o} \left(2\kappa_x x' + 2\kappa_y y' + \kappa_x' x + \kappa_y' y \right).
\end{aligned} \tag{2.37}$$

Equation (2.31) becomes with (2.35) through (2.37)

$$\begin{aligned}
& \mathbf{x}_\mathrm{o} \left(x'' + \kappa_x + \kappa_x^2 x + \kappa_x\kappa_y y - \frac{x'}{2\sigma'^2} \frac{\mathrm{d}\sigma'^2}{\mathrm{d}s} \right) \\
& + \mathbf{y}_\mathrm{o} \left(y'' + \kappa_y + \kappa_y^2 y + \kappa_x\kappa_y x - \frac{y'}{2\sigma'^2} \frac{\mathrm{d}\sigma'^2}{\mathrm{d}s} \right) \\
& - \mathbf{s}_\mathrm{o} \left(2\kappa_x x' + 2\kappa_y y' + \kappa_x' x + \kappa_y' y + \frac{1 + \kappa_x x + \kappa_y y}{\sigma'^2} \frac{\mathrm{d}\sigma'^2}{\mathrm{d}s} \right) \\
& = \frac{[c]e}{cp} \sigma' \left[\frac{\mathrm{d}\mathbf{S}}{\mathrm{d}s} \times \mathbf{B} \right].
\end{aligned} \tag{2.38}$$

This is the *general equation of motion* for a charged particles in a magnetic field $\mathbf{B}$. So far no approximations have been made. For practical use we may separate the individual components and get the differential equations for transverse motion

$$\begin{aligned}
& x'' - \kappa_x \left(1 + \kappa_x x + \kappa_y y \right) - \frac{1}{2} \frac{x'}{\sigma'^2} \frac{\mathrm{d}\sigma'^2}{\mathrm{d}s} \,, \\
& = \frac{[c]e}{cp} \sigma' \left[y' B_s - \left(1 + \kappa_x x + \kappa_y y \right) B_y \right] \\
& y'' - \kappa_y \left(1 + \kappa_x x + \kappa_y y \right) - \frac{1}{2} \frac{y'}{\sigma'^2} \frac{\mathrm{d}\sigma'^2}{\mathrm{d}s} \\
& = \frac{[c]e}{cp} \sigma' \left[\left(1 + \kappa_x x + \kappa_y y \right) B_x - x' B_s \right].
\end{aligned} \tag{2.39}$$

Chromatic effects originate from the momentum factor $\frac{e}{cp}$ which is different for particles of different energies. We expand this factor into a power series in δ

$$\frac{[c]e}{cp} = \frac{[c]e}{cp_\mathrm{o}} \left(1 - \delta + \delta^2 - \delta^3 + \ldots \right), \tag{2.40}$$

where $\delta = \Delta p/p_\mathrm{o}$ and $cp_\mathrm{o} = \beta E_\mathrm{o}$ is the ideal particle momentum. A further approximation is made when we expand σ' from (2.36) to third order in

x and y while restricting the discussion to paraxial rays with $x' \ll 1$ and $y' \ll 1$

$$\sigma' \approx 1 + \kappa_x\, x + \kappa_y\, y + \tfrac{1}{2}\left(x'^2 + y'^2\right)\left(1 - \kappa_x\, x - \kappa_y\, y\right) + \dots . \qquad (2.41)$$

Evaluating the derivative $\mathrm{d}\sigma'^2/\mathrm{d}s$ we obtain terms including second-order derivatives x'' and y''. Neglecting fourth-order terms x'' and y'' can be replaced by the unperturbed equations of motion $x'' + (\kappa_x^2 + k)x = 0$ and $y'' + (\kappa_y^2 + k)y = 0$. For the field components we insert in (2.39) expressions (2.25 – 27) while making use of (2.36,40,41). Keeping all terms up to third order in x, y, x', y' and δ we finally obtain equations of motion for a particle with charge e in an arbitrary magnetic field derivable from a scalar potential. For the horizontal plane this *general equation of motion* is expressed in (2.42).

$$
\begin{aligned}
x'' + (\kappa_x^2 + k)\, x =\; & \kappa_x\delta - \kappa_x\delta^2 + \kappa_x\delta^3 - (\underline{k} + \kappa_x\kappa_y)\, y \\
& - (\underline{m} + 2\kappa_x\underline{k} + 2\kappa_y k + 2\kappa_x^2\kappa_y)\, xy - \tfrac{1}{2}m\,(x^2 - y^2) \\
& - (\kappa_x^3 + 2\kappa_x k)\, x^2 - (\kappa_x\kappa_y^2 - \tfrac{1}{2}\kappa_x k + \tfrac{3}{2}\kappa_y\,\underline{k} - \tfrac{1}{2}\kappa_x'')\, y^2 \\
& + \tfrac{1}{2}\kappa_x\,(x'^2 - y'^2) + \kappa_x'\,(xx' + yy') + \kappa_y'\,(x'y - xy') + \kappa_y x'y' \\
& - \tfrac{1}{6}r\, x\,(x^2 - 3y^2) + \tfrac{1}{6}\,\underline{r}\, y\,(y^2 - 3x^2) \\
& + \tfrac{1}{12}\,(\kappa_y\underline{m} - 11\kappa_x m + 2\kappa_y^2 k - 10\kappa_x^2 k + k'' - \kappa_y\kappa_y'' \\
& \quad + \kappa_x\kappa_x'' - {\kappa_y'}^2 + {\kappa_x'}^2)\, x^3 \\
& - (2\kappa_x\underline{m} + \kappa_y m + \kappa_x^2\underline{k} + 2\kappa_x\kappa_y k)\, x^2 y \\
& + \tfrac{1}{4}\,(5\kappa_x m - 7\kappa_y\underline{m} + 6\kappa_x^2 k + k'' - \kappa_y\kappa_y'' - 2\kappa_y^2 k \\
& \quad + 5\kappa_x\kappa_x'' + {\kappa_x'}^2 - {\kappa_y'}^2 - \kappa_x\kappa_y\underline{k})\, xy^2 \\
& + \tfrac{1}{6}(10\kappa_x\kappa_y k + 8\kappa_y''\kappa_y + \kappa_x\underline{m} + 4\kappa_y^2\underline{k} + \underline{k''} \\
& \quad + 2\kappa_x'\kappa_y' + 5\kappa_y m)\, y^3 \\
& - (2\kappa_x^2 + \tfrac{3}{2}k)\, xx'^2 - (\kappa_x'\kappa_y + \kappa_x\kappa_y')\, xx'y - \kappa_x\kappa_x'\, x^2 x' \\
& - \tfrac{1}{2}\underline{k'}\, x^2 y' - \kappa_y\kappa_y'\, x'y^2 - \kappa_x\kappa_y\, xx'y' - \tfrac{1}{2}\,(\underline{k} + 3\kappa_x\kappa_y)\, x'^2 y \\
& + k'\, xyy' - \tfrac{1}{2}(k + \kappa_x^2)\, xy'^2 - (2\kappa_y^2 - \underline{k})\, x'yy' + \tfrac{1}{2}\underline{k'}\, y^2 y' - \tfrac{1}{2}\underline{k}\, yy'^2 \\
& + (2\kappa_x^2 + k)\, x\,\delta + (2\kappa_x\kappa_y + \underline{k})\, y\,\delta - \kappa_x'\, yy'\delta + \kappa_y'\, xy'\delta \\
& + \tfrac{1}{2}\kappa_x(x'^2 + y'^2)\delta + (\tfrac{3}{2}\kappa_y\underline{k} + \kappa_x\kappa_y^2 - \tfrac{1}{2}\kappa_x k - \tfrac{1}{2}\kappa_x'' - \tfrac{1}{2}m)\, y^2\delta \\
& + (\tfrac{1}{2}m + 2\kappa_x k + \kappa_x^3)\, x^2\delta + (\underline{m} + 2\kappa_x^2\kappa_y + 2\kappa_y k + 2\kappa_x\underline{k})\, xy\delta \\
& - (k + 2\kappa_x^2)\, x\delta^2 - (\underline{k} + 2\kappa_x\kappa_y)\, y\delta^2 + \mathcal{O}(4)\,.
\end{aligned} \qquad (2.42)
$$

In spite of our attempt to derive a general and accurate equation of motion, we note that some magnet boundaries are not correctly represented. The natural bending magnet is of the sector type and wedge or rectangular magnets require the introduction of additional corrections to the equations of motion which are not included here. This is also true for cases where a beam passes off center through a quadrupole, in which case theory assumes a combined function sector magnet and corrections must be applied to model correctly a quadrupole with parallel pole faces. The magnitude of such corrections is, however, in most cases very small. Equation (2.42) of motion shows an enormous complexity which, however, in real beam transport lines, becomes very much relaxed due to proper design and careful alignment of the magnets. Nonetheless (2.42), and (2.43) for the vertical plane, can be used as a reference to find and study particular perturbation terms. In a special beam transport line one or the other of these perturbation terms may become significant and can now be dealt with separately. This may be the case where strong multipole effects from magnet fringe fields cannot be avoided or because large beam sizes and divergences are important. The possible significance of any perturbation term must be evaluated for each beam transport system separately. In the vertical plane we get a very similar equation (2.43) which is to be expected since we have not yet introduced any asymmetry.

$$
\begin{aligned}
y'' + (\kappa_y^2 - k)y &= +\kappa_y\delta - \kappa_y\delta^2 + \kappa_y\delta^3 - (\underline{k} + \kappa_x\kappa_y)\,x \\
&+ (m - 2\kappa_y\underline{k} + 2\kappa_x k - 2\kappa_x\kappa_y^2)\,xy - \tfrac{1}{2}\,\underline{m}\,(x^2 - y^2) \\
&- (\kappa_y^3 - 2\kappa_y k)y^2 - (\kappa_x^2\kappa_y + \tfrac{1}{2}\kappa_y k + \tfrac{3}{2}\kappa_x\underline{k} - \tfrac{1}{2}\kappa_y'')\,x^2 \\
&- \tfrac{1}{2}\kappa_y(x'^2 - y'^2) + \kappa_y'(xx' + yy') - \kappa_x'\,(x'y - xy') + \kappa_x x'y' \\
&- \tfrac{1}{6}r\,y(y^2 - 3x^2) - \tfrac{1}{6}\underline{r}\,x(x^2 - 3y^2) \\
&- \tfrac{1}{12}(\kappa_x m - 11\kappa_y\underline{m} + 2\kappa_x^2 k - 10\kappa_y^2 k + k'' - \kappa_y\kappa_y'' \\
&+ \kappa_x\kappa_x'' + \kappa_x'^{\,2} - \kappa_y'^{\,2})\,y^3 \\
&+ (2\kappa_y m + \kappa_x\underline{m} - \kappa_y^2\underline{k} + 2\kappa_x\kappa_y k)\,xy^2 \\
&- \tfrac{1}{4}\,(5\kappa_y\underline{m} - 7\kappa_x m + 6\kappa_y^2 k + k'' + \kappa_x\kappa_x'' - 2\kappa_x^2 k \\
&- 5\kappa_y\kappa_y'' - \kappa_y'^{\,2} + \kappa_x'^{\,2} + \kappa_x\kappa_y\underline{k})\,x^2 y \\
&+ \tfrac{1}{6}(-10\kappa_x\kappa_y k + 8\kappa_x\kappa_x'' - \kappa_y m + 4\kappa_x^2\underline{k} + \underline{k}'' \qquad (2.43) \\
&+ 2\kappa_x'\kappa_y' - 5\kappa_x\underline{m})\,x^3 \\
&- (2\kappa_y^2 - \tfrac{1}{2}k - \underline{k})\,yy'^2 - (\kappa_x\kappa_y' + \kappa_x'\kappa_y)\,xyy' - \kappa_y\kappa_y'\,y^2 y' \\
&- \tfrac{1}{2}\underline{k}'\,x'y^2 - \kappa_x\kappa_x'\,x^2 y' - \kappa_x\kappa_y\,x'yy' - \tfrac{1}{2}\,(\underline{k} + 3\kappa_x\kappa_y)\,xy'^2 \\
&- k'\,xx'y + \tfrac{1}{2}(k - \kappa_y^2)x'^2 y - (2\kappa_x^2 + k)\,xx'y' + \tfrac{1}{2}\underline{k}'\,x^2 x' - \tfrac{1}{2}k\,xx'^2 \\
&+ (2\kappa_y^2 - k)\,y\delta + (2\kappa_x\kappa_y + \underline{k})\,x\delta - \kappa_y'xx'\delta + \kappa_x'\,x'y\delta \\
&+ \tfrac{1}{2}\kappa_y\,(x'^2 + y'^2)\,\delta + (\tfrac{3}{2}\kappa_x\underline{k} + \kappa_x^2\kappa_y + \tfrac{1}{2}\kappa_y k - \tfrac{1}{2}\kappa_y'' + \tfrac{1}{2}\underline{m})\cdot x^2\delta \\
&+ (-\tfrac{1}{2}\underline{m} - 2\kappa_y k + \kappa_y^3)\,y^2\delta - (m - 2\kappa_x\kappa_y^2 + 2\kappa_x k - 2\kappa_y\underline{k})\,xy\delta \\
&+ (k - 2\kappa_y^2)\,y\delta^2 - (\underline{k} + 2\kappa_x\kappa_y)\,x\delta^2 + \mathcal{O}(4)\,.
\end{aligned}
$$

In most beam transport lines the magnets are built in such a way that different functions like bending, focusing, etc., are not combined thus eliminating all terms that depend on those combinations like $\kappa_x \kappa_y$, $\kappa_x k$ or $m\kappa_x$ etc. As long as the terms on the right-hand sides are small we may apply perturbation methods to estimate the effects on the beam caused by these terms. It is interesting, however, to try to identify the perturbations with aberrations known from light optics.

Chromatic terms $\kappa_x \left(\delta - \delta^2 + \delta^3\right)$, for example, are constant perturbations for off momentum particles causing a shift of the equilibrium orbit which ideally is the trivial solution $x \equiv 0$ of the differential equation $x'' + (k + \kappa_x^2)x = 0$. Of course, this is not quite true since κ_x is not a constant but the general conclusion is still correct. This shift is equal to $\Delta x = \kappa_x(\delta - \delta^2 + \delta^3)/(k + \kappa_x^2)$ and is related to the *dispersion function* D by $D = \Delta x/\delta$. In light optics this corresponds to the dispersion of colors of a beam of white light (particle beam with finite energy spread) passing through a prism (bending magnet). We may also use a different interpretation for this term. Instead of a particle with an energy deviation δ in an ideal magnet κ_x we can interpret this term as the perturbation of a particle with the ideal energy by a magnetic field that deviates from the ideal value. In this case, we replace $\kappa_x \left(\delta - \delta^2 - \delta^3\right)$ by $-\Delta\kappa_x$ and the shift in the ideal orbit is then called an *orbit distortion*. Obviously, here and in the following paragraphs the interpretations are not limited to the horizontal plane alone but apply also to the vertical plane caused by similar perturbations. Terms proportional to x^2 cause *geometric aberrations* where the focal length depends on the amplitude x while terms involving x' lead to the well-known phenomenon of *astigmatism* or a combination of both aberrations. Additional terms depend on the particle parameters in both the vertical and horizontal plane and therefore lead to more complicated aberrations and *coupling*.

Terms depending also on the energy deviation δ, on the other hand, give rise to *chromatic aberrations* which are well known from light optics. Specifically, the term $(k + 2\kappa_x^2) \, x \, \delta$ is the source for the dependence of the focal length on the particle momentum. Some additional terms can be interpreted as combinations of aberrations described above.

It is interesting to write down the equations of motion for a pure quadrupole system where only $k \neq 0$ in which case (2.42) becomes

$$\begin{aligned}
x'' + kx &= kx \left(\delta - \delta^2 - \delta^3\right) - \tfrac{1}{12} \, k'' \, x \left(x^2 + 3y^2\right) \\
&\quad - \tfrac{3}{2} k \, x \, x'^2 + k \, x' \, yy' + k'xyy' + \mathcal{O}(4) \,.
\end{aligned} \tag{2.44}$$

We note that quadrupoles do not produce geometric perturbations of lower than third order. Only chromatic aberrations are caused in second order by quadrupole magnets.

2.3 Periodic Wiggler Magnets

Particular arrays or combinations of magnets can produce desirable results for a variety of applications. A specially useful device of this sort is a *wiggler magnet* which is composed of a series of short bending magnets with alternating field excitation. Such wiggler magnets are used for a variety of applications to either produce coherent or incoherent photon beams in electron accelerators, or to manipulate electron beam properties like beam emittance and energy spread. To compensate antidamping in a combined function synchrotron a wiggler magnet including a field gradient is used to modify the damping partition numbers [2.4]. In colliding-beam storage rings wiggler magnets are used to increase the beam emittance for maximum luminosity [2.5]. In other applications a very small beam emittance is desired as is the case in damping rings for linear colliders or synchrotron radiation sources which can be achieved by employing *damping wiggler* magnets in a different way [2.6] to be discussed in more detail in Chap. 9.

Wiggler magnets are generally designed as *flat wiggler magnet* [2.7] with field components only in one plane or as *helical wiggler magnets* [2.8 – 10] where the transverse field component rotates along the magnetic axis. In this discussion, we concentrate on flat wigglers which are used in growing numbers to generate, for example, intense beams of *synchrotron radiation* from electron beams, to manipulate beam parameters or to pump a *free electron laser*.

2.3.1 Wiggler Field Configuration

Whatever the application may be, the wiggler magnet deflects the electron beam transversely in an alternating fashion without introducing a net deflection on the beam as shown in Fig. 2.4.

Wiggler magnets are generally considered to be *insertion devices* which are not part of the basic magnet lattice but are installed in a magnet free straight section of the lattice. To minimize the effect of wiggler fields on the particle beam, the integrated magnetic field through the whole wiggler magnet must be zero

$$\int_{\text{wiggler}} B_\perp \, \mathrm{d}z = 0 \,. \tag{2.45}$$

Since a wiggler magnet is a straight device we use a fixed cartesian coordinate system (x, y, z) with the z-axis parallel to the wiggler axis to describe the wiggler field, rather than a curvilinear system that follows the oscillatory deflection of the reference path in the wiggler. The origin of the coordinate system is placed in the middle of one of the wiggler magnets. The whole magnet may be composed of N equal and symmetric pole pieces placed along the z-axis at a distance $\lambda_{\mathrm{p}}/2$ from pole center to pole center as depicted in Fig. 2.5. Each pair of adjacent wiggler poles forms one *wiggler period* with

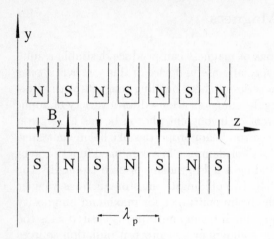

Fig. 2.4. Wiggler Magnet

a *wiggler period length* λ_p and the whole magnet is composed of $N/2$ periods. Since all periods are assumed to be the same and the beam deflection is compensated within each period no net beam deflection occurs for the complete magnet.

Upon closer inspection of the precise beam trajectory we observe a lateral displacement of the beam within a wiggler magnet composed of all equal magnet poles. To compensate this lateral beam displacement, the wiggler magnet should begin and end with only a half pole of length $\lambda_p/4$.

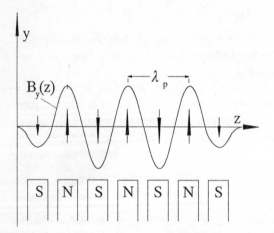

Fig. 2.5. Field distribution in a wiggler magnet

The individual magnets comprising a wiggler magnet are in general very short and the longitudinal field distribution differs considerable from a hard-edge model. In fact most of the field will be fringe fields. We consider only periodic fields which can be expanded into a Fourier series along the axis including a strong fundamental component with a period length λ_p and higher harmonics expressed by the ansatz [2.11]

$$B_y = B_o \sum_{n \geq o} b_{2n+1}(x, y) \cos[(2n + 1) k_p z], \tag{2.46}$$

where the wave number $k_p = 2\pi/\lambda_p$. The functions $b_i(x, y)$ describe the variation of the field amplitude orthogonal to the beam axis for the harmonic i. The content of higher harmonics is greatly influenced by the particular design of the wiggler magnet and the ratio of the period length to the pole gap aperture. For very long periods relative to the pole aperture the field profile approaches that of a hard-edge dipole field with a square field profile along the z-axis. For very short periods compared to the pole aperture, on the other hand, we find only a significant amplitude for the fundamental period and very small perturbations due to higher harmonics.

We may derive the magnetic field from Maxwell's equations based on a sinusoidal field along the axis. Each field harmonic may be determined separately due to the linear superposition of fields. To eliminate a dependence of the magnetic field on the horizontal variable x we assume a pole width which is large compared to the pole aperture and the fundamental field component, for example, is

$$B_y(y, z) = B_o b_1(y) \cos k_p z. \tag{2.47}$$

From Maxwell's curl equation $\nabla \times \mathbf{B} = 0$ we get $\frac{\partial B_z}{\partial y} = \frac{\partial B_y}{\partial z}$ and with (2.47) we have

$$\frac{\partial B_z}{\partial y} = \frac{\partial B_y}{\partial z} = -B_o b_1(y) k_p \sin k_p z. \tag{2.48}$$

Integration of (2.48) with respect to z gives for the vertical field component

$$B_y = -B_o k_p b_1(y) \int_o^z \sin k_p \bar{z} \, d\bar{z}. \tag{2.49}$$

We have not yet determined the y-dependence of the amplitude function $b_1(y)$. From $\nabla \cdot \mathbf{B} = 0$ and the independence of the field on the horizontal position we get with (2.47)

$$\frac{\partial B_z}{\partial z} = -B_o \frac{\partial b_1(y)}{\partial y} \cos k_p z. \tag{2.50}$$

Forming the second derivatives $\partial^2 B_z/(\partial y \, \partial z)$ from (2.48,50) we get for the amplitude function the differential equation

$$\frac{\partial^2 b_1(y)}{\partial y^2} = k_p^2 b_1(y), \tag{2.51}$$

which can be solved by the hyperbolic functions

$$b_1(y) = a \cosh k_p y + b \sinh k_p y. \tag{2.52}$$

Since the magnetic field is symmetric with respect to $y = 0$ and $b_1(0) = 1$ the coefficients are $a = 1$ and $b = 0$. Collecting all partial results, the *wiggler magnetic field* is finally determined by the components

$$
\begin{aligned}
B_x &= 0\,, \\
B_y &= B_\mathrm{o} \cosh k_\mathrm{p} y \, \cos k_\mathrm{p} z\,, \\
B_z &= -B_\mathrm{o} \sinh k_\mathrm{p} y \, \sin k_\mathrm{p} z\,,
\end{aligned}
\tag{2.53}
$$

where B_z is obtained by integration of (2.48) with respect to y.

The hyperbolic dependence of the field amplitude on the vertical position introduces higher-order field-errors which we determine by expanding the hyperbolic functions

$$
\begin{aligned}
\cosh k_\mathrm{p} y &= 1 + \frac{(k_\mathrm{p}\, y)^2}{2!} + \frac{(k_\mathrm{p}\, y)^4}{4!} + \frac{(k_\mathrm{p}\, y)^6}{6!} + \frac{(k_\mathrm{p}\, y)^8}{8!} + \cdots\,, \\
\sinh k_\mathrm{p} y &= +(k_\mathrm{p}\, y) + \frac{(k_\mathrm{p}\, y)^3}{3!} + \frac{(k_\mathrm{p}\, y)^5}{5!} + \frac{(k_\mathrm{p}\, y)^7}{7!} + \cdots\,.
\end{aligned}
\tag{2.54}
$$

Typically the vertical gap in a wiggler magnet is smaller than the period length or $y < \lambda_\mathrm{p}$. For larger apertures the field strength reduces drastically. Due to the fast convergence of the series expansions (2.54) only a few terms are required to obtain an accurate expression for the hyperbolic function within the wiggler aperture.

The expansion (2.54) displays the higher-order field components explicitly which, however, do not have the form of higher-order multipole fields and we cannot treat these fields just like any other multipole perturbation but must consider them separately.

To determine the path distortion due to the wiggler fields, we follow the reference trajectory through one quarter period starting at a symmetry plane in the middle of a pole. At the starting point $z = 0$ in the middle of a wiggler pole the beam direction is parallel to the reference trajectory and the deflection angle at a downstream point z is given by

$$
\begin{aligned}
\vartheta(z) &= \frac{[c]e}{cp} \int_\mathrm{o}^z B_y \, \mathrm{d}\bar{z} = \frac{[c]e}{cp} B_\mathrm{o} \cosh k_\mathrm{p} y \int_\mathrm{o}^z \cos k_\mathrm{p} \bar{z} \, \mathrm{d}\bar{z}\,, \\
&= -\frac{[c]e}{cp} B_\mathrm{o} \frac{1}{k_\mathrm{p}} \cosh k_\mathrm{p} y \, \sin k_\mathrm{p} z\,.
\end{aligned}
\tag{2.55}
$$

The maximum deflection angle is equal to the deflection angle for a quarter period or half a wiggler pole and is from (2.55) for $y = 0$ and $k_\mathrm{p} z = \pi/2$

$$
\theta = -\frac{[c]e}{cp} B_\mathrm{o} \frac{\lambda_\mathrm{p}}{2\pi}\,.
\tag{2.56}
$$

This deflection angle for a quarter period is used to define the *wiggler strength parameter*

$$K = \beta\gamma\theta = \frac{[c]\,e}{2\pi\,m\,c^2}\,B_\mathrm{o}\,\lambda_\mathrm{p} \qquad (2.57)$$

where $m\,c^2$ is the particle rest energy and γ the particle energy in units of the rest energy. In more practical units this strength parameter is

$$\boxed{K = 9.344\,B_\mathrm{o}(\mathrm{kG})\,\lambda_\mathrm{p}(\mathrm{m})\,.} \qquad (2.58)$$

The parameter K is a characteristic wiggler constant defining the *wiggler strength* and is not to be confused with the general focusing strength $K(s) = \kappa^2 + k(s)$.

2.3.2 Focusing in a Wiggler Magnet

The derivation of fringe field focusing in ordinary dipole magnets as discussed in [Ref. 2.1, Chap. 5] can be directly applied to wiggler magnets. The beam path in a wiggler magnet is generally not parallel to the reference trajectory z because of the transverse deflection in the wiggler field and follows a periodic sinusoidal form along the reference path. For this reason the field component B_z appears to the particle partially as a transverse field $B_\xi = B_z \tan\vartheta \approx B_z\,\vartheta$ where we use for a moment ξ as an auxiliary transverse coordinate normal to and in the plane of the actual wiggling beam path. We also assume that the wiggler deflection angle is small, $\vartheta \ll 1$. The field component B_ξ can be expressed with (2.53,55) more explicitly by

$$B_\xi = -\frac{[c]\,e}{cp}\,(B_\mathrm{o}\,\sin k_\mathrm{p}z)^2\,\frac{\sinh k_\mathrm{p}y\,\cosh k_\mathrm{p}y}{k_\mathrm{p}} \qquad (2.59)$$

and with the expansions (2.54) we have finally

$$B_\xi = -\frac{[c]\,e}{cp}\,(B_\mathrm{o}\,\sin k_\mathrm{p}z)^2\,(y + \tfrac{2}{3}\,k_\mathrm{p}^2\,y^3 + \ldots)\,. \qquad (2.60)$$

The term linear in y is similar to that found to produce vertical focusing in wedge magnets. Since the wiggler field appears quadratically in (2.59) $B_\xi(z) = B_\xi(-z)$ and $B_\xi(B_\mathrm{o}) = B_\xi(-B_\mathrm{o})$. In other words, the transverse field has the same sign along all wiggler poles independent of the polarity of the vertical main wiggler field. The integrated focusing field gradient per wiggler half pole is from (2.60)

$$g_y\,\ell = -\frac{[c]\,e}{cp}\,B_\mathrm{o}^2\int_0^{\lambda_\mathrm{p}/4} \sin^2 k_\mathrm{p}z\,\mathrm{d}z = -\frac{[c]\,e}{cp}\,\tfrac{1}{8}\,B_\mathrm{o}^2\,\lambda_\mathrm{p} \qquad (2.61)$$

where ℓ is the effective length of the focusing element. The integrated equivalent quadrupole strength or inverse focal length for each half pole with parallel entry and exit pole faces is

$$k_y \ell = -\frac{1}{f_y} = -\frac{1}{8}\left(\frac{[c]eB_o}{cp_o}\right)^2 \lambda_p = -\frac{\lambda_p}{8\rho_o^2}, \qquad (2.62)$$

where $1/\rho_o = \frac{[c]e}{cp}B_o$ is the inverse bending radius in the center of a wiggler pole at which point the field reaches the maximum value B_o. For N wiggler poles we have $2N$ times the focusing strength and the focal length of the total wiggler magnet of length $L_w = 1/2 N \lambda_p$ expressed in units of the wiggler strength parameter K becomes

$$\frac{1}{f_y} = \frac{K^2}{2\gamma^2}k_p^2 L_w. \qquad (2.63)$$

Tacitly, a rectangular form of the wiggler poles has been assumed (Fig. 2.6) and consistent with our sign convention we find that wiggler fringe fields cause focusing in the nondeflecting plane. Within the approximation used there is no corresponding focusing effect in the deflecting plane. This is the situation for most wiggler magnets or poles except for the first and last half pole where the beam enters the magnetic field normal to the pole face.

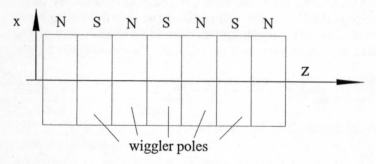

Fig. 2.6. Wiggler magnet with parallel pole end faces

A reason to possibly use wiggler magnets with rotated pole faces like wedge magnets originates from the fact that the wiggler focusing is asymmetric and not part of the lattice focusing and may therefore need to be compensated. For moderately strong wiggler fields the asymmetric focusing in both planes can mostly be compensated by small adjustments of lattice quadrupoles. The focusing effect of strong wiggler magnets, however, may generate a significant perturbation of the lattice focusing structure or create a situation where no stable solution for betatron functions exists anymore. The severity of this problem can be reduced by designing the wiggler poles as wedge magnets in such a way as to split the focusing equally between both the horizontal and vertical plane. In this case local correction can be applied efficiently in nearby lattice quadrupoles.

We will therefore discuss the focusing and transformation matrix through a wiggler pole in the case of arbitrary entry and exit angles. To derive the complete and general transformation matrices, we note that the whole wiggler field can be treated in the same way as the fringe field of ordinary magnets. The focal length of one half pole in the horizontal deflecting plane is from [Ref. 2.1, Sect. 5.3.2]

$$\frac{1}{f_x} = \int_0^{\lambda_p/4} \kappa'_x \, \eta \, \mathrm{d}\bar{s} + \kappa_{xo} \, \delta_f \,, \tag{2.64}$$

where the pole face rotation angle η has been assumed to be small and of the order of the wiggler deflection angle per pole (Fig. 2.7). With $\kappa_x = \kappa_{xo} \cos k_p z$ the field slope is

$$\kappa'_x = \kappa_{xo} \, k_p \sin k_p z \tag{2.65}$$

and after integration of (2.64) with $z \approx s$ the focal length for the focusing of a wiggler half pole is

$$\frac{1}{f_x} = \kappa_{xo} \left(\delta_f + \eta \right), \tag{2.66}$$

where δ_f is given by [Ref. 2.1, Eq. (5.68)] and is in the case of a wiggler magnet equal to the deflection angle of a half pole. In the case of rectangular wiggler pole $\eta = -\delta_f$ and the focusing in the deflecting plane vanishes as we would expect.

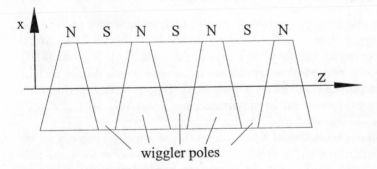

Fig. 2.7. Wiggler magnet with wedge shaped poles

In the nondeflecting plane [Ref. 2.1, Eq. (5.66)] applies and the focal length is for small angles η and δ

$$\frac{1}{f_y} = -\int_0^{\lambda_p/4} \kappa'_x \left[\eta + \delta(\bar{s}) \right] \mathrm{d}\bar{s} \,. \tag{2.67}$$

The focal length per wiggler half pole is after integration

$$\frac{1}{f_y} = -\kappa_{xo}\left(\eta + \delta_f\right) - \frac{\pi}{4}\,\kappa_{xo}\,\delta_f\,. \tag{2.68}$$

Here again setting $\eta = -\delta_f$ restores the result obtained in (2.63).

The focusing in each single wiggler pole is rather weak and we may apply thin lens approximation to derive the transformation matrices. For this we consider the focusing to occur in the middle of each wiggler pole with drift spaces of length $\lambda_p/4$ on each side. With $2/f$ being the focal length of a full pole in either the horizontal plane (2.66) or vertical plane (2.68) the transformation matrix for each wiggler pole is finally

$$
\begin{aligned}
\mathcal{M}_{\text{pole}} &= \begin{pmatrix} 1 & \lambda_p/4 \\ 0 & 1 \end{pmatrix} \begin{pmatrix} 1 & 0 \\ -2/f & 1 \end{pmatrix} \begin{pmatrix} 1 & \lambda_p/4 \\ 0 & 1 \end{pmatrix} \\
&= \begin{pmatrix} 1 - \frac{\lambda_p}{2f} & \frac{\lambda_p}{2}\left(1 - \frac{\lambda_p}{4f}\right) \\ -\frac{2}{f} & 1 - \frac{\lambda_p}{2f} \end{pmatrix} \approx \begin{pmatrix} 1 & \lambda_p/2 \\ -2/f & 1 \end{pmatrix},
\end{aligned}
\tag{2.69}
$$

where the approximation $\lambda_p \ll f$ was used. For a wiggler magnet of length $L_w = \frac{1}{2}N\lambda_p$ we have N poles and the total transformation matrix is

$$\mathcal{M}_{\text{wiggler}} = \mathcal{M}_{\text{pole}}^N\,. \tag{2.70}$$

This transformation matrix can be applied to each plane and any pole rotation angle η. Specifically, we set $\eta = -K/\gamma$ for a rectangular pole shape and $\eta = 0$ for pole rotations orthogonal to the path like in sector magnets.

2.3.3 Hard-Edge Model of Wiggler Magnets

Although the magnetic properties of wiggler magnets are well understood and easy to apply it is nonetheless often desirable to describe the effects of wiggler magnets in the form of hard-edge models. This is particularly true when special numerical programs are to be used which do not include the feature of properly modeling a sinusoidal wiggler field. On the other hand accurate field modeling is important since frequently strong wiggler magnets are to be inserted into a beam transport lattice.

For the proper modeling of linear properties of wiggler magnets we require three conditions to be fulfilled. The deflection angle for each pole should be the same as that for the equivalent hard-edge model. Similarly the edge focusing must be the same. Finally, like any other bending magnet in an electron circular accelerator, a wiggler magnet also contributes to quantum excitation and damping of the beam emittance and beam energy spread. The quantum excitation is in first approximation proportional to the third power of the curvature while the damping scales like the square of the curvature similar to focusing.

We consider now a wiggler field

$$B(z) = B_o \sin k_p z \tag{2.71}$$

and try to model the field for a half pole with parallel endpoles by a hard-edge magnet. Three conditions should be met. The deflection angle of the hard-edge model of length ℓ and field B must be the same as that for a wiggler half pole, or

$$\vartheta = \frac{\ell_h}{\rho_h} = \frac{[c]e}{cp_o} \int_{\text{halfpole}} B_y(z)\,dz = \frac{\lambda_p}{2\pi\,\rho_o}. \tag{2.72}$$

Here we use ρ_h for the bending radius of the equivalent hard-edge model and ρ_o is the bending radius for the peak wiggler field B_o. The edge focusing condition can be expressed by

$$\frac{1}{f} = \frac{\ell_h}{\rho_h^2} = \frac{1}{\rho_o^2} \int_{\text{halfpole}} \sin^2 k_p z\,dz = \frac{\lambda_p}{8\rho_o^2}. \tag{2.73}$$

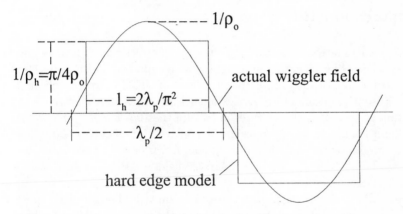

Fig. 2.8. Hard edge model for a wiggler magnet period

Modeling a wiggler field by a single hard-edge magnet for linear beam optics requires only two conditions to be met which can be done with the two parameters $B(z)$ and ℓ available. From (2.72,73) we get therefore the hard-edge magnet parameters (Fig. 2.8)

$$\rho_h = \frac{4}{\pi}\rho_o \qquad \text{and} \qquad \ell_h = \frac{2}{\pi^2}\lambda_p. \tag{2.74}$$

For a perfect modeling of the equilibrium energy spread and emittance due to quantum excitation in electron storage rings we would also like the cubic term to be the same

$$\frac{\ell_h}{\rho_h^3} \stackrel{?}{=} \frac{1}{\rho_o^3} \int_{\text{halfpole}} \sin^3 k_p z\,dz = \frac{\lambda_p}{3\pi\,\rho_o^3}. \tag{2.75}$$

Since we have no more free parameters available, we can at this pint only estimate the mismatch. With (2.73,74) we get from (2.75) the inequality

$$\frac{1}{3\pi} \neq \frac{\pi}{32}$$

which indicates that the quantum excitation from wiggler magnets is not correctly treated although the error is only about 8%. Similarly, one could decide that the quadratic and cubic terms must be equal while the deflection angle is let free. This would be a reasonable assumption since the total deflection angle of a wiggler is compensated anyway. In this case the deflection angle would be underestimated by about 8%. Where these mismatches are not significant, the simple hard-edge model (2.75) can be applied. For more accuracy the sinusoidal wiggler field must be segmented into smaller hard-edge magnets.

2.4 Superconducting Magnet

Superconducting materials are available in large quantities as thin wires and it has become possible to construct magnets with superconducting coils and fields exceeding significantly the saturation limit of 20 kG in iron dominated magnets. The magnetic-field properties in such magnets are determined by the location of current carrying wires, a feature that has become one of the most difficult engineering challenges in the construction of superconducting magnets. Any movement of wires during cool down or under the influence of great mechanical forces from the magnetic fields must be avoided. Generally, the superconducting material is embedded into a solid copper matrix to prevent thermal destruction in the event of a quench. The fact that most *superconducting magnets* have an iron collar on the outside of the coils does not affect our discussion since it mostly serve to keep the coils firmly in place and to shield the outside environment from excessive magnetic field. In this text we will not consider technical construction methods of superconducting magnets, but concentrate rather on the fundamental physical principles of generating multipole fields in conductor dominated magnets. These principles are obviously applicable to any ironless magnets, superconducting or other. More comprehensive accounts of superconducting magnet technology can be found, for example, in [2.12 – 14].

The determination of the magnetic field in such magnets is somewhat more complicated compared to that in iron dominated magnets. In the latter case, we performed an integration of Maxwell's curl equation along a path which is composed of a contribution from a known or desired field in the magnet aperture, while all other parts of the integration path do not contribute to the integral because either the field is very small like in unsaturated iron, or the path is orthogonal to the field lines. In iron free magnets, we must consider the whole path surrounding the current carrying coil.

Arbitrary multipole fields can be constructed in two somewhat related ways which we will discuss here in more detail. The first method uses two equal circular or elliptical conductors with the same current densities flowing in opposite directions. We consider such a pair of conductors to overlap, though not fully, as shown in Fig. 2.9 and note that there is no net current in the overlap region. The material can therefore be eliminated from this region providing the magnet *aperture* for the particle beam. We consider this simple case proposed first by *Rabi* [2.15] generating a pure dipole field. The magnetic field within a cylindrical conductor with uniform current density j is at radius r (Fig. 2.9)

$$H_\theta = \left[\frac{c}{4\pi}\right] \frac{2\pi}{c} j r$$

and in cartesian coordinates the field components are

$$H_x = -\left[\frac{c}{4\pi}\right] \frac{2\pi}{c} j y \quad \text{and} \quad H_y = \left[\frac{c}{4\pi}\right] \frac{2\pi}{c} j x . \tag{2.76}$$

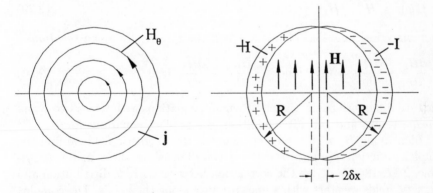

Fig. 2.9. Magnetic field in the overlap region of two cylindrical electrical conductors

Now we superimpose on this another identical conductor with the current running in the opposite direction $-j$. Since the electrical currents cancel, the fields cancel and there is no magnetic field. If however, both conductors are separated by a distance $\pm\delta x$ along the horizontal coordinate as shown in Fig. 2.9 a homogenous vertical field can be created in the overlap region. The horizontal field component H_x depends only on the vertical coordinate. The contributions of both conductors to H_x are therefore equal in magnitude but of opposite sign and cancel perfectly everywhere within the overlap of the conductors. The residual vertical field component from both conductors, on the other hand, is

$$H_y = \left[\frac{c}{4\pi}\right] \frac{2\pi}{c} j \left[(x + \delta x) - (x - \delta x)\right] = \left[\frac{c}{4\pi}\right] \frac{2\pi}{c} j \, 2 \, \delta x . \tag{2.77}$$

Within the area of the overlapping conductors, a pure vertical dipole field
has been generated that can be used for beam guidance after eliminating
the material in the overlap region where no net current flows. Other field
configurations can be obtained by similar superposition of elliptical current
carrying conductors [2.16].

The current density j is assumed to be constant and the current distribution just outside of the limits of the overlap region is

$$I(\varphi) = 2 j \delta x \cos\varphi. \tag{2.78}$$

Conversely, we can state that a cosine like current distribution on a circle produces a pure magnetic dipole field within that circle leading to the
second method of generating magnetic fields for beam transport systems
based on a specific current distribution along the periphery of the aperture.
The correlation between a particular current distribution and the resulting
higher-order multipole-fields can be derived in a formal and mathematical way. To do this it is convenient to use complex functions. A complex
function

$$f(w) = H_y + \mathrm{i} H_x \tag{2.79}$$

is an analytical functions of $w = x + \mathrm{i}y$ if the *Cauchy Riemann equations*

$$\frac{\partial H_y}{\partial x} - \frac{\partial H_x}{\partial y} = 0 \quad \text{and} \quad \frac{\partial H_y}{\partial y} + \frac{\partial H_x}{\partial x} = 0 \tag{2.80}$$

hold. We differentiate the Cauchy-Riemann equations with respect to x and
y, respectively, and get, after addition, the Laplace equations for H_x and
H_y. Identifying H_i with the components of the magnetic field we find the
complex function $f(w)$ to describe a two-dimensional field in the (x, y)-
plane [2.17, 18]. To derive the correlation between current distribution and
magnetic fields we start with a line current I along the z-axis. The complex
potential for the magnetic field generated by such a line current is

$$W = -\frac{2}{c} I \ln(w - v_{\mathrm{o}}), \tag{2.81}$$

where v_{o} is the location of the line current flowing in the z direction and
w the location of the field point in the complex (x, y)-plane. We may verify
the correctness of this potential by equating

$$H(w) = -\left[\frac{c}{4\pi}\right] \frac{\mathrm{d}W}{\mathrm{d}w} = \left[\frac{c}{4\pi}\right] \frac{2}{c} \frac{I}{w - v_{\mathrm{o}}} = (H_y + \mathrm{i} H_x), \tag{2.82}$$

where $w = x + \mathrm{i}y$ and $v_{\mathrm{o}} = x_{\mathrm{o}} + \mathrm{i}y_{\mathrm{o}}$. In cartesian coordinates such a line
current generates the well-known field components

$$H_y = \left[\frac{c}{4\pi}\right] \frac{2}{c} \frac{I}{R} \frac{x - x_{\mathrm{o}}}{R} \quad \text{and} \quad H_x = -\left[\frac{c}{4\pi}\right] \frac{2}{c} \frac{I}{R} \frac{y - y_{\mathrm{o}}}{R}$$

at a distance $R^2 = (x - x_\mathrm{o})^2 + (y - y_\mathrm{o})^2$ from a line current I. The complex function $f(w)$ has a first-order singularity at $w = v_\mathrm{o}$ and from *Cauchy's residue theorem* we find

$$\oint H(w)\, \mathrm{d}w = \left[\frac{c}{4\pi}\right] \frac{2}{c} \oint \frac{I}{w - v_\mathrm{o}}\, \mathrm{d}w = \left[\frac{c}{4\pi}\right] \mathrm{i} \frac{4\pi}{c}\, I, \qquad (2.83)$$

where $2I/c$ is the residue of the function $H(w)$. This can be applied to a number of line currents by adding up the r.h.s of all residues, or integrating a current distribution.

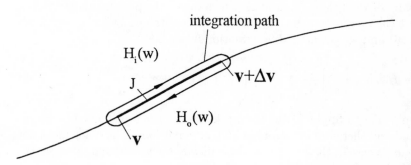

Fig. 2.10. Current sheet theorem [2.16]

Following *Beth* [2.16] we apply (2.83) to a current sheet along the line v, where $\mathrm{d}I/\mathrm{d}v = I(\varphi)$ is the line current density flowing perpendicular through the paper or x, y plane and parallel to the z-direction. A closed path leading clockwise and tightly around the current element $I(\varphi)\Delta v$ (Fig. 2.10) gives with (2.83)

$$H_\mathrm{i}(w) - H_\mathrm{o}(w) = -\mathrm{i} \left[\frac{c}{4\pi}\right] \frac{4\pi}{c}\, I(\varphi). \qquad (2.84)$$

We must consider here only the fields very close to the current sheet to eliminate the contributions to the fields from other current elements. Therefore the integration path surrounds closely the current element and H_i and H_o are the fields on either side of the current sheet for $w \to v$. This *current-sheet theorem* [2.16] will prove very useful in deriving magnetic fields from current distributions.

A cosine like current distribution on a circle produces a pure dipole field inside this circle. Generalizing this observation we study fields generated by higher-harmonic current distributions. Let the current be distributed along a circle of radius R

$$v = R\,\mathrm{e}^{\mathrm{i}\varphi}, \qquad (2.85)$$

where φ is the angle between the radius vector $\mathbf{R}$ and the positive x-coordinate. The azimuthal current distribution along the circle be

$$I(\varphi) = -\sum_{n=1}^{\infty} I_n \cos(n\varphi + \varphi_n) \qquad (2.86)$$

and replacing the trigonometric functions by exponential functions we get with (2.85)

$$\cos(n\varphi + \varphi_n) = \frac{1}{2}\left[\mathrm{e}^{\mathrm{i}\varphi_n}\left(\frac{v}{R}\right)^n + \mathrm{e}^{-\mathrm{i}\varphi_n}\left(\frac{R}{v}\right)^n\right]. \qquad (2.87)$$

An ansatz for the field configuration which meets the proper boundary condition at the location of the currents $w = v$ is obtained by inserting (2.87) into (2.86) and we get after differentiation from (2.83)

$$H_{\mathrm{i}} - H_{\mathrm{o}} = \mathrm{i}\left[\frac{c}{4\pi}\right]\frac{2\pi}{c}\sum_{n=1}^{\infty} n\frac{I_n}{R}\left[\mathrm{e}^{\mathrm{i}\varphi_n}\left(\frac{w}{R}\right)^{n-1} + \mathrm{e}^{-\mathrm{i}\varphi_n}\left(\frac{R}{w}\right)^{n+1}\right]. \qquad (2.88)$$

We would like to apply (2.88) to any location w not just at the current sheet and add therefore two more boundary conditions. The fields must not diverge anywhere in the (x, y)-plane specifically for $w \to 0$ and $w \to \infty$. We also exclude a constant field in all of the (x, y)-plane since it would require an infinite energy to establish. Application of these boundary conditions to (2.88) gives the fields

$$H_{\mathrm{i}}(w) = \mathrm{i}\left[\frac{c}{4\pi}\right]\frac{2\pi}{c}\sum_{n=1}^{\infty} n\frac{I_n}{R}\,\mathrm{e}^{\mathrm{i}\varphi_n}\left(\frac{w}{R}\right)^{n-1}$$

$$H_{\mathrm{o}}(w) = -\mathrm{i}\frac{2\pi}{c}\sum_{n=1}^{\infty} n\frac{I_n}{R}\,\mathrm{e}^{-\mathrm{i}\varphi_n}\left(\frac{R}{w}\right)^{n+1}. \qquad (2.89)$$

Both fields now meet proper boundary conditions for realistic fields at $w = 0$ and $w = \infty$ and are consistent with the current sheet theorem (2.84). Being interested only in the field inside the current distribution, we find with (2.82)

$$H_y + \mathrm{i}\,H_x = H_{\mathrm{i}}(w) = \mathrm{i}\left[\frac{c}{4\pi}\right]\frac{2\pi}{c}\sum_{n=1}^{\infty} n\frac{I_n}{R}\,\mathrm{e}^{\mathrm{i}\varphi_n}\left(\frac{x + \mathrm{i}y}{R}\right)^{n-1}, \qquad (2.90)$$

where $R^2 = (x - x_\mathrm{o})^2 + (y - y_\mathrm{o})^2$. The angle φ_n determines the orientation of the multipole fields and to be consistent with the definitions of conventional magnet fields we set $\varphi_n = 0$ for upright multipoles and $\varphi_n = \pi/2$ for rotated multipoles. An upright quadrupole field, for example, is defined by the imaginary term for $n = 2$ in (2.90)

$$H_{\mathrm{i}}(w) = \mathrm{i}\left[\frac{c}{4\pi}\right]\frac{4\pi}{c}\frac{I_2}{R^2}\,w = \mathrm{i}\,g\,w. \qquad (2.91)$$

For a rotated quadrupole the orientation $\varphi_2 = \pi/4$ and the field gradient is determined by

$$g = \left[\frac{c}{4\pi}\right] \frac{4\pi}{c} \frac{I_2}{R^2}. \tag{2.92}$$

The current distribution to create such a field is from (2.86)

$$I(\varphi) = -2\,I_2 \cos(2\varphi + \pi/4). \tag{2.93}$$

In a similar way current distributions can be derived for arbitrary higher-order field configurations including combinations of multipole fields . Proper selection of the orientation φ_n distinguishes between upright and rotated magnets. The derivation was made based on a current distribution on a circle but could have been based as well on an ellipse. We also chose to use only a thin current sheet. A "thick" current sheet can be represented by many thin current sheets and a linear superposition of fields . More detailed information about the techniques of building superconducting magnets exceeds the goals of this text and more specialized literature should be consulted, for example [2.14].

Problems

Problem 2.1. We may adopt thin lens approximation if the focal length of a quadrupole $f_o \gg \ell_q$, where ℓ_q is the quadrupole length. To quantify the error made by thin lens approximation define a dimensionless quantity R which depends only on its value for a thin quadrupole R_o. Plot R/R_o as a function of R_o. For what value of R_o does the difference exceed 10%? How long may a quadrupole be in units of the focal length before the difference between thin and thick lens approximation exceeds 1%?

Problem 2.2. We use generally a hard-edge model for bending and focusing magnets. We also note in (2.42,43) perturbation terms depending on the slope and curvature of the field distribution. Show in a mathematically rigorous way that these terms do not contribute to the linear transformation matrices of hard-edge model magnets although the field derivatives $\kappa' = \infty$, etc.

Problem 2.3. We generally ignore the effect of finite trajetory slopes in beam dynamics. On the other hand strong focusing is required, for example, at the collision point of a colliding-beam storage ring. Furthermore a beam of at least 10σ's must be accommodated for long beam life time. Derive a condition for the maximum slopes allowable such that the third-order term $3/2 kxx'^2$ is negligible. What is the contribution of the perturbation term

$\frac{1}{2}\kappa_x(x'^2 - y'^2)$ to beam dynamics? When does this contribution become serious?

Problem 2.4. Design an electrostatic quadrupole with an aperture radius of 3 cm which is strong enough to produce a tune split of $\delta\nu = 0.01$ between a counterrotating particle and antiparticle beam at an energy of your choice. Assume the quadrupole to be placed at a location with a betatron function of $\beta = 10$ m. How long must the quadrupole be if the electric field strength is to be limited to no more than 15 kV/cm?

Problem 2.5. The fringe fields of wiggler magnets contribute to some focusing. Derive an expression for the tune shift as a function of wiggler parameters assuming that the betatron function is symmetric but not constant along the wiggler magnet.

Problem 2.6. A wiggler magnet can be designed with wedge shaped poles which contributes to equal focusing in both planes. Derive an expression for the horizontal and vertical focal length as a function of the wiggler field and period length. What is the periodic solution for the betatron function?

Problem 2.7. Derive a hard-edge model for a wiggler half pole which meets all three matching conditions discussed in Sect. 2.3.3. Hint: To get the extra degree of freedom needed, split the half pole into two pieces.

Problem 2.8. Calculate and design the current distribution for a pure air coil or superconducting dipole magnet to produce a field of 50 kGauss in an aperture of 3 cm radius without exceeding an average current density of 100 A/mm^2.

Problem 2.9. Derive an expression for the current distribution to produce a combination of a dipole, quadrupole and sextupole field. Express the currents in terms of fields and field gradients.

Problem 2.10. Design a dipole magnet as proposed by *Rabi* with an aperture radius of 5 cm and a field of 1 kG. The separation of both circles should be no more than 20% of the radius. Calculate the required electrical current and the current density. To allow appropriate cooling the average current density should not exceed 5 A/mm^2. In case the magnet must be pulsed like a kicker magnet to stay below this thermal limit determine the maximum duty cycle. Calculate the stored field energy, inductance of the magnet and required voltage from the power supply if the rise time should be no more than 10% of the puls length.

3. Dynamics of Coupled Motion

Coupling between horizontal and vertical betatron oscillations plays an important role in beam dynamics. Since coupling is caused by linear as well as nonlinear fields, we observe the effects on beam characteristics in virtually any accelerator. In order to be able to introduce or eliminate coupling, whichever may be the case, in a controlled and predictable way, we need to understand the dynamics of coupling in more detail. In this chapter we will derive first the equations of motion for the two most general sources of coupling, the *solenoid field* and the field of a *rotated quadrupole*, solve the equations of motion and formulate modifications to beam dynamics parameters and functions of linear uncoupled motion.

3.1 Conjugate Trajectories

Lattice functions have been defined to express solutions to the equations of motion for individual trajectories. Conversely, there must be a way to express these lattice functions by the principal solutions of the equation of motion. This would enable us to determine lattice functions for coupled particle motion by integrating the equations of motion for two orthogonal trajectories. To do this we start from the differential equation of motion in normalized coordinates. A set of linearly independent principal solutions are given by

$$
\begin{aligned}
w_1(\varphi) &= \cos\nu\varphi, \\
w_2(\varphi) &= \sin\nu\varphi.
\end{aligned}
\tag{3.1}
$$

In regular coordinates with $u(s) = w\sqrt{\beta(s)}$, where $u(s)$ stands for $x(s)$ or $y(s)$ the *conjugate trajectories* are

$$
\boxed{
\begin{aligned}
u_1(s) &= \sqrt{\beta(s)}\,\cos\psi(s), \\
u_2(s) &= \sqrt{\beta(s)}\,\sin\psi(s),
\end{aligned}
}
\tag{3.2}
$$

and their derivatives

$$u_1'(s) = -\frac{\alpha(s)}{\sqrt{\beta(s)}} \cos\psi(s) - \frac{1}{\sqrt{\beta(s)}} \sin\psi(s),$$

$$u_2'(s) = -\frac{\alpha(s)}{\sqrt{\beta(s)}} \sin\psi(s) + \frac{1}{\sqrt{\beta(s)}} \cos\psi(s). \tag{3.3}$$

Using (3.2,3) all lattice functions can be expressed in terms of conjugate trajectories setting

$$\boxed{\begin{aligned} \beta(s) &= u_1^2(s) + u_2^2(s), \\ \alpha(s) &= -u_1(s)\,u_1'(s) - u_2(s)\,u_2'(s), \\ \gamma(s) &= u_1'^2(s) + u_2'^2(s). \end{aligned}} \tag{3.4}$$

The betatron phase advance $\Delta\psi = \psi - \psi_o$ between the point $s = 0$ and the point s can be derived from

$$\cos(\psi - \psi_o) = \cos\psi\,\cos\psi_o + \sin\psi\,\sin\psi_o,$$

where $\psi_o = \psi(0)$ and $\psi = \psi(s)$. With (3.2,4) we get

$$\cos\psi(s) = \frac{u_1(s)}{\sqrt{\beta(s)}} = \frac{u_1(s)}{\sqrt{u_1^2(s) + u_2^2(s)}} \tag{3.5}$$

and similarly,

$$\sin\psi(s) = \frac{u_2(s)}{\sqrt{\beta(s)}} = \frac{u_2(s)}{\sqrt{u_1^2(s) + u_2^2(s)}}. \tag{3.6}$$

The betatron phase advance then is given by

$$\boxed{\cos(\psi - \psi_o) = \frac{u_1\,u_{1o} + u_2\,u_{2o}}{\sqrt{u_1^2 + u_2^2}\,\sqrt{u_{1o}^2 + u_{2o}^2}},} \tag{3.7}$$

where $u_i = u_i(s)$ and $u_{io} = u_i(0)$. Finally, we can express the elements of the transformation matrix from $s = 0$ to s by

$$\mathcal{M}(s|0) = \begin{pmatrix} M_{11} & M_{12} \\ M_{21} & M_{22} \end{pmatrix} = \begin{pmatrix} u_1 u_{2o}' - u_2 u_{1o}' & u_{1o} u_2 - u_{2o} u_1 \\ u_{2o}' u_1' - u_{2o}' u_2' & u_{1o} u_2' - u_{2o} u_1' \end{pmatrix}. \tag{3.8}$$

The two linearly independent solutions (3.2) also can be used to define and characterize the phase space ellipse. At the start of a beam line we set $s = 0$ and $\psi(0) = 0$ and define an ellipse by the parametric vector equation

$$\mathbf{u}(0) = a\,[\mathbf{u}_1(0)\cos\phi + \mathbf{u}_2(0)\sin\phi], \tag{3.9}$$

where

$$\mathbf{u}(0) = \begin{pmatrix} u_o \\ u_o' \end{pmatrix} \qquad \text{and} \qquad \mathbf{u}_i(0) = \begin{pmatrix} u_{io} \\ u_{io}' \end{pmatrix}.$$

As the parameter ϕ varies over a period of 2π the vector follows the outline of an ellipse. To parametrize this ellipse we calculate the area enclosed by the phase ellipse. The area element is $dA = u_o' \, du_o$, from (3.9) we get

$$du_o = -a \left[u_{1o} \sin \phi + u_{2o} \cos \phi \right] d\phi$$

and the area enclosed by the ellipse is

$$A = -a^2 \, 2 \int\limits_o^\pi \left(u_{1o}' \cos \phi - u_{2o}' \sin \phi \right) \left(u_{1o} \sin \phi + u_{2o} \cos \phi \right) d\phi \,, \tag{3.10}$$

$$= a^2 \, \pi \left(u_{1o} u_{2o}' - u_{1o}' u_{2o} \right) = a^2 \, \pi \,,$$

since the expression in the brackets is the *Wronskian*, which we choose to normalize to unity. The Wronskian is an invariant of the motion and therefore the area of the phase ellipse along the beam transport line is preserved. For $a^2 = \epsilon$ the vector equation (3.9) describes the *phase ellipse* enclosing a beam with the emittance ϵ.

The formalism of conjugate trajectories has not produced any new insight into beam dynamics that we did not know before but it is an important tool for the discussion of coupled particle motion and provides a simple way to trace individual particles through complicated systems.

3.2 Particle Motion in a Solenoidal Field

General field equations have been derived in Chap. 2 with the only restriction that there be no solenoid fields, which allowed us to set $A_{00} = 0$ in (2.6), and concentrate on transverse fields only. Longitudinal fields like that produced in a *solenoid magnet* are used for very special purposes in beam transport systems and their effect on beam dynamics cannot be ignored.

We assume now that the lowest-order coefficient A_{00} in the potential (2.3) does not vanish

$$A_{00}(s) \neq 0 \,. \tag{3.11}$$

Longitudinal fields do not cause transverse deflection and we may therefore choose a cartesian coordinate system along such fields and set

$$\kappa_x = \kappa_y = 0 \,, \tag{3.12}$$

while the recursion formula (2.5) reduces to

$$A_{02} + A_{20} = -A_{00}'' \,. \tag{3.13}$$

For reasons of symmetry with respect to x and y we have $A_{02} = A_{20}$ and

$$A_{02} = A_{20} = -\tfrac{1}{2} A_{00}''. \tag{3.14}$$

With this the potential (2.3) for longitudinal fields is

$$V_s(x,y,s) = -\frac{cp}{e}\left[A_{00} - \tfrac{1}{4}A_{00}''(x^2+y^2)\right],$$
$$= -\frac{cp}{e}A_{00} + \tfrac{1}{4}\frac{cp}{e}A_{00}''r^2, \tag{3.15}$$

where we have made use of the rotational symmetry. The longitudinal field component becomes from (3.15) in linear approximation

$$B_s = +\frac{cp}{e}A_{00}' \tag{3.16}$$

and the transverse components

$$B_r = -\frac{cp}{e}\frac{1}{2}A_{00}''r = -\tfrac{1}{2}B_s'r, \tag{3.17}$$
$$B_\varphi = 0.$$

The azimuthal field component obviously vanishes because of symmetry. Radial field components appear whenever the longitudinal field strength varies as is the case in the fringe field region at the end of a solenoid shown in Fig. 3.1.

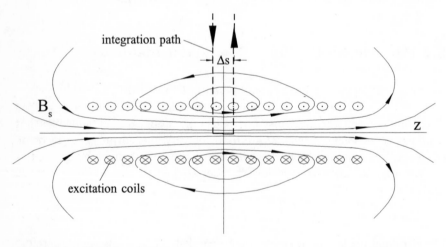

Fig. 3.1. Solenoid field

The strength B_s in the center of a long solenoid magnet can be calculated in the same way we determined dipole and higher-order fields utilizing *Stoke's theorem*. The integral $\oint \mathbf{B}\cdot\mathrm{ds}$ is performed along a path as indicated in Fig. 3.1. The only contribution to the integral comes from the integral along the field at the magnet axis. All other contributions vanish because the integration path cuts field lines at a right angle, where $\mathbf{B}\cdot\mathrm{ds} = 0$ or follows field lines to infinity where $B_s = 0$. We have therefore

$$\int \mathbf{B} \cdot \mathrm{d}s \;=\; B_\mathrm{s} \varDelta s \;=\; \frac{4\pi}{c} J \, \varDelta s \,, \tag{3.18}$$

where J is the total solenoid current per unit length. The *solenoid field strength* is therefore given by

$$B_\mathrm{s}(x=0, y=0) \;=\; \frac{4\pi}{c} J \,. \tag{3.19}$$

The total integrated radial field $\int B_r \mathrm{d}s$ can be evaluated from the central field for each of the fringe regions. We imagine a cylinder concentric with the solenoid axis and a radius r to extend from the solenoid center to a region well outside the solenoid. In the center of the solenoid a total magnetic flux of $\pi r^2 B_\mathrm{s}$ enters this cylinder. It is clear that along the infinitely long cylinder that flux will penetrate the surface of the cylinder through radial field components. We have therefore

$$\pi\, r^2\, B_\mathrm{s} \;=\; \int\limits_0^\infty 2\pi r\, B_r(r)\, \mathrm{d}s \,, \tag{3.20}$$

where we have set $s = 0$ at the center of the solenoid. The integrated radial field per fringe field is then

$$\int B_r(r)\, \mathrm{d}s \;=\; -\tfrac{1}{2}\, r\, B_\mathrm{s} \,. \tag{3.21}$$

The linear dependence of the integrated radial fields on the distance r from the axis constitutes linear focusing capabilities of solenoidal fringe fields, which is for example, used at *positron targets*. The positron source is generally a small piece of a heavy metal like tungsten placed in the path of a high energy electron beam. Through an electromagnetic cascade, positrons are generated and seem to emerge from a point like source into a large solid angle. If the target is placed in the center of a solenoid the radial positron motion couples with the longitudinal field to transfer the radial particle momentum into azimuthal momentum. At the end of the solenoid, the azimuthal motion couples with the radial field components of the fringe field to transfer azimuthal momentum into longitudinal momentum. In this idealized picture a divergent positron beam emerging from a small source area is transformed or focused into a quasi-parallel beam of larger cross section. Such a focusing device is called a $\lambda/4$-lens since the particles follow one quarter of a *helical trajectory* in the solenoid.

In other applications large volume solenoids are used as part of elementary particles detectors in high energy physics experiments performed at colliding-beam facilities. The strong influence of these solenoidal detector fields on beam dynamics in a storage ring must be compensated in most cases. In still other applications solenoid fields are used just to contain a particle beam within a small circular aperture like that along the axis of a linear accelerator.

Equation of motion: The equations of motion in a solenoid can be derived from (2.39), neglecting all transverse beam deflection

$$x'' - \frac{1}{2} \frac{x'}{\sigma'^2} \frac{d\sigma'^2}{ds} = \frac{e}{cp} \sigma' \left(y' B_s - B_y \right),$$

$$y'' - \frac{1}{2} \frac{y'}{\sigma'^2} \frac{d\sigma'^2}{ds} = \frac{e}{cp} \sigma' \left(B_x - x' B_s \right),$$

$$\tag{3.22}$$

where the field components derived from (3.17) are

$$\mathbf{B} = \left(-\tfrac{1}{2} B'_s x, \ -\tfrac{1}{2} B'_s y, \ B_s \right). \tag{3.23}$$

Following the same derivation as in Sect. 2.2 the general equations of motion in a solenoid field including up to third-order terms are

$$x'' = + \frac{e}{cp} B_s y' + \frac{1}{2} \frac{e}{cp} B'_s y$$
$$+ \frac{1}{4} \frac{e}{cp} \left(2x'^2 y' B_s + x'^2 y B'_s + 2y'^3 B_s + yy'^2 B'_s \right) + \mathcal{O}(4),$$

$$y'' = - \frac{e}{cp} B_s x' - \frac{1}{2} \frac{e}{cp} B'_s x$$
$$- \frac{1}{4} \frac{e}{cp} \left(2x' y'^2 B_s + xy'^2 B'_s + 2x'^3 B_s + xx'^2 B'_s \right) + \mathcal{O}(4).$$

$$\tag{3.24}$$

Considering only linear terms, the equations of motion in a solenoidal field are

$$x'' = + \frac{e}{cp} B_s y' + \frac{1}{2} \frac{e}{cp} B'_s y,$$

$$y'' = - \frac{e}{cp} B_s x' - \frac{1}{2} \frac{e}{cp} B'_s x,$$

$$\tag{3.25}$$

exhibiting clearly the coupling effected by solenoidal fields. In a uniform field, where $B'_s = 0$, the particle trajectory assumes the form of a helix parallel to the axis of the solenoid field.

The equations of motion (3.25) have been derived under the assumption of paraxial rays so that we can set $v \approx v_s$. In a solenoid field this approximation is not generally acceptable since we may use solenoid focusing for particles emerging from a target at large angles. We therefore replace all derivatives with respect to s by derivatives with respect to the time and use the particle velocity v, and replace $\frac{d}{ds} \rightarrow \frac{1}{v} \frac{d}{dt}$. In a uniform solenoid field the equations of motion are

$$\ddot{x} = + \frac{eB_s v}{cp} \dot{y} = \omega_L \dot{y},$$

$$\ddot{y} = - \frac{eB_s v}{cp} \dot{x} = -\omega_L \dot{x},$$

$$\tag{3.26}$$

where the *Larmor frequency* is defined by

$$\omega_{\mathrm{L}} = \frac{ec}{E} B_{\mathrm{s}},\tag{3.27}$$

and E is the total particle energy. Multiplying (3.26) by $\dot{x}$ and $\dot{y}$, respectively, and adding both equations we get $\mathrm{d}(\dot{x}^2 + \dot{y}^2)/\mathrm{d}t = 0$ or

$$\dot{x}^2 + \dot{y}^2 = v_t^2 = \mathrm{const}.\tag{3.28}$$

The transverse particle velocity v_t or total transverse momentum of the particle cp_t stays constant during the motion in a uniform solenoid field. For $\dot{x}_{\mathrm{o}} = 0$ and $\dot{y}_{\mathrm{o}} = v_t$ the transverse velocities can be expressed by

$$\begin{aligned} \dot{x} &= v_t \sin\omega_{\mathrm{L}} t, \\ \dot{y} &= v_t \cos\omega_{\mathrm{L}} t \end{aligned}\tag{3.29}$$

and the solutions of the equations of motion are

$$\begin{aligned} x(t) &= x_{\mathrm{o}} - \frac{v_t}{\omega_{\mathrm{L}}} \cos\omega_{\mathrm{L}} t, \\ y(t) &= y_{\mathrm{o}} + \frac{v_t}{\omega_{\mathrm{L}}} \sin\omega_{\mathrm{L}} t. \end{aligned}\tag{3.30}$$

The amplitude of the oscillating term in (3.30) is equal to the radius of the helical path

$$\rho_{\mathrm{h}} = \frac{cp_t}{eB_{\mathrm{s}}}.\tag{3.31}$$

The longitudinal motion is unaffected by the solenoid field and $\dot{v}_{\mathrm{s}} = 0$ as can be derived from the Lorentz equation since all transverse field components vanish and

$$s(t) = s_{\mathrm{o}} + v_s\, t.\tag{3.32}$$

The time to complete one period of the helix is

$$T = \frac{2\pi}{\omega_{\mathrm{L}}}\tag{3.33}$$

during which time the particle moves along the s-axis a distance

$$\Delta s = 2\pi \frac{cp_s}{eB_{\mathrm{s}}},\tag{3.34}$$

where p_s is the s-component of the particle momentum. A more general derivation of the equation of motion in a solenoid field will be discussed in Sect. 3.3.3.

3.3 Transverse Coupled Oscillations

In linear beam dynamics transverse motion of particles can be treated separately in the horizontal and vertical plane. This can be achieved by proper selection, design and alignment of beam transport magnets. Fabrication and alignment tolerances, however, will introduce, for example, rotated quadrupole components where only upright quadrupole fields were intended. The perturbation caused creates a coupling of both the horizontal and vertical oscillation and independent treatment is no longer accurate. Such linear coupling can be compensated in principle by additional rotated quadrupoles, but the beam dynamics for coupling effects must be known to perform a proper compensation.

3.3.1 Equations of Motion in Coupling Systems

The most generally used magnets that introduce coupling in beam transport systems are rotated quadrupoles and solenoid magnets and we will restrict our discussion of coupled beam dynamics to such magnets. Eqs. (2.42,43) include all coupling terms up to third order. Restricting beam dynamics to rotated quadrupole and solenoid fields defines the realm of *linear coupling*. The equations of motion in the presence of upright and rotated quadrupoles as well as solenoid fields are

$$
\begin{aligned}
x'' + kx &= -\underline{k}\,y + S\,y' + \tfrac{1}{2}\,S'\,y\,, \\
y'' - ky &= -\underline{k}\,x - S\,x' - \tfrac{1}{2}\,S'\,x\,,
\end{aligned}
\tag{3.35}
$$

where the solenoid field is expressed by

$$
S(s) = \frac{e}{cp}\,B_{\mathrm{s}}(s)\,.
\tag{3.36}
$$

In the following subsections we will derive separately the transformation through both rotated quadrupoles and solenoid magnets.

3.3.2 Coupled Beam Dynamics in Skew Quadrupoles

The distribution of rotated or *skew quadrupoles* and solenoid magnets is arbitrary and therefore no analytic solution can be expected for the differential equations (3.35). Similar to other beam line elements, we discuss solutions of the equations of motion for individual magnets only and assume that strength parameters within hard-edge model magnets stay constant. We discuss first the solutions of particle motion in skew quadrupoles alone and ignore solenoid fields. The equations of motion for skew quadrupoles are from (3.35)

$$
\begin{aligned}
x'' + \underline{k}\,y &= 0\,, \\
y'' + \underline{k}\,x &= 0\,.
\end{aligned}
\tag{3.37}
$$

These equations look very similar to the equations for ordinary upright quadrupoles except that the restoring forces now depend also on the particle amplitude in the other plane. We know the solution of the equation of motion for an upright focusing and defocusing quadrupole and will try to apply these solutions to (3.37). Combining the observation that each quadrupole is focusing in one plane and defocusing in the other with the apparent mixture of both planes for a skew quadrupole, we will try an ansatz for (3.37) which is made up of four principal solutions

$$
\begin{aligned}
x &= a \cos\varphi + \frac{b}{\sqrt{k}} \sin\varphi + c \cosh\varphi + \frac{d}{\sqrt{k}} \sinh\varphi, \\
y &= A \cos\varphi + \frac{B}{\sqrt{k}} \sin\varphi + C \cosh\varphi + \frac{D}{\sqrt{k}} \sinh\varphi,
\end{aligned}
\tag{3.38}
$$

where $\varphi = \sqrt{k}\, s$ and the variable s varies between zero and the full length of the quadrupole, $0 < s < \ell_q$. The coefficients $a, b, c, \ldots D$ must be determined to be consistent with the initial parameters of the trajectories (x_o, x'_o, y_o, y'_o). For $s = 0$ we get

$$
\begin{aligned}
x_o &= a + c, & y_o &= A + C, \\
x'_o &= b + d, & y'_o &= B + D.
\end{aligned}
\tag{3.39}
$$

Solutions (3.38) must be consistent with (3.37) from which we find

$$
\begin{aligned}
a &= A & c &= -C \\
b &= B & d &= -D.
\end{aligned}
\tag{3.40}
$$

From (3.39,40) we get finally the coefficients consistent with the initial conditions and the differential equations (3.37)

$$
\begin{aligned}
a &= A = \tfrac{1}{2}(x_o + y_o), & b &= B = \tfrac{1}{2}(x'_o + y'_o), \\
c &= -C = \tfrac{1}{2}(x_o - y_o), & d &= -D = \tfrac{1}{2}(x'_o - y'_o).
\end{aligned}
\tag{3.41}
$$

With these definitions the transformation through a skew quadrupole is

$$
\begin{pmatrix} x \\ x' \\ y \\ y' \end{pmatrix} = \mathbf{M}_{\mathrm{sq}} \begin{pmatrix} x_o \\ x'_o \\ y_o \\ y'_o \end{pmatrix},
\tag{3.42}
$$

where the transformation matrix $\mathbf{M}_{\mathrm{sq}}$ is the transformation matrix for a skew quadrupole,

$$
\mathbf{M}_{\mathrm{sq}}(s|0) =
\tag{3.43}
$$

$$
\frac{1}{2}
\begin{pmatrix}
\mathcal{C}^+(\varphi) & \frac{1}{\sqrt{k}}\mathcal{S}^+(\varphi) & \mathcal{C}^-(\varphi) & \frac{1}{\sqrt{k}}\mathcal{S}^-(\varphi) \\
-\sqrt{k}\,\mathcal{S}^-(\varphi) & \mathcal{C}^+(\varphi) & -\sqrt{k}\,\mathcal{S}^+(\varphi) & \mathcal{C}^-(\varphi) \\
\mathcal{C}^-(\varphi) & \frac{1}{\sqrt{k}}\mathcal{S}^-(\varphi) & \mathcal{C}^+(\varphi) & \frac{1}{\sqrt{k}}\mathcal{S}^+(\varphi) \\
-\sqrt{k}\,\mathcal{S}^+(\varphi) & \mathcal{C}^-(\varphi) & -\sqrt{k}\,\mathcal{S}^-(\varphi) & \mathcal{C}^+(\varphi)
\end{pmatrix},
$$

with $\mathcal{C}^\pm(\varphi) = \cos\varphi \pm \cosh\varphi$ and $\mathcal{S}^\pm(\varphi) = \sin\varphi \pm \sinh\varphi$ and $\varphi = \sqrt{k}\, s$.

This transformation matrix is quite elaborate and becomes useful only for numerical calculations on computers. We employ again *thin lens approximation* where the quadrupole length vanishes, $\ell_{sq} \to 0$, in such a way as to preserve the integrated magnet strength or the focal length f. The matrix (3.43) then reduces to the simple form

$$
\mathcal{M}_{sq}(0|\ell_{sq}) = \begin{pmatrix} 1 & \ell_{sq} & 0 & 0 \\ 0 & 1 & -1/f & 0 \\ 0 & 0 & 1 & \ell_{sq} \\ -1/f & 0 & 0 & 1 \end{pmatrix},
\tag{3.44}
$$

where the focal length is defined as $f^{-1} = \underline{k}\,\ell_{sq}$. Note that we have not set $\ell_{sq} = 0$ but retained the linear terms in ℓ_{sq}, which is a more appropriate thin-lens approximation for weak skew quadrupoles of finite length. Along the diagonal the transformation matrix looks like a drift space of length ℓ_{sq} while the off-diagonal elements describe the focusing of the thin skew quadrupole.

3.3.3 Equations of Motion in a Solenoid Magnet

The solutions of the equations of motion for a solenoid magnet are more complex since we must now include terms that depend on the slope of the particle trajectories as well. Ignoring skew quadrupoles the differential equations of motion in a *solenoid magnet* becomes from (3.35)

$$
\begin{aligned}
x'' - S(s)\,y' - \tfrac{1}{2}\,S'(s)\,y &= 0, \\
y'' + S(s)\,x' + \tfrac{1}{2}\,S'(s)\,x &= 0.
\end{aligned}
\tag{3.45}
$$

Strong coupling between both planes is obvious and the variation of coordinates in one plane depends entirely on the coordinates in the other plane. We note a high degree of symmetry in the equations in the sense that both coordinates change similar as a function of the other coordinates. This suggests that a rotation of the coordinate system may help simplify the solution of the differential equations. We will therefore try such a coordinate rotation in complex notation by defining

$$
R = (x + \mathrm{i}\,y)\,\mathrm{e}^{-\mathrm{i}\,\phi(s)},
\tag{3.46}
$$

where the rotation angle ϕ may be a function of the independent variable s. From (3.45) a single differential equation can be formed in complex notation

$$
(x + \mathrm{i}y)'' + \mathrm{i}\,S(s)\,(x+\mathrm{i}y)' + \mathrm{i}\,\tfrac{1}{2}\,S'(s)\,(x+\mathrm{i}y) = 0.
\tag{3.47}
$$

The rotation (3.46) can now be applied directly to (3.47) and with

$$
(x + \mathrm{i}y)' = R'\,\mathrm{e}^{\mathrm{i}\phi} + \mathrm{i}\phi'\,R\,\mathrm{e}^{\mathrm{i}\phi}
$$

and

$$(x + iy)'' = R'' e^{i\phi} + 2 i \phi' R' e^{i\phi} + i \phi'' R e^{i\phi} - \phi'^2 R e^{i\phi}$$

$$R'' - [S(s) \phi' + \phi'^2] R$$
$$+ i 2 [\phi' + \tfrac{1}{2} S(s)] R' + i [\phi'' + \tfrac{1}{2} S'(s)] R = 0. \tag{3.48}$$

At this point the introduction of the coordinate rotation allows to greatly simplify (3.48) by assuming a continuous rotation along the beam axis with a rotation angle defined by

$$\phi(s) = -\tfrac{1}{2} \int_{s_o}^{s} S(\bar{s}) \, d\bar{s} \tag{3.49}$$

where the solenoid field starts at s_o. We are able to eliminate two terms in the differential equation (3.48). Since a positive solenoid field generates Lorentz forces that deflect the particles onto counter clockwise spiraling trajectories, we have included the negative sign in (3.49) to remain consistent with our sign convention. From (3.49) we have $\phi' = -\tfrac{1}{2} S(s)$ and $\phi'' = -\tfrac{1}{2} S'(s)$, which after insertion into (3.48) results in the simple equation of motion

$$R'' + \tfrac{1}{4} S^2(s) R = 0. \tag{3.50}$$

With $R = v + iw$ we finally get two uncoupled equations

$$v'' + \tfrac{1}{4} S^2(s) v = 0,$$
$$w'' + \tfrac{1}{4} S^2(s) w = 0. \tag{3.51}$$

By the introduction of coordinate rotation we have reduced the coupled differential equations (3.45) to the form of uncoupled equations of motion exhibiting focusing in both planes. At the entrance to the solenoid field $\phi = 0$ and therefore $v_o = x_o$ and $w_o = y_o$. To determine the particle motion through the solenoid field of length L_s we simply follow the particle coordinates (v, w) through the solenoid as if it were a quadrupole of strength $k_s = \tfrac{1}{4} S^2(L_s)$ followed by a rotation of the coordinate system by the angle $-\phi(L_s)$ to get back to cartesian coordinates (x, y).

3.3.4 Transformation Matrix for a Solenoid Magnet

Similar to the transformation through quadrupoles and other beam transport magnets we may formulate a transformation matrix for a solenoid magnet. Instead of uncoupled 2×2 transformation matrices, however, we must use 4×4 matrices to include coupling effects. Each coordinate now depends on the initial values of all coordinates, $x(s) = x(x_o, x'_o, y_o, y'_o)$, etc. The transformation through a solenoid is performed in two steps in which the first is the solution of (3.51) in the form of the matrix $\mathcal{M}_s$, and the second is a coordinate rotation introduced through the matrix $\mathcal{M}_r$. The total transformation is therefore

$$\begin{pmatrix} x \\ x' \\ y \\ y' \end{pmatrix} = \mathcal{M}_r \, \mathcal{M}_s \begin{pmatrix} x_o \\ x'_o \\ y_o \\ y'_o \end{pmatrix}. \tag{3.52}$$

In analogy to the transformation through an upright quadrupole, we get from (3.51) the transformation matrix $\mathcal{M}_s$ from the beginning of the solenoid field at s_o to a point s inside the solenoid magnet. The strength parameter in this case is $1/4 S^2$ assumed to be constant along the length of the magnet and the transformation matrix becomes then

$$\mathcal{M}_s(s_o|s) = \begin{pmatrix} \cos\phi & \frac{2}{S}\sin\phi & 0 & 0 \\ -\frac{S}{2}\sin\phi & \cos\phi & 0 & 0 \\ 0 & 0 & \cos\phi & \frac{2}{S}\sin\phi \\ 0 & 0 & -\frac{S}{2}\sin\phi & \cos\phi \end{pmatrix}, \tag{3.53}$$

where $\phi = 1/2 S\, s$. The next step is to introduce the coordinate rotation $\mathcal{M}_r$ which we derive from the vector equation

$$(\mathbf{x} + \mathrm{i}\,\mathbf{y}) = (\mathbf{v} + \mathrm{i}\,\mathbf{w})\, \mathrm{e}^{-\mathrm{i}\phi(s)}, \tag{3.54}$$

where the vectors are defined like $\mathbf{x} = (x, x')$, etc. Note that the value of the rotation angle ϕ is proportional to the strength parameter and the sign of the solenoid field defines the orientation of the coordinate rotation. Fortunately, we need not keep track of the sign since the components of the focusing matrix $\mathcal{M}_s$ are even functions of s and do not depend on the direction of the solenoid field.

By separating (3.54) into its real and imaginary part and applying *Euler's identity* $\mathrm{e}^{\alpha} = \cos\alpha + \mathrm{i}\sin\alpha$, we get for the rotation matrix at the point s within the solenoid magnet

$$\mathcal{M}_r = \begin{pmatrix} \cos\phi & 0 & \sin\phi & 0 \\ -\frac{S}{2}\sin\phi & \cos\phi & \frac{S}{2}\cos\phi & \sin\phi \\ -\sin\phi & 0 & \cos\phi & 0 \\ -\frac{S}{2}\cos\phi & -\sin\phi & -\frac{S}{2}\sin\phi & \cos\phi \end{pmatrix}. \tag{3.55}$$

The total transformation matrix for a solenoid magnet from $s_o = 0$ to s finally is the product of (3.53) and (3.55)

$$\mathcal{M}_{\mathrm{sol}}(0|s < L) =$$
$$\begin{pmatrix} \cos^2\phi & \frac{1}{S}\sin 2\phi & \sin\phi\cos\phi & \frac{2}{S}\sin^2\phi \\ -S\sin\phi\cos\phi & \cos 2\phi & \frac{S}{2}\cos 2\phi & \sin 2\phi \\ -\sin\phi\cos\phi & -\frac{2}{S}\sin^2\phi & \cos^2\phi & \frac{1}{S}\sin 2\phi \\ -\frac{S}{2}\cos 2\phi & -\sin 2\phi & -S\sin\phi\cos\phi & \cos 2\phi \end{pmatrix}. \tag{3.56}$$

This transformation matrix is correct for any point s inside the solenoid magnet but caution must be taken when we try to apply this transformation matrix for the whole solenoid by setting $s = L_s$. The result would be inaccurate because of a discontinuity caused by the solenoid fringe field. Only the focusing matrix $\mathcal{M}_s$ for the whole solenoid becomes a simple extension of (3.53) to the end of the solenoid by setting $\phi(L_s) = \Phi = \frac{1}{2}SL_s$.

Due to the solenoid fringe field, which in the hard-edge approximation adopted here is a thin slice, the rotation matrix exhibits a discontinuity. For $s = L_s + \epsilon$, where $\epsilon \to 0$ the phase is $\phi(L_s) = \Phi$ but the solenoid strength is now zero, $S = 0$. Therefore, the rotation matrix assumes the form

$$
\mathcal{M}_r = \begin{pmatrix}
\cos\Phi & 0 & \sin\Phi & 0 \\
0 & \cos\Phi & 0 & \sin\Phi \\
-\sin\Phi & 0 & \cos\Phi & 0 \\
0 & -\sin\Phi & 0 & \cos\Phi
\end{pmatrix}. \tag{3.57}
$$

After multiplication of (3.53) with (3.57) and setting $\phi = \Phi$ the transformation matrix for a complete solenoid magnet is finally

$$
\mathcal{M}_{sol}(0|L) =
$$
$$
\begin{pmatrix}
\cos^2\Phi & \frac{1}{S}\sin 2\Phi & \sin\Phi\cos\Phi & \frac{2}{S}\sin^2\Phi \\
-\frac{S}{2}\sin\Phi\cos\Phi & \cos^2\Phi & -\frac{S}{2}\sin^2\Phi & \sin\Phi\cos\Phi \\
-\sin\Phi\cos\Phi & -\frac{2}{S}\sin^2\Phi & \cos^2\Phi & \frac{1}{S}\sin 2\Phi \\
\frac{S}{2}\sin^2\Phi & -\sin\Phi\cos\Phi & -\frac{S}{2}\sin\Phi\cos\Phi & \cos^2\Phi
\end{pmatrix}. \tag{3.58}
$$

Comparing matrices (3.56,58) we find no continuous transition between both matrices since only one matrix includes the effect of the fringe field. In reality the fringe field is not a thin-lens and therefore a continuous transition between both matrices can be derived but to stay consistent with the rest of this chapter, we assume for our discussions hard-edge magnet models.

From the matrix (3.53) some special properties of particle trajectories in a solenoid can be derived. For $\Phi = \pi/2$ a parallel beam becomes focused to a point at the magnet axis. A trajectory entering a solenoid with the strength $\Phi = \frac{1}{2}SL = \pi/2$ at say y_o will follow half a period of a spiraling trajectory with a radius $\rho = y_o/2$ and exit the solenoid at $x = y = 0$. Similarly, a beam emerging from a point source on the axis and at the start of the solenoid field will have been focused to a parallel beam at the end of the solenoid. Such a solenoid is used to focus, for example, a divergent positron beam emerging from the target source and is called a $\lambda/4$-*lens* or *quarter-wavelength solenoid* for obvious reasons.

The focusing properties of the whole solenoid are most notable when the field strength is weak and the focal length is long compared to the length of the solenoid. In this case, the focal length can be taken immediately from the M_{21} and M_{43} element of the transformation matrix as we did for quadrupoles and other focusing devices

$$\frac{1}{f_x} = M_{21} = -\tfrac{1}{2}\, S \sin\phi \cos\phi,$$

$$\frac{1}{f_y} = M_{43} = -\tfrac{1}{2}\, S \sin\phi \cos\phi.$$

(3.59)

In contrast to quadrupole magnets the focal length of a solenoid magnet is the same in both planes and is in thin-lens approximation

$$\frac{1}{f_{\mathrm{sol}}} = \tfrac{1}{4}\, S^2\, L_{\mathrm{s}} = \tfrac{1}{4}\left(\frac{e}{cp}\right)^2 B_{\mathrm{s}}^2\, L_{\mathrm{s}}\,.$$

(3.60)

The focal length is always positive and a solenoid will therefore always be focusing independent of the sign of the field or the sign of the particle charge. Transformation matrices have been derived for the two most important coupling magnets in beam transport systems, the skew quadrupole and the solenoid magnet, which allows us now to employ linear beam dynamics in full generality including linear coupling. Using (4 × 4)-transformation matrices any particle trajectory can be described whether coupling magnets are included or not. Specifically, we may use this formalism to incorporate compensating schemes when strongly coupling magnets must be included in a particular beam transport line.

3.3.5 Betatron Functions for Coupled Motion

For the linear uncoupled motion of particles in electromagnetic fields we have derived powerful mathematical methods to describe the dynamics of single particles as well as that of a beam composed of a large number of particles. Specifically, the concept of phase space to describe a beam at a particular location and the ability to transform this phase space from one point of the beam transport line to another allow us to design beam transport systems with predictable results. These theories derived for particle motion in one degree of freedom can be expanded to describe coupled motion in both the horizontal and vertical plane.

Ripken [3.1] first developed a complete theory of coupled betatron oscillations and of particle motion in four-dimensional phase space. In our discussion of coupled betatron motion and phase space transformation we will closely follow his theory. The basic idea hinges on the fact that the differential equations of motion provide the required number of independent solutions, two for oscillations in one plane and four for coupled motion in two planes, to define a two- or four-dimensional ellipsoid which serves as the boundary in phase space for the beam enclosed by it. Since the transformations in beam dynamics are symplectic, we can rely on invariants of the motion which are the basis for the determination of beam characteristics at any point along the beam transport line if we only know such parameters at one particular point.

Before we discuss coupled motion in more detail it might be useful to recollect some salient features of linear beam dynamics. In Sect. 3.1 the concept of conjugate trajectories was introduced, which can be used to define a phase ellipse at $s = 0$ in parametric form. Due to the symplecticity of the transformations we find the area of the phase ellipse to be a constant of motion and we may describe the phase ellipse at any point s along the beam line by

$$\mathbf{v}(s) = \sqrt{\epsilon}\left[\mathbf{v}_1(s) \cos \vartheta + \mathbf{v}_2(s) \sin \vartheta\right], \tag{3.61}$$

where the vector functions $\mathbf{v}_i(s) = (v_i, v_i')$ are two independent solutions of the equation of motion. The Wronskian is a constant of motion normalized to unity in which case the phase ellipse (3.9) has the area $A = \pi\epsilon$, where ϵ is the beam emittance for the beam enclosed by the ellipse. The solutions are of the form

$$\begin{aligned} v_1(s) &= \sqrt{\beta(s)} \cos \phi(s), \\ v_2(s) &= \sqrt{\beta(s)} \sin \phi(s), \end{aligned} \tag{3.62}$$

and forming the Wronskian we find

$$\beta\phi' = 1$$

as we would expect.

Expanding this theory of beam dynamics to coupled motion we note that the Wronskian is equivalent to the Lagrange invariant for symplectic transformations.

$$\mathbf{v}_i \, \boldsymbol{\mathcal{S}} \, \mathbf{v}_k = \text{const}, \tag{3.63}$$

where $\mathbf{v}_i$ and $\mathbf{v}_k$ are independent solution vectors $\mathbf{v}_i = (x_i, p_{xi}, y_i, p_{yi})$ of the differential equations of motion. For uncoupled motion (3.63) is equal to the Wronskian since $p_{xi} = p_o x_i'$ for purely transverse magnetic fields. The general relation between both coordinates can be expressed by the transformation

$$\begin{pmatrix} x \\ p_x \\ y \\ p_y \end{pmatrix} = \boldsymbol{\mathcal{S}} \cdot \begin{pmatrix} x \\ x' \\ y \\ y' \end{pmatrix} = \begin{pmatrix} 1 & 0 & 0 & 0 \\ 0 & 1 & +\tfrac{1}{2}S & 0 \\ 0 & 0 & 1 & 0 \\ -\tfrac{1}{2}S & 0 & 0 & 1 \end{pmatrix} \begin{pmatrix} x \\ x' \\ y \\ y' \end{pmatrix} \tag{3.64}$$

considering solenoid fields as the only longitudinal magnetic fields. Analogous to (3.61) we use now four independent solution vectors

$$\mathbf{v}_i = \begin{pmatrix} x_i \\ x_i' \\ y_i \\ y_i' \end{pmatrix}, \qquad \text{where} \quad i = 1, 2, 3, 4 \tag{3.65}$$

normalized by the conditions

$$\mathbf{v}_1 \, \boldsymbol{\mathcal{S}} \, \mathbf{v}_2 = 1 \qquad \text{and} \qquad \mathbf{v}_3 \, \boldsymbol{\mathcal{S}} \, \mathbf{v}_4 = 1 \, , \tag{3.66}$$

to define the location of a particle in four-dimensional phase space. Since the solutions are transformed by area preserving or symplectic transformations we describe the phase ellipse by

$$\begin{aligned}
\mathbf{v}(s) &= \sqrt{\epsilon_{\mathrm{I}}} \left[\mathbf{v}_1(s) \cos \vartheta_{\mathrm{I}} + \mathbf{v}_2(s) \sin \vartheta_{\mathrm{I}} \right] \cos \chi \\
&+ \sqrt{\epsilon_{\mathrm{II}}} \left[\mathbf{v}_3(s) \cos \vartheta_{\mathrm{II}} - \mathbf{v}_4(s) \sin \vartheta_{\mathrm{II}} \right] \sin \chi \, .
\end{aligned} \tag{3.67}$$

As the independent variables $\chi, \vartheta_{\mathrm{I}}$ and ϑ_{II} vary from 0 to 2π the vector $\mathbf{v}$ covers all points on the surface of the four-dimensional ellipsoid while the shape of the ellipse varies along the beam line consistent with the variation of the vector functions $\mathbf{v}_i$. In this ansatz we chose two modes of oscillations indicated by the index I and II. If the oscillations were uncoupled, we would identify mode-I with the horizontal oscillation and mode-II with the vertical motion and (3.67) would still hold with $\chi = 0$ having only horizontal non-vanishing components while $\mathbf{v}_{3,4}$ contain nonzero components only in the vertical plane for $\chi = \pi/2$. For independent solutions $\mathbf{v}_i$ of coupled motion, we try an ansatz

$$\begin{aligned}
x_1(s) &= \sqrt{\beta_{x_{\mathrm{I}}}(s)} \, \cos \phi_{x_{\mathrm{I}}}(s) \, , & y_1(s) &= \sqrt{\beta_{y_{\mathrm{I}}}(s)} \, \cos \phi_{y_{\mathrm{I}}}(s) \, , \\
x_2(s) &= \sqrt{\beta_{x_{\mathrm{I}}}(s)} \, \sin \phi_{x_{\mathrm{I}}}(s) \, , & y_2(s) &= \sqrt{\beta_{y_{\mathrm{I}}}(s)} \, \sin \phi_{y_{\mathrm{I}}}(s) \, , \\
x_3(s) &= \sqrt{\beta_{x_{\mathrm{II}}}(s)} \, \cos \phi_{x_{\mathrm{II}}}(s) \, , & y_3(s) &= \sqrt{\beta_{y_{\mathrm{II}}}(s)} \, \cos \phi_{y_{\mathrm{II}}}(s) \, , \\
x_4(s) &= \sqrt{\beta_{x_{\mathrm{II}}}(s)} \, \sin \phi_{x_{\mathrm{II}}}(s) \, , & y_4(s) &= \sqrt{\beta_{y_{\mathrm{II}}}(s)} \, \sin \phi_{y_{\mathrm{II}}}(s) \, ,
\end{aligned} \tag{3.68}$$

which is consistent with the earlier definitions of *conjugate trajectories*. In Sect. 3.1 we defined conjugate trajectories to be independent solutions normalized to the same phase ellipse and developed relationships between these trajectories and betatron functions. These relationships can be expanded to coupled motion by defining betatron functions for both modes of oscillations similar to (3.4)

$$\begin{aligned}
\beta_{x_{\mathrm{I}}} &= x_1^2 + x_2^2 \, , & \beta_{x_{\mathrm{II}}} &= x_3^2 + x_4^2 \, , \\
\beta_{y_{\mathrm{I}}} &= y_1^2 + y_2^2 \, , & \beta_{y_{\mathrm{II}}} &= y_3^2 + y_4^2 \, .
\end{aligned} \tag{3.69}$$

The phase functions can be defined similar to (3.5) by

$$\begin{aligned}
\cos \phi_{x_{\mathrm{I}}} &= \frac{x_1}{\sqrt{x_1^2 + x_2^2}} \, , & \cos \phi_{x_{\mathrm{II}}} &= \frac{x_3}{\sqrt{x_3^2 + x_4^2}} \, , \\
\cos \phi_{y_{\mathrm{I}}} &= \frac{y_1}{\sqrt{y_1^2 + y_2^2}} \, , & \cos \phi_{y_{\mathrm{II}}} &= \frac{y_3}{\sqrt{y_3^2 + y_4^2}} \, .
\end{aligned} \tag{3.70}$$

All other lattice functions can be defined in a similar way. By following the conjugate trajectories (3.68) and utilizing the (4×4)-transformation matrices including coupling effects we are able to determine the betatron functions at any point along the coupled beam transport line. To correlate parameters of the four-dimensional phase ellipse with quantities that can be measured, we write the solutions in the form

$$x_1(s) = \sqrt{\beta_{x_\mathrm{I}}(s)} \cos\phi_{x_\mathrm{I}}(s), \qquad x_2(s) = \sqrt{\beta_{x_\mathrm{I}}(s)} \sin\phi_{x_\mathrm{I}}(s),$$
$$x_1'(s) = \sqrt{\gamma_{x_\mathrm{I}}(s)} \cos\psi_{x_\mathrm{I}}(s), \qquad x_2'(s) = \sqrt{\gamma_{x_\mathrm{I}}(s)} \sin\psi_{x_\mathrm{I}}(s), \tag{3.71}$$

and similar for all other solutions. Comparing the second equation (3.71) with the derivative of the first equation we find the definitions

$$\gamma_{x_\mathrm{I}} = \frac{\beta_{x_\mathrm{I}}^2 \phi_{x_\mathrm{I}}'^2 + \alpha_{x_\mathrm{I}}^2}{\beta_{x_\mathrm{I}}} = \frac{1 + \alpha_{x_\mathrm{I}}^2}{\beta_{x_\mathrm{I}}} \tag{3.72}$$

and

$$\psi_{x_\mathrm{I}} = \phi_{x_\mathrm{I}} - \arctan\frac{1}{\alpha_{x_\mathrm{I}}}. \tag{3.73}$$

The other parameters γ_{x_II}, etc. are defined similarly and the phase ellipse (3.67) can now be expressed by the four-dimensional vector

$$\mathbf{v}(s) = \sqrt{\epsilon_\mathrm{I}} \begin{pmatrix} \sqrt{\beta_{x_\mathrm{I}}} \cos(\phi_{x_\mathrm{I}} + \vartheta_\mathrm{I}) \\ \sqrt{\gamma_{x_\mathrm{I}}} \cos(\psi_{x_\mathrm{I}} + \vartheta_\mathrm{I}) \\ \sqrt{\beta_{y_\mathrm{I}}} \cos(\phi_{y_\mathrm{I}} + \vartheta_\mathrm{I}) \\ \sqrt{\gamma_{y_\mathrm{I}}} \cos(\psi_{y_\mathrm{I}} + \vartheta_\mathrm{I}) \end{pmatrix} \cos\chi$$
$$+ \sqrt{\epsilon_\mathrm{II}} \begin{pmatrix} \sqrt{\beta_{x_\mathrm{II}}} \cos(\phi_{x_\mathrm{II}} + \vartheta_\mathrm{II}) \\ \sqrt{\gamma_{x_\mathrm{II}}} \cos(\psi_{x_\mathrm{II}} + \vartheta_\mathrm{II}) \\ \sqrt{\beta_{y_\mathrm{II}}} \cos(\phi_{y_\mathrm{II}} + \vartheta_\mathrm{II}) \\ \sqrt{\gamma_{y_\mathrm{II}}} \cos(\psi_{y_\mathrm{II}} + \vartheta_\mathrm{II}) \end{pmatrix} \sin\chi \tag{3.74}$$

This vector covers all points on the surface of the four-dimensional ellipsoid as $\chi, \vartheta_\mathrm{I}$ and ϑ_II vary independently from 0 to π. For one-dimensional oscillations we know from the definition of the phase ellipse that the product $\sqrt{\epsilon_u} \sqrt{\beta_u}$ is equal to the beam size or beam envelope E_u and $\sqrt{\epsilon_u} \sqrt{\gamma_u}$ equal to the angular beam envelope A_u, where $u = x$ or y. These definitions of beam envelopes can be generalized to coupled motion but we find from (3.74) that the envelopes have two contributions. Each point on the phase ellipse for an uncoupled beam appears now expanded into an ellipse with an area $\pi\epsilon_\mathrm{II}$ as shown in Fig. 3.2.

In a real beam transport line we are not able to observe experimentally the four-dimensional phase ellipse. By methods of emittance measurements, however, we may determine the area for the projection of the four-dimensional ellipsoid onto the $(x - x')$, the $(y - y')$ or the $(x - y)$-plane.

To do that we note in (3.74) that the maximum amplitude of a particle in the u-plane occurs for $\phi_{u\mathrm{I,II}} = -\vartheta_{u\mathrm{I,II}}$ and a projection angle χ given by

$$\sin^2 \chi = \frac{\epsilon_{u\mathrm{II}} \beta_{u\mathrm{II}}}{E_u}, \tag{3.75}$$

where $u = x$ or y and the beam envelope for coupled motion is given by

$$E_u = \sqrt{\epsilon_{u\mathrm{I}}\beta_{u\mathrm{I}} + \epsilon_{u\mathrm{II}}\beta_{u\mathrm{II}}}. \tag{3.76}$$

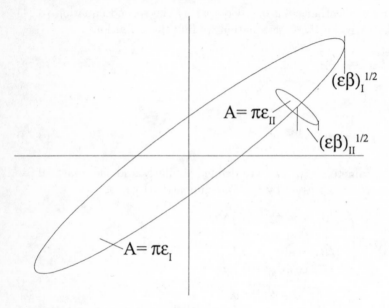

Fig. 3.2. Phase space ellipse for coupled motion

Similarly, we get from the second component of (3.74) the angular envelope

$$A_u = \sqrt{\epsilon_{u\mathrm{I}}\gamma_{u\mathrm{I}} + \epsilon_{u\mathrm{II}}\gamma_{u\mathrm{II}}} \tag{3.77}$$

for $\psi_{u\mathrm{I,II}} = -\vartheta_{u\mathrm{I,II}}$ and a projection angle given by

$$\sin^2 \chi = \frac{\epsilon_{u\mathrm{II}} \beta_{u\mathrm{II}}}{A_u}. \tag{3.78}$$

To completely determine the phase ellipse we calculate also the slope x' for the particle at $x = E_x$ which is the slope of the envelope E'. Taking the derivative of (3.76) we get

$$E'_u = -\frac{\epsilon_{u\mathrm{I}}\alpha_{u\mathrm{I}} + \epsilon_{u\mathrm{II}}\alpha_{u\mathrm{II}}}{\sqrt{\epsilon_{u\mathrm{I}}\beta_{u\mathrm{I}} + \epsilon_{u\mathrm{II}}\beta_{u\mathrm{II}}}}. \tag{3.79}$$

Expressing the equation of the phase ellipse in terms of these envelope definitions we get

$$A_u^2 \, u^2 - 2 \, E_u' \, E_u \, uu' + E_u^2 \, u'^2 \; = \; \epsilon_u^2 \,, \tag{3.80}$$

and inserting $u = E_u$ and $u' = E_u'$ into (3.80) we get for the emittance of the projection ellipse

$$\epsilon_u \; = \; E_u \sqrt{A_u^2 - E_u'^{\,2}} \,. \tag{3.81}$$

The envelope functions can be measured noting that $E^2 = \sigma_{11}, A^2 = \sigma_{22}$ and $EE' = -\sigma_{12}$ where the σ_{ij} are elements of the *beam matrix*. Because of the deformation of the four-dimensional phase ellipse through transformations, we cannot expect that the projection is a constant of motion and the projected emittance is therefore of limited use.

A more important and obvious projection is that onto the (x, y)-plane which shows the actual beam cross section under the influence of coupling. For this projection we use the first and third equation in (3.74) and find an elliptical beam cross section. The spatial envelopes E_x and E_y have been derived before in (3.76) and become here

$$\begin{aligned} E_x &\; = \; \sqrt{\epsilon_{x\mathrm{I}}\beta_{x\mathrm{I}} + \epsilon_{x\mathrm{II}}\beta_{x\mathrm{II}}} \,, \\ E_y &\; = \; \sqrt{\epsilon_{y\mathrm{I}}\beta_{y\mathrm{I}} + \epsilon_{y\mathrm{II}}\beta_{y\mathrm{II}}} \,. \end{aligned} \tag{3.82}$$

The y-coordinate for E_x, which we denote by E_{xy}, can be derived from the third equation in (3.74) noting that now $\vartheta_{y\mathrm{I,II}} = -\phi_{x\mathrm{I,II}}$, χ is given by (3.75) and

$$E_{xy} \; = \; \frac{\epsilon_{\mathrm{I}} \, \sqrt{\beta_{x\mathrm{I}}\beta_{y\mathrm{I}}} \, \cos \Delta\phi_{\mathrm{I}} + \epsilon_{\mathrm{II}} \, \sqrt{\beta_{x\mathrm{II}}\beta_{y\mathrm{II}}} \, \cos \Delta\phi_{\mathrm{II}}}{\sqrt{\epsilon_{x\mathrm{I}}\beta_{x\mathrm{I}} + \epsilon_{x\mathrm{II}}\beta_{x\mathrm{II}}}} \,, \tag{3.83}$$

where $\Delta\phi_{\mathrm{I,II}} = \phi_{x\mathrm{I,II}} - \phi_{y\mathrm{I,II}}$.

For $E_{xy} \neq 0$ we find the beam cross section tilted due to coupling. The tilt angle ψ of the ellipse is determined by

$$\tan 2\psi \; = \; \frac{2 \, E_x \, E_{xy}}{E_x^2 - E_y^2}$$

or more explicitly

$$\boxed{\tan 2\psi \; = \; 2 \, \frac{\epsilon_{\mathrm{I}} \, \sqrt{\beta_{x\mathrm{I}}\beta_{y\mathrm{I}}} \, \cos \Delta\phi_{\mathrm{I}} + \epsilon_{\mathrm{II}} \, \sqrt{\beta_{x\mathrm{II}}\beta_{y\mathrm{II}}} \, \cos \Delta\phi_{\mathrm{II}}}{\epsilon_{x\mathrm{I}} \, \Delta\beta_{\mathrm{I}} + \epsilon_{x\mathrm{II}} \, \Delta\beta_{\mathrm{II}}} \,.} \tag{3.84}$$

The beam cross section of a coupled beam is tilted as can be directly observed, for example, through a light monitor which images the beam cross section by the emission of synchrotron light. This rotation vanishes as we would expect for vanishing coupling when $\beta_{x\mathrm{II}} \to 0$ and $\beta_{y\mathrm{I}} \to 0$. The tilt angle is not a constant of motion and therefore different tilt angles can be observed at different points along a beam transport line.

We have discussed *Ripken's* theory [3.1] of coupled betatron motion which allows the formulation of beam dynamics for arbitrary strength of coupling. The concept of conjugate trajectories and transformation matrices through skew quadrupoles and solenoid magnets are the basic tools required to determine coupled betatron functions and the tilt of the beam cross section.

Problems

Problem 3.1. Consider a point source of particles (e.g. a positron conversion target) on the axis of a solenoidal field. Determine the solenoid parameters for which the particles would exit the solenoid as a parallel beam. Such a solenoid is also called a $\lambda/4$-lens, why? Let the positron momentum be 10 MeV/c. What is the maximum solid angle accepted from the target that can be focused to a beam of radius $r = 1$ cm? What is the exit angle of a particle which emerges from the target at a radius of 1mm? Express the transformation of this $\lambda/4$-lens in matrix formulation

Problem 3.2. Determine the length and field strength of a solenoid magnet which is able to rotate a flat 10 GeV beam by $90°$.

Problem 3.3. Choose a FODO lattice for a circular accelerator and insert at a symmetry point a thin rotated quadrupole. Calculate the tilt of the beam cross section at this point as a function of the strength of the rotated quadrupole. Place the same skew quadrupole in the middle of a FODO half cell and determine if the rotation of the beam aspect ratio at the symmetry point requires a stronger or a weaker field. Explain why.

Problem 3.4. Prove the validity of (3.69) and (3.70).

Problem 3.5. Assume two cells of a symmetric FODO lattice and determine the betatron functions for a phase advance of $90°$ per cell. Now introduce a rotational misalignment of the first quadrupole by an angle α which generates coupling of the horizontal and vertical betatron oscillations.

a) Calculate and plot the perturbed betatron functions β_{I} and β_{II} and compare with the unperturbed solution.

b) If the beam emittances are $\epsilon_{\mathrm{I}} = \epsilon_{\mathrm{II}} = 1$mm-mrad what is the beam aspect ratio and beam rotation at the end of cell one and two with and without the rotation of the first quadrupole?

4. Higher-Order Perturbations

The rules of linear beam dynamics allow the design of beam transport systems with virtually any desired beam characteristics. Whether such characteristics actually can be achieved depends greatly on our ability or lack thereof to control the source and magnitude of perturbations. Only the lowest-order perturbation terms were discussed in [4.1] in the realm of linear, paraxial beam dynamics. With the continued sophistication of accelerator design and increased demand on beam quality it becomes more and more important to also consider higher-order *magnetic field perturbations* as well as *kinematic perturbation* terms.

The effects of such terms in beam-transport lines, for example, may compromise the integrity of a carefully prepared very low emittance beam for linear colliders or may contribute to nonlinear distortion of the chromaticity in circular accelerators and associated reduced beam stability. Studying nonlinear effects we will not only consider nonlinear fields but also the effects of linear fields errors in higher order, whether it be higher-order perturbation terms or higher-order approximations for the equations of motion. The sources and physical nature of perturbative effects must be understood to determine limits to beam parameters and to design correcting measures. In this chapter we will consider and discuss some of the most relevant nonlinear perturbations.

4.1 Kinematic Perturbation Terms

Perturbations of beam dynamics not only occur when there are magnetic field and alignment errors present. During the derivation of the general equation of motion in Chap. 2 we encountered in addition to general multipole fields a large number of *kinematic perturbation* terms or *higher-order field perturbations* which appear even for ideal magnets and alignment. Generally, such terms become significant for small circular accelerators or wherever beams are deflected in high fields generating bending radii of order unity or less. If, in addition, the beam sizes are large the importance of such perturbations is further aggravated. In many cases well-known aberration phenomena from light optics can be recognized.

Of the general equations of motion (2.42,43) we consider terms up to third order for ideal linear upright magnets and get the equation of motion in the horizontal and deflecting plane

$$
\begin{aligned}
x'' + (\kappa^2 + k)\,x &= -\kappa_x^3\,x^2 + 2\kappa_x k\,x^2 + \left(\tfrac{1}{2}\kappa_x k + \tfrac{1}{2}\kappa_x''\right)y^2 \\
&\quad - \tfrac{1}{2}\kappa_x(x'^2 - y'^2) + \kappa_x'\left(xx' + yy'\right) \\
&\quad - \tfrac{1}{12}\left(2\kappa_x\kappa_x'' + \kappa_x^2 - \kappa_x'^2 + k'' - 11\kappa_x^2 k\right)x^3 + \left(2\kappa_x^2 + \tfrac{3}{2}k\right)xx'^2 \\
&\quad + \tfrac{1}{4}\left(2\kappa_x\kappa_x'' - \kappa_x'^2 + 5\kappa_x^2 k - 3\kappa_x^2 k + k''\right)xy^2 \\
&\quad + k'\,xyy' - \tfrac{1}{2}k\,xy'^2 + kx'yy' \\
&\quad + \kappa_x\delta + (k + 2\kappa_x^2)\,x\delta + (\kappa_x^3 + 2\kappa_x k)\,x^2\delta + \tfrac{1}{2}\kappa_x\,x'^2\delta \\
&\quad - \tfrac{1}{2}(\kappa_x k + \kappa_x'')\,y^2\delta - \kappa_x\delta^2 - (k + 2\kappa_x^2)\,x\delta^2 + \kappa_x\delta^3 + \mathcal{O}(4)\,.
\end{aligned}
\tag{4.1}
$$

In the nondeflecting or vertical plane the equation of motion is

$$
\begin{aligned}
y'' - k\,y &= +2\kappa_x k\,xy + \kappa_x'\left(xy' - x'y\right) + \kappa_x\,x'y' \\
&\quad + \tfrac{1}{12}\left(2\kappa_x'^2\kappa_x'' + \kappa_x'^2 - k'' - \kappa_x^2 k\right)y^3 \\
&\quad - \kappa_x\kappa_x'\,y^2 y' - 2\kappa_x^2\,xx'y' - k'\,xx'y \\
&\quad + \tfrac{1}{4}\left(2\kappa_x\kappa_x'' + \kappa_x'^2 - k'' - 3\kappa_x^2 k\right)x^2 y \\
&\quad + \tfrac{3}{2}k\,yy'^2 + \tfrac{1}{2}k\,x'^2 y - kxx'y' \\
&\quad + \kappa_x'\,x'y\delta - k\,y\delta + k\,y\delta^2 + \mathcal{O}(4)\,.
\end{aligned}
\tag{4.2}
$$

It is quite clear from these equations that most perturbations become significant only for large amplitudes and oblique particle trajectories or for very strong magnets. The lowest-order quadrupole perturbations are of third order in the oscillation amplitudes and therefore become significant only for large beam sizes. Second-order perturbations occur only in combined-function magnets and for particle trajectories traversing a quadrupole at large amplitudes or offsets from the axis. Associated with the main fields and perturbation terms are also chromatic variations thereof and wherever beams with a large energy spread must be transported such perturbation terms become important. Specifically, the quadrupole terms $kx\delta$ and $ky\delta$ determine the chromatic aberration of the focusing system and play a significant role in the transport of beams with large momentum spread. In most cases of beam dynamics, all except the linear chromatic terms can be neglected.

Evaluating the effect of perturbations on a particle beam, we must carefully select the proper boundary conditions for bending magnets. Only for sector magnets is the field boundary normal to the reference path and occurs therefore at the same location s independent of the amplitude. Generally, this is not true and we must adjust the integration limits according to the particle oscillation amplitudes x and y and actual magnet boundary just as we did in the derivation of linear transformation matrices for rectangular or wedge magnets.

4.2 Control of the Central Beam Path

In circular accelerators the equilibrium orbit depends on all perturbations and orbit corrections in the ring. Not so in an open beam transport system where the distortions of the beam at a location s depend only on the upstream perturbations. This is actually also the case in a circular accelerator during injection trials while the beam is steered around the ring for the first turn and has not yet found a closed orbit. Only after one turn the errors as well as the corrections repeat and the particle starts to oscillate about the equilibrium orbit. The results of this discussion can therefore also be applied to large circular accelerators in support of establishing the first turn.

Beam-path correction schemes which are different from closed orbit corrections in circular accelerators are required to steer a beam along a prescribed beam path. On first thought one might just install a number of beam position monitors, preferably about four monitors per betatron wave length, and steer the beam to the center of each monitor with an upstream steering magnet. This method is efficient for short beam lines but may fail for long systems due to an intrinsic instability of this correction scheme.

To review the nature of this instability, we discuss the path of a beam through a very long beam-transport system as one might encounter in the design of linear colliders or for the first turn of a very large circular accelerator. Since there is an infinite variety of beam-transport lattices we also restrict the discussion to a FODO *channel*. This restriction, however, does not invalidate the application of the derived results to other lattice types. The advantage of a FODO channel is only that analytical methods can be applied to develop an optimized correction scheme. Such an optimum correction can then be applied to any lattice by numeric methods.

In linear collider systems, for example, such a long periodic focusing lattice is required both along the linear accelerator and along beam-transport lines from the linac to the collision point to prevent degradation of the beam quality or beam loss. The linear beam dynamics is straightforward and most of the beam dynamics efforts must be concentrated on the control of errors and their effects on the beam. In this section we will derive characteristics of the beam-path distortion due to dipole field errors.

The distortion of the beam path due to one kick caused by a localized dipole field error develops exactly like the sine-like principal solution. With θ the kick angle at a point along the transport line where the betatron phase is ψ_θ we get at any point s downstream with $\psi(s) > \psi_\theta$ the path distortion

$$\Delta u(s) = \sqrt{\beta(s)} \sqrt{\beta_\theta} \sin[\psi(s) - \psi_\theta] \, \theta \,. \tag{4.3}$$

The distortion depends on the value for the betatron function both at the observation point and at the point of the dipole field error. Consequently, it is important to minimize the probability for such errors at high-beta points. The actual distortion measured at a particular point also depends on the phase difference $[\psi(s) - \psi_\theta]$ because the particle starts a sinusoidal oscillation at the location of the dipole field error.

This oscillatory feature of the perturbation is constructively utilized in the design of a particle *spectrometer* where we are interested in the momentum of particles emerging from a point source like a fixed target hit by high energy particles. We consider the target as the perturbation which deflects particles from the direction of the incoming path. The emerging particles have a wide distribution in scattering angles and momentum and it is interesting for the study of scattering processes to measure the intensity distribution as a function of these parameters.

Downstream from the target we assume a beam-transport system composed of bending magnets and quadrupoles with focusing properties consistent with the desired measurements. To measure the angular distribution for elastic scattering the particle detectors are placed at $\psi(s) - \psi_\theta = \pi/2 \cdot (2n+1)$ where φ_θ is the phase at the target and n is a positive integer.

In case of inelastic scattering events we are interested to also measure the loss of momentum. All particles with the same momentum are expected to arrive at the same point of the particle detector independent of the starting angle from the target. In this case we place detectors also at the location $\psi(s) - \psi_\theta = \pi n$, which is an image point of the target. To measure the momentum distribution we add a bending magnet into the beam line which generates a dispersion and spreads out the particle beam according to the momentum distribution for easy recording by particle detectors.

Launching error: At the start of the transport line the beam may get launched with an error in angle or position. The angle error is equivalent to a kick at the launch point causing a distortion of the beam path as determined by (4.3). A position error at the launch point develops like a cosine-like trajectory and the distortion of the beam path at the point s is given by

$$\Delta u(s) = \frac{\sqrt{\beta(s)}}{\sqrt{\beta_\mathrm{o}}} \cos[\psi(s) - \psi_\mathrm{o}] \, \Delta u_\mathrm{o} \,, \tag{4.4}$$

where $\beta_\mathrm{o}, \psi_\mathrm{o}$ and Δu_o are taken at the launch point. The ratio $\sqrt{\beta(s)/\beta_\mathrm{o}}$ is called the magnification of the beam-transport line since it determines the distortion Δu or beam size as a function of the distortion Δu_o or beam size at the starting point, respectively.

Statistical alignment and field error: In a long beam-transport line with many focusing elements the error of the beam position or angle at the end of the beam line depends on all errors. We have therefore instead of (4.3) with $\mathrm{d}\theta = \Delta \frac{1}{\rho} \, \mathrm{d}\sigma$

$$\Delta u(s) = \sqrt{\beta(s)} \int\limits_{s_0}^{s} \sqrt{\beta(\sigma)} \, \Delta\frac{1}{\rho}(s) \, \sin[\psi(s) - \psi(\sigma)] \, d\sigma \,. \tag{4.5}$$

In most cases, however, the errors can be assumed to be localized in a thin magnet and we may set $\int \Delta\frac{1}{\rho}(s) ds = \theta_i$. After replacing the integral in (4.5) with a sum over all kicks the path distortion is

$$\Delta u(s) = \sqrt{\beta(s)} \sum_{i} \sqrt{\beta(s_i)} \, \theta_i \, \sin[\psi(s) - \psi_i] \,. \tag{4.6}$$

In case of a statistical distribution of the errors the probable displacement of the beam position is given by

$$\sigma_u(s) = \sqrt{\beta(s)} \sqrt{\beta_{\mathrm{avg}}} \, \sigma_\theta \sqrt{\tfrac{1}{2} N_\theta} \,, \tag{4.7}$$

where σ_u is the rms value of the path displacement along the beam line, β_{avg} the average value of the betatron function at places, where errors occur, σ_θ the rms value of θ and N_θ is the total number of errors. We also made use of the statistical distribution of the error locations ψ_i and have $\langle \sin^2(\psi(s) - \psi_i) \rangle = 1/2$. The errors θ_i can be dipole field errors but more important they can be the result of misaligned quadrupoles with $\sigma_\theta = k\ell \, \sigma_{\mathrm{q}}$ where $f = 1/(k\ell)$ is the focal length of the quadrupoles and σ_{q} the rms value of the transverse quadrupole misalignment.

Beam-transport lines which are particular sensitive to statistical misalignment errors are part of the *Stanford Linear Collider* (SLC) [4.2] and of future linear colliders. In these projects an extremely small beam size is desired and only strong focusing in the transport line can prevent a dilution of the beam emittance. In the SLC approximately 500 bending magnets with a strong quadrupole component form a FODO channel and are used to transport particles from the linear accelerator to the collision point where high energy physics experiments are performed. The focal length of these magnets is about $f \approx 3.0$ m and the average value of the betatron function is $\beta_{\mathrm{avg}} \approx 5.3$ m, since the phase advance per cell is $\phi = 108°$ and $L \approx 2.5$ m. To obtain a large luminosity for high energy experiments, the beam is focused down to about $\sigma \approx 1.4 \, \mu$m at the collision point, where $\beta(s) = 0.005$ m. From (4.7) we get with these parameters and $\sigma_\theta = 0.33 \, \sigma_{\mathrm{q}}$ for the probable displacement of the beam center at the collision point

$$\sigma_{\mathrm{u}} = 0.85 \, \sigma_{\mathrm{q}} \,, \tag{4.8}$$

where σ_{q} is the rms misalignment error of the quadrupoles. Extreme accuracy of the quadrupole position is required to aim the beam onto a target of 1.4 μm radius. In fact, the rms quadrupole alignment must be precise to better than $\sigma_{\mathrm{q}} = 1.6 \, \mu$m. It is clear that on an absolute scale this precision cannot be achieved with known alignment techniques. An orbit correction system is required to correct for the unavoidable quadrupole misalignments.

This is possible for static misalignments but not for dynamic movement, due to ground vibrations. For a detailed discussion of ground vibrations consult the review article by *Fischer* [4.3].

Path distortion and acceleration: In a beam-transport line which also contains accelerating structures, the effect of perturbations at low energies on the path distortion downstream are modified by adiabatic damping. The contribution of the kick angle θ_i occurring at the energy γ_i in (4.6) to the path distortion is reduced at the energy γ by the adiabatic damping factor $\sqrt{\gamma_i/\gamma}$.

It is interesting to note that the contribution of path distortions in a FODO channel is independent of the acceleration. To show this we assume a FODO channel made up of equal strength quadrupoles. The distance between quadrupoles increases proportional to the beam energy and as a consequence so does the betatron function. Since the beam emittance scales like $1/\gamma$ it is obvious that the beam envelope $\sqrt{\epsilon\beta}$ is the same in each quadrupole independent of the energy. Now we assume statistical quadrupole misalignments of the same magnitude along the channel. Normalizing to the contribution of the path distortion from an error at a location where the beam energy is γ_o, we find $\beta(s)$ to scale like γ/γ_o, β_i like γ_i/γ_o and the quadrupole strength or kick angle θ_i like γ_o/γ_i. Adding the adiabatic damping factor γ_i/γ we find the individual contribution to path distortions to be independent of energy. We may therefore apply the results of this discussion to any FODO channel with or without acceleration. This is even true for any periodic lattice if we only sort the different quadrupole families into separate sums in (4.6).

Correction of path distortions: To correct a static distortion of the beam path we use a set of beam position monitors and a set of correction magnets. If we have an alternating sequence of correctors and monitors it might seem to be sufficient to correct the beam path at every monitor using the previous correction magnet. In long beam lines this procedure may, however, lead to a resonant build up of distortions rather than a correction. While the beam position is corrected at every monitor the direction of the beam path at the monitors is not, leading to ever increasing distortions between the monitors and to divergent corrector strengths. We may prove this statement assuming a FODO channel with monitors and correctors distributed as shown in Fig. 4.1. The mathematics can be simplified greatly without loss of generality if we assume correctors and monitors to be located in the middle of the quadrupoles. We also assume that the path distortion to be caused by quadrupole misalignments only and that the monitors be located in every second quadrupole of the FODO channel only such that the betatron function is the same for all monitors and correctors. Both assumptions are not fundamental but simplify the formalism. The path be corrected perfectly at the monitor M_i and therefore the path distortion at the monitor

M_{i+1} is the combination of all errors θ_j between the monitors M_i and M_{i+1} and a contribution from the slope of the path at the monitor M_i. The path distortion at the monitor M_{i+1} is therefore

$$\Delta u_{i+1} = \beta \sum_j \theta_j \sin(\psi_{i+1} - \psi_j) + \beta \Delta u_i' \sin \Delta\psi \,,$$

$$= \beta S_{i+1} + \beta \Delta u_i' \sin \Delta\psi \,, \tag{4.9}$$

and the slope

$$\Delta u_{i+1}' = \sum_j \theta_j \cos(\psi_{i+1} - \psi_j) + \Delta u_i' \cos \Delta\psi \,,$$

$$= C_{i+1} + \Delta u_i' \cos \Delta\psi \,, \tag{4.10}$$

where $\Delta\psi = \psi_{i+1} - \psi_i$ is the phase difference between adjacent monitors and S_{i+1}, C_{i+1} obvious abbreviations.

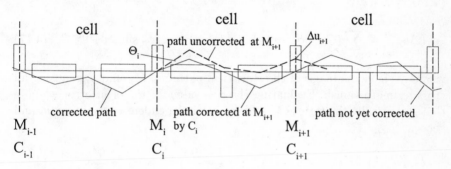

Fig. 4.1. Beam-path correction

The betatron phase at the monitor M_{i+1} is ψ_{i+1} and $\psi(s_j) = \psi_j$ is the phase at the location of the errors between monitors M_i and M_{i+1}. Now we correct the path distortion with the corrector C_i and get for the displacement at the monitor M_{i+1}

$$\Delta_{\mathrm{c}} u_i = \beta \Theta_i \sin \Delta\psi_{\mathrm{c}} \,, \tag{4.11}$$

where Θ_i is the beam deflection angle by the corrector and $\Delta\psi_{\mathrm{c}}$ the phase between the corrector C_i and monitor M_{i+1}. The corrector field is adjusted to move the beam orbit such as to compensate the orbit distortion at the monitor M_{i+1} and the sum of (4.9,11) is therefore

$$0 = \beta S_{i+1} + \beta \Delta u_i' \sin \Delta\psi + \beta \Theta_i \sin \Delta\psi_{\mathrm{c}} \,. \tag{4.12}$$

The slope of the beam path at M_{i+1} is the combination of the yet uncorrected effects of errors and correctors

$$\Delta u'_{i+1} = C_{i+1} + \Delta u'_i \cos \Delta \psi + \Theta_i \cos \Delta \psi_c. \tag{4.13}$$

Solving (4.12) for the corrector angle we get

$$\Theta_i = \frac{-S_{i+1} + \Delta u'_i \sin \Delta \psi}{\sin \Delta \psi_c} \tag{4.14}$$

and inserted into (4.13) the slope of the beam path at the monitor M_{i+1} is

$$\Delta u'_{i+1} = C_{i+1} + \Delta u'_i \cos \Delta \psi - (S_{i+1} + \Delta u'_i \sin \Delta \psi) \tan \Delta \psi_c. \tag{4.15}$$

Evaluating this recursion formula we find for the path direction at monitor M_n

$$\Delta u'_n = \sum_{m=1}^{n} (C_m - S_m \cot \Delta \psi_c) (\cos \Delta \psi + \sin \psi \cot \Delta \psi_c)^{n-m}. \tag{4.16}$$

The expectation value for the path distortion is derived from the square of (4.16), which gives

$$(\Delta u'_n)^2 = \sum_{m=1}^{n} (C_m - S_m \cot \Delta \psi_c)^2 (\cos \Delta \psi + \Delta \psi_c \sin \Delta \psi)^{2(n-m)}, \tag{4.17}$$

where we have made use of the statistical nature of C_m and S_m causing all cross terms to vanish. Evaluating the rms value for the orbit error we note that statistically the terms C_m and S_m are independent of the index m and get

$$\langle (\Delta u'_n)^2 \rangle = \langle (C - S \cot \Delta \psi_c)^2 \rangle \frac{1 - H^{2m}}{1 - H^2} \tag{4.18}$$

with

$$H = \cos \Delta \psi + \sin \Delta \psi \cot \Delta \psi_c = \frac{\sin \Delta}{\sin \Delta \psi_c}$$

and $\Delta = \Delta \psi_c + \Delta \psi$. For large numbers of magnets (4.18) leads to

$$\langle (\Delta u'_n)^2 \rangle = \langle (C - S \cot \Delta \psi_c)^2 \rangle \cdot \begin{cases} H^{2m-2} & \text{for } H \gg 1, \\ m & \text{for } H \approx 1, \\ \frac{1}{1-H^2} & \text{for } H \ll 1. \end{cases} \tag{4.19}$$

The result of a simple path correction as assumed depends greatly on the position of the corrector magnet. If the corrector C_i is located at the monitor M_i in a 90°-FODO lattice the simple correction scheme works well since $H^2 \ll 1$. On the other hand if the corrector C_i is close to the monitor M_{i+1} then $\Delta \psi_c \approx 0$ and the correction scheme quickly diverges. Even in the middle position we have a slow divergence proportional to the total number m. This case is important for linear collider beam lines. Here the beam energies are very high and therefore long corrector magnets are incorporated into the quadrupoles which fill all space between the monitors. The effective position of the corrector is therefore in the middle between the monitors. A correct derivation for these extended correctors reveals the same behavior.

A more sophisticated *path correction* scheme is required to prevent a divergent behavior. If the beam is observed in two consecutive monitors it is possible to correct in one of the two monitors position and angle. To do this, we assume the layout of Fig. 4.2 to correct both, the position, Δu_2, and the angle $\Delta u_2'$ at the monitor M_2 with the use of the two correctors C_1 and C_2.

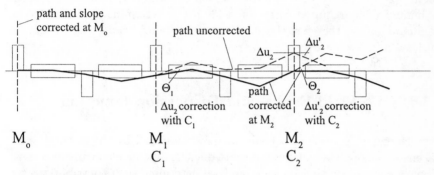

Fig. 4.2. Complete beam-path correction scheme

The corrected path and path direction at the monitor M_2 is determined by

$$\begin{aligned} \Delta u_2 = 0 = \Delta u_o + b\,\Theta_1 + B\,\Theta_2 , \\ \Delta u_2' = 0 = \Delta u_o' + d\,\Theta_1 + D\,\Theta_2 , \end{aligned} \tag{4.20}$$

where we used elements of the transformation matrices $M(M_2|C_1) = \begin{pmatrix} a & b \\ c & d \end{pmatrix}$ from the first corrector at C_1 to the monitor M_2 and $M(M_2|C_2) = \begin{pmatrix} A & B \\ C & D \end{pmatrix}$ for the second corrector at C_2 to the same monitor. The angles Θ_i are the deflection angles of the corrector magnets. The slope $\Delta u_2'$ can be determined only through two position measurements at say M_1 and M_2. We must assume that there are no errors between M_1 and M_2 to obtain a meaningful relationship between the orbit position at these monitors. From the transformation matrix $M(M_2|M_1) = \begin{pmatrix} m_{11} & m_{12} \\ m_{21} & m_{22} \end{pmatrix}$ between both monitors we get for the orbit position in the second monitor $\Delta u_2 = m_{11}\Delta u_1 + m_{12}\Delta u_1'$, which we solve for the slope

$$\Delta u_1' = \frac{\Delta u_2 - m_{11}\,\Delta u_o}{m_{12}} . \tag{4.21}$$

All quantities on the r.h.s. can be measured or are known. With (4.20) we can solve for the two corrector strengths

$$\Theta_1 = \frac{D\,\Delta u_1 - B\Delta u_1'}{d\,B - b\,D}\,,$$

$$\Theta_2 = -\frac{d\Delta u_1 - b\Delta u_1'}{d\,B - b\,D} \qquad\qquad (4.22)$$

It is easy to express the determinant by lattice functions

$$dB - bD = \sqrt{\beta_1^c\,\beta_2^c}\,\sin(\psi_1 - \psi_2)\,, \qquad\qquad (4.23)$$

where the phases are taken at the locations of the correctors. Obviously, we minimize the required corrector fields if the determinant of (4.22) is a maximum and we therefore choose the phase distance between correctors to be close to 90° when $\psi_1 - \psi_2 = \pi/2$.

4.3 Dipole Field Errors and Dispersion Function

The dispersion function of a beam line is determined by the strength and placement of dipole magnets. As a consequence, dipole field errors also contribute to the dispersion function and we determine such contributions to the dispersion function due to dipole field errors. First, we note from the general expression for the linear dispersion function that the effect of dipole errors adds linearly to the dispersion function by virtue of the linearity of the equation. We may therefore calculate separately the effect of dipole errors and add the results to the ideal solution for the dispersion function.

Self compensation of perturbations: The linear superposition of individual dipole contributions to the dispersion function can be used in a constructive way. Any contribution to the dispersion function by a short magnet can be eliminated again by a similar magnet located 180° in betatron phase downstream from the first magnet. If the betatron function at the location of both magnets is the same, the magnet strengths are the same too. For quantitative evaluation we assume two dipole errors introducing a beam deflection by the angles θ_1 and θ_2 at locations with betatron functions of β_1 and β_2 and betatron phases ψ_1 and ψ_2, respectively. Since the dispersion function or fractions thereof evolve like a sine like function, we find for the variation of the dispersion function at a phase $\psi(s) \geq \psi_2$

$$\Delta D(s) = \theta_1\sqrt{\beta\,\beta_1}\,\sin\left[\psi(s) - \psi_1\right] + \theta_2\sqrt{\beta\,\beta_2}\,\sin\left[\psi(s) - \psi_2\right]. \qquad (4.24)$$

For the particular case where $\theta_1 = \theta_2$ and $\beta_1 = \beta_2$ we find

$$\Delta D(s) = 0 \quad \text{for} \quad \psi_2 - \psi_1 = (2n+1)\,\pi. \qquad\qquad (4.25)$$

If $\theta_1 = -\theta_2$

$$\Delta D(s) = 0 \quad \text{for} \quad \psi_2 - \psi_1 = 2n\pi\,, \tag{4.26}$$

where n is an integer. This property of the dispersion function can be used in periodic lattices if for example a vertical displacement of the beam line is desired. In this case we would like to deflect the beam at one point vertically and as soon as the beam reaches the desired elevation a second dipole magnet deflects the beam by the same angle but opposite sign to level the beam line parallel to the horizontal plane. In an arbitrary lattice such a beam displacement can be accomplished without creating a residual dispersion outside the beam deflecting section if we place two bending magnets at locations separated by a betatron phase of 2π. Similarly, a deflection in the same direction by two dipole magnets does not create a finite dispersion outside the deflecting section if both dipoles are separated by a betatron phase of $(2n+1)\pi$. This feature is important to simplify beam-transport lattices since no additional quadrupoles are needed to match the dispersion function.

Sometimes it is necessary to deflect the beam in both the horizontal and vertical direction. This can be done in a straightforward way by a sequence of horizontal and vertical bending sections leading, however, in general to long beam lines. In a more compact version we combine the beam deflection in both planes within one magnet. To obtain some vertical deflection for example in an otherwise horizontally deflecting beam line, we may rotate a section of the beam line about the beam axis at the start of the desired vertical deflection. Some of the horizontal deflection is thereby transformed into the vertical plane. At the start of such a section we introduce by the rotation of the coordinate system a sudden change in all lattice functions. Specifically, a purely horizontal dispersion function is coupled partly into a vertical dispersion. If we rotate the beam line and coordinate system back at a betatron phase of $2n\pi$ downstream from the first rotation, the coupling of the dispersion function as well as that of other lattice functions is completely restored. For that to work without further matching, however, we require that the rotated part of the beam line has a phase advance of $2n\pi$ in both planes as, for example, a symmetric FODO lattice would have.

Perturbations in open transport lines: While these properties are useful for specific applications, general beam dynamics requires that we discuss the effects of errors on the dispersion function in a more general way. To this purpose we use the general equation of motion (2.42) or (2.43) up to linear terms in δ and add constant perturbation terms from [Ref. 4.1, Eq. (7.22) or (7.23)]. In the following discussion we use only the horizontal equation of motion, but the results can be immediately applied to the vertical plane as well. The equation of motion with only linear chromatic terms and a quadratic sextupole term is

$$x'' + (\kappa_x^2 + k)x = kx\delta - \tfrac{1}{2}mx^2(1 - \delta) - \Delta\kappa_x(1 - \delta) + \mathcal{O}(2)\,. \qquad (4.27)$$

We observe two classes of perturbation terms, the ordinary chromatic terms and those due to field errors. Taking advantage of the linearity of the solution we decompose the particle position into four components

$$x = x_\beta + x_c + \eta_x\,\delta + v_x\,\delta\,, \qquad (4.28)$$

where x_β is the betatron motion, x_c the distorted beam path or orbit, η_x the ideal dispersion function and v_x the perturbation of the dispersion that derives from field errors. The individual parts of the solution then are determined by the following set of differential equations:

$$x_\beta'' + (\kappa_x^2 + k)\,x_\beta = -\tfrac{1}{2}mx_\beta^2 + mx_\beta x_c\,, \qquad (4.29)$$

$$x_c'' + (\kappa_x^2 + k)\,x_c = -\Delta\kappa_x - \tfrac{1}{2}mx_c^2\,, \qquad (4.30)$$

$$\eta_x'' + (\kappa_x^2 + k)\,\eta_x = \kappa_x\,, \qquad (4.31)$$

$$v_x'' + (\kappa_x^2 + k)\,v_x = kx_c - mx_c\eta_x + \Delta\kappa_x + \tfrac{1}{2}mx_c^2\,. \qquad (4.32)$$

In the ansatz (4.28) we have ignored the energy dependence of the betatron function since it will be treated separately as an aberration and has no impact on the dispersion. We have solved (4.29,30,31) before and concentrate therefore on the solution of (4.32). Obviously, the field errors cause distortions of the beam path x_c which in turn cause additional variations of the dispersion function. The principal solutions are

$$C(s) = \sqrt{\beta(s)/\beta_o}\,\cos(\psi(s) - \psi_o)\,,$$

$$S(s) = \sqrt{\beta(s)\beta_o}\,\sin(\psi(s) - \psi_o)\,,$$

and the Greens function becomes

$$G(s,\sigma) = S(s)\,C(\sigma) - S(\sigma)\,C(s) = \sqrt{\beta(s)\beta(\sigma)}\,\sin\left[\psi(s) - \psi(\sigma)\right]\,.$$

With this the solution of (4.32) is

$$v_x(s) = -\,x_c(s) \qquad (4.33)$$

$$+ \sqrt{\beta_x(s)}\int_o^s (k - m\eta_x)\,\sqrt{\beta_x(\sigma)}\,x_c(\sigma)\,\sin[\psi_x(s) - \psi_x(\sigma)]\,\mathrm{d}\sigma\,.$$

Here we have split off the solution for the two last perturbation terms in (4.32) which, apart from the sign, is exactly the orbit distortion (4.30). In a closed lattice we look for a periodic solution of (4.33), which can be written in the form

$$v_x(s) = -\,x_c(s) + \frac{\sqrt{\beta_x(s)}}{2\sin\pi\nu_x}\int_s^{s+L_p} (k - m\eta_x)\,\sqrt{\beta_x(\sigma)} \qquad (4.34)$$

$$\times\,x_c(\sigma)\,\cos\nu_x[\varphi_x(s) - \varphi_x(\sigma) + \pi]\,\mathrm{d}\sigma\,,$$

where $x_c(s)$ is the periodic solution for the distorted orbit. In the vertical plane we have exactly the same solution except for a change in sign for some terms

$$v_y(s) = -y_c(s) - \frac{\sqrt{\beta_y(s)}}{2\sin\pi\nu_y} \int_s^{s+lp} (k - m\eta_x)\sqrt{\beta_y(\sigma)}$$

$$\times y_c(\sigma)\cos\nu_y[\varphi_y(s) - \varphi_y(\sigma) + \pi]\,d\sigma. \tag{4.35}$$

For reasons of generality we have included here sextupoles to permit chromatic corrections in long curved beam lines with bending magnets. The slight asymmetry due to the term $m\eta_x$ in the vertical plane derives from the fact that in real accelerators only one orientation of the sextupoles is used. Due to this orientation the perturbation in the horizontal plane is $-\frac{1}{2}mx^2(1-\delta)$ and in the vertical plane $mxy(1-\delta)$. In both cases we get the term $m\eta_x$ in the solution integrals.

Again we may ask how this result varies as we add acceleration to such a transport line. Earlier in this section we found that the path distortion is independent of acceleration under certain periodic conditions. By the same arguments we can show that the distortion of the dispersions (4.34) and (4.35) are also independent of acceleration and the result of this discussion can therefore be applied to any periodic focusing channel.

4.4 Dispersion Function in Higher Order

The first-order change in the reference path for off energy particles is proportional to the relative momentum error. The proportionality factor is a function of the position s and is called the *dispersion function*. This result is true only in linear beam dynamics. We will now derive chromatic effects on the reference path in higher order to allow a more detailed determination of the chromatic stability criteria. The linear differential equation for the normalized dispersion function is

$$\frac{d^2 w_o}{d\varphi^2} + \nu^2 w_o = \nu^2 \beta^{3/2} \frac{1}{\rho} = \nu^2 F(\varphi), \tag{4.36}$$

where φ is the betatron phase, $w_o = \eta_o/\sqrt{\beta}$, $\beta(s)$ the betatron function and $\eta(s)$ the dispersion function. The periodic solution of (4.36) is called the normalized dispersion function

$$w_o(\varphi) = \sum_{n=-\infty}^{+\infty} \frac{\nu^2 F_n\, e^{in\varphi}}{\nu^2 - n^2}, \tag{4.37}$$

where

$$F(\varphi) = \beta^{\frac{3}{2}} \frac{1}{\rho} = \sum F_n \, e^{in\varphi} . \tag{4.38}$$

This linear solution includes only the lowest-order chromatic error term from the bending magnets and we must therefore include higher-order chromatic terms into the differential equation of motion. To do that we use the general differential equation of motion while ignoring all coupling terms

$$
\begin{aligned}
x'' + (\kappa^2 + k)\, x = &+ \kappa\delta - \kappa\delta^2 + \kappa\delta^3 - \tfrac{1}{2}\, m\, (1 - \delta)\, x^2 + \kappa'\, xx' \\
&- (\kappa^3 + 2\kappa k)\, (1 - \delta)\, x^2 + \tfrac{1}{2}\, \kappa\, (1 - \delta)\, x'^2 \\
&+ (2\kappa^2 + k)\, x\, \delta - (k + 2\kappa^2)\, x\delta^2 + \mathcal{O}(4) ,
\end{aligned}
\tag{4.39}
$$

where $\kappa = 1/\rho$. We are only interested in the chromatic solution with vanishing betatron oscillation amplitudes and insert for the particle position therefore

$$x_\eta = \eta_{\rm o}\, \delta + \eta_1\, \delta^2 + \eta_2\, \delta^3 + \mathcal{O}(4) . \tag{4.40}$$

Due to the principle of linear superposition we get separate differential equations for each component η_i by collecting on the right-hand side terms of equal power in δ. For the terms linear in δ, we find the well-known differential equation for the dispersion function

$$\eta_{\rm o}'' + K(s)\, \eta_{\rm o} = \kappa = \sum_n R_{n{\rm o}} \, e^{in\varphi} , \tag{4.41}$$

where we also express the perturbations by their Fourier expansions. The terms quadratic in δ form the differential equation

$$
\begin{aligned}
\eta_1'' + K(s)\, \eta_1 &= -\kappa - \tfrac{1}{2} m \eta_{\rm o}^2 + (\kappa^3 + 2\kappa k)\, \eta_{\rm o}^2 + \tfrac{1}{2}\, \kappa\, \eta_{\rm o}'^2 \\
&\quad + \kappa' \eta_{\rm o} \eta_{\rm o}' + (2\kappa^2 + k)\, \eta_{\rm o} \\
&= -\sum_n R_{n{\rm o}} \, e^{in\varphi} + \sum_n R_{1n} \, e^{in\varphi} ,
\end{aligned}
\tag{4.42}
$$

and terms cubic in δ are determined by

$$
\begin{aligned}
\eta_2'' + K(s)\, \eta_2 &= +\kappa + \tfrac{1}{2} m \eta_{\rm o}^2 + (\kappa^3 + 2\kappa k)\, \eta_{\rm o}^2 - \tfrac{1}{2}\kappa\eta_{\rm o}'^2 \\
&\quad - \kappa' \eta_{\rm o} \eta_{\rm o}' - (2\kappa^2 + k)\, \eta_{\rm o} \\
&\quad - m \eta_{\rm o} \eta_1 - 2\, (\kappa^3 + 2\kappa k)\, \eta_{\rm o} \eta_1 + (2\kappa^2 + k)\, \eta_1 \\
&\quad + \kappa \eta_{\rm o}' \eta_1' + \kappa'\, (\eta_{\rm o} \eta_1' + \eta_{\rm o}' \eta_1) + \kappa'\, \eta_{\rm o} \eta_{\rm o}' \\
&= +\sum_n R_{n{\rm o}} \, e^{in\varphi} - \sum_n R_{1n} \, e^{in\varphi} + \sum_n R_{2n} e^{in\varphi} .
\end{aligned}
\tag{4.43}
$$

We note that the higher-order dispersion functions are composed of the negative lower-order solutions plus an additional perturbation. After transformation of these differential equations into normalized variables, $w = \eta/\sqrt{\beta}$, etc., we get with $j = 0, 1, 2$ differential equations of the form

$$\ddot{w}_j(\varphi) + \nu_o^2\, w_j(\varphi) \;=\; \nu_o^2\, \beta^{3/2}\, F(s) \;=\; \sum_{m=0}^{m=j} \sum_{n=-\infty}^{n=\infty} (-1)^{m+j} F_{mn}\, \mathrm{e}^{in\varphi}\,, \quad (4.44)$$

where we have expressed the periodic perturbation on the r.h.s. by an expanded Fourier series which in addition to the perturbation terms includes the factor $\nu_o^2\, \beta^{3/2}$. Noting that the dispersion functions $w_j(\varphi)$ are periodic, we try the ansatz

$$w_j(\varphi) \;=\; \sum_n w_{jn}\, \mathrm{e}^{in\varphi}\,. \quad (4.45)$$

Inserting (4.45) into (4.44) we can solve for the individual Fourier coefficients w_{jn} by virtue of the orthogonality of the exponential functions $\mathrm{e}^{in\varphi}$ and get for the dispersion functions up to second order

$$w_o(\varphi) \;=\; +\sum_n \frac{F_{on}\, \mathrm{e}^{in\varphi}}{\nu^2 - n^2}\,,$$

$$w_1(\varphi) \;=\; -\sum_n \frac{F_{on}\, \mathrm{e}^{in\varphi}}{\nu^2 - n^2} + \sum_n \frac{F_{1n}\, \mathrm{e}^{in\varphi}}{\nu^2 - n^2}\,, \qquad (4.46)$$

$$w_2(\varphi) \;=\; +\sum_n \frac{F_{on}\, \mathrm{e}^{in\varphi}}{\nu^2 - n^2} - \sum_n \frac{F_{1n}\, \mathrm{e}^{in\varphi}}{\nu^2 - n^2} + \sum_n \frac{F_{2n}\, \mathrm{e}^{in\varphi}}{\nu^2 - n^2}\,.$$

The solutions of the higher-order differential equations have the same integer-resonance behavior as the linear solution for the dispersion function. The higher-order corrections will become important for lattices where strong sextupoles are required in which cases the sextupole terms may be the major perturbations to be considered. Other perturbation terms depend mostly on the curvature κ in the bending magnets and, therefore, maybe small for large rings or beam-transport lines with weak bending magnets.

4.4.1 Chromaticity in Higher Approximation

So far we have used only quadrupole and sextupole fields to define and calculate the chromaticity. From the general equations of motion we know, however, that many more perturbation terms act just like sextupoles and therefore cannot be omitted without further discussion. To derive the relevant equations of motion from (2.42,43) we set $x = x_\beta + \eta_x\delta$ and $y = y_\beta + \eta_y\delta$ where we keep for generality the symmetry between vertical and horizontal plane. Neglecting, however, coupling terms we get with perturbations quadratic in (x, y, δ) but at most linear in δ and after separating the dispersion function a differential equation of the form ($u = x_\beta$ or y_β)

$$u_\beta'' + K\, u_\beta \;=\; -\Delta K\, u_\beta\delta - \Delta L\, u_\beta'\, \delta\,, \qquad (4.47)$$

where

$$K_x = \kappa_x^2 + k, \tag{4.48}$$

$$K_y = \kappa_y^2 - k, \tag{4.49}$$

$$-\Delta K_x = 2\kappa_x^2 + k - (m + 2\kappa_x^3 + 4\kappa_x k)\,\eta_x$$
$$- (\underline{m} + 2\kappa_x k + 2\kappa_y k)\,\eta_y + \kappa_x'\eta_x' - \kappa_y'\eta_y', \tag{4.50}$$

$$-\Delta K_y = 2\kappa_y^2 - k + (m - 2\kappa_y \underline{k} + 2\kappa_x k)\,\eta_x$$
$$+ (\underline{m} - 2\kappa_y^3 + 4\kappa_y k)\,\eta_y - \kappa_x'\eta_x' + \kappa_y'\eta_y', \tag{4.51}$$

$$-\Delta L_x = -\Delta L_y = +\kappa_x'\eta_x + \kappa_y'\eta_y + \kappa_x\eta_x' + \kappa_y\eta_y',$$

$$= +\frac{\mathrm{d}}{\mathrm{d}s}\,(\kappa_x\eta_x + \kappa_y\eta_y). \tag{4.52}$$

The perturbation terms (4.47) depend on the betatron oscillation amplitude as well as on the slope of the betatron motion. If by some transformation we succeed in transforming (4.47) into an equation with terms proportional only to u we obtain immediately the chromaticity. We try a transformation of the form $u = v \cdot f(s)$ where $f(s)$ is a still to be determined function of s. With $u' = v'f + vf'$ and $u'' = v''f + 2v'f' + vf''$ (4.47) becomes

$$v''f + 2v'f' + vf'' + Kvf + \Delta K\,vf\delta + \Delta L\,v'f\delta + \Delta L\,vf'\delta = 0. \tag{4.53}$$

Now we introduce a condition defining the function f such that in (4.53) the coefficients of v' vanish. This occurs if

$$2f' = -\Delta L\,f\delta. \tag{4.54}$$

To first order in δ this equation can be solved by

$$f = 1 + \tfrac{1}{2}\,\delta\,(\kappa_x\,\eta_x + \kappa_y\,\eta_y) \tag{4.55}$$

and (4.53) becomes

$$v'' + \left[K + (f'' + \delta\Delta K)\right]v = 0. \tag{4.56}$$

The chromaticity in this case is $\xi = \frac{1}{4\pi}\oint(\frac{f''}{\delta} + \Delta K)\beta \mathrm{d}s$ which becomes with $\frac{2f''}{\delta} = \frac{\mathrm{d}^2}{\mathrm{d}s^2}(\kappa_x\eta_x + \kappa_y\eta_y)$ and (4.50)

$$\xi_x = \frac{1}{4\pi}\oint\left(\frac{f''}{\delta} + \Delta K_x\right)\beta_x\,\mathrm{d}s = \frac{1}{4\pi}\oint \tfrac{1}{2}\frac{\mathrm{d}^2}{\mathrm{d}s^2}(\kappa_x\eta_x + \kappa_y\eta_y)\,\beta_x\,\mathrm{d}s$$
$$- \frac{1}{4\pi}\oint\beta_x\left[(2\kappa_x^2 + k) + \kappa_x'\eta_x' + \kappa_y'\eta_y'\right] \tag{4.57}$$
$$- (m + 2\kappa_x^3 + 4\kappa_x k)\,\eta_x - (\underline{m} + 2\kappa_x \underline{k} + 2\kappa_y k)\,\eta_y]\,\mathrm{d}s.$$

The first integral can be integrated twice by parts to give $\oint \tfrac{1}{2}(\kappa_x\eta_x + \kappa_y\eta_y)\,\beta''\mathrm{d}s$ and with $\beta'' = \gamma_x - K\beta$, (4.57) gives the *horizontal chromaticity*

$$\xi_x = \frac{1}{4\pi} \oint \left[-(2\kappa_x^2 + k) - \kappa_x' \eta_x' - \kappa_y' \eta_y' \right.$$
$$+ \left. (m + 2\kappa_x^3 + 4\kappa_x k)\, \eta_x + (\underline{m} + 2\kappa_x \underline{k} + 2\kappa_y k)\, \eta_y \right] \beta_x \mathrm{d}s \qquad (4.58)$$
$$+ \frac{1}{4\pi} \oint (\kappa_x \eta_x + \kappa_y \eta_y)\, \gamma_x \mathrm{d}s \,.$$

A similar expression can be derived for the *vertical chromaticity*

$$\xi_y = \frac{1}{4\pi} \oint \left[(2\kappa_y^2 + k) + \kappa_x' \eta_x' + \kappa_y' \eta_y' \right.$$
$$- \left. (m + 2\kappa_x \underline{k} + 3\kappa_y k)\, \eta_y - (\underline{m} + \kappa_y^3 + 3\kappa_y k)\, \eta_y \right] \beta_y \mathrm{d}s \qquad (4.59)$$
$$+ \frac{1}{4\pi} \oint (\kappa_x \eta_x + \kappa_y \eta_y)\, \gamma_y \mathrm{d}s \,.$$

In deriving the chromaticity we used the usual curvilinear coordinate system for which the sector magnet is the natural bending magnet. For rectangular or wedge magnets the chromaticity must be determined from (4.59) by taking the edge focusing into account. Generally, this is done by applying a delta function focusing at the edges of dipole magnets with a focal length of

$$\frac{1}{f_x} = \frac{1}{\rho} \tan \epsilon \int \delta(s_\mathrm{edge})\, \mathrm{d}s \,. \qquad (4.60)$$

Similarly, we proceed with all other terms which include focusing.

The chromaticity can be determined experimentally simply by measuring the tunes for a beam circulating with a momentum slightly different from the lattice reference momentum. In an electron ring, this is generally not possible since any momentum deviation of the beam is automatically corrected by radiation damping within a short time. To sustain an electron beam at a momentum different from the reference energy, we must change the frequency of the accelerating cavity. Due to the mechanics of phase focusing, a particle beam follows such an orbit that the particle revolution time in the ring is an integer multiple of the rf oscillation period in the accelerating cavity. By proper adjustment of the rf frequency the beam orbit is centered along the ideal orbit and the beam momentum is equal to the ideal momentum as determined by the actual magnetic fields.

Now if the rf frequency is raised, for example, the oscillation period becomes shorter and the revolution time for the beam must become shorter too. This is accomplished only if the beam momentum is changed in such a way that the particles now follow a new orbit that is shorter than the ideal reference orbit. Such orbits exist for particles with momentum less than the reference momentum. The relation between revolution time and momentum deviation is a lattice property expressed by the momentum compaction which we write now in the form

$$\frac{\Delta f_{\rm rf}}{f_{\rm rf}} = -\eta_{\rm c}\frac{\Delta cp}{cp_{\rm o}}.\tag{4.61}$$

Through the knowledge of the lattice and momentum compaction we can relate a relative change in the rf frequency to a change in the beam momentum. Measurement of the tune change due to the momentum change determines immediately the chromaticity.

4.4.2 Nonlinear Chromaticity

The chromaticity of a circular accelerator is defined as the linear rate of change of the tunes with the relative energy deviation δ. With the increased amount of focusing that is being applied in modern circular accelerators, especially in storage rings, to obtain specific particle beam properties like very high energies in large rings or a small emittance, the linear chromaticity term is no longer sufficient to describe the chromatic dynamics of particle motion. Quadratic and cubic terms in δ must be considered to avoid severe stability problems for particles with energy error. Correcting the chromaticity with only two families of sextupoles as described in (Ref. [4.1], Sect. 7.5.2) we would indeed correct the linear chromaticity but the nonlinear chromaticity may be too severe to permit stable beam operation.

We derive the nonlinear chromaticity from the equation of motion expressed in normalized coordinates and including up to third-order chromatic focusing terms

$$\ddot{w} + \nu_{\rm oo}^2\, w = p(\varphi)\, w = \left(a\,\delta + b\,\delta^2 + c\,\delta^3\right)w,\tag{4.62}$$

where the coefficients a, b, c are the perturbation functions up to third order in δ and linear in the amplitude w, and where ν_{oo} is the unperturbed tune. This equation defines nonlinear terms for the chromaticity which have been solved for the quadratic term [4.4] and for the cubic term [4.5 – 6]. While second and third-order terms become significant in modern circular accelerators, higher-order terms can be recognized by numerical particle tracking but are generally insignificant.

Since the perturbations in (4.62) are periodic for a circular accelerator we may Fourier expand the coefficients

$$a(\varphi) = a_{\rm o} + \sum_{n\neq 0} a_n\, e^{in\varphi},$$

$$b(\varphi) = b_{\rm o} + \sum_{n\neq 0} b_n\, e^{in\varphi},\tag{4.63}$$

$$c(\varphi) = c_{\rm o} + \sum_{n\neq 0} c_n\, e^{in\varphi}.$$

From the lowest-order harmonics of the perturbations we get immediately the first approximation of *nonlinear chromaticities*

$$\nu_{\rm o}^2 = \nu_{\rm oo}^2 - a_{\rm o}\,\delta - b_{\rm o}\,\delta^2 - c_{\rm o}\delta^3 \tag{4.64}$$

or

$$\nu_{\rm o}^2 = \nu_{\rm oo}^2 - \frac{1}{2\pi}\int\limits_{\rm o}^{2\pi} \tilde{p}(\varphi)\,{\rm d}\varphi. \tag{4.65}$$

With this definition we reduce the equation of motion (4.62) to

$$\ddot{w} + \nu_{\rm o}^2 w = \Bigg(\delta \sum_{n>0} 2a_n \,\cos n\varphi + \delta^2 \sum_{n>0} 2b_n \,\cos n\varphi$$
$$+\delta^3 \sum_{n>0} 2c_n \,\cos n\varphi\Bigg)\, w \;=\; \delta\, p(\varphi)\, w\,. \tag{4.66}$$

The remaining perturbation terms look oscillatory and therefore seem not to contribute to an energy dependent tune shift. In higher-order approximation, however, we find indeed resonant terms which do not vanish but contribute to a systematic tune shift. Such higher-order tune shifts cannot be ignored in all cases and therefore an analytical expression for this chromatic tune shift will be derived. To solve the differential equation (4.66) we consider the r.h.s. as a small perturbation with δ serving as the smallness parameter. Mathematical methods for a solution have been developed and are based on a power series in δ. We apply this method to both the cosine and sine like principal solution and try the ansatz

$$C(\varphi) = \sum_{n\geq0} C_n(\varphi)\,\delta^n \qquad \text{and} \qquad S(\varphi) = \sum_{n\geq0} S_n(\varphi)\,\delta^n\,. \tag{4.67}$$

Concentrating first on the cosine like solution we insert (4.67) into (4.66) and sort for same powers in δ noting that each term must vanish separately to make the ansatz valid for all values of δ. The result is a set of differential equations for the individual solution terms

$$\ddot{C}_{\rm o} + \nu_{\rm o}^2 C_{\rm o} = 0\,,$$
$$\ddot{C}_1 + \nu_{\rm o}^2 C_1 = p(\varphi)\,C_{\rm o}\,,$$
$$\cdots\cdots$$
$$\ddot{C}_n + \nu_{\rm o}^2 C_n = p(\varphi)\,C_{n-1}\,. \tag{4.68}$$

These are defining equations for the functions $C_{\rm o}, C_1, \ldots C_n$ with $C_i = C_i(\varphi)$ and each function depending on a lower-order solution. The lowest-order solutions are the principal solutions of the unperturbed motion

$$C_{\rm o}(\varphi) = \cos\nu_{\rm o}\,\varphi \qquad \text{and} \qquad S_{\rm o}(\varphi) = \frac{1}{\nu_{\rm o}} \sin\nu_{\rm o}\,\varphi\,. \tag{4.69}$$

The differential equations (4.68) can be solved with the Green's Function method which we have applied earlier to deal with perturbation terms. All successive solutions can now be derived from the unperturbed solutions through

$$C_{n+1}(\varphi) = \frac{1}{\nu_{\mathrm{o}}} \int_{\mathrm{o}}^{\varphi} p(\alpha) \sin[\nu_{\mathrm{o}}(\alpha - \varphi)] C_n(\alpha) \, d\alpha,$$

$$S_{n+1}(\varphi) = \frac{1}{\nu_{\mathrm{o}}} \int_{\mathrm{o}}^{\varphi} p(\alpha) \sin[\nu_{\mathrm{o}}(\alpha - \varphi)] S_n(\alpha) \, d\alpha.$$

(4.70)

With the unperturbed solution C_{o} we get for C_1

$$C_1(\varphi) = \frac{1}{\nu_{\mathrm{o}}} \int_{\mathrm{o}}^{\varphi} p(\alpha) \sin[\nu_{\mathrm{o}}(\alpha - \varphi)] \cos(\nu_{\mathrm{o}}\alpha) \, d\alpha$$

(4.71)

and utilizing this solution C_2 becomes

$$C_2(\varphi) = \frac{1}{\nu_{\mathrm{o}}^2} \int_{\mathrm{o}}^{\varphi} p(\alpha) \sin[\nu_{\mathrm{o}}(\alpha - \varphi)] \cos(\nu_{\mathrm{o}}\alpha) \, d\alpha$$
$$\times \int_{\mathrm{o}}^{\alpha} p(\beta) \sin[\nu_{\mathrm{o}}(\beta - \alpha)] \cos(\nu_{\mathrm{o}}\beta) \, d\beta.$$

(4.72)

Further solutions are derived following this procedure although the formulas get quickly rather elaborate. With the cosine and sine like solutions we can formulate the transformation matrix for the whole ring

$$\mathbf{M} = \begin{pmatrix} C(2\pi) & S(2\pi) \\ C'(2\pi) & S'(2\pi) \end{pmatrix}$$

(4.73)

and applying *Floquet's theorem*, the tune of the circular accelerator can be determined from the trace of the transformation matrix

$$2\cos 2\pi\nu = C(2\pi) + S'(2\pi),$$

(4.74)

where $S' = \mathrm{d}s/\mathrm{d}\varphi$. With the ansatz (4.67) this becomes

$$2\cos 2\pi\nu = \sum_{n \geq 0} C_n(2\pi) \delta^n + \sum_{n \geq 0} S'_n(2\pi) \delta^n.$$

(4.75)

Retaining only up to third-order terms in δ we finally get after some manipulations with (4.70)

$$\cos 2\pi\nu = \cos 2\pi\nu_{\mathrm{o}} - \frac{1}{2\nu_{\mathrm{o}}} \sin 2\pi\nu_{\mathrm{o}} \int_{\mathrm{o}}^{2\pi} p(\alpha) \, d\alpha$$

$$+ \frac{1}{2\nu_{\mathrm{o}}^2} \int_{\mathrm{o}}^{2\pi} \int_{\mathrm{o}}^{\alpha} p(\alpha) \, p(\beta) \sin[\nu_{\mathrm{o}}(\alpha - \beta - 2\pi)]$$
$$\times \sin[\nu_{\mathrm{o}}(\beta - \alpha)] \, d\beta \, d\alpha$$

$$+ \frac{1}{2\nu_{\mathrm{o}}^3} \int_{\mathrm{o}}^{2\pi} \int_{\mathrm{o}}^{\alpha} \int_{\mathrm{o}}^{\beta} p(\alpha) \, p(\beta) \, p(\gamma) \sin[\nu_{\mathrm{o}}(\alpha - \gamma - 2\pi)]$$
$$\times \sin[\nu_{\mathrm{o}}(\beta - \alpha)] \sin[\nu_{\mathrm{o}}(\gamma - \beta)] \, d\gamma \, d\beta \, d\alpha.$$

(4.76)

These integrals can be evaluated analytically and (4.76) becomes after some fairly lengthy but straightforward manipulations

$$
\begin{aligned}
\cos 2\pi\nu \;=\; & \cos 2\pi\nu_{\mathrm{o}} \\
& - \delta^2 \left(\frac{\pi \sin 2\pi\nu_{\mathrm{o}}}{2\nu_{\mathrm{o}}} \sum_{n>0} \frac{a_n^2}{n^2 - 4\nu_{\mathrm{o}}^2} \right) \\
& - \delta^3 \left(\frac{\pi \sin 2\pi\nu_{\mathrm{o}}}{\nu_{\mathrm{o}}} \sum_{n>0} \frac{a_n\, b_n}{n^2 - 4\nu_{\mathrm{o}}^2} \right) \\
& - \delta^3 \left\{ \frac{\pi \sin 2\pi\nu_{\mathrm{o}}}{4\nu_{\mathrm{o}}} \sum_{s>0}\sum_{t>0} \frac{a_s\, a_t}{t^2 - 4\nu_{\mathrm{o}}^2} \right. \\
& \left. \times \left[a_{s+t} \frac{1 + \frac{4\nu_{\mathrm{o}}^2}{t(s+t)}}{(s+t)^2 - 4\nu_{\mathrm{o}}^2} + a_{|s-t|} \frac{1 - \frac{4\nu_{\mathrm{o}}^2}{t(s-t)}}{(s+t)^2 - 4\nu_{\mathrm{o}}^2} \right] \right\} + \mathcal{O}(\delta^4).
\end{aligned}
\tag{4.77}
$$

This expression defines the chromatic tune shift up to third order. Note that the tune ν_{o} is not the unperturbed tune but already includes the lowest-order approximation of the chromaticity (4.64). The relevant perturbations here are linear in the betatron amplitude and drive therefore half-integer resonances as is obvious from (4.77). The main contribution to the perturbation observed here are from the quadrupole and sextupole terms

$$
p(\varphi) = \nu_{\mathrm{o}}^2 \beta^2 (k - m\eta_x)(\delta - \delta^2 + \delta^3 \dots).
\tag{4.78}
$$

In large storage rings the nonlinear chromaticity becomes quite significant as demonstrated in Fig. 4.3. Here the tune variation with energy in the storage ring PEP is shown both for the case where only two families of sextupoles are used to compensate for the natural chromaticities [4.5]. Since in this ring an energy acceptance of at least $\pm 1\%$ is required, we conclude from Fig. 4.3 that insufficient stability is available because of the nonlinear chromaticity terms shifting the tunes for off-momentum particles to an integer resonance within the desired energy acceptance.

For circular accelerators or rings with a large natural chromaticity it is important to include in the calculation of the nonlinear chromaticity higher-order terms of the dispersion function η_x. Following the discussion in Sect. 4.3 we set in (4.78)

$$
\eta_x(\varphi) = \eta_{xo} + \eta_1\, \delta + \eta_2\, \delta^2 + \dots
$$

and find the Fourier components a_n and b_n in (4.77) defined by

$$
\begin{aligned}
\nu_{\mathrm{o}}^2 \beta^2 (k - m\,\eta_{xo}) &= \sum_{n\geq 0} 2\, a_n \cos n\varphi, \\
-\nu_{\mathrm{o}}^2 \beta^2 (k - m\,\eta_{xo} + m\,\eta_1) &= \sum_{n\geq 0} 2\, b_n \cos n\varphi.
\end{aligned}
\tag{4.79}
$$

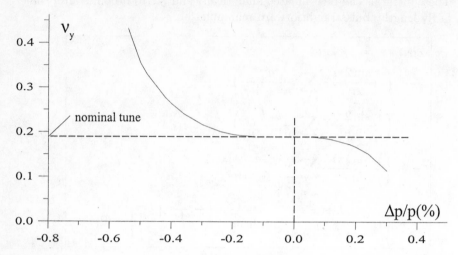

Fig. 4.3. Variation of the vertical tune with energy in the storage ring PEP if the chromaticities are corrected by only two families of sextupoles

Nonlinear energy terms in the η-function can sometimes become quite significant and must be included to improve the accuracy of analytical expressions for the nonlinear chromaticity. In such cases more sophisticated methods of chromaticity correction are required to control nonlinear chromaticities as well. One procedure is to distribute sextupoles in more than two families while keeping their total strength to retain the desired chromaticity. Using more than two families of sextupoles allows us to manipulate the strength of specific harmonics a_n such as to minimize the nonlinear chromaticities. Specifically, we note in (4.77) that the quadratic chromaticity term originates mainly from the resonant term $\frac{a_n^2}{n^2-4\nu_o^2}$. This term can be minimized by a proper distribution of sextupoles suppressing the nth-harmonic of the *chromaticity function* $\nu^2\beta^2(k-m\eta)$. Special computer programs like PATRICIA [4.5] calculate the contribution of each sextupole to the Fourier coefficients a_n and provide thereby the information required to select optimum sextupole locations and field strength to minimize quadratic and cubic chromaticities.

4.5 Perturbation Methods in Beam Dynamics

In previous sections of this chapter, mathematical procedures have been developed to evaluate the effect of specific perturbations on beam dynamics parameters. It is the nature of perturbations that they are unknown and certain assumptions as to their magnitude and distribution have to be made.

Perturbations can be systematic, statistical but periodic or just statistical and all can have a systematic or statistical time dependence.

Systematic perturbations in most cases become known through careful magnetic measurements and evaluation of the environment of the beam line. By construction magnet parameters may be all within statistical tolerances but systematically off the design values. This is commonly the case for the actual magnet length. Such deviations generally are of no consequences since the assumed magnet length in the design of a beam-transport line is arbitrary within limits. After the effective length of any magnet type to be included in a beam line is determined by magnetic measurements, beam optics calculations need to be repeated to reflect the actual magnet length. Similarly, deviations of the field due to systematic errors in the magnet gap or bore radius can be cancelled by experimental calibration of the fields with respect to the excitation current. Left are then only small statistical errors in the strength and length within each magnet type.

One of the most prominent systematic perturbation is an energy error a particle may have with respect to the ideal energy. We have treated this perturbation in much detail leading to dispersion or η-functions and chromaticities.

Other sources of systematic field errors come from the magnetic field of ion pumps or RF klystrons, from earth magnetic field, and current carrying cables along the beam line. The latter source can be substantial and requires some care in the choice of the direction the electrical current flows such that the sum of currents in all cables is mostly if not completely compensated. Further sources of systematic field perturbations originate from the vacuum chamber if the permeability of the chamber or welding material is too high, if eddy currents exist in cycling accelerators or due to persistent currents in superconducting materials which are generated just like eddy currents during the turn on procedure. All these effects are basically accessible to measurements and compensatory measures in contrast to statistical perturbations as a result of fabrication tolerances.

4.5.1 Periodic Distribution of Statistical Perturbations

Whatever statistical perturbations exist in circular accelerators, we know that these perturbations are periodic, with the ring circumference being the period length. The perturbation can therefore always be expressed by a Fourier series. The equation of motion in the presence of, for example, dipole field errors is in normalized coordinates

$$\ddot{w} + \nu_o^2 w = -\nu_o^2 \beta^{3/2} \Delta \frac{1}{\rho}. \tag{4.80}$$

The perturbation $\beta^{3/2} \Delta \frac{1}{\rho}$ is periodic and can be expressed by the Fourier series

$$\beta^{3/2}\Delta\frac{1}{\rho} = \sum_n F_n e^{in\varphi}, \tag{4.81}$$

where $\nu_o\varphi$ is the betatron phase and the Fourier harmonics F_n are given by

$$F_n = -\frac{1}{2\pi} \oint \left[-\beta^{1/2}(\sigma) \cdot \Delta\frac{1}{\rho}(\sigma) \right] e^{-in\varphi(\sigma)} d\sigma. \tag{4.82}$$

The location of the errors is not known and we may therefore only calculate the expectation value for the perturbation by multiplying (4.82) with its complex conjugate. In doing so, we note that each localized dipole perturbation deflects the beam by an angle θ and replace therefore the integral in (4.82) by a sum over all perturbations. With $\int \Delta\frac{1}{\rho} d\sigma \approx \theta$ we get for $F_n F_n^* = |F_n|^2$

$$|F_n|^2 = \frac{1}{4\pi^2} \left[\sum_k \beta_k \theta_k^2 + \sum_{k \neq j} \sqrt{\beta_k \beta_j}\, \theta_k \theta_j\, e^{-in(\varphi_k - \varphi_j)} \right], \tag{4.83}$$

where β_k is the betatron function at the location of the dipole perturbation. The second sum in (4.83) vanishes in general, since the phases for the perturbations are randomly distributed.

For large circular accelerators composed of a regular lattice unit like FODO cells we may proceed further in the derivation of the effects of perturbations which then lets us determine the field and alignment tolerances of magnets. For simplicity, we assume that the lattice magnets are the source of dipole perturbations and that the betatron functions are the same at all magnets. Equation (4.83) then becomes

$$|F_n|^2 = \frac{1}{4\pi^2} N_m \beta_m \sigma_\theta^2, \tag{4.84}$$

where σ_θ is the expectation value for the statistical deflection angle due to dipole perturbations. In a little more sophisticated approach, we would separate all magnets into groups with the same strength and betatron function and (4.84) becomes

$$|F_n|^2 = \frac{1}{4\pi^2} \sum_m N_m \beta_m \sigma_{\theta,m}^2, \tag{4.85}$$

where the sum is taken over all groups of similar perturbations and N_m is the number of perturbations within the group m. In a pure FODO lattice, for example, obvious groups would be all QF's, all QD's and all bending magnets. From now on we will, however, not distinguish between such groups anymore to simplify the discussion.

Periodic dipole perturbations cause a periodic orbit distortion which is from (4.80)

$$w(\varphi) = -\sum_n \frac{\nu_o^2 F_n}{(\nu_o^2 - n^2)} e^{in\varphi}. \tag{4.86}$$

The expectation value for the orbit distortion is obtained by multiplying (4.86) with it's complex conjugate and we get with $w(\varphi) = u(s)/\sqrt{\beta}$

$$u\,u^* = \beta(s)\,\nu^4\,|F_n|^2 \sum_{n=-\infty}^{+\infty} \frac{e^{in\varphi}}{(\nu^2-n^2)^2} \sum_{m=-\infty}^{+\infty} \frac{e^{-im\varphi}}{(\nu^2-m^2)^2}\,. \qquad (4.87)$$

The sums can be replaced by $-\pi\,\frac{\cos\nu(\pi-\varphi)}{\nu\sin\nu\pi}$ and we get finally for the expectation value of the orbit distortion σ_u at locations with a betatron function $\beta(s)$

$$\boxed{\sigma_u^2 = \beta\,\frac{N\,\bar\beta\,\sigma_\theta^2}{8\sin^2\pi\nu}\,,} \qquad (4.88)$$

where $\bar\beta$ is the average betatron function at the locations of perturbations and β the betatron function at the locations of observation. This result is in full agreement with the result [Ref. 4.1, Eq. (7.30)] for misaligned quadrupoles, where $\sigma_\theta = \sigma_q/f$, σ_q the statistical quadrupole misalignment and f the focal length of the quadrupole.

This procedure is not restricted to dipole errors only but can be applied to most any perturbation occurring in a circular accelerator. For this we determine which quantity we would like to investigate, be it the tunes, the chromaticity, perturbation of the dispersion functions, or any other beam parameter. Variation of expressions for such quantities due to variations of magnet parameters and squaring such variation we get the perturbation of the quantity under investigation. Generally, perturbation terms of order n in normalized coordinates are expressed by

$$P_n(s) = \nu^2\,\beta^{3/2}\,\beta^{n/2}\,p_n(s)\,w^n\,. \qquad (4.89)$$

Because the perturbations are assumed to be small, we may replace the oscillation amplitudes w^n in the perturbation term by their principle unperturbed solutions. Considering that the beam position w is a composite of, for example, betatron oscillation w_β, orbit distortion w_o, and energy error w_η we set

$$w = w_\beta + w_o + w_\eta \qquad (4.90)$$

and note that any higher-order perturbation contributes to the orbit, the eta-function, the tunes, betatron functions, and other beam parameters. Orbit distortions in sextupoles of strength m, for example, produce the perturbations

$$P_2(s) = \tfrac{1}{2}\,\nu^2\,\beta^{5/2}\,m\,w^2 \qquad (4.91)$$

which for $w_\eta = 0$ can be decomposed into three components

$$P_o(s) = \tfrac{1}{2} \nu^2 \beta^{5/2} m w_o^2,$$
$$P_1(s) = \nu^2 \beta^{5/2} m w_\beta w_o, \qquad\qquad (4.92)$$
$$P_2(s) = \tfrac{1}{2} \nu^2 \beta^{5/2} m w_\beta^2.$$

The perturbation P_o causes an orbit distortion and since the perturbations are randomly distributed the contribution to the orbit according to (4.88) is

$$\sigma_u^2 = \beta_u \frac{N_s \beta_{us} \sigma_\theta^2}{8 \sin^2 \pi \nu_u}, \qquad\qquad (4.93)$$

where N_s is the number of sextupoles, β_{us} the value of the betatron function and σ_u the rms orbit distortion at the sextupoles, $\sigma_\theta = \frac{1}{2} m \sigma_u^2 \ell_s$ and ℓ_s is the effective sextupole length. In cases of very strong sextupoles iteration methods must be applied since the orbit perturbation depends on the orbit. Similarly, we could have set $w_o = 0$ to calculate the perturbation of the η-function due to sextupole magnets.

The linear perturbation in (4.92) causes a statistical tune shift and a perturbation of the betatron function. Following the derivation of tune shifts in [Ref. 4.1, Sect. 7.3.1], we find the expectation value for the tune shift to be

$$\langle \delta^2 \nu \rangle = \frac{1}{16\pi^2} \sum_k \beta_k m_k \ell_k \langle u_o^2 \rangle_k, \qquad\qquad (4.94)$$

where $\sigma_u^2 = \langle u_o^2 \rangle$ is the random misalignment of the sextupole magnets or random orbit distortions in the sextupoles.

We find the interesting result, that sextupoles contribute to a tune error only if there is a finite orbit distortion or misalignment u_o, while a finite betatron oscillation amplitude of a particle in the same sextupoles does not contribute to a tune shift. Similarly, we may use the effects of systematic errors derived in [Ref. 4.1, Chap. 7] to get expressions for the probable variation of the betatron function due to gradient errors from sextupoles.

In the approximation of small perturbations, we are able to determine the expectation value for the effect of statistical errors on a particular beam parameter or lattice function. This formalism is used particularly when we are interested to define tolerances for magnetic field quality and magnet alignment by calculating backwards from the allowable perturbation of beam parameters to the magnitude of the errors. Some specific statistical effects will be discussed in subsequent sections.

Periodic perturbations in circular accelerators: Alignment and field errors in circular accelerators not only cause a distortion of the orbit but also a perturbation of the η-functions. Although these perturbations occur in both the horizontal and vertical plane, we will discuss only the effect in the vertical plane. While the derivations are the same for both planes the errors contribute only to a small perturbation of the already existing horizontal η-function while the ideal vertical η-function vanishes, and therefore the perturbation can contribute a large variation of beam parameters. This is specifically true for electron storage ring where the vertical beam emittance is very small and a finite vertical η-function may increase this emittance considerably.

Similar to (4.27) we use the equation of motion

$$y'' - k\,y \;=\; +\Delta\kappa_y - \Delta\kappa_y\,\delta - ky\delta + mxy \tag{4.95}$$

with the decomposition

$$y \;=\; y_{\mathrm{c}} + v_y\,\delta \tag{4.96}$$

and get in normalized coordinates $\tilde{y} = y/\sqrt{\beta_y}$, while ignoring the betatron motion, the differential equations for the orbit distortion $\tilde{y}_{\mathrm{c}}$

$$\ddot{\tilde{y}}_{\mathrm{c}} + \nu_y^2\,\tilde{y}_{\mathrm{c}} \;=\; +\nu_y^2\,\beta^{3/2}\left(\Delta\kappa_y + mx_{\mathrm{c}}y_{\mathrm{c}}\right) \tag{4.97}$$

and for the perturbation of the η-function $\tilde{v}_y = v_y/\sqrt{\beta_y}$

$$\ddot{\tilde{v}}_y + \nu_y^2\,\tilde{v}_y \;=\; -\nu_y^2\,\beta_y^{3/2}\,\Delta\kappa_y + \nu_y^2\,\beta_y^2\,(k - m\eta_x)\,\tilde{y}_{\mathrm{c}}\,. \tag{4.98}$$

First we note in a linear lattice where $m = 0$ that the differential equations for both the closed orbit distortion and the η-function perturbation are the same except for a sign in the perturbation. In analogy to (4.88)

$$\langle v_y^2(s)\rangle \;=\; \frac{\beta(s)}{8\sin^2\pi\nu}\sum_i\langle\theta_{y,i}^2\,\beta_i\rangle\,. \tag{4.99}$$

The perturbation of the η-function becomes more complicated in strong focusing lattices, where the chromaticity is corrected by sextupole fields. In this case, we note that all perturbation terms on the r.h.s. are periodic and we express them in Fourier series

$$\nu^2\,\beta_y^{3/2}\,\Delta\kappa_y \;=\; \sum_{n=-\infty}^{n=+\infty} F_n\,\mathrm{e}^{in\varphi} \tag{4.100}$$

with

$$F_n \;=\; \frac{\nu^2}{2\pi}\int_0^{2\pi}\beta^{3/2}\,\Delta\kappa_y\,\mathrm{e}^{-in\tau}\,\mathrm{d}\tau$$

and

$$\nu^2\,\beta_y^2\,(k - m\eta_x) \;=\; \sum_{n=-\infty}^{n=+\infty} A_n\,\mathrm{e}^{in\varphi} \tag{4.101}$$

with

$$A_n = \frac{\nu^2}{2\pi} \int_o^{2\pi} \beta^2 \left(k - m\eta_x\right) e^{-in\tau} \, d\tau \, .$$

We also make use of the periodicity of the perturbation of the η-function and set

$$\tilde{v}_y = \sum_{n=-\infty}^{n=+\infty} E_n \, e^{in\varphi} \, . \tag{4.102}$$

Inserting (4.100 – 102) into (4.98) we get

$$\sum_n [(\nu^2 - n^2) \, E_n + F_n] \, e^{in\varphi} - \sum_{m,r} \frac{A_m \, F_r}{\nu^2 - n^2} \, e^{i(m+r)\varphi} \; = \; 0 \, .$$

Noting that this equation must be true for all phases φ all terms with the same exponential factor must vanish separately and we may solve for the harmonics of the η-function

$$E_n \; = \; -\frac{F_n}{\nu^2 - n^2} + \sum_r \frac{A_{n-r} \, F_r}{(\nu^2 - n^2) \, (\nu^2 - r^2)} \, . \tag{4.103}$$

The perturbation of the η-function is therefore

$$\tilde{v}_y(\varphi) \; = \; -\tilde{y}_c(\varphi) + \sum_{n,r} \frac{A_{n-r} \, F_r}{(\nu^2 - n^2) \, (\nu^2 - r^2)} \, e^{in\varphi} \, , \tag{4.104}$$

where we have made use of the periodic solution of the closed orbit

$$\tilde{y}_c(\varphi) \; = \; \sum_n \frac{F_n}{\nu^2 - n^2} \, e^{in\varphi} \, .$$

We extract from the double sum on the r.h.s. of (4.104) all terms with $n = r$ and get from those terms the expression $A_o \sum_n \frac{F_n}{(\nu^2 - n^2)^2} \, e^{in\varphi}$. The coefficient A_o, however, is just the natural chromaticity $A_o = 2\xi_o/\nu$ and the perturbation of the η-function is from (4.104)

$$\tilde{v}_y(\varphi) \; = \; -\tilde{y}_c(\varphi) + \frac{2\xi_y}{\nu_y} \sum_n \frac{F_n \, e^{in\varphi}}{(\nu^2 - n^2)^2} + \sum_{n \neq r} \frac{A_{n-r} \, F_r \, e^{in\varphi}}{(\nu^2 - n^2)(\nu^2 - r^2)} \, . \tag{4.105}$$

By correcting the orbit distortion and compensating the chromaticity, we are able to greatly reduce the perturbation of the vertical η-function. All terms with $r = 0$ vanish for a truly random distribution of misalignment errors since $F_o = 0$. Taking the quadrupole lattice as fixed we find the remaining terms to depend mainly on the distribution of the orbit correction F_r and sextupole positions A_i. For any given sextupole distribution the orbit correction must be done such as to eliminate as much as possible all harmonics of the orbit in the vicinity of the tunes $r \not\approx \nu_y$ and to center the corrected orbit such that $F_o = 0$.

Furthermore, we note that some care in the distribution of the sextupoles must be exercised. While this distribution is irrelevant for the mere correction of the natural chromaticities, higher harmonics of the *chromaticity function* must be held under control as well. The remaining double sum is generally rather small since the resonance terms have been eliminated and either $\nu - n$ or $\nu - r$ is large. However, in very large rings or very strong focusing rings this contribution to the perturbation of the η-function may still be significant.

4.5.2 Statistical Methods to Evaluate Perturbations

In an open beam-transport line the perturbation effect at a particular point depends only on the upstream perturbations. Since perturbations cannot change the position but only the slope of particle trajectories, we merely transform the random kick angle θ_k from the location of the perturbation to the observation point. Adding all perturbations upstream of the observation point we get

$$
\begin{aligned}
u(s) &= \sqrt{\beta(s)} \sum_{\substack{k \\ \psi_k < \psi(s)}} \sqrt{\beta_k}\, \sin(\psi(s) - \psi_k)\, \theta_k\,, \\
u'(s) &= \frac{1}{\sqrt{\beta(s)}} \sum_{\substack{k \\ \psi_k < \psi(s)}} \sqrt{\beta_k}\, \cos(\psi(s) - \psi_k)\, \theta_k\,.
\end{aligned}
\tag{4.106}
$$

The expectation value for the position of the beam center at the observation point becomes from the first equation (4.106) noting the complete independence of the perturbations

$$
\sigma_u(s) = \sqrt{\beta(s)}\, \tfrac{1}{2}\, N\, \sqrt{\bar{\beta}}\, \sigma_\theta\,.
\tag{4.107}
$$

Random variations of the beam position are customarily corrected by special steering magnets if such correction is required at all. In long beam-transport systems like those required in linear colliders a mere correction of the beam position at the collision point, for example, may not be acceptable. Specifically, nonlinear perturbations lead to an incoherent increase of the beam size which can greatly reduce the usefulness of the colliding-beam system. In the next subsection we will therefore discuss general perturbations in beam-transport lines and their effect on the beam cross section.

Control of beam size in transport lines: For the transport of beams with a very small beam size or beam emittance like in linear collider facilities we are specially concerned about the impact of any kind of perturbation on the integrity of a small beam emittance. Errors can disturb the beam size in many ways. We have discussed already the effect of dipole errors on the dispersion. The distortion of the dispersion causes an increase in the beam size due to the energy spread in the beam. Quadrupole field errors affect the value of betatron functions and therefore the beam size. Vertical orbit distortions in sextupoles give rise to vertical – horizontal coupling. In this section we will try to evaluate these effects on the beam size.

We use the equations of motion (2.43,44) up to second order in x, y and δ, and assume the curvature to be small of the order or less than (x, y, δ). This is a proper assumption for high-energy beam transport lines like in linear colliders. For lower-energy beam lines very often this assumption is still correct and where a better approximation is needed more perturbation terms must be considered. For the horizontal plane we get

$$x'' + (\kappa_x^2 + k)\, x = \kappa_x \delta - \kappa_x \delta^2 - \underline{k} y - \tfrac{1}{2} m\, (x^2 - y^2)(1 - \delta)$$
$$- \Delta \kappa_x\, (1 - \delta) + kx\delta + \Delta kx\, (1 - \delta) \tag{4.108}$$
$$+ \mathcal{O}(3)$$

and for the vertical plane

$$y'' - \underline{k} y = \kappa_y\, \delta - \kappa_y\, \delta^2 - \underline{k} x + mxy(1 - \delta) - ky\delta$$
$$- \Delta \kappa_y\, (1 - \delta) - \Delta ky(1 - \delta) + \mathcal{O}(3)\,. \tag{4.109}$$

In these equations rotated magnets $(\kappa_y, \underline{k}, \underline{m})$ are included as small quantities because rotational alignment errors of upright magnets cause rotated error fields although no rotated magnets per se are used in the beam line. For the solution of (4.108, 109) we try the ansatz

$$x = x_\beta + x_c + \eta_x \delta + v_x \delta + w_x \delta^2\,,$$
$$y = y_\beta + y_c + \eta_y \delta + v_y \delta + w_y \delta^2\,. \tag{4.110}$$

Here we define (x_β, y_β) as the betatron oscillations, (x_c, y_c) the orbit distortions, (η_x, η_y) the dispersion function, (v_x, v_y) the perturbations of the dispersion functions due to magnetic field errors, and (w_x, w_y) the first-order chromatic perturbation of the dispersion functions $(\eta_{\text{tot}} = \eta + v + w\delta + \ldots)$. This ansatz leads to the following differential equations in the horizontal plane where we assume the bending radii to be large and κ_x, κ_y are therefore treated as small quantities

$$x_\beta'' + kx_\beta = -\underline{k}\, y_\beta - \tfrac{1}{2} m\, (x_\beta^2 - y_\beta^2) - m\, (x_\beta x_c - y_\beta y_c) + \Delta kx_\beta\,, \tag{4.111}$$

$$x_c'' + kx_c = -\Delta \kappa_x + \Delta kx_c - \underline{k}\, y_c - \tfrac{1}{2}m(x_c^2 - y_c^2)\,, \tag{4.112}$$

$$\eta_x'' + k\eta_x = +\kappa_x\,, \tag{4.113}$$

$$v_x'' + kv_x = -\underline{k}\, v_y - m\, (x_\beta + x_c)\, (\eta_x + v_x) + m\, (y_\beta + y_c)\, (\eta_y + v_y)$$
$$+ \Delta k(x_c + x_\beta) + \Delta \kappa_x + \Delta k\, (\eta_x + v_x) \tag{4.114}$$
$$+ kx_\beta + kx_c + \tfrac{1}{2}m(x_c^2 - y_c^2) + \underline{k} y_c\,,$$

$$w_x'' + k\, w_x = -\kappa_x - \tfrac{1}{2}m(\eta_x^2 + 2\eta_x v_x - v_y^2) + k(\eta_x + v_x)$$
$$+ m\, (x_c \eta_x + x_c v_x - y_c v_y) + (\eta_x + v_x)x_\beta - v_y y_\beta\,. \tag{4.115}$$

Similarly, we get for the vertical plane

$$y_\beta'' - k y_\beta = -\underline{k} x_\beta + m\, x_\beta y_\beta - \Delta k y_\beta + m\,(x_c y_\beta + x_\beta y_c)\,, \tag{4.116}$$

$$y_c'' - k y_c = -\Delta \kappa_y - \Delta k y_c - \underline{k} x_c + m\, x_c y_c\,, \tag{4.117}$$

$$\eta_y'' - k \eta_y = +\kappa_y\,, \tag{4.118}$$

$$\begin{aligned} v_y'' - k v_y = &+ \Delta \kappa_y - \underline{k}(\eta_y + v_y) + m\,(x_\beta + x_c)\,(\eta_y + v_y) \\ &+ m\,(\eta_x + v_x)\,(y_\beta + y_c) + \Delta k(y_\beta + y_c) + \underline{k} x_c \\ &- k(y_\beta + y_c) - \Delta k(\eta_y + v_y) - m x_c y_c\,, \end{aligned} \tag{4.119}$$

$$\begin{aligned} w_y'' - k w_y = &- \kappa_y + k(\eta_y + v_y) \\ &+ m\,(\eta_x \eta_y + \eta_x v_y + v_x \eta_y + v_x v_y)\,. \end{aligned} \tag{4.120}$$

The solution of all these differential equations is, if not already known, straightforward. We consider every perturbation to be localized as a thin element causing just a kick which propagates along the beam line. If β_j is the betatron function at the observation point and β_i that at the point of the perturbation P_i the solution of (4.111 – 120) have the form

$$u_j = \sqrt{\beta_j} \sum_i \sqrt{\beta_i}\,\sin\psi_{ji} \int P_i\,\mathrm{d}s\,. \tag{4.121}$$

The kick due to the perturbation is $\theta_i = \int P_i \mathrm{d}s$, where the integral is taken along the perturbation assumed to be short. To simplify the equations to follow we define the length $\ell_i = \theta_i/\langle P_i\rangle$. Since most errors derive from real magnets, this length is identical with that of the magnet causing the perturbation and $\psi_{ji} = \psi_j - \psi_i$ is the betatron phase between perturbation and observation point. A closer look at (4.111 – 121) shows that many perturbations depend on the solution requiring an iterative solution process. Here we will, however, concentrate only an the first iteration.

Ignoring coupling terms we have in (4.111) two types of perturbations, statistically distributed focusing errors Δk and geometric aberration effects due to sextupoles. We assume here that the beam line is chromatically corrected by the use of self-correcting achromats in which case the term $\frac{1}{2}m(x_\beta^2 - y_\beta^2)$ is self-canceling. The expectation value for the betatron oscillation amplitude due to errors, setting $\langle P_i\rangle = P_i$, is then

$$x_\beta^2(s) = \beta_x(s)\sum_i \beta_{xi}(P_i \ell_i)^2 \sin^2\psi_{ji}$$

or

$$\langle x_\beta^2(s)\rangle = \beta_x(s)\overline{\beta}_x\,\langle \underline{k}^2 y_\beta^2 + \Delta k^2 x_\beta^2 + m^2(x_\beta^2 x_c^2 + y_\beta^2 y_c^2)\rangle \tfrac{1}{2} N_M \ell^2\,,$$

where $\overline{\beta}_x$ is the average value of the betatron functions at the errors, N_M the number of perturbed magnets and ℓ the magnet length. With $\underline{k} = k\alpha$, where α is the rotational error, we get

$$\langle x_\beta^2(s) \rangle = \tfrac{1}{2} \beta_x(s)\,\overline{\beta}_x\, N_M k^2 \ell^2$$
$$\times \left[\sigma_\alpha^2\, \sigma_y^2 + \sigma_k^2 \sigma_x^2 + \frac{m^2}{k^2} (\sigma_x^2\, \sigma_{yc}^2 - \sigma_y^2 \sigma_{yc}^2) \right]. \tag{4.122}$$

We have assumed the errors to have a gaussian error distribution with standard width σ. Therefore, $\sigma_\alpha^2 = \langle \alpha^2 \rangle$, $\sigma_k^2 = \langle (\Delta k / k)^2 \rangle, \sigma_{xc} = \langle x_c^2 \rangle$, etc., and σ_y, σ_x the standard beam size for the gaussian particle distribution. Since $\langle x_\beta^2(s) \rangle / \beta(s) = \Delta \epsilon_x$ is the increase in beam emittance and $\sigma_x^2 = \epsilon_x \overline{\beta}_x, \sigma_y^2 = \epsilon_y \overline{\beta}_y$ we get for a round beam for which $\epsilon_x = \epsilon_y$ and the average values for the betatron functions are the same $(\overline{\beta}_x = \overline{\beta}_y)$

$$\boxed{\frac{\Delta \epsilon_x}{\epsilon_x} = \tfrac{1}{2} \overline{\beta}^2\, N_M\, k^2 \ell^2 \left[\sigma_\alpha^2 + \sigma_k^2 + \frac{m^2}{k^2} (\sigma_{xc}^2 + \sigma_{yc}^2) \right].} \tag{4.123}$$

To keep the perturbation of the beam small the alignment (σ_α) and magnet field quality (σ_k) must be good and the focusing weak which, however, for other reasons is not desirable. For a chromatically corrected beam line we have $k/m = \overline{\eta}_x$ which can be used in (4.123). The perturbation of the vertical beam emittance follows exactly the same results because we used a round beam.

The expectation value for the shift of the beam path is derived from (4.112, 117) with (4.121) in a similar way as for the betatron oscillations

$$\langle x_c^2(s) \rangle = \tfrac{1}{2} \beta_x(s)\, \overline{\beta}_x\, N_M\, \ell^2 \left[\langle \Delta \kappa_x^2 \rangle + k^2 \sigma_k^2 \langle x_c^2 \rangle + k^2\, \sigma_\alpha^2 \langle y_c^2 \rangle \right]. \tag{4.124}$$

This expression for the path distortion, however, is not to be used to calculate the perturbation of the dispersion. In any properly operated beam line one expects this path distortion to be corrected leading to a smaller residual value depending on the correction scheme applied and the resolution of the monitors. With some effort the path can be corrected to better than 1mm rms which should be used to evaluate path dependent perturbation terms in (4.111 – 4.120). In the vertical plane we get

$$\langle y_c^2(s) \rangle = \tfrac{1}{2} \beta_y(s)\, \overline{\beta}_y\, N_M\, \ell^2 \left[\langle \Delta \kappa_y^2 \rangle + k^2 \sigma_\alpha^2 \langle x_c^2 \rangle + k^2 \sigma_k^2 \langle y_c^2 \rangle \right]. \tag{4.125}$$

The perturbation of the dispersion is with (4.114,121)

$$v_x(s) = -x_c(s) + \sqrt{\beta_x(s)} \sum_i \sqrt{\beta_{xi}}\, P_{xi}\, \ell_i\, \sin \psi_{xji}. \tag{4.126}$$

In (4.114) we note the appearance of the same perturbation terms as for the path distortion apart from the sign and we therefore separate that solution in (4.126). The perturbations left are then

$$P_{xi} = (k - m\,\eta_x)\,(x_\beta + x_c) + m\,(y_\beta + y_c)\,\eta_y + \Delta k \eta_x + \dots. \tag{4.127}$$

It should be noted that in this derivation the betatron phase ψ_{ji} does not depend on the energy since the chromaticity is corrected. Without this assumption, we would get another contribution to v_x from the beam-path distortion. We also note that the chromaticity factor $(k - m\eta_x)$ can to first order be set to zero for chromatically corrected beam lines. The expectation value for the distortion of the dispersion is finally given by

$$\langle v_x^2(s)\rangle = \langle x_c^2(s)\rangle \tag{4.128}$$
$$+ \tfrac{1}{2}\beta_x(s)\,\overline{\beta}_x\,N_M\,\ell^2\left[\langle \Delta k^2\rangle \overline{\eta}_x{}^2 + m^2\,\overline{\eta}_y{}^2\langle y_\beta^2\rangle + m^2\,\overline{\eta}_y{}^2\,\langle y_c^2\rangle\right]$$

or with some manipulation

$$\langle v_x^2(s)\rangle = \langle x_c^2(s)\rangle \tag{4.129}$$
$$+ \tfrac{1}{2}\beta_x(s)\,\overline{\beta}_x\,N_M\,k^2\,\ell^2\left[\sigma_k^2\,\overline{\eta}_x{}^2 + \frac{\overline{\eta}_y{}^2}{\overline{\eta}_x{}^2}\left(\overline{\beta}_y\,\epsilon_y + \sigma_{yc}^2\right)\right].$$

The perturbation of the dispersion function is mainly caused by quadrupole field errors while the second term vanishes for a plane beam line where $\eta_y = 0$. In principle, the perturbation can be corrected if a specific path distortion is introduced which would compensate the perturbation at the point s as can be seen from (4.126). In the vertical plane we proceed just the same and get instead of (4.126)

$$v_y(s) = -y_c(s) + \sqrt{\beta_y(s)}\sum_i \sqrt{\beta_{yi}}\,P_{yi}\,\ell_i\,\sin\psi_{yji} \tag{4.130}$$

with

$$P_{yi} = (k - m\eta_x)\,(y_\beta + y_c) + mv_x(y_\beta + y_c) \tag{4.131}$$
$$+ m\,(\eta_y + v_y\,)\,(x_\beta + x_c) - \Delta k(\eta_y + v_y) - \underline{k}\,(\eta_y + v_y).$$

Again due to chromaticity correction we have $(k - m\eta_x) \approx 0$ and get for the expectation value of v_y in first approximation with $v_y \equiv 0$ in (4.131)

$$\langle v_y^2(s)\rangle = \langle y_c^2(s)\rangle + \tfrac{1}{2}\beta_y(s)\,\overline{\beta}_y\,N_M\,k^2\,\ell^2 \tag{4.132}$$
$$\times\left[\left(\sigma_k^2 + \sigma_\alpha^2 + \frac{\langle x_c^2\rangle}{\overline{\eta}_x^2}\right)\overline{\eta}_y^2 + \frac{\overline{v}_x^2}{\overline{\eta}_x^2}\left(\beta_y\epsilon_y + \langle y_c^2\rangle\right)\right].$$

For a plane beam line where $\eta_y \equiv 0$ we clearly need to go through a further iteration to include the dispersion perturbation which is large compared to η_y. In this approximation, we also set $y_c(s) = 0$ and get

$$\langle v_y^2(s)\rangle_i = \frac{\overline{v}_x^2}{\overline{\eta}_x^2}\left(\beta_y\epsilon_y + \langle y_c^2\rangle\right). \tag{4.133}$$

This used in a second iteration gives finally for the variation of the vertical dispersion function due to field and alignment errors

$$\langle v_y^2(s)\rangle_2 = \langle y_c^2(s)\rangle + \tfrac{1}{2}\beta_y(s)\,\bar{\beta}_y\,N_M\,k^2\,\ell^2\left(\sigma_k^2 + \sigma_\alpha^2 + \frac{\langle x_c^2\rangle}{\bar{\eta}_x^2}\right) \qquad (4.134)$$

$$\times\left[(\bar{\eta}_y^2 + \bar{v}_y^2) + \frac{\bar{v}_y^2}{\bar{\eta}_x^2}\,\beta_x\epsilon_x + \frac{\bar{v}_x^2}{\bar{\eta}_x^2}\,(\beta_y\epsilon_y + \langle y_c^2\rangle)\right].$$

This second-order dispersion due to dipole field errors is generally small but becomes significant in linear-collider facilities where extremely small beam emittances must be preserved along beam lines up to the collision point.

Problems

Problem 4.1.   Consider a compact storage ring with a circumference of 9 m, $\bar{\beta}_x = \bar{\beta}_y = 1.2$ m and beam emittances $\epsilon_x = 10$ mm mrad and $\epsilon_y = 5$ mm mrad. The arcs consist of two 180° sector bending magnets ($\rho = 1.1$ m) with a vertical aperture of $g = 10$ cm. The bending magnet field drops off linearly in the fringe region which we assume to be one vertical aperture long. Determine the five strongest nonchromatic perturbation terms for the bending magnets. Which terms contribute to the orbit and which to the focusing or the tune? What is the average orbit distortion and tune shift? Are these effects the same for all particles in the beam?

Problem 4.2.   Use the perturbation term $P_2(s)$ in (4.92) and show that pure betatron oscillations in sextupoles do not cause a tune shift in first approximation. Identify the approximation made which may lead to a tune shift in higher order.

Problem 4.3.   Consider a long, straight beam-transport line for a beam with and emittance of $\epsilon = 10^{-12}$ rad-m from the end of a 500 GeV linear collider linac toward the collision point. Use a FODO channel with $\beta_{\max} = 5$ m and determine lateral, rotational and strength tolerances for the FODO cell quadrupoles to prevent the beam emittance from dilution of more than 10%.

5. Hamiltonian Nonlinear Beam Dynamics

Deviations from linear beam dynamics in the form of perturbations and aberrations play an important role in accelerator physics. Beam parameters, quality and stability are determined by our ability to correct and control such perturbations. Hamiltonian formulation of nonlinear beam dynamics allows us to study, understand and quantify the effects of geometric and chromatic aberrations in higher order than discussed so far. Based on this understanding we may develop correction mechanisms to achieve more and more sophisticated beam performance. We will first discuss higher-order beam dynamics as an extension to the linear matrix formulation followed by specific discussions on aberrations. Finally, we develop the *Hamiltonian perturbation theory* for particle beam dynamics in accelerator systems.

5.1 Higher-Order Beam Dynamics

Chromatic and geometric aberrations appear specifically in strong focusing transport systems designed to preserve carefully prepared beam characteristics. As a consequence of correcting *chromatic aberrations* by sextupole magnets, nonlinear *geometric aberrations* are introduced. The effects of both types of aberrations on beam stability must be discussed in a formal way. Based on quantitative expressions for aberrations, we will be able to determine criteria for stability of a particle beam.

5.1.1 Multipole Errors

The general equations of motion (2.42,43) exhibit an abundance of driving terms which depend on second or higher-order transverse particle coordinates (x, x', y, y') or linear and higher-order momentum errors δ. *Magnet alignment* and *field errors* add another multiplicity to these perturbation terms. Although the designers of accelerator lattices and beam guidance magnets take great care to minimize undesired field components and avoid focusing systems that can lead to large transverse particle deviations from the reference orbit, we cannot completely ignore such perturbation terms.

In previous chapters we have discussed the effect of some of these terms and have derived among other effects such basic beam dynamics features as the dispersion function, orbit distortions, chromaticity and tune shifts as a consequence of particle momentum errors or magnet alignment and field errors. More general tools are required to determine the effect of any arbitrary driving term on the particle trajectories. In developing such tools we will assume a careful design of the accelerator under study in layout and components so that the driving terms on the r.h.s. of (2.42,43) can be treated truly as perturbations. This may not be appropriate in all circumstances in which cases numerical methods need to be applied. For the vast majority of accelerator physics applications it is, however, appropriate to treat these higher-order terms as perturbations.

This assumption simplifies greatly the mathematical complexity. Foremost, we can still assume that the general equations of motion are linear differential equations. We may therefore continue to treat every perturbation term separately as we have done so before and use the unperturbed solutions for the amplitude factors in the perturbation terms. The perturbations are reduced to functions of the location along the beam line s and the relative momentum error δ only and such differential equations can be solved analytically as we will see. Summing all solutions for the individual perturbations finally leads to the composite solution of the equation of motion in the approximation of small errors.

The differential equations of motion (2.42,43) can be expressed in a short form by

$$u'' + K(s)\,u \;=\; \sum_{\mu,\nu,\sigma,\rho,\tau \geq 0} p_{\mu\nu\sigma\rho\tau}(s)\,x^{\mu}\,x'^{\nu}\,y^{\sigma}\,y'^{\rho}\,\delta^{\tau}\,, \tag{5.1}$$

where $u = x$ or $u = y$ and the quantities $p_{\mu\nu\sigma\rho\tau}(s)$ represent the coefficients of perturbation terms. The same form of equation can be used for the vertical plane but we will restrict the discussion to only one plane neglecting coupling effects.

Some of the *perturbation terms* $p_{\mu\nu\sigma\rho\tau}$ can be related to aberrations known from geometrical light optics. Linear particle beam dynamics and gaussian geometric light optics works only for paraxial beams where the light rays or particle trajectories are close to the optical axis or reference path. Large deviations in amplitude, as well as fast variations of the amplitudes or large slopes, create aberrations in the imaging process leading to distortions of the image known as spherical aberrations, coma, distortions, curvature and astigmatism. While corrections of such aberrations are desired, the means to achieve corrections in particle beam dynamics are different from those used in light optics. Much of the theory of particle beam dynamics is devoted to diagnose the effects of aberrations on particle beams and to develop and apply such corrections.

The transverse amplitude x can be separated into its components which under the assumptions made are independent from each other

$$x \;=\; x_{\beta} + x_{\mathrm{o}} + x_{\delta} + \sum x_{\mu\nu\sigma\rho\tau}\,. \tag{5.2}$$

The first three components of solution (5.2) have been derived earlier and are associated with specific lowest order perturbation terms:

$x_\beta(s)$ is the *betatron oscillation amplitude* and general solution of the homogeneous differential equation of motion with vanishing perturbations $p_{\mu\nu\sigma\rho\tau} = 0$ for all indices.

$x_o(s)$ is the *orbit distortion* and is a special solution caused by amplitude and momentum independent perturbation terms like dipole field errors or displacements of quadrupoles or higher multipoles causing a dipole-field error. The relevant perturbations are characterized by $\mu = \nu = \sigma = \rho = \tau = 0$ but otherwise arbitrary values for the perturbation p_{ooooo}. Note that in the limit $p_{ooooo} \to 0$ we get the ideal reference path or reference orbit $x_o(s) = 0$.

$x_\delta(s)$ is the chromatic equilibrium orbit for particles with an energy different from the ideal reference energy, $\delta \neq 0$, and differs from the reference orbit with or without distortion $x_o(s)$ by the amount $x_\delta(s)$ which is proportional to the dispersion function $\eta(s)$ and the relative momentum deviation δ. In this case $\mu = \nu = \sigma = \rho = 0$ and $\tau = 1$.

All other solutions $x_{\mu\nu\sigma\rho\tau}$ are related to remaining higher-order perturbations. The perturbation term p_{1oooo}, for example, acts just like a quadrupole and may be nothing else but a quadrupole field error causing a tune shift and a variation in the betatron oscillations. Other terms, like p_{ooloo} can be correlated with linear coupling or with chromaticity if $p_{1ooo1} \neq 0$. Sextupole terms p_{2oooo} are used to compensate chromaticities, in which case the amplitude factor x^2 is expressed by the betatron motion and chromatic displacement

$$x^2 \approx (x_\beta + x_\delta)^2 = (x_\beta + \eta\,\delta)^2 \approx 2\,\eta\,x_\beta\,\delta\,. \tag{5.3}$$

The x_β^2-term, which we neglected while compensating the chromaticity, is the source for geometric aberrations due to sextupolar fields becoming strong for large oscillation amplitudes and the $\eta^2\delta^2$-term contributes to higher-order solution of the η-function. We seem to make arbitrary choices about which perturbations to include in the analysis. Generally therefore only such perturbations are included in the discussion which are most appropriate to the problem to be investigated and solved. If, for example, we are only interested in the orbit distortion x_o, we ignore in lowest order of approximation the betatron oscillation x_β and all chromatic and higher-order terms. Should, however, chromatic variations of the orbit be of interest one would evaluate the corresponding component separately. On the other hand if we want to calculate the chromatic variation of betatron oscillations, we need to include the betatron oscillation amplitudes as well as the off momentum orbit x_δ.

In treating higher-order perturbations we make an effort to include all perturbations that contribute to a specific aberration to be studied or to define the order of approximation used if higher-order terms are to be ignored. A careful inspection of all perturbation terms close to the order of approximation desired is prudent to ensure that no significant term is missed. Conversely such an inspection might very well reveal correction possibilities. An example is the effect of chromaticity which is generated by quadrupole field errors for off momentum particles but can be compensated only by sextupole fields at locations where the dispersion function is finite. Here the problem is corrected by fields of a different order from those causing the chromaticity.

To become more quantitative we discuss the analytical solution of (5.1). Since in our approximation this solution is the sum of all partial solutions for each individual perturbation term, the problem is solved if we find a general solution for an arbitrary perturbation. The solution of, for example, the horizontal equation of motion

$$x'' + K(s)\, x \; = \; p_{\mu\nu\sigma\rho\tau}\, x^\mu\, x'^\nu\, y^\sigma\, y'^\rho\, \delta^\tau \tag{5.4}$$

can proceed in two steps. First we replace the oscillation amplitudes on the r.h.s. by their most significant components

$$
\begin{aligned}
x^\mu &\to (x_\beta + x_{\rm o} + x_\delta)^\mu\,, & y^\sigma &\to (y_\beta + y_{\rm o} + y_\delta)^\sigma\,, \\
x'^\nu &\to (x'_\beta + x'_{\rm o} + x'_\delta)^\nu\,, & y'^\rho &\to (y'_\beta + y'_{\rm o} + y'_\delta)^\rho\,.
\end{aligned}
\tag{5.5}
$$

As discussed before, in a particular situation only those of these components are eventually retained that are significant to the problem. Since most accelerators are constructed in the horizontal plane we may set the vertical dispersion $y_\delta = 0$. The decomposition (5.5) is inserted into the r.h.s of (5.4) and again only terms significant for the particular problem and to the approximation desired are retained. The solution $x_{\mu\nu\sigma\rho\tau}$ can be further broken down into components each relating to only one individual perturbation term. Whatever number of perturbation terms we decide to keep, the basic differential equation for the perturbation is of the form

$$P'' + K(s)\, P \; = \; p\,(x_\beta, x'_\beta, x_{\rm o}, x'_{\rm o}, x_\delta, x'_\delta, y_\beta, y'_\beta, y_{\rm o}, y'_{\rm o}, y_\delta, y'_\delta, \delta, s)\,. \tag{5.6}$$

The second step in the solution process is to derive the actual solution of (5.6). Demonstrating the principle of the solution process we restrict the r.h.s. to linear and quadratic, uncoupled terms. In addition orbit distortions and vertical dispersion are ignored. These assumptions reduce the complexity of formulas without limiting the application of the solution process. We use the principal solutions on the r.h.s. of the simplified equation (5.6) and set for the betatron oscillations

$$
\begin{aligned}
x_\beta &= C_x(s)\, x_{\beta {\rm o}} + S_x(s)\, x'_{\beta {\rm o}}\,, \\
x'_\beta &= C'_x(s)\, x_{\beta {\rm o}} + S'_x(s)\, x'_{\beta {\rm o}}\,,
\end{aligned}
\tag{5.7}
$$

and for the off momentum orbit

$$
\begin{aligned}
x_\delta &= C_x(s)\, x_{\delta {\rm o}} + S_x(s)\, x'_{\delta {\rm o}} + D_x(s)\, \delta_{\rm o}\,, \\
x'_\delta &= C'_x(s)\, x_{\delta {\rm o}} + S'_x(s)\, x'_{\delta {\rm o}} + D'_x(s)\, \delta_{\rm o}\,.
\end{aligned}
\tag{5.8}
$$

As long as we do not include acceleration or energy losses the particle energy stays constant and $\delta_o = \delta(s) = $ const. The principal solutions depend on the initial conditions $x_{\beta o}, x'_{\beta o}, x_{\delta o}, \dots$ etc. at $s = 0$ and (5.6) becomes

$$P(s)'' + K(s)\,P(s) \;=\; p(x_{\beta o}, x'_{\beta o}, x_{\delta o}, x'_{\delta o}, \delta_o, s)\,. \tag{5.9}$$

This is exactly the form which has been discussed in [Ref. 5.1, Sect. 4.8.3] with solution

$$P(s) \;=\; \int_o^s p(\tilde{s})\,G(s, \tilde{s})\,\mathrm{d}\tilde{s} \tag{5.10}$$

and the Green's function defined by

$$G(s, \tilde{s}) \;=\; S(s)\,C(\tilde{s}) - C(s)\,S(\tilde{s})\,. \tag{5.11}$$

Following these steps we may calculate, at least in principle, the perturbations $P(s)$ for any arbitrary higher-order driving term $p(s)$. In praxis, however, even principal solutions of particle trajectories in composite beam transport systems can be expressed only in terms of the betatron functions. Since the betatron functions cannot be expressed in a convenient analytical form, we are unable to express the integral (5.10) in closed analytical form and must therefore employ numerical methods.

5.1.2 Nonlinear Matrix Formalism

In linear beam dynamics this difficulty has been circumvented by the introduction of transformation matrices, a principle which can be used also for beam transport systems including higher-order perturbation terms [5.2 − 4] . The solution of (5.1) can be expressed by (5.10) in terms of initial conditions. Similar to discussions in the context of linear beam dynamics we solve (5.9) for individual lattice elements only, where $K(s) = $ const. In this case (5.10) can be solved for any piecewise constant perturbation along a beam line. Each solution depends on initial conditions at the beginning of the magnetic element and the total solution can be expressed in the form

$$
\begin{aligned}
x(s) &= c_{11o}\,x_o + c_{12o}\,x'_o + c_{13o}\,\delta_o + c_{111}\,x_o^2 + c_{112}\,x_o x'_o + \dots, \\
x'(s) &= c_{21o}\,x_o + c_{22o}\,x'_o + c_{23o}\,\delta_o + c_{211}\,x_o^2 + c_{212}\,x_o x'_o + \dots,
\end{aligned}
\tag{5.12}
$$

where the coefficients c_{ijk} are functions of s. The nomenclature of the indices becomes obvious if we set $x_1 = x$, $x_2 = x'$, and $x_3 = \delta$. The coefficient c_{ijk} then determines the effect of the perturbation term $x_j x_k$ on the variable x_i. In operator notation we may write

$$c_{ijk} \;=\; \langle x_i | x_{jo} x_{ko} \rangle\,. \tag{5.13}$$

The first-order coefficients are the principal solutions

$$c_{110}(s) = C(s), \qquad c_{210}(s) = C'(s),$$
$$c_{120}(s) = S(s), \qquad c_{220}(s) = S'(s), \qquad (5.14)$$
$$c_{130}(s) = D(s), \qquad c_{230}(s) = D'(s).$$

Before continuing with the solution process, we note that the variation of the oscillation amplitudes (x', y') are expressed in a curvilinear coordinate system generally used in beam dynamics. This definition, however, is not identical to the intuitive assumption that the slope x' of the particle trajectory is equal to the angle Θ between the trajectory and reference orbit. In a curvilinear coordinate system the slope $x' = \mathrm{d}x/\mathrm{d}s$ is a function of the amplitude x. To clarify the transformation, we define angles between the trajectory and the reference orbit by

$$\frac{\mathrm{d}x}{\mathrm{d}\sigma} = \Theta \qquad \text{and} \qquad \frac{\mathrm{d}y}{\mathrm{d}\sigma} = \Phi, \qquad (5.15)$$

where with the curvature $\kappa = 1/\rho$

$$\mathrm{d}\sigma = (1 + \kappa x)\,\mathrm{d}s. \qquad (5.16)$$

In linear beam dynamics there is no numerical difference between x' and Θ which is a second-order effect nor is there a difference in straight parts of a beam transport line where $\kappa = 0$. The relation between both definitions is from (5.15,16)

$$\Theta = \frac{x'}{1 + \kappa x} \qquad \text{and} \qquad \Phi = \frac{y'}{1 + \kappa x}, \qquad (5.17)$$

where $x' = \mathrm{d}x/\mathrm{d}s$ and $y' = \mathrm{d}y/\mathrm{d}s$. We will use these definitions and formulate second-order transformation matrices in a cartesian coordinate system (x, y, z). Following *Brown's* notation [5.2] we may express the nonlinear solutions of (5.4) in the general form

$$u_i = \sum_{j=1}^{3} c_{ijo}\, u_{jo} + \sum_{\substack{j=1 \\ k=1}}^{3} T_{ijk}(s)\, u_{jo} u_{ko}, \qquad (5.18)$$

with

$$(u_1, u_2, u_3) = (x, \Theta, \delta), \qquad (5.19)$$

where σ is the length of the particle trajectory. *Nonlinear transformation coefficients* T_{ijk} are defined similar to coefficients c_{ijk} in (5.13) by

$$T_{ijk} = \langle u_i | u_{jo} u_{ko} \rangle, \qquad (5.20)$$

where the coordinates are defined by (5.19). In linear approximation both coefficients are numerically the same and we have

$$\begin{pmatrix} c_{11o} & c_{12o} & c_{13o} \\ c_{21o} & c_{22o} & c_{23o} \\ c_{31o} & c_{32o} & c_{33o} \end{pmatrix} = \begin{pmatrix} C(s) & S(s) & D(s) \\ C'(s) & S'(s) & D'(s) \\ 0 & 0 & 1 \end{pmatrix}. \tag{5.21}$$

Earlier in this section we decided to ignore coupling effects which could be included easily in (5.18) if we set for example $x_4 = y$ and $x_5 = y'$ and expand the summation in (5.18) to five indices. For simplicity, however, we will continue to ignore coupling.

The equations of motion (2.42,43) are expressed in curvilinear coordinates and solving (5.10) results in coefficients c_{ijk} which are different from the coefficients T_{ijk} if one or more variables are derivatives with respect to s. In the equations of motion all derivatives are transformed like (5.17) generating a Θ-term as well as an $x\Theta$-term. If, for example, we were interested in the perturbations to the particle amplitude x caused by perturbations proportional to $x_o\Theta_o$ we are looking for the coefficient $T_{112} = \langle x|x_o\Theta_o\rangle$. Collecting from (2.42) only second-order perturbation terms proportional to xx' we find

$$x = c_{112}\,x_o\,x'_o = c_{112}\,x_o\,\Theta_o + \mathcal{O}(3)\,. \tag{5.22}$$

An additional second-order contribution appears as a spill over from the linear transformation

$$x = c_{12o}\,x'_o = c_{12o}\,(1 + \kappa_x x_o)\,\Theta_o = c_{12o}\,\Theta_o + c_{12o}\,\kappa_x\,x_o\,\Theta_o\,. \tag{5.23}$$

Collecting all $x_o\Theta_o$-terms we get finally

$$T_{112} = c_{112} + \kappa_x\,c_{12o} = c_{112} + \kappa_x\,S(s)\,. \tag{5.24}$$

To derive a coefficient like $T_{212} = \langle\Theta|x_o\Theta_o\rangle$ we also have to transform the derivative of the particle trajectory at the end of the magnetic element. First we look for all contributions to x' from $x_ox'_o$-terms which originate from $x' = c_{220}x'_o + c_{212}x_ox'_o$. Setting in the first term $x'_o = \Theta_o(1 + \kappa_x x_o)$ and in the second term $x_ox'_o \approx x_o\Theta_o$ we get with $c_{22o} = S'(s)$ and keeping again only second-order terms

$$x' = [c_{212} + \kappa_x\,S'(s)]\,x_o\Theta_o\,.$$

On the l.h.s. we replace x' by $\Theta(1 + \kappa_x\,x) = \Theta + \kappa_x\,x\Theta$ and using the principal solutions we get

$$x\Theta \approx (C_x\,x_o + S_x\,\Theta_o)\,(C'_x\,x_o + S'_x\,\Theta_o) = (C_x S'_x + C'_x S_x)\,x_o\Theta_o$$

keeping only the $x_o\Theta_o$-terms. Collecting all results the second-order coefficient for this perturbation becomes

$$T_{212} = \langle\Theta|x_o\Theta_o\rangle = c_{212} + \kappa_x S'_x(s) - \kappa_x(C_x S'_x + C'_x S_x)\,. \tag{5.25}$$

In a similar way we can derive all second-order coefficients T_{ijk}. Equations (5.18) define the transformation of the particle coordinates in second order through a particular magnetic element. For the transformation to quadratic terms we may ignore the third order difference between the coefficients c_{ijk} and T_{ijk} and get

$$
\begin{aligned}
x^2 &= \left(C_x x_o + S_x x'_o + D_x \delta_o \right)^2 , \\
xx' &= \left(C_x x_o + S_x x'_o + D_x \delta_o \right) \left(C'_x x_o + S'_x x'_o + D'_x \delta_o \right) , \\
x\delta &= \left(C_x x_o + S_x x'_o + D_x \delta_o \right) \delta_o ,
\end{aligned}
\tag{5.26}
$$

$$\vdots \quad \text{etc..}$$

All transformation equations can now be expressed in matrix form after correctly ordering equations and coefficients and a general second-order transformation matrix can be formulated [5.4] in the form

$$
\begin{pmatrix} x \\ \Theta \\ \delta \\ x^2 \\ x\Theta \\ x\delta \\ \Theta^2 \\ \Theta\delta \\ \delta^2 \end{pmatrix}
= \boldsymbol{M}
\begin{pmatrix} x \\ \Theta_o \\ \delta_o \\ x_o^2 \\ x_o \Theta_o \\ x_o \delta_o \\ \Theta_o^2 \\ \Theta_o \delta_o \\ \delta_o^2 \end{pmatrix} ,
\tag{5.27}
$$

where we have ignored the y-plane. The *second-order transformation matrix* is then

$$
\boldsymbol{M} =
\tag{5.28}
$$

$$
\begin{pmatrix}
C & S & D & T_{111} & T_{112} & T_{116} & T_{122} & T_{126} & T_{166} \\
C' & S' & D' & T_{211} & T_{212} & T_{216} & T_{222} & T_{226} & T_{266} \\
0 & 0 & 1 & 0 & 0 & 0 & 0 & 0 & 0 \\
0 & 0 & 0 & C^2 & 2CS & 2CD & S^2 & 2SD & D^2 \\
0 & 0 & 0 & CC' & CS'+C'S & CD'+C'D & SS' & SD'+S'D & DD' \\
0 & 0 & 0 & 0 & 0 & C & 0 & S & D \\
0 & 0 & 0 & C'^2 & 2C'S' & 2C'D & S'^2 & 2S'D' & D'^2 \\
0 & 0 & 0 & 0 & 0 & C' & 0 & S' & D' \\
0 & 0 & 0 & 0 & 0 & 0 & 0 & 0 & 1
\end{pmatrix}
$$

with $C = C_x, S = S_x, \ldots$ etc.

A similar equation can be derived for the vertical plane. If coupling effects are to be included the matrix could be further expanded to include also such terms. While the matrix elements must be determined individually for each magnetic element in the beam transport system, we may in analogy to linear beam dynamics multiply a series of such matrices to obtain the transformation matrix through the whole composite beam transport line. As a matter of fact the transformation matrix has the same appearance as (5.27) for a single magnet or a composite beam transport line and the magnitude of the nonlinear matrix elements will be representative of imaging errors like spherical and chromatic aberrations.

To complete the derivation of *second-order transformation matrices* we derive, as an example, an expression of the matrix element T_{111} from the equation of motion (2.42). To obtain all x_o^2-terms we look in (2.42) for perturbation terms proportional to x^2, xx' and x'^2, replace these amplitude factors by principal solutions (5.7) and collect only terms quadratic in x_o to get the relevant perturbation

$$p(s) = [-(\tfrac{1}{2} m + 2\kappa_x k + \kappa_x^3) C_x^2 x_o^2 + \tfrac{1}{2} \kappa_x C_x'^2 x_o^2 + \kappa_x' C_x C_x'] x_o^2. \quad (5.29)$$

First, we recollect that the theory of nonlinear transformation matrices is based on the constancy of magnet strength parameters and therefore we must set $\kappa_x' = 0$. Where this is an undue simplification like in magnet fringe fields one could approximate the smooth variation of κ_x by a step function. Inserting (5.29) into (5.10) we get for the second-order matrix element

$$c_{111} = T_{111} = -(\tfrac{1}{2} m + 2\kappa_x k + \kappa_x^3) \int_o^s C_x^2(\tilde{s}) \, G(s, \tilde{s}) \, \mathrm{d}\tilde{s}$$

$$\quad (5.30)$$

$$- \tfrac{1}{2} \kappa_x \int_o^s C_x'^2 \, G(s, \tilde{s}) \, \mathrm{d}\tilde{s}.$$

The integrants are powers of trigonometric functions and can be evaluated analytically. In a similar way we may now derive any second-order matrix element of interest. A complete list of all second order matrix elements can be found in [5.2].

This formalism is valuable whenever the effect of second-order perturbations must be evaluated for particular particle trajectories. Specifically, it is suitable for nonlinear beam simulation studies where a large number of particles representing the beam are to be traced through nonlinear focusing systems to determine, for example, the particle distribution and its deviation from linear beam dynamics at a focal point. This formalism is included in the program TRANSPORT [5.5] allowing the determination of the coefficients T_{ijk} for any beam transport line and providing fitting routines to eliminate such coefficients by proper adjustment and placement of nonlinear elements like sextupoles.

5.2 Aberrations

From light optics we are familiar with the occurrence of *aberrations* which cause the distortion of optical images. We have repeatedly noticed the similarity of particle beam optics with geometric or paraxial light optics and it is therefore not surprising that there is also a similarity in imaging errors. Aberrations in particle beam optics can cause severe stability problems and must therefore be controlled.

We distinguish two classes of aberrations, *geometric aberrations* and for off momentum particles *chromatic aberrations*. The geometric aberrations become significant as the amplitude of betatron oscillations increases while chromatic aberration results from the variation of the optical system parameters for different colors of the light rays or in our case for different particle energies. For the discussion of salient features of aberration in particle beam optics we study the equation of motion in the horizontal plane and include only bending magnets, quadrupoles and sextupole magnets. The equation of motion in this case becomes in normalized coordinates $w = x/\sqrt{\beta_x}$

$$\ddot{w} + \nu_{\rm o}^2 w = \nu_{\rm o}^2 \beta^{2/3} \kappa_{\rm o}\, \delta + \nu_{\rm o}^2 \beta^2 k_{\rm o}\, w\, \delta - \tfrac{1}{2} \nu_{\rm o}^2 \beta^{5/2} m_{\rm o}\, w^2\,, \qquad (5.31)$$

where $\beta = \beta_x$.

The particle deviation w from the ideal orbit is composed of two contributions, the betatron oscillation amplitude w_β and the shift in the equilibrium orbit for particles with a relative momentum error δ. This orbit shift w_δ is determined by the normalized dispersion function at the location of interest, $w_\delta = \tilde{\eta}\,\delta = \frac{\eta}{\sqrt{\beta}}\,\delta$, and the particle position can be expressed by the composition

$$w = w_\beta + w_\delta = w_\beta + \tilde{\eta}\delta. \qquad (5.32)$$

Inserting (5.32) into (5.31) and employing the principle of linear superposition (5.31) can be separated into two differential equations, one for the betatron motion and one for the dispersion function neglecting quadratic or higher-order terms in δ. The differential equation for the dispersion function is then

$$\ddot{\tilde{\eta}} + \nu_{\rm o}^2 \tilde{\eta} = \nu_{\rm o}^2 \beta^{1/2} \kappa + \nu_{\rm o}^2 \beta^2 k\, \tilde{\eta}\, \delta - \tfrac{1}{2} \nu_{\rm o}^2 \beta^{5/2} m\, \tilde{\eta}^2\, \delta\,, \qquad (5.33)$$

which has been solved earlier in Sect. 4.3.

All other terms include the betatron oscillation w_β contributing therefore to aberrations of betatron oscillations expressed by the differential equation

$$\ddot{w}_\beta + \nu_{\rm o}^2 w_\beta = \nu_{\rm o}^2 \beta^2 k\, w_\beta\, \delta - \nu_{\rm o}^2 \beta^2 m\, \eta\, w_\beta\, \delta - \tfrac{1}{2} \nu_{\rm o}^2 \beta^{5/2} m\, w_\beta^2\,. \qquad (5.34)$$

The third term in (5.34) is of geometric nature causing a perturbation of beam dynamics at large betatron oscillation amplitudes and, as will be discussed in Sect. 5.3, also gives rise to an amplitude dependent tune shift. This term appears as an isolated term in second order and no local compensation scheme is possible. *Geometric aberrations* must therefore be expected whenever sextupole magnets are used to compensate for chromatic aberrations.

The first two terms in (5.34) represent the natural chromaticity and the compensation by sextupole magnets, respectively. Whenever it is possible to compensate the chromaticity at the location where it occurs both terms would cancel for $m\eta = k$. Since the strength changes sign for both magnets going from one plane to the other the compensation is correct in both planes. This method of chromaticity correction is quite effective in long beam transport systems with many equal lattice cells. An example of such a correction scheme are the beam transport lines from the SLAC linear accelerator to the collision point of the *Stanford Linear Collider*, SLC, [5.6]. This transport line consists of a dense sequence of strong magnets forming a combined function FODO channel (for parameters see example #2 in [Ref. 5.1, Tab. 6.1). In these magnets dipole, quadrupole and sextupole components are combined in the pole profile and the chromaticity compensation occurs locally.

This method of compensation, however, does not work generally in circular accelerators because of special design criteria which often require some parts of the accelerator to be dispersion free and the chromaticity created by the quadrupoles in these sections must then be corrected elsewhere in the lattice. Consequently both chromaticity terms in (5.34) do not cancel anymore locally and can be adjusted to cancel only globally.

The consequence of these less than perfect chromaticity correction schemes is the occurrence of aberrations through higher-order effects. We get a deeper insight for the effects of these aberrations in a circular accelerator by noting that the coefficients of the betatron oscillation amplitude w_β for both chromatic perturbations are periodic functions in a circular accelerator and can therefore be expanded into a Fourier series. Only nonoscillatory terms of these expansions cancel if the chromaticity is corrected while all other higher harmonics still appear as chromatic aberrations.

5.2.1 Geometric Aberrations

Geometric perturbations from sextupole fields scale proportional to the square of the betatron oscillation amplitude leading to a loss of stability for particles oscillating at large amplitudes. From the third perturbation term in (5.34) we expect this limit to occur at smaller amplitudes in circular accelerators where either the betatron functions are generally large or where the focusing and therefore the chromaticity and required sextupole correction is strong or where the tunes are large. Most generally this occurs in large proton and electron colliding-beam storage rings or in electron storage rings with strong focusing.

Compensation of nonlinear perturbations: In most older circular accelerators the chromaticity is small and can, if at all, be corrected by two families of sextupoles. Although in principle only two sextupole magnets for the whole ring are required for *chromaticity compensation*, this is in most cases impractical since the strength of the sextupoles becomes too large exceeding technical limits or leading to loss of beam stability because of intolerable geometric aberrations. For chromaticity compensation we generally choose a more even distribution of sextupoles around the ring and connect them into two families compensating the horizontal and vertical chromaticity, respectively. This scheme is adequate for most small and not too strong focusing circular accelerators. Where beam stability suffers from geometric aberrations more sophisticated sextupole correction schemes must be utilized.

To analyze the geometric aberrations due to sextupoles and develop correction schemes we follow a particle along a beam line including sextupoles. Here we understand a beam line to be an open system from a starting point to an image point at the end or one full circumference of a circular accelerator. Following any particle through the beam line and ignoring for the moment nonlinear fields we expect the particle to move along an ellipse in phase space as shown in Fig. 5.1. Travelling through the complete beam line of phase advance $\Psi = 2\pi\nu_{\mathrm{o}}$ a particle moves for ν_{o} revolutions along the phase ellipse in Fig. 5.1.

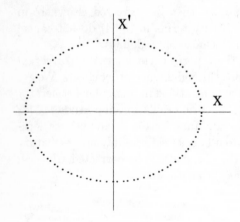

Fig. 5.1. Linear particle motion in phase space

Including nonlinear perturbations due to, for example, sextupole magnets the phase space trajectory becomes distorted from the elliptical form as shown in Fig. 5.2. An arbitrary distribution of sextupoles along a beam line can cause large variations of the betatron oscillation amplitude leading to possible loss of particles on the vacuum chamber wall even if the motion is stable in principle. The PEP storage ring [5.7] was the first storage ring to require a more sophisticated sextupole correction [5.8] beyond the mere compensation of the two chromaticities because geometric aberrations caused by specific design goals were too strong to give sufficient beam stability. Chromaticity correction with only two families of sextupoles in PEP would have produced large amplitude dependent tune shifts leading to reduced beam stability.

Such a situation can be greatly improved with additional sextupole families [5.8] to minimize the effect of these nonlinear perturbation. Although individual perturbations may not be reduced much by this method the sum of all perturbations can be compensated to reduce the overall perturbation to a tolerable level.

In this sextupole correction scheme the location and strength of the individual sextupoles are selected such as to minimize the perturbation of the particle motion in phase space at the end of the beam transport line. Although this correction scheme seems to work in not too extreme cases it is not sufficient to guarantee beam stability. This scheme works only for one amplitude due to the nonlinearity of the problem and in cases where sextupole fields are no longer small perturbations we must expect a degradation of this compensation scheme for larger amplitudes. As the example of PEP shows, however, an improvement of beam stability can be achieved beyond that obtained by a simple two family chromaticity correction. Clearly, a more formal analysis of the perturbation and derivation of appropriate correction schemes are desirable.

Sextupoles separated by a $-\mathcal{I}$ transformation: A chromaticity correction scheme that overcomes this amplitude dependent aberration has been proposed by *Brown* and *Servranckx* [5.9]. In this scheme possible sextupole locations are identified in pairs along the beam transport line such that each pair is separated by a negative unity transformation

$$-\mathcal{I} = \begin{pmatrix} -1 & 0 & 0 & \\ 0 & -1 & 0 & 0 \\ 0 & 0 & -1 & 0 \\ 0 & 0 & 0 & -1 \end{pmatrix} . \tag{5.35}$$

Placing sextupoles of equal strength at these two locations we get an additive contribution to the chromaticity correction. The effect of geometric aberrations, however, is canceled for all particle oscillation amplitudes. This can be seen if we calculate the transformation matrix through the first sextupole, the $-\mathcal{I}$ section, and then through the second sextupole. The sextupoles are assumed to be thin magnets inflicting kicks on particle trajectories by the amount

$$\Delta x' = -\tfrac{1}{2} m_o \ell_s \left(x^2 - y^2 \right),$$

and

$$\Delta y' = -m_o \ell_s \, xy \,,$$

where ℓ_s is the sextupole length. We form a 4×4 transformation matrix through a thin sextupole and get

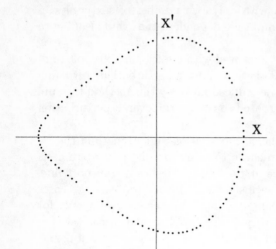

Fig. 5.2. Typical phase space motion in the presence of nonlinear fields

$$
\begin{pmatrix} x \\ x' \\ y \\ y' \end{pmatrix} = \mathcal{M}_s(x_o, y_o) \begin{pmatrix} x_o \\ x'_o \\ y_o \\ y'_o \end{pmatrix}
$$

$$
= \begin{pmatrix} 1 & 0 & 0 & 0 \\ -1/2 m_o \ell_s x_o & 1 & 1/2 m_o \ell_s y_o & 0 \\ 0 & 0 & 1 & 0 \\ 0 & 0 & m_o \ell_s x_o & 1 \end{pmatrix} \begin{pmatrix} x_o \\ x'_o \\ y_o \\ y'_o \end{pmatrix} .
$$

(5.36)

To evaluate the complete transformation we note that in the first sextupole the particle coordinates are (x_o, y_o) and become after the $-\mathcal{I}$-*transformation* in the second sextupole $(-x_o, -y_o)$. The transformation matrix through the complete unit is therefore

$$
\mathcal{M}_t = \mathcal{M}_s(x_o, y_o) \cdot (-\mathcal{I}) \cdot \mathcal{M}_s(-x_o, -y_o) = -\mathcal{I} .
$$

(5.37)

Independent of the oscillation amplitude we observe a complete cancellation of geometric aberrations in both the horizontal and vertical plane. This correction scheme has been applied successfully to the final focus system of the *Stanford Linear Collider* [5.10] where chromatic as well as geometric aberrations must be controlled and compensated to high accuracy to allow the focusing of a beam to a spot size at the collision point of only a few micrometer.

The effectiveness of this correction scheme and its limitations in circular accelerators has been analyzed in more detail by *Emery* [5.11] and we will discuss some of his findings. As an example, we use strong focusing FODO cells for an extremely low emittance electron storage ring [5.11] and investigate the beam parameters along this lattice. Any other lattice could be used as well since the characteristics of aberrations is not lattice dependent although the magnitude may be. The particular FODO lattice under

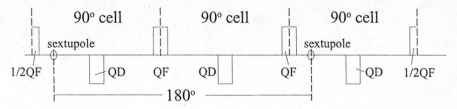

Fig. 5.3. FODO lattice and chromaticity correction

discussion as shown in Fig. 5.3, is a thin lens lattice with 90° cells, a distance between quadrupoles of $L_q = 3.6$ m and an integrated half quadrupole strength of $(k\ell_q)^{-1} = \sqrt{2}\,L_q$. The horizontal and vertical betatron functions at the symmetry points are 12.29 and 2.1088 m respectively. Three FODO cells are shown in Fig. 5.3 including one pair of sextupoles separated by 180° in betatron phase space.

We choose a phase ellipse for an emittance of $\epsilon = 200$ mm mrad which is an upright ellipse at the beginning of the FODO lattice, Fig. 5.4a. Due to quadrupole focusing the ellipse becomes tilted at the entrance to the first sextupole, Fig. 5.4b. The thin lens sextupole introduces a significant angular perturbation, Fig. 5.4c, leading to large lateral aberrations in the quadrupole QF, Fig. 5.4d. At the entrance to the second sextupole the distorted phase ellipse is rotated by 180° and all aberrations are compensated again by this sextupole, Fig. 5.4e. Finally, the phase ellipse at the end of the third FODO cell is again an upright ellipse with no distortions left, Fig. 5.4f. The range of stability therefore extends to infinitely large amplitudes ignoring any other detrimental effects.

The compensation of aberrations works as long as the phase advance between sextupoles is exactly 180°. A shift of the second sextupole by a few degrees or a quadrupole error resulting in a similar phase error between the sextupole pair would greatly reduce the compensation. In Fig. 5.5 the evolution of the phase ellipse from Fig. 5.4 is repeated but now with a phase advance between the sextupole pair of only 175°. A distortion of the phase ellipse due to aberrations can be observed which may build up to instability as the particles pass through many similar cells. Emery has analyzed numerically this degradation of stability and finds empirically the maximum stable betatron amplitude to scale with the phase error like $\Delta\varphi^{-0.52}$ [5.11]. The sensitivity to phase errors together with unavoidable quadrupole field errors and orbit errors in sextupoles can significantly reduce the effectiveness of this compensation scheme.

The single most detrimental arrangement of sextupoles compared to the perfect compensation of aberrations is to interleave sextupoles which means to place other sextupoles between two pairs of compensating sextupoles

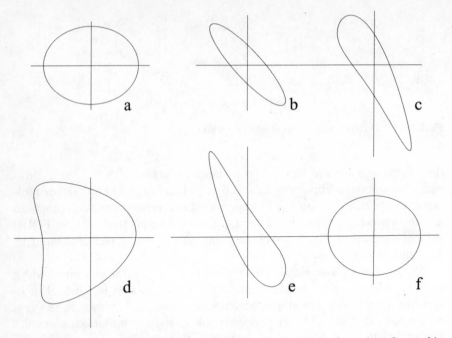

Fig. 5.4. Phase ellipses along a FODO channel including nonlinear aberrations due to thin sextupole magnets separated by exactly 180° in betatron phase

[5.12]. Such interleaved sextupoles introduce amplitude dependent phase shifts leading to phase errors and reduced compensation of aberrations. This limitation to compensate aberrations is present even in a case without apparent interleaved sextupoles as shown in Fig. 5.6 for the following reason.

The assumption of thin magnets is sometimes convenient but, as Emery points out, can lead to erroneous results. For technically realistic solutions, we must allow the sextupoles to assume a finite length and find, as a consequence, a loss of complete compensation for geometric aberrations because sextupoles of finite length are only one particular case of interleaved sextupole arrangements. If we consider the sextupoles made up of thin slices we still find that each slice of the first sextupole has a corresponding slice exactly 180° away in the second sextupoles. However, other slices are interleaved between such ideal pairs of thin slices. In Fig. 5.6 the sequence of phase ellipses from Fig. 5.4 is repeated with the only difference of using now a finite length of 0.3 m for the sextupoles. From the last phase ellipse it becomes clear that the aberrations are not perfectly compensated as was the case for thin sextupoles. Although the $-\mathcal{I}$-transformation scheme to eliminate geometric aberrations is not perfectly effective for real beam lines it is still prudent to arrange sextupoles in that way to minimize aberrations and apply additional corrections.

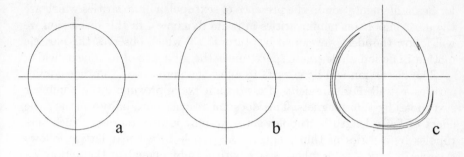

Fig. 5.5. Thin sextupole magnets separated by 175° in betatron phase space. The unperturbed phase ellipse **a** becomes slightly perturbed **b** at the end of the first triple FODO cell (Fig. 5.3) and more so after passing through many such triplets **c**

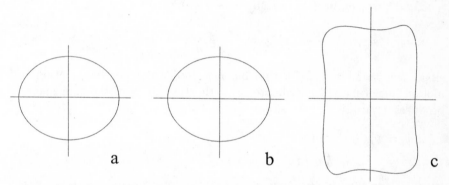

Fig. 5.6. Phase ellipses along a FODO channel including nonlinear aberrations due to finite length sextupole magnets placed exactly 180° apart. Phase ellipse **a** transforms to **b** after one FODO triplet cell and to **c** after passage through many such cells

5.2.2 Filamentation of Phase Space

Some distortion of the unperturbed trajectory in phase space due to aberrations is inconsequential to beam stability as long as this distortion does not build up and starts growing indefinitely. A finite or infinite growth of the beam emittance enclosed within a particular particle trajectory in phase space may at first seem impossible since we deal with macroscopic, nondissipating magnetic fields where Liouville's theorem must hold. Indeed numerical simulations indicate that the total phase space occupied by the beam does not increase but an originally elliptical boundary in phase space can grow, for example, tentacles like a spiral galaxy leading to larger beam sizes without actually increasing the phase space density. This phenomenon is called *filamentation* of the phase space and evolves like shown in Fig. 5.7.

For particle beams this filamentation is as undesirable as an increase in beam emittance or beam loss. We will therefore try to derive the causes for beam filamentation in the presence of sextupole nonlinearities which are the most important nonlinearities in beam dynamics. In this discussion we will follow the ideas developed by *Autin* [5.13] which observes the particle motion in action angle phase space under the influence of nonlinear fields.

The discussion will be general to allow the application to other nonlinearities as well. For simplicity of expression, we approximate the nonlinear sextupoles by thin magnets. This does not restrict our ability to study the effect of finite length sextupoles since we may always represent such sextupoles by a series of thin magnets. A particle in a linear lattice follows a circle in action angle phase space with a radius equal to the action J_o. The appearance of a nonlinearity along the particle trajectory will introduce an amplitude variation ΔJ to the action which is from (1.48) for both the horizontal and vertical plane

$$\begin{aligned}
\Delta J_x &= \nu_{xo}\, w\, \Delta w + \frac{1}{\nu_{xo}}\, \dot{w}\, \Delta \dot{w} = \frac{1}{\nu_{xo}}\, \dot{w}\, \Delta \dot{w}, \\
\Delta J_y &= \nu_{yo}\, v\, \Delta v + \frac{1}{\nu_{yo}}\, \dot{v}\, \Delta \dot{v} = \frac{1}{\nu_{yo}}\, \dot{v}\, \Delta \dot{v},
\end{aligned} \tag{5.38}$$

since $\Delta w = \Delta v = 0$ for a thin magnet. Integration of the equations of motion in normalized coordinates over the "length" ℓ of the thin magnet produces a variation of the slopes

$$\begin{aligned}
\Delta \dot{w} &= \nu_{xo}\, \sqrt{\beta_x}\, \tfrac{1}{2} m\ell \left(x^2 - y^2\right), \\
\Delta \dot{v} &= -\nu_{yo}\, \sqrt{\beta_y}\, m\ell\, xy.
\end{aligned} \tag{5.39}$$

We insert (5.39) into (5.38) and get with (1.43) after linearization of the trigonometric functions a variation of the action

$$\begin{aligned}
\Delta J_x &= \frac{m\ell}{4} \sqrt{\frac{2 J_x \beta_x}{\nu_{xo}}} \left\{ \left(J_x \beta_x - 2 J_y \beta_y \frac{\nu_x}{\nu_y}\right) \sin \psi_x + J_x \beta_x \sin 3\psi_x \right. \\
&\quad \left. - J_y \beta_y \frac{\nu_x}{\nu_y} \left[\sin(\psi_x + 2\psi_y) + \sin(\psi_x - 2\psi_y)\right] \right\}, \tag{5.40}
\end{aligned}$$

$$\Delta J_y = \frac{m\ell}{2} \sqrt{\frac{2 J_x \beta_x}{\nu_{xo}}} J_y \beta_y \left[\sin(\psi_x + 2\psi_y) - \sin(\psi_x - 2\psi_y)\right].$$

Since the action is proportional to the beam emittance, (5.40) allow us to study the evolution of beam filamentation over time. The increased action from (5.40) is due to the effect of one nonlinear sextupole magnet and we obtain the total growth of the action by summing over all turns and all sextupoles. To sum over all turns we note that the phases in the trigonometric functions increase by $2\pi\nu_{o,x,y}$ every turn and we have for the case of a single sextupole after an infinite number of turns expressions of the form

$$\sum_{n=0}^{\infty} \sin[(\psi_{xj} + 2\pi\nu_{xo}\, n) + 2(\psi_{yj} + 2\pi\nu_{yo}\, n)],$$

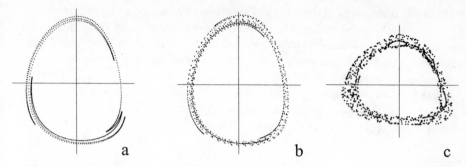

Fig. 5.7. Filamentation of phase space after passage through an increasing number of FODO cells

where ψ_{xj} and ψ_{yj} are the phases at the location of the sextupole j. Such sums of trigonometric functions are best solved in the form of exponential functions. In this case the sine function terms are equivalent to the imaginary part of the exponential functions

$$e^{i(\psi_{xj}+2\psi_{yj})} e^{i2\pi(\nu_{xo}+2\nu_{yo})n} .$$

The second factor forms an infinite geometric series and the imaginary part of the sum is therefore

$$\text{Im} \, \frac{e^{i(\psi_{xj}+2\psi_{yj})}}{1 - e^{i2\pi(\nu_{xo}+2\nu_{yo})}} = \frac{\cos[(\psi_{xj} - \pi\nu_{xo}) + 2(\psi_{yj} - \pi\nu_{yo})]}{2 \sin[\pi(\nu_{xo} + 2\nu_{yo})]} . \tag{5.41}$$

This solution has clearly resonant character leading to an indefinite increase of the action if $\nu_{xo} + 2\nu_{yo}$ is an integer. Similar results occur for the other three terms and *Autin's* method to observe the evolution of the action coordinate over many turns allows us to identify four resonances driven by sextupolar fields which can lead to particle loss and loss of beam stability if not compensated. Resonant growth of the apparent beam emittance occurs according to (5.40) for

$$\begin{aligned} \nu_{xo} &= q_1, &\text{or}&& \nu_{xo} + 2\nu_{yo} &= q_3, \\ 3\nu_{xo} &= q_2, &\text{or}&& \nu_{xo} - 2\nu_{yo} &= q_4, \end{aligned} \tag{5.42}$$

where q_i are integers. In addition to the expected integer and third integer resonance in the horizontal plane, we find also two third order coupling resonances in both planes where the sum resonance leads to beam loss while the difference resonance only initiates an exchange of the horizontal and vertical emittances. The asymmetry is not fundamental and is the result of our choice to use only upright sextupole fields.

So far we have studied the effect of one sextupole on particle motion. Since no particular assumption was made as to the location and strength of this sextupole, we conclude that any other sextupole in the ring would drive

the same resonances and we obtain the beam dynamics under the influence
of all sextupoles by adding the individual contributions. In the expressions
of this section we have tacitly assumed that the beam is observed at the
phase $\psi_{xo,yo} = 0$. If this is not the desired location of observation the phases
ψ_{xj} need to be replaced by $\psi_{xj} - \psi_{xo}$, etc., where the phases $\psi_{xj,yj}$ define the
location of the sextupole j. Considering all sextupoles in a circular lattice we
sum over all such sextupoles and get, as an example, for the sum resonance
used in the derivation above from (5.40)

$$\Delta J_{x,\nu_x + 2\nu_y} = -\sum_j \frac{m_j \ell_j}{4} \sqrt{\frac{2J_x \beta_{xj}}{\nu_{xo}}} J_y \beta_{yj} \frac{\nu_x}{\nu_y} \sin(\psi_{xj} + 2\psi_{yj}). \quad (5.43)$$

Similar expressions exist for other resonant terms. Equation (5.43) indicates
a possibility to reduce the severity of driving terms for the four resonances.
Sextupoles are primarily inserted into the lattice where the dispersion function is nonzero to compensate for chromaticities. Given sufficient flexibility
these sextupoles can be arranged to avoid driving these resonances. Additional sextupoles may be located in dispersion free sections and adjusted to
compensate or at least minimize the four resonance driving terms without
affecting the chromaticity correction. The perturbation ΔJ is minimized by
distributing the sextupoles such that the resonant driving terms in (5.40)
are as small as possible. This is accomplished by *harmonic correction* which
is the process of minimization of expressions

$$\sum_j m_j \ell_j \beta_x^{3/2} e^{i\psi_{xj}} \rightarrow 0, \quad (5.44)$$

$$\sum_j m_j \ell_j \beta_x^{3/2} e^{i3\psi_{xj}} \rightarrow 0, \quad (5.45)$$

$$\sum_j m_j \ell_j \beta_x^{1/2} \beta_y e^{i\psi_{xj}} \rightarrow 0, \quad (5.46)$$

$$\sum_j m_j \ell_j \beta_x^{1/2} \beta_y e^{i(\psi_{xj} + 2\psi_{yj})} \rightarrow 0, \quad (5.47)$$

$$\sum_j m_j \ell_j \beta_x^{1/2} \beta_y e^{i(\psi_{xj} - 2\psi_{yj})} \rightarrow 0. \quad (5.48)$$

The perturbations of the action variables in (5.40) cancel perfectly if we insert sextupoles in pairs at locations which are separated by a $-\mathbf{I}$ *transformation* as discussed previously in this section. The distribution of sextupoles in
pairs is therefore a particular solution to (5.44) for the elimination of beam
filamentation and specially suited for highly periodic lattices while (5.44 –
48) provide more flexibility to achieve similar results in general lattices and
sextupole magnets of finite length.

Cancellation of resonant terms does not completely eliminate all aberrations caused by sextupole fields. Because of the existence of the nonlinear
sextupole fields the phases ψ_j depend on the particle amplitude and resonant
driving terms are therefore canceled only to first order. For large amplitudes

we expect increasing deviation from the perfect cancellation leading eventually to beam filamentation and beam instability. Maximum stable oscillation amplitudes in (x, y)-space due to nonlinear fields form the *dynamic aperture* which is to be distinguished from the physical aperture of the vacuum chamber. This dynamic aperture is determined by numerical tracking of particles. Given sufficiently large physical apertures in terms of linear beam dynamics to meet particular design specifications, including some margin for safety, the goal of correcting nonlinear aberrations is to extend the dynamic aperture beyond the physical aperture. Methods discussed above to increase the dynamic aperture have been applied successfully to a variety of particle storage rings, especially by *Autin* [5.13 – 14], to the antiproton cooling ring ACOL where a particularly large dynamic aperture is required.

5.2.3 Chromatic Aberrations

Correction of natural chromaticities is not a complete correction of all chromatic aberrations. For sensitive lattices nonlinear chromatic perturbation terms must be included. Both linear as well as nonlinear chromatic perturbations have been discussed in detail in Sect. 4.3. Such terms lead primarily to gradient errors and therefore the sextupole distribution must be chosen such that driving terms for half integer resonances are minimized. Together with tune shifts due to gradient field errors we observe also a variation of the betatron function. Chromatic gradient errors in the presence of sextupole fields are

$$p_1(s) = (k - m\eta)\,\delta \tag{5.49}$$

and the resulting variation of the betatron function has been derived in [Ref. 5.1, Sect. 7.3.3]. For the perturbation (5.49) the linear variation of the betatron function with momentum is from [Ref. 5.1, Eq. (7.71)]

$$\frac{\Delta\beta(s)}{\beta_\mathrm{o}} = \frac{\delta}{2\sin 2\pi\nu_\mathrm{o}} \int_s^{s+L} \beta\,(k - m\eta)\,\cos[2\nu_\mathrm{o}\,(\varphi_s - \varphi_\sigma + 2\pi)]\mathrm{d}\sigma\,, \tag{5.50}$$

where L is the length of the superperiod, $\varphi_s = \varphi(s)$ and $\varphi_\sigma = \varphi(\sigma)$. The same result can be expressed in the form of a Fourier expansion for N_s superperiods in a ring lattice by

$$\frac{\Delta\beta}{\beta} = \delta\,\frac{\nu_\mathrm{o}}{4\pi} \sum_q \frac{F_q\,\mathrm{e}^{iN_\mathrm{s}q\varphi}}{\nu_\mathrm{o}^2 - (N_\mathrm{s}q/2)^2}\,, \tag{5.51}$$

where

$$F_q = \frac{\nu_\mathrm{o}}{2\pi} \int_\mathrm{o}^{2\pi} \beta^2\,(k - m\eta)\,\mathrm{e}^{iN_\mathrm{s}q\varphi}\,\mathrm{d}\varphi\,. \tag{5.52}$$

Both expressions exhibit the presence of half integer resonances and we must expect the area of beam stability in phase space to be reduced for off momentum particles because of the increased strength of the resonances. Obviously, this perturbation does not appear in cases where the chromaticity is corrected locally so that $(k - m\eta) \equiv 0$ but few such cases exist. To minimize the perturbation of the betatron function, we look for sextupole distributions such that the Fourier harmonics are as small as possible by eliminating excessive "fighting" between sextupoles and by minimizing the resonant harmonic $q = 2\nu_o$. Overall, however, it is not possible to eliminate this *beta beat* completely. With a finite number of sextupoles the beta beat can be adjusted to zero only at a finite number of points along the beam line.

In colliding-beam storage rings, for example, we have specially sensitive sections just adjacent to the collision points. To maximize the luminosity the lattice is designed to produce small values of the betatron functions at the collision points and consequently large values in the adjacent quadrupoles. In order not to further increase the betatron functions there and make the lattice more sensitive to errors, one might choose to seek sextupole distributions such that the beta beat vanishes at the collision point and its vicinity.

Having taken care of chromatic gradient errors we are left with the variation of geometric aberrations as a function of particle momentum. Specifically, resonance patterns vary and become distorted as the particle momentum is changed. Generally this should not cause a problem as long as the dynamic aperture can be optimized to exceed the physical aperture. A momentum error will introduce only a small variation to the dynamic aperture as determined from geometric aberrations for on momentum particles only. If, however, the dynamic aperture is limited by some higher-order resonances even a small momentum change can cause a big difference in the stable phase space area.

Analytical methods are useful to minimize detrimental effects of geometric and chromatic aberrations due to nonlinear magnetic fields. We have seen how by careful distribution of the chromaticity correcting sextupoles, resonant beam emittance blow up and excessive beating of the betatron functions for off momentum particles can be avoided or at least minimized within the approximations used. In Sect. 5.3, we will also find that sextupolar fields can produce strong tune shifts for larger amplitudes leading eventually to instability at nearby resonances. Here again a correct distribution of sextupoles will have a significant stabilizing effect. Although there are a number of different destabilizing effects, we note that they are driven by only a few third order resonances. Specifically, in large circular lattices a sufficient number of sextupoles and locations for additional sextupoles are available for an optimized correction scheme. In small rings such flexibility often does not exist and therefore the sophistication of chromaticity correction is limited. Fortunately, in smaller rings the chromaticity is much smaller and some of the higher-order aberrations discussed above are very small and need not be compensated. Specifically, the amplitude dependent tune shift is generally negligible in small rings while it is this effect which limits the dynamic aperture in most cases of large circular accelerators.

The optimization of sextupole distribution requires extensive analysis of the linear lattice and time consuming albeit straightforward numerical calculations which are best left for computer programs. The program HARMON [5.15] has been developed particularly with this optimization procedure in mind. Input requires a matched linear lattice with possible locations for sextupoles. The program then adjusts sextupole strengths such as to correct the chromaticities to the desired values while minimizing aberrations.

For open beam transport lines the program TRANSPORT [5.5], based on a second-order transformation theory [5.3], provides a powerful way to eliminate any specified second-order aberration. A special second-order matching routine allows to match sextupole strengths to meet desired values for second-order aberrations at a specified point along the lattice. Since the results are broken down to reflect the contribution of all second-order transformation elements, it is easy to identify the most detrimental.

In trying to solve aberration problems in beam dynamics we are, however, mindful of approximations made and terms neglected for lack of mathematical tools to solve analytically the complete nonlinear dynamics in realistic accelerators. The design goals for circular accelerators become more and more demanding on our ability to control nonlinear aberrations. On one hand the required cross sectional area in the vicinity of the ideal orbit for a stable beam remains generally constant for most designs but the degree of aberrations is increased in an attempt to reach very special beam characteristics. As a consequence, the nonlinear perturbations become stronger and the limits of dynamic aperture occur for smaller amplitudes compared to less demanding lattices and require more and more sophisticated methods of chromaticity correction and control of nonlinear perturbations.

5.2.4 Particle Tracking

No mathematical methods are available yet to calculate analytically the limits of the dynamic aperture for any but the most simple lattices. High order of approximations are required to treat strong aberrations in modern circular accelerator designs. The most efficient way to determine beam stability characteristics for a particular lattice design is to perform numerical *particle tracking* studies.

Aberrations in open beam transport lines have been studied since the early days of strong focusing and second-order transformation matrices have been derived by *Brown* [5.2 – 4] to quantify such aberrations. To study the effect of aberrations on the focusing properties along an arbitrary beam line, the program DECAY TURTLE [5.16] has been developed. This program traces a large number of particles distributed over a specified range of initial coordinates through the nonlinear beam transport system and provides an accurate particle distribution in real or phase space at the desired location of the beam line.

Circular accelerators are basically not different from very long periodic beam transport lines and therefore programs like DECAY TURTLE could be applied here as well. In many modern accelerator designs, however, second-order simulation of aberrations is not sufficient anymore and additional simulation features like the inclusion of synchrotron oscillations are required which are not part of programs specially designed for beam transport lines. A number of tracking programs have been developed meeting such needs.

Perturbations of localized nonlinear fields on a particle trajectory are easy to calculate and tracking programs follow single particles along their path incorporating any nonlinear perturbation encountered. Since most nonlinear fields are small, we may use thin lens approximation and passage of a particle through a nonlinear field of any order inflicts therefore only a deflection on the particle trajectory. During the course of tracking the deflections of all nonlinearities encountered are accumulated for a large number of turns and beam stability or instability is judged by the particle surviving the tracking or not, respectively. The basic effects of nonlinear fields in numerical tracking programs are therefore reduced to what actually happens to particles travelling through such fields producing results in an efficient way. Of course from an intellectual point of view such programs are not completely satisfactory since they serve only as tools providing little direct insight into actual causes for limitations to the dynamic aperture and instability.

The general approach to accelerator design is to develop first a lattice in linear approximation meeting the desired design goals followed by an analytical approach to include chromaticity correcting sextupoles in an optimized distribution. Further information about beam stability and *dynamic aperture* can at this point only be obtained from numerical tracking studies. Examples of widely used computer programs to perform such tracking studies are in historical order PATRICIA [5.8], MARYLIE [5.17], RACETRACK [5.18], PATPET [5.19].

Tracking programs generally require as input an optimized linear lattice and perform then particle tracking for single particles as well as for a large number of particles simulating a full beam. Nonlinear fields of any order can be included as thin lenses in the form of isolated multipole magnets like sextupoles or a multipole errors of regular lattice magnets. The multipole errors can be chosen to be systematic or statistical and the particle momentum may have a fixed offset or may be oscillating about the ideal momentum due to synchrotron oscillations.

Results of such computer studies contribute information about particle dynamics which is not available otherwise. The motion of single particles in phase space can be observed together with an analysis of the frequency spectrum of the particle under the influence of all nonlinear fields included and at any desired momentum deviation.

Further information for the dynamics of particle motion can be obtained from the frequency spectrum of the oscillation. An example of this is shown in Fig. 5.8 as a function of oscillation amplitudes. For small amplitudes we notice only the fundamental horizontal betatron frequency ν_x. As the oscillation amplitude is increased this basic frequency is shifted toward lower values while more frequencies appear. We note the appearance of higher harmonics of ν_x due to the nonlinear nature of motion.

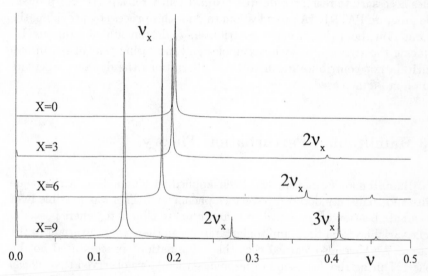

Fig. 5.8. Frequency spectrum for betatron oscillations with increasing amplitudes (X) as determined by particle tracking with PATRICIA

The motion of a particle in phase space and its frequency spectrum as a result of particle tracking can give significant insight into the dynamics of a single particle. For the proper operation of an accelerator, however, we also need to know the overall stability of the particle beam. To this purpose we define initial coordinates of a large number of particles distributed evenly over a cross section normal to the direction of particle propagation to be tested for stability. All particles are then tracked for many turns and the surviving particles are displayed over the original cross section at the beginning of the tracking thus defining the area of stability or *dynamic aperture*.

Having pursued analytical as well as numerical computations to determine the dynamic aperture we are still faced with the question, what to do if the analytical optimization of sextupole distribution will not yield a dynamic aperture as large as required. In such a case special features of tracking programs may be employed to maximize the dynamic aperture.

Adjusting the sextupole distribution due to analytical recipes does not tell us the magnitude of improvement to the dynamic aperture for each compensation because each of the resonant terms has its weighting factor which is unknown to us. In extreme cases where further improvement of the dynamic aperture is desired, we may therefore rely on scientifically less satisfactory applications of trial and error methods to get access to analytically unreachable corrections. By varying the strength of a sextupole, one at a time, the dynamic aperture is determined through particle tracking and any variation that produces an increase in the dynamic aperture is accepted as the new sextupole strength. As stated this method lacks scientific merit but is sometimes necessary to reach the desired beam stability for a particular project. The program PATRICIA provides some aid for this process by providing the information about the strongest contributors to detrimental aberrations. By reducing the strength of such sextupoles and increasing that of sextupoles which do not contribute much to these aberrations the dynamic aperture can often be increased.

5.3 Hamiltonian Perturbation Theory

The Hamiltonian formalism has been applied in Chap. 1 to derive tune shifts and to discuss detailed resonance phenomena. This was possible by a careful application of *canonical transformation* to eliminate, where possible, cyclic variables from the Hamiltonian and obtain thereby an invariant of the motion. We have also learned that this "elimination" process need not be perfect. During the discussion of resonance theory, we observed that *slowly varying terms* function almost like *cyclic variables* giving us important information about the stability of the motion.

During the discussion of the resonance theory, we tried to transform perturbation terms to a higher order in oscillation amplitude than required by the approximation desired and where this was possible we would then ignore such higher-order fast-oscillating terms. This procedure was successful for all terms but resonant terms. In this section we will ignore resonant terms and concentrate on higher-order terms which we have ignored so far [5.21]. By application of a canonical identity transformation we try to separate from fast oscillating terms those which vary only slowly. To that goal we start from the *nonlinear Hamiltonian* (1.53)

$$H = \nu_0 J + p_n(\varphi) J^{n/2} \cos^n \psi.$$

Fast-oscillating terms can be transformed to a higher order by a *canonical transformation* which can be derived from the *generating function*

$$G_1 = \psi J_1 + w(\psi, \varphi) J_1^{n/2}, \tag{5.53}$$

where the function $w(\psi, \varphi)$ is an arbitrary but periodic function in ψ and φ which we will determine later. From (5.53) we get for the new angle variable ψ_1 and the old action variable J

$$
\begin{aligned}
\psi_1 &= \frac{\partial G_1}{\partial J_1} = \psi + \tfrac{n}{2}\, w(\psi, \varphi)\, J^{n/2-1}, \\
J &= \frac{\partial G_1}{\partial \psi} = J_1 + \frac{\partial w}{\partial \psi}\, J_1^{n/2},
\end{aligned}
\tag{5.54}
$$

and the new Hamiltonian is

$$
H_1 = H + \frac{\partial G_1}{\partial \varphi} = H + \frac{\partial w(\psi, \varphi)}{\partial \varphi}\, J_1^{n/2}.
\tag{5.55}
$$

We replace now the old variables (ψ, J) in the Hamiltonian by the new variables (ψ_1, J_1) and expand the nth root

$$
J^{n/2} = \left(J_1 + \frac{\partial w}{\partial \psi}\, J_1^{n/2} \right)^{n/2} = J_1^{n/2} + \frac{n}{2}\frac{\partial w}{\partial \psi}\, J_1^{n-1} + \ldots .
\tag{5.56}
$$

With (5.54,56) the Hamiltonian (5.55) becomes

$$
\begin{aligned}
H_1 &= \nu_o J_1 + J_1^{n/2} \left[\nu_o \frac{\partial w}{\partial \psi} + p_n(\varphi) \cos^n \psi + \frac{\partial w}{\partial \varphi} \right] \\
&\quad + J_1^{n-1} \left[\frac{n}{2} p_n(\varphi) \cos^n \psi \frac{\partial w}{\partial \psi} \right] + \mathcal{O}\left(J_1^{n+1/2} \right).
\end{aligned}
\tag{5.57}
$$

We have neglected all terms of order $n + 1/2$ or higher in the amplitude J as well as quadratic terms in $w(\psi, \varphi)$ or derivations thereof. We still must express all terms of the Hamiltonian in the new variables and define therefore the function

$$
Q(\psi, \varphi) = \nu_o \frac{\partial w}{\partial \psi} + p_n(\varphi) \cos^n \psi + \frac{\partial w}{\partial \varphi}.
\tag{5.58}
$$

In the approximation of small perturbations we have $\psi_1 \approx \psi$ or $\psi_1 = \psi + \Delta\psi$ and may expand (5.58) like

$$
\begin{aligned}
Q(\psi_1, \varphi) &= Q(\psi, \varphi) + \frac{\partial Q}{\partial \psi}\, \Delta\psi, \\
&= Q(\psi, \varphi) + \frac{n}{2}\, w(\psi_1, \varphi)\, J_1^{n/2-1} \frac{\partial Q}{\partial \psi},
\end{aligned}
\tag{5.59}
$$

where we used the first equation of (5.54). The Hamiltonian can be greatly simplified if we make full use of the periodic but otherwise arbitrary function $w(\psi_1, \varphi)$. With (5.59) we obtain from (5.57)

$$
\begin{aligned}
H_1 &= \nu_o J_1 + J_1^{n/2}\, Q(\psi_1, \varphi) \\
&\quad + \frac{n}{2} J_1^{n-1} \left[p_n(\varphi) \cos^n \psi_1 \frac{\partial w}{\partial \psi} - w(\psi, \varphi) \frac{\partial Q}{\partial \psi} \right] + \ldots
\end{aligned}
\tag{5.60}
$$

and we will derive the condition that

$$Q(\psi, \varphi) = 0. \tag{5.61}$$

First we set

$$\cos^n \psi_1 = \sum_{m=-n}^{n} a_{nm}\, e^{im\psi_1} \tag{5.62}$$

and try an expansion of $w(\psi_1, \varphi)$ in a similar way

$$w(\psi_1, \varphi) = \sum_{m=-n}^{n} w_m(\varphi)\, e^{im\,(\psi_1 - \nu_o \varphi)}. \tag{5.63}$$

This definition of the function w obviously is still periodic in ψ and φ as long as $w_m(\varphi)$ is periodic. With

$$\frac{\partial w}{\partial \psi_1} = \sum_{m=-n}^{n} w_m(\varphi)\, i\,m\, e^{im(\psi_1 - \nu_o \varphi)}$$

and

$$\frac{\partial w}{\partial \varphi} = \sum_{m=-n}^{n} \left[\frac{\partial w_m}{\partial \varphi} - i\nu_o m w_m(\varphi) \right] e^{im(\psi_1 - \nu_o \varphi)}$$

we get instead of (5.58)

$$Q(\psi_1, \varphi) \approx Q(\psi, \varphi) = i\nu_o \sum_{m=-n}^{n} m w_m\, e^{im(\psi_1 - \nu_o \varphi)} \tag{5.64}$$

$$+ p_n(\varphi) \sum_{m=-n}^{n} a_{nm} e^{im\psi_1} + \sum_{m=-n}^{n} \left(\frac{\partial w_m}{\partial \varphi} - i\nu_o m w_m \right) e^{im(\psi_1 - \nu_o \varphi)} = 0$$

noting from (5.59) that the difference $\Delta Q = Q(\psi_1, \varphi) - Q(\psi, \varphi)$ contributes nothing to the term of order $J_1^{n/2}$ for $n > 2$. The imaginary terms cancel and we get

$$Q(\psi_1, \varphi) \approx p_n(\varphi) \sum_{m=-n}^{n} a_{nm}\, e^{im\psi} + \sum_{m=-n}^{n} \frac{\partial w_m}{\partial \varphi}\, e^{im(\psi - \nu_o \varphi)} = 0. \tag{5.65}$$

This equation must be true for all values of φ and therefore the individual terms of the sums must vanish independently

$$p_n(\varphi)\, a_{nm} + \frac{\partial w_m}{\partial \varphi}\, e^{-im\,\nu_o \varphi} = 0 \tag{5.66}$$

for all values of m. After integration we have

$$w_m(\varphi) = w_{mo} - a_{nm} \int_o^\varphi p_n(\phi)\, e^{im\nu_o\phi}\, d\phi \tag{5.67}$$

and since the coefficients $w_m(\varphi)$ must be periodic, $w_m(\varphi) = w_m(\varphi + \frac{2\pi}{N})$, where N is the superperiodicity, we are able to eventually determine the function $w(\psi_1, \varphi)$. With

$$w_m(\varphi)\, e^{im(\psi_1 - \nu_o\varphi)} = w_m\left(\varphi + \frac{2\pi}{N}\right) e^{im(\psi_1 - \nu_o\varphi - \frac{2\pi}{N}\nu_o)}$$

and (5.67) we have

$$w_{mo}\, e^{im(\psi - \nu_o\varphi)} - a_{nm}\, e^{im(\psi - \nu_o\varphi)} \int_o^\varphi p_n(\bar\phi)\, e^{im\nu_o\bar\phi}\, d\bar\phi$$

$$= e^{im(\psi - \nu_o\varphi - \frac{2\pi}{N}\nu_o)} (w_{mo} - a_{nm}) \int_o^{\varphi + \frac{2\pi}{N}} p_n(\bar\phi)\, e^{im\nu_o\bar\phi}\, d\bar\phi.$$

Solving for w_{mo} we get

$$w_{mo}(1 - e^{im\frac{2\pi}{N}\nu_o}) = a_{nm} \int_o^{2\pi/N} p_n(\bar\phi)\, e^{im\nu_o\bar\phi}\, d\bar\phi. \tag{5.68}$$

A solution for w_{mo} exists only if there are no perturbations and $p(\varphi) \equiv 0$ or if $(1 - e^{im\frac{2\pi}{N}\nu_o}) \neq 0$. In other words we require the condition

$$m\nu_o \neq qN, \tag{5.69}$$

where q is an integer number. The canonical transformation (5.53) leads to the condition (5.61) only if the particle oscillation frequency is off resonance. We have therefore the result that all nonresonant perturbation terms can be transformed to higher-order terms in the oscillation amplitudes while the resonant terms lead to phenomena discussed in Sect. 1.2. From (5.68) we derive w_{mo}, obtain the function $w_m(\varphi)$ from (5.67) and finally the function $w(\psi_1, \varphi)$ from (5.63). Since $Q(\psi_1, \varphi) = 0$ we get from (5.60) the Hamiltonian

$$H_1 = \nu_o J_1 + \frac{n}{2} J_1^{n-1} \left[p_n(\varphi) \cos^n \psi_1 \frac{\partial w}{\partial \psi_1} - w \frac{\partial Q}{\partial \psi_1} \right] + \dots. \tag{5.70}$$

Nonresonant terms appear only in order J_1^{n-1}. As long as such terms can be considered small we conclude that the particle dynamics is determined by the linear tune ν_o, a tune shift or tune spread caused by perturbations and resonances. Note that the Hamiltonian (5.70) is not the complete form but addresses only the nonresonant case of particle dynamics while the resonant case of the Hamiltonian has been derived earlier.

We will now continue to evaluate (5.70) and note that the product

$$w(\psi_1, \varphi) \frac{\partial Q(\psi_1, \varphi)}{\partial \psi_1} = 0$$

in this approximation and get

$$T(\psi, \varphi) = \tfrac{n}{2} \, p_n(\varphi) \, \cos^n \psi \, \frac{\partial w}{\partial \psi} \,, \tag{5.71}$$

where we have dropped the index on ψ and set from now on $\psi_1 = \psi$ which is not to be confused with the variable ψ used before the transformation (5.53). Using the Fourier spectrum for the perturbations and summing over all but resonant terms $q \neq q_r$ we get from (5.68)

$$w_{mo}(1 - \mathrm{e}^{\mathrm{i}m\frac{2\pi}{N}\nu_o}) = a_{nm} \sum_{q \neq q_r} \int_0^{2\pi/N} p_{nq} \mathrm{e}^{\mathrm{i}(m\nu_o - qN)\varphi} \, \mathrm{d}\varphi \,,$$

$$= a_{nm} \sum_{q \neq q_r} p_{nq} \frac{\mathrm{e}^{\mathrm{i}m\nu_o \frac{2\pi}{N}} - 1}{\mathrm{i}\,(m\nu_o - qN)} \,, \tag{5.72}$$

or

$$w_{mo} = \mathrm{i}\, a_{nm} \sum_{q \neq q_r} \frac{p_{nq}}{m\nu_o - qN} \,. \tag{5.73}$$

Note that we have excluded in the sums the resonant terms $q = q_r$ where $m_r \nu_o - q_r N = 0$. These resonant terms include also terms $q = 0$ which do not cause resonances of the ordinary type but lead to tune shifts and tune spreads. After insertion into (5.67) and some manipulations we find

$$w_m(\varphi) = \mathrm{i}\, a_{nm} \sum_{q \neq q_r} \frac{p_{nq}}{m\nu_o - qN} - a_{nm} \sum_{q \neq q_r} \int_0^\varphi p_{nq} \, \mathrm{e}^{\mathrm{i}(m\nu_o - qN)\phi} \, \mathrm{d}\phi \,,$$

$$= \mathrm{i}\, a_{nm} \sum_{q \neq q_r} p_{nq} \frac{\mathrm{e}^{\mathrm{i}(m\nu_o - qN)\varphi}}{m\nu_o - qN} \,, \tag{5.74}$$

and with (5.63)

$$w(\psi, \varphi) = \mathrm{i} \sum_{m=-n}^{n} \sum_{q \neq q_r} \frac{a_{nm} p_{nq}}{m\nu_o - qN} \mathrm{e}^{\mathrm{i}m\psi} \, \mathrm{e}^{-\mathrm{i}qN\varphi} \,. \tag{5.75}$$

From (5.71) we get with (5.62) and (5.75)

$$T(\psi, \varphi) = \mathrm{i}\frac{n}{2} \sum_{q \neq q_r} p_{nq} \, \mathrm{e}^{-\mathrm{i}qN\varphi} \sum_{m=-n}^{n} a_{nm} \, \mathrm{e}^{\mathrm{i}m\psi} \, m \, w(\psi, \varphi) \,. \tag{5.76}$$

This function $T(\psi, \varphi)$ is periodic in ψ and φ and we may apply a Fourier expansion like

$$T(\psi, \varphi) = \sum_{r} \sum_{s \neq \frac{r\nu_o}{N}} T_{rs} \, \mathrm{e}^{\mathrm{i}(r\psi - sN\varphi)} \,, \tag{5.77}$$

where the coefficients T_{rs} are determined by

$$T_{rs} = \frac{N}{4\pi^2} \int\limits_0^{2\pi} e^{-ir\psi}\, d\psi \int\limits_0^{2\pi/N} e^{isN\varphi}\, T(\psi,\varphi)\, d\varphi. \tag{5.78}$$

To evaluate (5.78) it is most convenient to perform the integration with respect to the betatron phase ψ before we introduce the expansions with respect to φ. Using (5.62,63,71) we get from (5.78) after some reordering

$$T_{rs} = i\frac{nN}{4\pi} \sum_{m=-n}^{n} m \int_0^{2\pi} \sum_{j=-n}^{n} \frac{a_{nj}}{2\pi}\, e^{i(j+m-r)\psi}\, d\psi$$

$$\times \int_0^{2\pi/N} p_n(\varphi)\, w_m(\varphi)\, e^{i(m\nu_{\mathrm{o}}-sN)\varphi}\, d\varphi.$$

The integral with respect to ψ is zero for all values $j + m - r \neq 0$ and therefore equal to $a_{n,r-m}$

$$T_{rs} = i\frac{nN}{4\pi} \sum_{m=-n}^{m} m\, a_{m,r-m} \int\limits_0^{2\pi/N} p_n(\varphi)\, w_m(\varphi)\, e^{-i(m\nu_{\mathrm{o}}-sN)\varphi}\, d\varphi. \tag{5.79}$$

Expressing the perturbation $p_n(\varphi)$ by its Fourier expansion and replacing $w_m(\varphi)$ by (5.74), (5.79) becomes

$$T_{rs} = -\frac{n}{2} \sum_{m=-n}^{n} m\, a_{m,r-m}\, a_{n,m} \sum_{q \neq q_{\mathrm{r}}} \frac{p_{n,s-q}\, p_{n,q}}{m\nu_{\mathrm{o}} - qN}. \tag{5.80}$$

With this expression we have fully defined the function $T(\psi,\varphi)$ and obtain for the non-resonant Hamiltonian (5.70)

$$H = \nu_{\mathrm{o}} J + J^{n-1} \sum_{r} \sum_{s \neq \frac{r}{N}\nu_{\mathrm{o}}} T_{rs}\, e^{i(r\psi - sN\varphi)}. \tag{5.81}$$

We note in this result a higher-order amplitude dependent tune spread which has a constant contribution T_{oo} as well as oscillatory contributions.

Successive application of appropriate canonical transformations has lead us to derive detailed insight into the dynamics of particle motion in the presence of perturbations. Of course every time we applied a canonical transformation of variables it was in the hope of obtaining a cyclic variable. Except for the first transformation to action angle variables, this was not completely successful. However, we were able to extract from perturbation terms depending on both action and angle variables such elements that do not depend on the angle variable. As a result, we are now able to determine to a high order of approximation shifts in the betatron frequency caused by perturbations as well as the occurrence and nature of resonances.

Careful approximations and simplifications had to be made to keep the mathematical formulation manageable. Specifically we had to restrict the perturbation theory in this section to one order of multipole perturbation and we did not address effects of coupling between horizontal and vertical betatron oscillations.

From a more practical view point one might ask to what extend this higher-order perturbation theory is relevant for the design of particle accelerators. Is the approximation sufficient or is it more detailed than needed? As it turns out so often in physics we find the development of accelerator design to go hand in hand with the theoretical understanding of particle dynamics. Accelerators constructed up to the late sixties were designed with moderate focusing and low chromaticities requiring no or only very weak sextupole magnets. In contrast more modern accelerators require much stronger sextupole fields to correct for the chromaticities and as a consequence, the effects of perturbations, in this case third-order perturbations, become more and more important. The ability to control the effects of such perturbations actually limits the performance of particle accelerators. For example, colliding-beam storage rings the strongly nonlinear fields introduced by the beam-beam effect limit the attainable luminosity while a lower limit on the attainable beam emittance for synchrotron light sources or damping rings is determined by strong sextupole fields.

5.3.1 Tune Shift in Higher Order

In Sect. 1.2 and (1.59) we found the appearance of tune shifts due to even order multipole perturbations only. The third-order sextupole fields therefore would not affect the tunes. This was true within the degree of approximation used at that point. In this section, however, we have derived higher-order tune shifts and should therefore discuss again the effect of sextupolar fields on the tune.

Before we evaluate the sextupole terms, however, we like to determine the contribution of a quadrupole perturbation to the *higher-order tune shift*. In lower order we have derived earlier a coherent tune shift for the whole beam. We use (5.80) and calculate T_{oo} for $n = 2$

$$T_{oo} = -\sum_{q \neq q_r} p_{2,q}\, p_{2,-q} \sum_{m=-2}^{2} \frac{m\, a_{2,m}\, a_{2,-m}}{m\nu_o - qN} . \tag{5.82}$$

With $4\,a_{2,2} = a_{2,-2} = 2\,a_{2,0} = 1$ and $a_{2,1} = a_{2,-1} = 0$ the term in the bracket becomes

$$\frac{-2}{-2\nu_o - qN} + \frac{2}{2\nu_o - qN} = \frac{4qN}{(2\nu_o)^2 - (qN)^2}$$

and (5.82) is simplified to

$$T_{oo} = -\sum_{q \neq q_r} p_{2,q}\, p_{2,-q} \frac{4qN}{(2\nu_o)^2 - (qN)^2}. \tag{5.83}$$

In this summation we note the appearance of the index q in pairs as a positive and a negative value. Each such pair cancels and therefore

$$T_{oo,2} = 0, \tag{5.84}$$

where the index$_2$ indicates that this coefficient was evaluated for a second-order quadrupole field. This result is not surprising since all quadrupole fields contribute directly to the tune and formally a quadrupole field perturbation cannot be distinguished from a "real" quadrupole field.

In a similar way we derive the T_{oo} coefficient for a third-order multipole or a sextupolar field. From (5.80) we get

$$T_{oo,3} = -\frac{3}{2} \sum_{q \neq q_r} p_{3,q}\, p_{3,-q} \sum_{m=-3}^{3} \frac{m\, a_{3,m}\, a_{3,-m}}{m\nu_o - qN}. \tag{5.85}$$

Since $\cos^3 \psi$ is an even function we have $a_{3,m} = a_{3,-m}$, $a_{3,1} = 3/8$ and $a_{3,3} = 1/8$. The second sum in (5.85) becomes now

$$\frac{1}{64}\left(\frac{3}{3\nu_o + qN} + \frac{q}{\nu_o + qN} + \frac{q}{\nu_o - qN} + \frac{3}{3\nu_o - qN} \right)$$
$$= \frac{1}{64}\left[\frac{18\nu_o}{\nu_o^2 - (qN)^2} + \frac{18\nu_o}{(3\nu_o)^2 - (qN)^2} \right],$$

and after separating out the terms for $q = 0$ (5.85) becomes

$$T_{oo,3} = -\frac{15}{32\nu_o}\, p_{3,0}^2$$
$$-\frac{27\nu_o}{64} \sum_{q \neq q_r} p_{3,q}\, p_{3,-q} \left[\frac{1}{\nu_o^2 - (qN)^2} + \frac{1}{(3\nu_o)^2 - (qN)^2} \right]. \tag{5.86}$$

This expression in general is nonzero and we found, therefore, that sextupole fields indeed, contribute to a tune shift although in a high order of approximation. This tune shift can actually become very significant for strong sextupoles and for tunes close to an integer or third integer resonances. Although we have excluded resonances, $q = q_r$, terms close to resonances become important. Obviously, the tunes should be chosen such as to minimize both terms in the bracket of (5.86). This can be achieved with $\nu_o = qN + \frac{1}{2}N$ and $3\nu_o = rN + \frac{1}{2}N$ where q and r are integers. Eliminating ν_o from both equations we get the condition $3q - r + 1 = 0$ or $r = 3q + 1$. With this we finally get from the two tune conditions the relation $2\nu_o = (2q + 1)N$ or

$$\nu_{opt} = \frac{2q+1}{2}\, N. \tag{5.87}$$

Of course, an additional way to minimize the tune shift is to arrange the sextupole distribution in such a way as to reduce strong harmonics in (5.86). In summary we find for the non-resonant Hamiltonian in the presence of sextupole fields.

$$H_3 = \nu_0 J + T_{oo,3} J^2 + \text{higher order terms} \tag{5.88}$$

and the betatron oscillation frequency or tune is given by

$$\nu = \nu_0 + 2 T_{oo,3} J. \tag{5.89}$$

In this higher-order approximation of beam dynamics we find that sextupole fields cause an *amplitude dependent tune shift* in contrast to our earlier first-order conclusion

$$\frac{\Delta\nu}{\nu_0} = \frac{\nu - \nu_0}{\nu_0} = T_{oo,3}\left(\gamma u^2 + 2\alpha u u' + \beta u'^2\right) = T_{oo,3}\,\epsilon, \tag{5.90}$$

where we have used (1.49) with ϵ the emittance of a single particle oscillating with a maximum amplitude $a^2 = \beta\epsilon$. We have shown through higher-order perturbation theory that odd order nonlinear fields like sextupole fields, can produce amplitude dependent tune shifts which in the case of sextupole fields are proportional to the square of the betatron oscillation amplitude and therefore similar to the tune shift caused by octupole fields. In a beam where particles have different betatron oscillation amplitudes this tune shift leads to a tune spread for the whole beam.

In practical accelerator designs requiring strong sextupoles for chromaticity correction it is mostly this tune shift which moves large amplitude particles onto a resonance thus limiting the dynamic aperture. Since this tune shift is driven by the integer and third-order resonance, it is imperative in such cases to arrange the sextupoles such as to minimize this driving term for geometric aberration.

Problems

Problem 5.1.   Derive the expression for the second-order matrix element T_{166} and give a physical interpretation for each term. Perform the integration in (5.30) and show that $T_{111} = -(\frac{1}{2}m + 2\kappa_x k + \kappa_x^3)\frac{1}{3}\left[S(\dot{s}) + \frac{D(s)}{\kappa_x}\right]$.

Problem 5.2.   Expand the second-order transformation matrix to include path length terms relevant for the design of an isochronous beam transport system and derive expressions for the matrix elements. Which elements must be adjusted and how would you do this? Which parameters would you observe to control your adjustment?

Problem 5.3.   Consider a 180° spectrometer [Ref. 5.1, Fig. 5.32] for particles with a central momentum of 300 MeV and use a maximum dipole field of 10 kGauss. Calculate the spread of the image points due to second-order perturbations. Plot the distribution of image points for $\delta = 0$ and $\delta = 1\%$ as a function of the trajectory angle x'_o. What is the maximum allowable range of angles x'_o to achieve an energy resolution of 0.1%?

Problem 5.4.   Consider a large circular accelerator made of many FODO cells with a phase advance of 90° per cell. Locate chromaticity correcting sextupoles in the center of each quadrupole and calculate the magnitude for two of the five expressions (5.44 – 48). Now place sextupole pairs 180° apart and calculate the same two expressions for the new sextupole distribution.

Problem 5.5.   Sextupoles are used to compensate for chromatic aberrations at the expense of geometric aberrations. Derive a condition for which the geometric aberration has become as large as the original chromatic aberration. What is the average perturbation of geometric aberrations on the betatron motion? Try to formulate a "rule of thumb" stability criteria for the maximum sextupole strength. Is it better to place a chromaticity correcting sextupole at a high beta location (weak sextupole) or at a low beta location (weak aberration)?

Problem 5.6.   Consider both sextupole distributions of problem 5.4 and form a phasor diagram of one of expressions (5.44 – 48) for the first four or more FODO cells. Discuss desirable features of the phasor diagram and explain why the $-\mathcal{I}$ correction scheme works well. A phasor diagram is constructed by adding vectorially each term of an expression (5.44 – 48) going along a beam line.

Problem 5.7.   The higher-order chromaticity of a lattice may include a strong quadratic term. What dependence on energy would one expect in this case for the beta beat? Why? Can your finding be generalized to higher-order terms?

Problem 5.8.   Use the lattice of problem 5.4 and determine the tunes of the ring. Are the tunes the best choices for the superperiodicity of the ring to avoid first, second and third-order resonances driving aberrations? How would you go about improving the situation?

Problem 5.9.   Consider the lattice of problem 5.4 and locate chromaticity correcting sextupoles in the center of quadrupoles such that they conform with the $-\mathcal{I}$-transformation scheme. Derive an expression for the tune shift with amplitude as a function of a phase error between sextupole pairs. How precise, for example, must the phase advance be controlled in this lattice in order for the tune shift with amplitude parameter $(\Delta\nu_x/\nu_x)/\epsilon_x$ not to exceed a value of 1000 m^{-1}?

Problem 5.10.   Consider the lattice of problem 5.4 and place one octupole, for convenience, in the center of a quadrupole to generate a tune spread of $\Delta\nu = 0.02$ for Landau damping in a beam with an emittance of $\epsilon = 0.1$ mm-mrad. Determine the fourth order stability diagram and derive the stability limit. Will a 10σ beam survive? Does your result change with the number of octupoles employed and how?

6. Charged Particle Acceleration

Particle acceleration by rf fields has been discussed, for example, in considerable detail in [6.1,2] and [6.3] where relationships between longitudinal phase oscillation and beam stability are derived and discussed. The accelerating fields were assumed to be available in resonant cavities, but we ignored the conditions that must be met to generate such fields and ensure positive energy transfer to the particle beam. In this chapter we will discuss relevant characteristics of rf cavities and study the interaction of the rf generator with accelerating cavity and beam. Conditions will be derived which must be met for beam stability followed by discussions on synchrotron oscillation as well as energy acceptance in higher order. Finally, we combine transverse focusing and acceleration by formulating beam dynamics in a FODO channel along a linear accelerator while particles are accelerated.

It is not the intention here to develop a detailed derivation of rf theory but we will restrict ourselves to such aspects which are of importance for particle accelerator physics. Considerable limits on the application of accelerator principles are imposed by technical limitations in various accelerator systems as, for example, the rf system and it is therefore useful for the accelerator designer to have a basic knowledge of such limits.

6.1 Accelerating Fields in Resonant rf Cavities

Commonly high frequency rf fields are used to accelerate charged particles and the interaction of such electromagnetic waves with charged particles has been discussed in [6.3] together with the derivation of *synchronization conditions* to obtain continuous particle acceleration. In doing so plane rf waves have been used ignoring the fact that such fields do not have electrical field components in the direction of particle and wave propagation. Although this assumption has not made the results obtained so far obsolete, a satisfactory description of the *wave-particle interaction* must include the establishment of appropriate field configurations.

Electromagnetic waves useful for particle acceleration must exhibit field components in the direction of particle propagation which in our coordinate system is the z-direction. The synchronization condition can be achieved in two ways. An electromagnetic wave travels along the direction of the desired particle acceleration in such a way that the phase velocity is equal to the velocity of the particle. In this case a particle starting, say, at the crest of the wave where the field strength is largest, would be continuously accelerated at the maximum rate as it moves along with the wave. Another way of particle acceleration occurs from electromagnetic fields in rf cavities placed at particular locations along the particle path. In this case the phase velocity of the wave is irrelevant. For positive *particle acceleration* the phase of the electromagnetic field must be adjusted such that the integrated acceleration is positive, while the particle passes through the cavity. Obviously, if the velocity of the particle or the length of the cavity is such that it takes several oscillation periods for a particle to traverse the cavity no efficient acceleration is possible unless the electromagnetic wave travels along with the particle beam.

6.1.1 Wave Equation

To generate electromagnetic field components in the direction of wave propagation we cannot use free plane waves, but must apply specific boundary conditions by properly placing conducting metallic surfaces to modify the electromagnetic wave into the desired form. The theory of electromagnetic waves, waveguides and modes is well established and we repeat here only those aspects relevant to particle acceleration. For more detailed reading consult, for example, references [6.4 – 8]. *Maxwell's equations* for our application in a charge free environment are

$$\nabla(\epsilon\,\mathbf{E}) = 0\,, \qquad \nabla\times\mathbf{E} = -[\mu_o c]\frac{\mu}{c}\frac{d\mathbf{H}}{dt}\,,$$
$$\nabla(\mu\,\mathbf{H}) = 0\,, \qquad \nabla\times\mathbf{H} = [\epsilon_o c]\frac{\epsilon}{c}\frac{d\mathbf{E}}{dt}\,, \tag{6.1}$$

and we look for solutions in the form of rf fields oscillating with frequency ω. A uniform medium is assumed which need not be a vacuum but may have a *dielectric constant* ϵ and a *magnetic permeability* μ. Maxwell's curl equations become then

$$\nabla\times\mathbf{E} = -\mathrm{i}\,[\mu_o c]\,\omega\,\frac{\mu}{c}\,\mathbf{H}\,,$$
$$\nabla\times\mathbf{H} = \mathrm{i}\,[\epsilon_o c]\,\omega\,\frac{\epsilon}{c}\,\mathbf{E}\,. \tag{6.2}$$

Eliminating the magnetic field strength from both equations and using the vector relation $\nabla\times(\nabla\times\mathbf{a}) = \nabla(\nabla\mathbf{a}) - \nabla^2\mathbf{a}$, we get the well-known *wave equations*

$$\nabla^2\mathbf{E} + \mu\epsilon\frac{\omega^2}{c^2}\,\mathbf{E} = 0\,,$$
$$\nabla^2\mathbf{H} + \mu\epsilon\frac{\omega^2}{c^2}\,\mathbf{H} = 0\,. \tag{6.3}$$

In the case of a plane wave propagating along the z-axis the transverse partial derivatives vanish

$$\frac{\partial}{\partial x} = \frac{\partial}{\partial y} = 0, \tag{6.4}$$

since field parameters of a plane wave do not vary transverse to the direction of propagation. The differential equation (6.3) for say the vertical electrical field component then becomes

$$\left(\frac{\partial^2}{\partial z^2} + \mu\epsilon \frac{\omega^2}{c^2} \right) E_y = 0 \tag{6.5}$$

with the solution

$$E_y = E_{oy}\, e^{i(\omega t - kz)}, \tag{6.6}$$

where for nontrivial solutions $E_{oy} \neq 0$ and the wave number

$$k = \sqrt{\mu\epsilon}\,\frac{\omega}{c}. \tag{6.7}$$

For real values of the wave number k the solutions of (6.3) describe waves propagating with the *phase velocity*

$$v_{\mathrm{ph}} = \frac{z}{t} = \frac{c}{\sqrt{\mu\epsilon}} \leq c. \tag{6.8}$$

An imaginary component of k, on the other hand, would lead to an exponential damping term for the fields, a situation that occurs, for example, in a conducting surface layer where the fields decay exponentially over a distance of the skin depth. In the absence of conducting boundary conditions the wave number is real and describes propagating waves of arbitrary frequencies. As has been noted before, however, such plane waves lack electrical field components in the direction of propagation because Maxwell's curl equation for the magnetic field strength vanishes $(\nabla \times \mathbf{H})_z = 0$ due to (6.4). Consequently, the E_z-component of the electrical field vanishes as well and there is no acceleration of charged particles in the direction of motion. In the following subsection we will therefore introduce appropriate conditions to obtain from (6.3) waves with longitudinal field components.

6.1.2 Waveguide Modes

Significant modification of wave patterns can be obtained from the proximity of metallic boundaries. To demonstrate this we evaluate the electromagnetic field of a wave propagating along the axis of a rectangular metallic pipe or *rectangular waveguide* as shown in Fig. 6.1. Since we are interested in getting a finite value for the z-component of the electrical field we try the ansatz

$$E_z = \psi_x(x) \cdot \psi_y(y) \cdot \psi_z(z) \tag{6.9}$$

and look for boundary conditions that are required to obtain nonvanishing longitudinal fields. Insertion into the first of (6.3) gives

$$\frac{\psi_x''(x)}{\psi_x(x)} + \frac{\psi_y''(y)}{\psi_y(y)} + \frac{\psi_z''(z)}{\psi_z(z)} = -\mu\epsilon\frac{\omega^2}{c^2} \tag{6.10}$$

where the r.h.s. is a constant while the functions $\psi_u(u)$ are functions of the variable $u = x, y,$ or z alone. In order that this equation be true for all values of the coordinates, the ratios $\frac{\psi_u''(u)}{\psi_u(u)}$ must be constant and we may write (6.10) in the form

$$k_x^2 + k_y^2 + k_z^2 = -\mu\epsilon\frac{\omega^2}{c^2} = -k^2, \tag{6.11}$$

where the parameters k_u are constants.

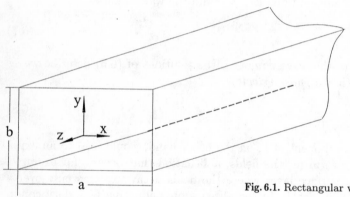

Fig. 6.1. Rectangular waveguide

Differentiating (6.9) twice with respect to z results in the differential equation for the z-component of the electrical field

$$\frac{d^2 E_z}{dz^2} = k_z^2 E_z, \tag{6.12}$$

which can be solved readily. The wavenumber k must be real for propagating waves and with

$$k_c^2 = -k_x^2 - k_y^2, \tag{6.13}$$

we get

$$k_z = i\sqrt{k^2 - k_c^2}. \tag{6.14}$$

The solution (6.9) of the wave equation for the z-component of the electrical field is then finally

$$E_z = E_{oz}\,\psi_x(x)\,\psi_y(y)\,e^{i(\omega t - k_z z)}, \tag{6.15}$$

where

$$-k_z^2 = k^2 - k_c^2 = \mu\epsilon\frac{\omega^2}{c^2} - k_c^2. \tag{6.16}$$

The nature of the parameters in this equation will determine if the wave fields are useful for acceleration of charged particles. First we note that the *phase velocity* is given by

$$v_{\text{ph}} = \frac{\omega}{k_z} = \frac{\omega}{\sqrt{k^2 - k_c^2}}. \tag{6.17}$$

An electromagnetic wave in a rectangular metallic pipe is propagating only if the phase velocity is real or $k > k_c$ and the quantity k_c is therefore called the *cutoff wave number*. For frequencies with a wave number less than the cutoff value the phase velocity becomes imaginary and the wave decays exponentially like $\exp\left(-\sqrt{|k^2 - k_c^2|}z\right)$.

Conducting boundaries modify electromagnetic waves in such a way that finite longitudinal electric field components can be produced, which at least in principle, can be used for particle acceleration. Although we have found solutions seemingly suitable for particle acceleration, we cannot use such an electromagnetic wave propagating, for example, in a smooth rectangular pipe to accelerate particles. Inserting (6.11) into (6.17), the phase velocity of a traveling waveguide mode in a rectangular pipe becomes

$$\boxed{v_{\text{ph}} = \frac{c}{\sqrt{\mu\epsilon}\sqrt{1 - (k_c/k)^2}}} \tag{6.18}$$

and is with $k > k_c$ and $\mu \approx \epsilon \approx 1$ in vacuum or air larger than the velocity of light. As a consequence, there can be no net acceleration since the wave rolls over the particles, which cannot move faster than the speed of light. This problem occurs in a smooth pipe of any cross section. We must therefore seek for modifications of a smooth pipe in such a way that the phase velocity is reduced or to allow a standing wave pattern, in which case the phase velocity does not matter anymore. The former situation occurs for *traveling wave* linac structures, while the latter is specially suited for single *accelerating cavities*, which we will discuss here in more detail.

For a standing wave pattern $k_z = 0$ or $k = k_c$ and we get with (6.11,13) *the cutoff frequency*

$$\omega_c = \frac{ck_c}{\sqrt{\mu\epsilon}}. \tag{6.19}$$

A *waveguide wavelength* can be defined by

$$\lambda_g = \frac{2\pi}{\sqrt{k^2 - k_c^2}} \tag{6.20}$$

which is always longer than the *free space wavelength* $\lambda = 2\pi/k$ and we get from (6.11) with (6.13)

$$\frac{1}{\lambda^2} = \frac{1}{\lambda_g^2} + \frac{1}{\lambda_c^2},$$ (6.21)

where $\lambda_c = 2\pi/k_c$.

To complete the solution (6.15) for transverse dimensions we apply boundary conditions to the amplitude functions ψ_x and ψ_y. The rectangular waveguide be centered along the z-axis with a width a in the x-direction and a height b in the y-direction (Fig. 6.1). Since the tangential component of the electrical field must vanish at conducting surfaces, the boundary conditions are

$$\begin{aligned} \psi_x(x) &= 0 \quad \text{for} \quad x = \pm\tfrac{1}{2}\,a, \\ \psi_y(y) &= 0 \quad \text{for} \quad y = \pm\tfrac{1}{2}\,b. \end{aligned}$$ (6.22)

The solutions must be cosine functions to meet these boundary conditions and the complete solution (6.15) for the *longitudinal electric field* can be expressed by

$$E_z = E_o \cos\frac{m\pi x}{a} \cos\frac{n\pi y}{b}\, \mathrm{e}^{\mathrm{i}\left(\omega t - k_z z\right)},$$ (6.23)

where $m \geq 1$ and $n \geq 1$ are arbitrary integers. The trigonometric functions are eigenfunctions of the differential equation (6.12) with the boundary conditions (6.22) and the integers m and n are eigenvalues. In a similar way we get an expression for the z-component of the magnetic field strength H_z. The boundary conditions require that the tangential magnetic field component at a conducting surface is the same inside and outside the conductor which is equivalent to the requirement that

$$\left.\frac{\partial H_z}{\partial x}\right|_{x=\pm 1/2a} = 0 \quad \text{and} \quad \left.\frac{\partial H_z}{\partial y}\right|_{y=\pm 1/2b} = 0.$$ (6.24)

These boundary conditions can be met by sine functions and the z-component of the magnetic field strength is therefore in analogy to (6.23) given by

$$H_z = H_o \sin\frac{m\pi x}{a} \sin\frac{n\pi y}{b}\, \mathrm{e}^{\mathrm{i}(\omega t - k_z z)}.$$ (6.25)

The cutoff frequency is the same for both the electrical and magnetic field component and is closely related to the dimension of the wave guide. With the definition (6.13) the cutoff frequency can be determined from

$$\boxed{k_c^2 = -k_x^2 - k_y^2 = \left(\frac{m\pi}{a}\right)^2 + \left(\frac{n\pi}{b}\right)^2.}$$ (6.26)

All information necessary to complete the determination of field components have been collected. Using (6.2,23) the component equations are with $\frac{\partial}{\partial z} = -ik_z$

$$[\mu_o c]\,\mu\frac{\omega}{c}\,H_x = i\frac{\partial E_z}{\partial y} - k_z E_y\,, \qquad [\epsilon_o c]\,\epsilon\frac{\omega}{c}\,E_x = -i\frac{\partial H_z}{\partial y} + k_z H_y\,,$$

$$[\mu_o c]\,\mu\frac{\omega}{c}\,H_y = k_z E_x - i\frac{\partial E_z}{\partial x}\,, \qquad [\epsilon_o c]\,\epsilon\frac{\omega}{c}\,E_y = -k_z H_x + i\frac{\partial H_z}{\partial x}\,, \qquad (6.27)$$

$$[\mu_o c]\,\mu\frac{\omega}{c}\,H_z = i\frac{\partial E_y}{\partial x} - i\frac{\partial E_x}{\partial y}\,, \qquad [\epsilon_o c]\,\epsilon\frac{\omega}{c}\,E_z = -i\frac{\partial H_y}{\partial x} + i\frac{\partial H_x}{\partial y}\,.$$

From the first four equations we may derive expressions for the transverse field components E_x, E_y, H_x, H_y as functions of the known z-components

$$E_x = \frac{i}{k_z^2 - \mu\epsilon\frac{\omega^2}{c^2}}\left(k_z\frac{\partial E_z}{\partial x} + [\mu_o c]\,\mu\frac{\omega}{c}\frac{\partial H_z}{\partial y}\right),$$

$$E_y = \frac{i}{k_z^2 - \mu\epsilon\frac{\omega^2}{c^2}}\left(k_z\frac{\partial E_z}{\partial y} - [\mu_o c]\,\mu\frac{\omega}{c}\frac{\partial H_z}{\partial x}\right),$$

$$H_x = \frac{i}{k_z^2 - \mu\epsilon\frac{\omega^2}{c^2}}\left([\epsilon_o c]\,\epsilon\frac{\omega}{c}\frac{\partial E_z}{\partial y} - k_z\frac{\partial H_z}{\partial x}\right), \qquad (6.28)$$

$$H_y = \frac{i}{k_z^2 - \mu\epsilon\frac{\omega^2}{c^2}}\left([\epsilon_o c]\,\epsilon\frac{\omega}{c}\frac{\partial E_z}{\partial x} + k_z\frac{\partial H_z}{\partial y}\right),$$

where

$$k_z^2 = \mu\epsilon\left(\frac{\omega}{c}\right)^2 - \left(\frac{m\pi}{a}\right)^2 - \left(\frac{n\pi}{b}\right)^2. \qquad (6.29)$$

By application of proper boundary conditions at the conducting surfaces of a rectangular waveguide we have derived expressions for the z-component of the electromagnetic fields and are able to formulate the other field components in terms of the z-component. Two fundamentally different field configurations can be distinguished depending on whether we choose E_z or H_z to vanish. All field configurations, for which $E_z = 0$, form the class of *transverse electrical modes* or short TE-*modes*. Similarly, all fields for which $H_z = 0$, form the class of *transverse magnetic modes* or short TM-*modes*. Each class of modes consists of all modes obtained by varying the integers m and n. The particular choice of these mode integers is commonly included in the mode nomenclature and we speak therefore of TM$_{mn}$ or TE$_{mn}$-modes.

For the remainder of this chapter we will concentrate only on the transverse magnetic or TM-modes, since TE-modes are useless for particle acceleration. The lowest order TM-mode is the TM$_{11}$-mode producing the z-component of the electrical field, which is maximum along the z-axis of the rectangular waveguide and falls off from there like a cosine function to reach zero at the metallic surfaces. Such a mode would be useful for particle acceleration if it were not for the phase velocity being larger than the speed of light. In the next subsection we will see how this mode may be used anyway. The next higher mode, the TM_{21}-mode would have a similar distribution in the vertical plane but exhibits a node along the z-axis. Because the field is zero along the axis, we get acceleration only if the particle beam passes off-axis through the waveguide in which case we may as well build only one half of the waveguide and excite it to a TM_{11}-mode.

Before we continue the discussion on field configurations we note that electromagnetic waves with frequencies above cutoff frequency propagate along the axis of the rectangular waveguide. Electromagnetic energy travels along the waveguide with a velocity known as the *group velocity* defined by

$$v_g = \frac{d\omega}{dk_g}.$$ (6.30)

The frequency of this electromagnetic wave is from (6.21) with (6.11,19,20)

$$\omega = \sqrt{\frac{c^2 k_g^2}{\mu\epsilon} + \omega_c^2}$$ (6.31)

and the group velocity

$$\boxed{v_g = \frac{d\omega}{dk_g} = \frac{c}{\sqrt{\mu\epsilon}}\sqrt{1 - \frac{k_c^2}{k^2}} < c.}$$ (6.32)

In contrast to the phase velocity, the group velocity is always less than the speed of light as it should be.

6.1.3 rf Cavities

These waveguide modes are not yet ready for particle acceleration, because of excessive phase velocities. This obstacle for successful particle acceleration can be solved by considering two waves traveling in opposite directions on the same axis of a waveguide. Both fields have the form (6.23) and the superposition of both fields is

$$E_z = 2 E_o \cos\frac{m\pi x}{a} \cos\frac{n\pi y}{b} \cos\frac{p\pi z}{d} e^{i\omega t},$$ (6.33)

where p is an integer and with (6.20)

$$d = \frac{p\pi}{k_z} = \frac{p\lambda_g}{2}.$$ (6.34)

The superposition of two equal but opposite waves form a *standing wave* with nodes half a waveguide wavelength apart. Closing off the waveguide at such node points with a metallic surface fulfills automatically all boundary conditions. The resulting rectangular box forms a resonant cavity enclosing a standing electromagnetic wave which in principle, can be used for particle acceleration since the phase velocity has become irrelevant. In analogy to the waveguide mode nomenclature, we extend the nomenclature to cavities by adding a third index for the eigenvalue p. The lowest cavity mode is then the TM_{110}-mode. The indices m and n cannot be zero because of the boundary conditions for E_z. For $p = 0$ we find E_z to be constant along the whole length of the cavity varying only with x and y. The boundary conditions are met automatically at the end caps since with $p = 0$ also $k = 0$ and the transverse field components vanish everywhere. The electrical field configuration for the TM_{110}-mode consists therefore of a finite E_z-component being constant only along z and falling off transversely from a maximum value along the z-axis in the center of the cavity to zero at the walls.

In practical applications boxes with rectangular boundaries are rarely used as accelerating cavities. Round cylindrical boxes similar to *pill boxes* are the more common form of cavities since it is easier to obtain the mechanical precision required for high performance cavities (Fig. 6.2). The derivation of the field configuration is similar to that for rectangular waveguides although now the wave equation (6.3) is expressed in cylindrical coordinates (r, φ, z) and we get for the z-component of the electrical field

$$\frac{\partial^2 E_z}{\partial r^2} + \frac{1}{r}\frac{\partial E_z}{\partial r} + \frac{1}{r^2}\frac{\partial^2 E_z}{\partial \varphi^2} + \frac{\partial^2 E_z}{\partial z^2} + \mu\epsilon\frac{\omega^2}{c^2}E_z = 0. \tag{6.35}$$

In a stationary configuration the field is expected to be periodic in φ while the z-dependence is the same as for rectangular waveguides. Using the derivatives $\frac{\partial}{\partial \varphi} = im$, where m is again an eigenvalue, and $\frac{\partial}{\partial z} = -ik_z$, we get from (6.35)

$$\frac{\partial^2 E_z}{\partial r^2} + \frac{1}{r}\frac{\partial E_z}{\partial r} + \left(k_c^2 - \frac{m^2}{r^2}\right)E_z = 0 \tag{6.36}$$

and $k_c^2 = k_z^2 - \mu\epsilon\frac{\omega^2}{c^2}$ consistent with its previous definition. This differential equation can be solved with Bessel functions in the form [6.9]

$$E_z = E_o\,J_m(k_c r)\,e^{i(\omega t - m\varphi - k_z z)}, \tag{6.37}$$

which must meet the boundary condition $E_z = 0$ for $r = a$ where a is the radius of the cylindrical waveguide. The location of the cylindrical boundaries are determined by the roots of Bessel functions of order m. For the lowest order $m = 0$ the first root is at

$$k_c a_1 = 2.405 \qquad \text{or at a radius} \qquad a_1 = \frac{2.405}{k_c}. \tag{6.38}$$

By adding endcaps at $z = \pm\, 1/2\, d$ standing waves are established again and with $i\,k_z = \frac{p\pi}{d}$ and (6.16)

$$k_c^2 = \mu\epsilon\frac{\omega^2}{c^2} - \frac{p^2\pi^2}{d^2}. \tag{6.39}$$

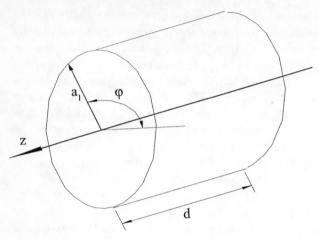

Fig. 6.2. Cylindrical resonant cavity (pill box cavity)

Solving for the *resonance frequency* ω of the lowest order or $\mathrm{TM_{olo}}$-mode, we get with (6.38), $m = 0$ and $p = 0$

$$\omega_{\mathrm{olo}} = \frac{c}{\sqrt{\mu\epsilon}} \frac{2.405}{a_1} \tag{6.40}$$

and the z-component of the electrical field is

$$E_z = E_{z,\mathrm{olo}} \, \mathrm{J_o}\left(2.405 \frac{r}{a_1} \right) e^{i\omega_{\mathrm{olo}}t} . \tag{6.41}$$

The resonance frequency is inversely proportional to the diameter of the cavity and to keep the size of accelerating cavities manageable, short wave radio frequencies are chosen. For electron linear accelerators a wavelength of $\lambda = 10$ cm is often used corresponding to a frequency of 2997.93 MHz and a cavity radius of $a_1 = 3.38$ cm. For storage rings a common frequency is 499.65 MHz or $\lambda = 60$ cm and the radius of the resonance cavity is $a_1 = 22.97$ cm. The size of the cavities is in both cases quite reasonable. For much lower rf frequencies the size of a resonant cavity becomes large. Where such low frequencies are desired the diameter of a cavity can be reduced at the expense of efficiency by loading it with magnetic material like ferrite with a permeability $\mu > 1$ as indicated by (6.40). This technique also allows the change of the resonant frequency during acceleration to synchronize with low energy protons, for example, which have not yet reached relativistic energies. To keep the rf frequency synchronized with the revolution frequency, the permeability of the magnetic material in the cavity can be changed by an external electrical current.

The nomenclature for different modes is similar to that for rectangular waveguides and cavities. The eigenvalues are equal to the number of field maxima in φ, r and z and are indicated as indices in this order. The TM_{010}-*mode* therefore exhibits only a radial variation of field strength independent of φ and z. Similar to rectangular waveguides we can derive the other field components from (6.3) and get

$$E_r = \frac{i}{k_z^2 - \mu\,\epsilon\,\frac{\omega^2}{c^2}}\left(k_z\frac{\partial E_z}{\partial r} + i\,[\mu_o c]\,\mu\frac{\omega}{c}\frac{m}{r}H_z\right),$$

$$E_\varphi = \frac{i}{k_z^2 - \mu\,\epsilon\,\frac{\omega^2}{c^2}}\left(i\,k_z\frac{m}{r}E_z - [\mu_o]\,\mu\frac{\omega}{c}\frac{\partial H_z}{\partial r}\right),$$

$$H_r = \frac{-i}{k_z^2 - \mu\,\epsilon\,\frac{\omega^2}{c^2}}\left(i\,[\epsilon_o]\,\epsilon\frac{\omega}{c}\frac{m}{r}E_z - k_z\frac{\partial H_z}{\partial r}\right),$$

$$H_\varphi = \frac{i}{k_z^2 - \mu\,\epsilon\,\frac{\omega^2}{c^2}}\left([\epsilon_o]\,\epsilon\frac{\omega}{c}\frac{\partial E_z}{\partial r} + i\,k_z\frac{m}{r}H_z\right).$$

(6.42)

Again we distinguish TM-modes and TE-modes but continue to consider only TM-modes for particle acceleration. The z-component for the magnetic field strength H_z can be derived similar to (6.25) if needed. Electrical fields in such a cavity have all the necessary properties for particle acceleration. Small openings along the z-axis allow the beam to pass through the cavity and gain energy from the accelerating field. Cylindrical cavities can be excited in many different modes with different frequencies. For particle acceleration the dimensions of the cavity are chosen such that at least one resonant frequency satisfies the synchronicity condition of the beam line or circular accelerator. In general this is the frequency of the TM_{010}-mode which is also called the *fundamental cavity frequency*.

From the expressions (6.42) we conclude that the lowest order TM-mode does not include transverse electrical field components since $k_z = 0$. In addition the radial component of the magnetic field vanishes since $m = 0$ and the azimuthal component becomes with (6.41)

$$H_\varphi = -i\,[\epsilon_o c]\sqrt{\frac{\epsilon}{\mu}}\,E_{z,010}\,J_1\left(2.405\,\frac{r}{a_1}\right)e^{i\omega_{010}t}.$$

(6.43)

The kinetic energy gained in such a cavity can be obtained by integrating the time dependent field along the particle path. The cavity center be located at $z = 0$ and a particle enters the cavity at $z = -d/2$ at which moment the rf field may have the phase δ. The electric field along the z-axis has the form $E_z = E_{zo}\sin(\omega t + \delta)$ and we get for the *kinetic energy gain* of a particle passing through the cavity with velocity v

$$\Delta E_{\mathrm{kin}} = e\int_{-d/2}^{d/2}E_z\,\mathrm{d}z = eE_{zo}\int_{-d/2}^{d/2}\sin\left(\omega\frac{z}{v} + \delta\right)\mathrm{d}z.$$

(6.44)

In general, the change in the particle velocity is small during passage of one rf cavity and the integral is a maximum for $\delta = \pi/2$ when the field reaches a maximum at the moment the particle is half way through the cavity. Defining an accelerating cavity voltage $\widehat{V}_{rf} = E_{zo}\,d = E_{o1o}\,d$ the kinetic energy gain is

$$\Delta E_{\text{kin}} = e\,\widehat{V}_{rf}\,\frac{\sin\frac{\omega d}{2v}}{\frac{\omega d}{2v}} = e\widehat{V}_{\text{cy}}\,, \qquad (6.45)$$

where the *transit-time factor*

$$T = \frac{\sin\frac{\omega d}{2v}}{\frac{\omega d}{2v}} \qquad (6.46)$$

and the maximum *effective cavity voltage* $\widehat{V}_{\text{cy}}$ seen by particles. The transit-time factor provides the correction on the particle acceleration due to the time variation of the field while the particles traverse the cavity. In a resonant pill box cavity, as shown in Fig. 6.3a, we have $d = \lambda/2$ and the transit-time factor for a particle traveling approximately at the speed of light is

$$T_{\text{pillbox}} = \frac{2}{\pi} < 1\,. \qquad (6.47)$$

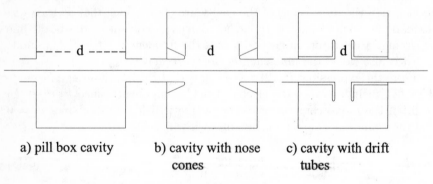

a) pill box cavity   b) cavity with nose   c) cavity with drift
                             cones   tubes

Fig. 6.3. Resonant cavities with drift tubes (schematic)

As the cavity length or the active accelerating gap in the cavity is reduced, the transient time factor can be slightly increased. The simple pill box cavity may be modified by adding nose cones (Fig. 6.3b) or by adding drift tubes at the entrance and exit of the cavity as shown in Fig. 6.3c. In this case the parameter d in (6.46) is the active accelerating gap.

6.1.4 Cavity Losses and Shunt Impedance

Radio frequency fields can be enclosed within conducting surfaces only because electrical surface currents are induced by these fields which provide
the shielding effect. For a perfect conductor with infinite surface conductivity these currents would be lossless and the excitation of such a cavity would
persist indefinitely. This situation is achieved to a considerable degree, albeit not perfect, in superconducting cavities. In warm cavities constructed
of copper or aluminum the finite resistance of the material causes surface
currents to produce heating losses leading to a depletion of field energy. To
sustain a steady field in the cavity, radio frequency power must be supplied
continuously. The surface currents in the conducting cavity boundaries can
be derived from Maxwell's curl equation or *Ampere's law* (6.2) as it is also
called. In cylindrical coordinates this vector equation becomes for the lowest
order TM-mode in component form

$$-\frac{\partial H_\varphi}{\partial z} = \left[\frac{c}{4\pi}\right] \frac{4\pi}{c} j_r ,$$

$$0 = j_\varphi , \tag{6.48}$$

$$\frac{H_\varphi}{r} + \frac{\partial H_\varphi}{\partial r} = \left[\frac{c}{4\pi}\right] \frac{4\pi}{c} j_z + \mathrm{i}[\epsilon_0 c] \,\omega\, \frac{\epsilon}{c}\, E_z .$$

Because we do not consider perfectly but only well conducting boundaries,
we expect fields and surface currents to penetrate somewhat into the conducting material. The depth of penetration of fields and surface currents
into the conductor is well-known as the *skin depth* [6.4]

$$\delta_{\mathrm{s}} = [\sqrt{4\pi\epsilon_0}] \frac{c}{\sqrt{4\pi}} \sqrt{\frac{2}{\mu_{\mathrm{w}}\omega\sigma_{\mathrm{w}}}} , \tag{6.49}$$

where σ_{w} is the conductivity of the cavity wall and μ_{w} the permeability of
the wall material. The azimuthal magnetic field component induces surface
currents in the cylindrical walls as well as in the end caps. In both cases
the magnetic field decays within a skin depth from the surface inside the
conductor. The first equation (6.48) applies to the end caps and the integral
through the skin depth is

$$\int_S^{S+\delta_{\mathrm{s}}} \frac{\partial H_\varphi(r)}{\partial z}\, \mathrm{d}z \approx H_\varphi(r)\,|_S^{S+\delta_{\mathrm{s}}} \approx -H_\varphi(r,S) , \tag{6.50}$$

since $H_\varphi(r, S + \delta_{\mathrm{s}}) \approx 0$ just under the surface S of the wall. We integrate
also the third equation (6.48) at the cylindrical walls and get for the first
term $\int H_\varphi/r\, \mathrm{d}r \approx H_\varphi\delta_{\mathrm{s}}/a_1$, which is very small, while the second term has
a form similar to (6.50). The electrical term E_z vanishes because of the
boundary condition and the surface current densities are therefore related
to the magnetic fields by

$$j_z \, \delta_s \; = \; \left[\frac{4\pi}{c} \right] \frac{c}{4\pi} \, H_\varphi(a_1, z) \,,$$

$$j_r \, \delta_s \; = \; \left[\frac{4\pi}{c} \right] \frac{c}{4\pi} \, H_\varphi(r, \pm \tfrac{1}{2} \, d) \,. \tag{6.51}$$

The *cavity losses* per unit wall surface area are given by

$$\frac{\mathrm{d}P_{\mathrm{cy}}}{\mathrm{d}S} \; = \; R_s \, j_s^2 \,, \tag{6.52}$$

where j_s is the surface current and R_s is the *surface resistance* given by

$$R_s \; = \; \frac{[4\pi\epsilon_0]}{\sigma_{\mathrm{w}} \delta_s} \,. \tag{6.53}$$

With $j_s = j_{r,z} \delta_s$, (6.43,49) and after integration of (6.52) over all inside surfaces of the cavity the energy loss rate becomes

$$P_{\mathrm{cy}} \; = \; [4\pi\epsilon_0] \, \frac{\omega \, \delta_s \, \epsilon}{16\pi} \, \frac{\mu_{\mathrm{w}}}{\mu} \, E_{\mathrm{o1o}}^2 \int\limits_{S} \mathrm{J}_1^2 \left(2.405 \frac{r}{a_1} \right) \mathrm{d}S \,, \tag{6.54}$$

where ϵ and μ is the dielectric constant and permeability of the material inside the cavity, respectively. Evaluating the integral over all surfaces we get for the cylindrical wall the integral value $2\pi \, a_1 d \, J_1^2(2.405)$. For each of the two end caps the integral $\int_0^{a_1} \mathrm{J}_1^2(2.405 \frac{r}{a_1}) \, r \, \mathrm{d}r$ must be evaluated and is from integration tables [6.10]

$$\int\limits_{0}^{a_1} \mathrm{J}_1^2 \left(2.405 \frac{r}{a_1} \right) r \, \mathrm{d}r \; = \; \tfrac{1}{2} \, a_1^2 \, \mathrm{J}_1^2(2.405) \,. \tag{6.55}$$

The total *cavity wall losses* become finally with $V_{\mathrm{cy}} = E_{\mathrm{o1o}} \, d$

$$\boxed{ P_{\mathrm{cy}} \; = \; [4\pi\epsilon_0] \, \frac{\omega \delta_s \epsilon}{8} \, \frac{\mu_{\mathrm{w}}}{\mu} \, V_{\mathrm{cy}} \, J_1^2(2.405) \, \frac{a_1 \, (a_1 + d)}{d^2} \,. } \tag{6.56}$$

It is convenient to separate fixed cavity parameters from adjustable parameters. Once the cavity is constructed the only adjustable parameter is the strength of the electrical field E_{o1o}, which is closely related to the effective cavity voltage. Expressing the cavity losses in terms of an impedance we get from (6.56)

$$P_{\mathrm{cy}} \; = \; \frac{V_{\mathrm{cy}}^2}{2 R_{\mathrm{cy}}} \,, \tag{6.57}$$

where the cavity *shunt impedance* including transient time is defined by [1]

---

[1] The shunt impedance is defined in the literature sometimes by $P_{\mathrm{cy}} = V_{\mathrm{cy}}^2/R_s$ in which case the numerical value of the shunt impedance is larger by a factor of two.

$$R_s = [4\pi\epsilon_o] \frac{4}{\omega\delta_s\epsilon} \frac{\mu}{\mu_w} \frac{d^2}{a_1(a_1+d)} \frac{1}{J_1^2(2.405)} \frac{\sin\frac{\omega d}{2v}}{\frac{\omega d}{2v}} . \qquad (6.58)$$

In accelerator design we prefer sometimes to use the shunt impedance per unit length or the *specific shunt impedance*. The required length depends on the accelerating voltage needed and the rf power available. With

$$r_{cy} = 2R_s/d \qquad (6.59)$$

the *cavity losses* are instead of (6.57)

$$P_{cy} = \frac{V_{cy}^2}{r_s L_{cy}}, \qquad (6.60)$$

where L_{cy} is the total length of all cavities producing the voltage V_{cy}. The specific shunt impedance is proportional to the square root of the rf frequency $r_s \propto \sqrt{\omega}$ favoring high values for the frequency. A practical limit is reached for very high frequencies when the cavity apertures become too small for the particle beam to pass through or when the size of the cavities prevents an efficient cooling of wall losses.

As an example, we calculate from (6.58) the shunt impedance for a pill box cavity designed for a resonance frequency of 358 MHz. The wavelength is $\lambda = 85$ cm, the cavity length $d = 42.5$ cm and the cavity radius $a_1 = 32.535$ cm. Constructing this cavity from aluminum the skin depth would be $\delta_s = 4.44\,\mu$m and the specific shunt impedance becomes $r_s = 15.2$ MΩ/m while the measured value for this cavity as shown in Fig. 6.4 is 18.0 MΩ/m. The difference is due to two competing effects. The open aperture along the axis for the beam has the tendency to reduce the shunt impedance while the nose cones being a part of the actual cavity increase the transient time factor and thereby the effective shunt impedance (6.58). The simple example of a pill box cavity produces rather accurate results, however, for more precise estimates computer programs have been developed to calculate the mode frequencies and shunt impedances for all modes in arbitrary rotationally symmetric cavities (for example, SUPERFISH [6.11] or URMEL [6.12]). More sophisticated three-dimensional programs are becoming available (for example MAFIA [6.12]) to simulate rf properties of arbitrary forms of cavities.

The *specific shunt impedance* for a pill box cavity can be expressed in a simple form as a function of the rf frequency only and is for realistic cavities approximately

$$r_s(\text{M}\Omega/\text{m}) \approx 1.28\sqrt{f_{rf}\,(\text{MHz})} \qquad \text{for copper and}$$
$$r_s(\text{M}\Omega/\text{m}) \approx 1.06\sqrt{f_{rf}\,(\text{MHz})} \qquad \text{for aluminum}. \qquad (6.61)$$

The shunt impedance should be maximum in order to minimize cavity losses for a given acceleration. Since the interior of the cavity must be

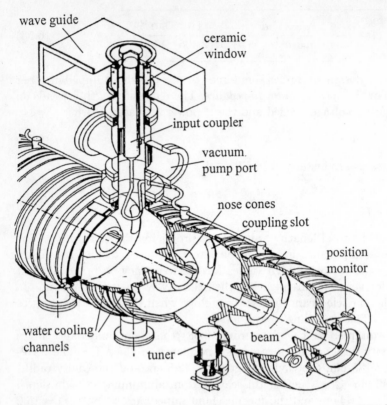

Fig. 6.4. Resonant accelerating cavity at 358 MHz for the storage ring PEP [6.13]

evacuated $\mu = \epsilon = 1$ and $\mu_{\mathrm{w}} = 1$ because we do not consider magnetic materials to construct the cavity. The only adjustable design parameters left are the skin depth and the transient time factor. The skin depth can be minimized by using well conducting materials like copper or aluminum. A typical design for such a cavity consisting of five cells [6.13] is shown in Fig. 6.4.

To derive the *quality factor* of the cavity the energy W stored in the electromagnetic field within the cavity must be calculated. The field energy is the volume integral of the square of the electrical or magnetic field and we have in case of a $\mathrm{TM_{olo}}$-mode with $W = [4\pi\epsilon_{\mathrm{o}}] \frac{\epsilon}{8\pi} \int_V E_z^2 \, \mathrm{d}V$ and (6.41) for the *stored cavity energy*

$$ W = [4\pi\epsilon_{\mathrm{o}}] \frac{\epsilon}{8} E_{\mathrm{olo}}^2 \, d \, a_1^2 \, J_1^2(2.405) \,. \tag{6.62} $$

The *quality factor* Q of a resonator is defined as the ratio of the stored energy to the energy loss per radian

$$ \boxed{Q = 2\pi \frac{\text{stored energy}}{\text{energy loss/cycle}} = \omega \frac{W}{P_{\mathrm{cy}}} \,,} \tag{6.63} $$

or with (6.56,62)

$$Q = \frac{d_s}{\delta} \frac{\mu_w}{\mu} \frac{a_1}{a_1 + d}. \tag{6.64}$$

The quality factor determines the *cavity time constant* since the fields decay exponentially like $e^{-t/\tau_{cy}}$ due to wall losses, where τ_{cy} is the *cavity time constant* and the decay rate of the stored energy in the cavity is

$$\frac{dW}{dt} = -\frac{2}{\tau_{cy}} W. \tag{6.65}$$

The change in the stored energy is equal to the cavity losses P_{cy} and the cavity time constant is from (6.63)

$$\tau_{cy} = \frac{2W}{P_{cy}} = \frac{2Q}{\omega}, \tag{6.66}$$

which is also called the *cavity filling time* because it describes the build up time of fields in a cavity following a step function application of rf power.

6.1.5 Determination of rf Parameters

A variety of rf parameters has to be chosen for a circular accelerator. Some parameters relate directly to beam stability criteria and are therefore easy to determine. Other parameters have less of an impact on beam stability and are often determined by nonphysical criteria like availability and economics. Before rf parameters can be determined a few accelerator and lattice parameters must be known. Specifically, we need to know the desired minimum and maximum beam energy, the beam current, the circumference of the ring, the momentum compaction factor, and the bending radius of the magnets. Further, we make a choice of the maximum desired rate of particle acceleration per turn or determine the energy loss per turn to synchrotron radiation which needs to be compensated. During the following discussion we assume that these parameters are known.

rf frequency: One of the most prominent parameters for rf accelerating systems is the *rf frequency* of the electromagnetic fields. For highly relativistic beams there is no fundamental reason for a particular choice of the rf frequency and can therefore be selected on technical and economic grounds. The *rf frequency* must, however, be an integer multiple, the *harmonic number*, of the particle revolution frequency. The harmonic number can be any integer from a beam stability point of view. In specific cases, the harmonic number need to be a multiple of a smaller number. Considering, for example, a colliding beam facility with N_{IP} collision points an optimum harmonic number is divisible by $N_{IP}/2$. In this case $N_{IP}/2$ bunches could be filled in each of the two counter rotating beams leading to a maximum collision rate. Other such considerations may require the harmonic number to contain additional factors. In general, most flexibility is obtained if the harmonic number is divisible by many small factors.

Within these considerations the harmonic number can be chosen from a large range of rf frequencies without generally affecting beam stability. Given complete freedom of choice, however, a low frequency is preferable to a high frequency. For low rf frequencies the bunch length is longer and electromagnetic interaction with the beam environment is reduced since high frequency modes are not excited significantly. A longer bunch length also reduces the particle density in the bunch and thereby potentially troublesome *intra beam scattering* [6.14, 15]. In proton and heavy ion beams a longer bunch length leads to a reduced *space charge tune shift* and therefore allows to accelerate a higher beam intensity. For these reasons lower frequency systems are used mostly in low energy ($\gamma < 100$) circular accelerators. The downside of low rf frequencies is the fact that the accelerating cavities become large and less efficient and rf sources are limited in power.

The size of circular accelerators imposes a lower limit on the rf frequency since the synchronicity condition requires that the rf frequency be at least equal to the revolution frequency in which case the harmonic number is equal to unity. A higher harmonic number to accommodate more than a single particle bunch further increases the required rf frequency. Most electron and very high energy proton accelerators operate at rf frequencies of a few hundred MHz, while lower frequencies are preferred for ion or medium energy proton accelerators. For some applications it is critical to obtain short particle bunches which is much easier to achieve with a high rf frequency. The appropriate choice of the rf frequency therefore dependents much on the desired parameters for the particular application and is mostly chosen as a compromise between competing requirements including economic considerations like cost and availability.

Synchronous phase and rf voltage: The most common use of an rf system is for acceleration while particles pass through a resonant cavity at the moment when the voltage reaches the crest of the rf wave and particles gain a kinetic energy equivalent to the full cavity voltage. This is the general accelerating mode in linear accelerators. In circular accelerators, however, the principle of *phase focusing* requires that particles be accelerated off the crest at a *synchronous phase* ψ_s, where the effective accelerating voltage is $V_a = \hat{V}_{cy} \sin \psi_s$. The peak rf voltage $\hat{V}_{cy}$ and the synchronous phase are determined by the desired momentum acceptance and acceleration per turn.

The *momentum acceptance* of a circular accelerator has been derived in [6.3], is proportional to the square root of the cavity voltage and must be adjusted for the larger of several momentum acceptance requirements. To successfully inject a beam into a circular accelerator the voltage must be sufficiently large to accept the finite momentum spread in the injected beam. In addition any phase spread or timing error of the incoming beam translates into momentum errors due to synchrotron oscillations. For acceleration of a high intensity beam an additional allowance to the rf voltage must be made to compensate beam loading, which will be discussed later in more detail.

After injection into a circular accelerator an electron beam may change considerably its momentum spread due to quantum excitation as a result of emitting synchrotron radiation. This momentum spread has a gaussian distribution and to assure long beam lifetime the momentum acceptance must be large enough to contain several standard deviations. In proton and heavy ion accelerators some phase space manipulation may be required during the injection process which contributes another lower limit for the required rf voltage. In general, there are a number of requirements that determine the ultimate momentum acceptance of an accelerator and the most stringent requirement may very well be different for different accelerator designs and applications. Generally, circular accelerators are designed for a momentum acceptance of a few percent.

6.2 Beam-Cavity Interaction

The proper operation of the rf system in a particle accelerator depends more than any other component on the detailed interaction with the particle beam. This results from the observation that a particle beam can induce fields in the accelerating cavities of significant magnitude compared to the generator produced voltages and we may therefore not neglect the presence of the particle beam. This phenomenon is called *beam loading* and can place severe restrictions on the beam current that can be accelerated. In this section main features of such interaction and stability conditions for most efficient and stable particle acceleration will be discussed.

6.2.1 Coupling Between rf Field and Particles

In our discussions about particle acceleration we have tacitly assumed that particles would gain energy from the fields in accelerating cavities merely by meeting the synchronicity conditions. This is true for a weak particle beam which has no significant effect on the fields within the cavity. As we try, however, to accelerate an intense beam the actual accelerating fields become modified by the presence of considerable electrical charges from the particle beam. This *beam loading* can be significant and ultimately limits the maximum beam intensity that can be sustained.

The phenomenon of beam loading will be defined and characterized in this section leading to conditions and parameters to assure positive energy flow from the rf power source to the beam. Fundamental consideration to this discussion are the principles of energy conservation and linear superposition of fields which allow us to study field components from one source independent of fields generated by other sources. Specifically, we may treat beam induced fields separately from fields generated by rf power sources.

Externally driven accelerating cavity: Accelerator cavities can be described as *damped oscillators* with *external excitation*. Damping occurs due to energy losses in the walls of the cavity and transfer of energy to the particle beam while an external rf power source is connected to the cavity to sustain the rf fields. Many features of an accelerating cavity can be expressed in well-known terms of a damped, externally excited *harmonic oscillator* which is described in the form

$$\ddot{x} + 2\alpha\dot{x} + \omega_o^2 x = D\,e^{i\omega t}, \tag{6.67}$$

where α is the *damping decrement*, ω_o the unperturbed oscillator frequency, ω the frequency and D the amplitude of the *external driving force*. The equilibrium solution can be expressed in the form $x = A\,e^{i\omega t}$, where the complex amplitude A is determined after insertion of this ansatz into (6.67)

$$A = \frac{D}{\omega_o^2 - \omega^2 + i2\alpha\omega} = a\,e^{i\Psi}. \tag{6.68}$$

The angle Ψ is the phase shift between the external excitation and the oscillator and the amplitude $a = \mathrm{Re}(A)$ is from (6.68)

$$a = \frac{D}{\sqrt{(\omega_o^2 - \omega^2)^2 + 4\alpha^2\omega^2}}. \tag{6.69}$$

Plotting the oscillation amplitude a as a function of the excitation frequency ω we get the *resonance curve* for the oscillator as shown in Fig. 6.5. The resonance frequency at which the oscillator reaches the maximum amplitude depends on the damping and is

$$\omega_r = \sqrt{\omega_o^2 - 2\alpha^2}. \tag{6.70}$$

For an undamped oscillator the resonance amplitude becomes infinite but is finite whenever there is damping. The oscillator can be excited within a finite distance from the resonance frequency and the width of the resonance curve at half maximum amplitude is

$$\Delta\omega_{1/2} \approx \pm 2\sqrt{3}\,\alpha \quad \text{for} \quad \alpha \ll \omega_r. \tag{6.71}$$

If there were no external excitation to sustain the oscillation, the amplitude would decay like $a \propto e^{-\alpha t}$. The energy of the oscillator scales like $W \propto A^2$ and the energy loss per unit time $P = -dW/dt = 2\alpha W$ which can be used to determine the *quality factor* of this oscillator as defined in (6.63)

$$Q = \frac{\omega_r}{2\alpha}. \tag{6.72}$$

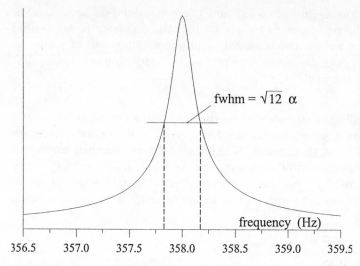

Fig. 6.5. Resonance curve for a damped oscillator

The quality factor is reduced as damping increases. For the case of an accelerating cavity, we expect therefore a higher Q-value called the *unloaded* Q_o when there is no beam, and a reduced quality factor called *loaded Q* when there is a beam extracting energy from the cavity. The time constant for the decay of oscillation amplitudes or the *cavity damping time* is

$$t_{\mathrm{d}} = \frac{1}{\alpha} = \frac{2Q}{\omega_{\mathrm{r}}} \tag{6.73}$$

which is the same as the cavity filling time (6.66) and the field amplitude decays to $1/\mathrm{e}$ during Q/π oscillations. Coming back to the equation of motion (6.67) for this oscillator, we have the solution

$$x(t) = a\,\mathrm{e}^{\mathrm{i}(\omega t + \Psi)} \tag{6.74}$$

noting that the oscillator assumes the same frequency as the external excitation but is out of synchronism by the phase Ψ. The magnitude and sign of this phase shift depends on the excitation frequency and can be derived from (6.68) in the form

$$\omega_{\mathrm{r}}^2 - \omega^2 + \mathrm{i}\,2\alpha\omega = \frac{1}{a}\,\mathrm{e}^{\mathrm{i}\Psi} = \frac{1}{a}\left(\cos\Psi - \mathrm{i}\sin\Psi\right).$$

Both the real and imaginary parts must separately be equal and we get for the phase shift between excitation and oscillator

$$\cot\Psi = \frac{\omega^2 - \omega_{\mathrm{r}}^2}{2\alpha\omega} \approx 2Q\,\frac{\omega - \omega_{\mathrm{r}}}{\omega_{\mathrm{r}}}, \tag{6.75}$$

where we have made use of (6.72) and the approximation $\omega \approx \omega_r$. For excitation at the resonance frequency we find the oscillator to lag behind the driving force by $\pi/2$ and is almost in phase or totally out of phase for very low or very high frequencies, respectively. In rf jargon this phase shift is called the *tuning angle*.

Network modelling of an accelerating cavity: In an accelerating cavity similar situations exist as just discussed and we will use therefore characteristic parameters and terminology of externally driven, damped oscillators in our further discussions of rf systems. Electrically, an accelerating cavity can be represented by a parallel resonant circuit (Fig. 6.6) which is driven by an external rf current source I_g and image currents of the particle beam I_b.

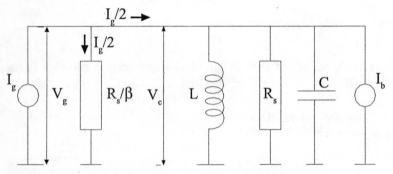

Fig. 6.6. Network model for an rf generator and an accelerating cavity

The amount of rf power available from the generator in the accelerating cavity depends greatly on the relative impedance of cavity and generator. Both have to be matched to assure optimum power transfer. To derive conditions for that we define the *internal impedance* of the current source or *rf generator* in terms of the *shunt impedance* R_s of an empty cavity as defined in (6.57)

$$R_g = \frac{R_s}{\beta},\tag{6.76}$$

where β is the *coupling coefficient* still to be defined. This coefficient depends on the actual hardware of the coupling arrangement for the rf power from the generator at the entrance to the cavity and quantifies the generator impedance as seen from the cavity in units of the *cavity shunt impedance* R_s (Fig. 6.6). Since this coupling coefficient depends on the hardware, we need to specify the desired operating condition to determine the proper adjustment of the coupling during assembly. This adjustment is done by either rotating a loop coupler (Fig. 6.4) with respect to the cavity axis or adjustment of the aperture in case of capacitive coupling through a hole.

The inductance L and capacitance C form a *parallel resonant circuit* with the *resonant frequency*

$$\omega_r = \frac{1}{\sqrt{LC}}. \tag{6.77}$$

The rf power available at the cavity from the generator is

$$P_g = \tfrac{1}{2} Y_L V_g^2, \tag{6.78}$$

where Y_L is the loaded *cavity admittance* including energy transfer to the beam and V_g is the generator voltage. Unless otherwise noted, the voltages, currents and power used in this section are the maximum values of these oscillating quantities. At resonance where all reactive power vanishes we use the generator current I_g and network admittance $Y = Y_g + Y_L$ to replace the *generator voltage*

$$V_g = \frac{I_g}{Y} = \frac{I_g}{Y_g + Y_L}$$

and get after insertion into (6.78) the *generator power* in the form

$$P_g = \tfrac{1}{2} \frac{Y_L}{|Y_g + Y_L|^2} I_g^2. \tag{6.79}$$

Noting that the *generator power* has a maximum, which can be determined from $\partial P_g / \partial Y_L = 0$, we obtain the well-known result that the rf power transfer from the generator becomes a maximum if the load is matched to the internal impedance of the generator by adjusting

$$Y_L = Y_g \quad \text{or} \quad R_L = \frac{R_s}{\beta} \tag{6.80}$$

replacing the admittances by the respective impedances. The maximum available rf power is therefore with $Y_g = \beta / R_s$

$$P_g = \frac{1}{8} \frac{R_s}{\beta} I_g^2. \tag{6.81}$$

To calculate the *quality factor* for a cavity, we note the stored energy is $W = 1/2\,CV^2$ and the energy loss rate $P_{cy} = 1/2\,V^2/R$. Using the definition (6.63) the unloaded quality factor becomes with $R = R_s$

$$Q_o = \omega_r C R_s, \tag{6.82}$$

and the quality factor for the total circuit as seen by the beam is with

$$\frac{1}{R_b} = \frac{\beta}{R_s} + \frac{1}{R_s} = \frac{1+\beta}{R_s} \tag{6.83}$$

and (6.82)

$$Q = \omega_r C R_b = \frac{Q_o}{1 + \beta}. \tag{6.84}$$

Off resonance the generator voltage and current are no more in phase. The phase difference can be derived from the *complex impedance* of the network which is the same seen from the generator as well as seen from the beam

$$\frac{1}{Z} = \frac{1}{R_b} + i\omega C + \frac{1}{i\omega L}. \tag{6.85}$$

The complex impedance becomes with (6.77,84) from (6.85)

$$\boxed{\frac{1}{Z} = \frac{1}{R_b}\left(1 + iQ\frac{\omega^2 - \omega_r^2}{\omega_r\,\omega}\right)} \tag{6.86}$$

and with $I_g = V_g/Z$ the generator current is

$$I_g = \frac{V_g}{R_b}\left(1 + iQ\frac{\omega^2 - \omega_r^2}{\omega_r\,\omega}\right) = \frac{V_g}{R_b}\left(1 - i\tan\Psi\right). \tag{6.87}$$

Close to resonance the *tuning angle* becomes from (6.87) with $\omega \approx \omega_o$

$$\tan\Psi \approx -Q\frac{\omega^2 - \omega_r^2}{\omega_r\,\omega} \approx -2Q\frac{\omega - \omega_r}{\omega_r} \tag{6.88}$$

in agreement with (6.75) except for a phase shift of -90°, which is introduced to be consistent with the definition of the synchronous phase. The variation of the tuning angle is shown in Fig. 6.7 as a function of the generator frequency. From (6.87) the generator voltage at the cavity is finally

$$V_g = \frac{R_b\,I_g}{1 - i\tan\Psi} = R_b\,I_g\,\cos\Psi\,e^{i\Psi}. \tag{6.89}$$

At frequencies below the resonance frequency the tuning angle is positive and therefore the generator current lags the voltage by the phase Ψ. This case is also called *inductive detuning* since the impedance looks mainly inductive. Conversely, the detuning is called *capacitive detuning* because the impedance looks mostly capacitive for frequencies above resonance frequency.

A bunched particle beam passing through a cavity acts as a current just like the generator current and therefore the same relationships with respect to beam induced voltages exist. In case of capacitive detuning, for example, the beam induced voltage V_b lags in phase behind the beam current I_b.

The effective accelerating voltage in the cavity is a composition of the *generator voltage*, the *induced voltage*, and the phase relationships between themselves and relative to the particle beam. To assure a stable beam, the resulting *cavity voltage* must meet the requirements of particle acceleration

tuning angle (deg)

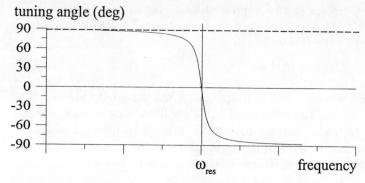

Fig. 6.7. Tuning angle Ψ as a function of the generator frequency

to compensate, for example, lost energy into synchrotron radiation. We determine the conditions for that by deriving first the generator voltage V_{gr} at resonance and without beam loading while voltage and current are in phase. From Fig. 6.6 we get

$$V_{\mathrm{gr}} = \frac{I_{\mathrm{g}}}{Y_{\mathrm{L}} + Y_{\mathrm{g}}} = \frac{I_{\mathrm{g}}}{\frac{1}{R_{\mathrm{o}}} + \frac{\beta}{R_{\mathrm{o}}}} = \frac{R_{\mathrm{o}} I_{\mathrm{g}}}{1 + \beta} \tag{6.90}$$

and with (6.81) the generator voltage at resonance becomes

$$V_{\mathrm{gr}} = \frac{2\sqrt{2\beta}}{1 + \beta} \sqrt{P_{\mathrm{g}} R_{\mathrm{s}}}. \tag{6.91}$$

The generator voltage at the cavity is therefore with (6.89)

$$V_{\mathrm{g}} = V_{\mathrm{gr}} \cos\Psi\, \mathrm{e}^{\mathrm{i}\Psi}. \tag{6.92}$$

This is the cavity voltage seen by a negligibly small beam and can be adjusted to meet beam stability requirements by varying the tuning angle Ψ and rf power P_{g}.

6.2.2 Beam Loading and rf System

For more substantial beam currents the effect of *beam loading* must be included to obtain the effective cavity voltage. Similar to the derivation of the generator voltage in a cavity, we may derive the induced voltage from the beam current passing through that cavity. Since there is no fundamental difference between generator and beam current, the *induced voltage* is in analogy to (6.92)

$$V_{\mathrm{b}} = -V_{\mathrm{br}} \cos\Psi\, \mathrm{e}^{\mathrm{i}\Psi}, \tag{6.93}$$

where the negative sign indicates that the induced voltage is decelerating the beam. The particle distribution in the beam occurs in bunches and the beam current therefore can be expressed by a Fourier series. Here we are only interested in the harmonic I_h of the beam current and find for bunches short compared to the rf wavelength

$$I_h = 2 I_b, \tag{6.94}$$

where I_b is the average beam current and h the *harmonic number*. The approximation for short bunches with $\ell \ll \lambda_{rf}$ holds as long as $\sin k_{rf}\ell \approx k_{rf}\ell$ with $k_{rf} = 2\pi/\lambda_{rf}$. For longer bunches the factor 2 is modified as can be derived from an appropriate Fourier expansion. At the resonance frequency $\omega_r = h\omega_o$ the beam induced voltage in the cavity is with (6.83)

$$V_{br} = \frac{R_s I_h}{1+\beta} = \frac{2R_s I_b}{1+\beta}. \tag{6.95}$$

The resulting cavity voltage is the superposition of both voltages, the generator and the induced voltage. This superposition, including appropriate phase factors, is often represented in a *phasor diagram*. In such a diagram a complex quantity $\tilde{z}$ is represented by a vector of length $|\tilde{z}|$ with the horizontal and vertical components being the real and imaginary part of $\tilde{z}$, respectively. The phase of this vector increases counter clockwise and is given by $\tan\varphi = \mathrm{Im}\{\tilde{z}\}/\mathrm{Re}\{\tilde{z}\}$. In an application to rf parameters we represent voltages and currents by vectors with a length equal to the magnitude of voltage or current and a counter clockwise rotation of the vector by the phase angle φ.

The particle beam current can be chosen as the reference being parallel to the real axis and we obtain from the quantities derived so far the phasor diagram as shown in Fig. 6.8. First we determine the relationships between individual vectors and phases and then the correct adjustments of variable rf parameters. In Fig. 6.8 the generator current is assumed to have the still to be determined phase ϑ with respect to the beam current while the generator voltage and beam induced voltage lag by the phase Ψ behind the beam current. The resulting cavity voltage $\tilde{\mathbf{V}}_{cy}$ is the phasor addition of both voltages $\tilde{\mathbf{V}}_g + \tilde{\mathbf{V}}_b$ as shown in Fig. 6.8.

The adjustment of the rf system must now be performed in such a way as to provide the desired gain in *kinetic energy* $U_o = e V_{cy} \sin\psi_s$ where V_{cy} is the maximum value of the cavity voltage and ψ_s the *synchronous phase*. To maximize the energy flow from the generator to the cavity the load must be matched such that it appears to the generator purely resistive. This is achieved by adjusting the phase ψ_g to get the cavity voltage V_{cy} and generator current I_g in phase which occurs for

$$\psi_g = \tfrac{1}{2}\pi - \psi_s \tag{6.96}$$

as shown in Fig. 6.9. Obviously, this is only true for a specific value of the beam current. General operation will deviate from this value and therefore we often match to the maximum desired beam current. For lower currents,

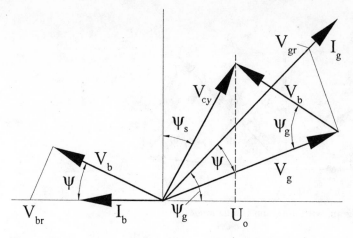

Fig. 6.8. Phasor diagram for an accelerating cavity and arbitrary tuning angle

where the energy transfer is not optimum anymore, some loss of efficiency is acceptable.

The *tuning angle* adjustment for optimum matching can be derived from Fig. 6.9 and applying the law of sines we have with (6.96)

$$\frac{V_{br} \cos \Psi_m}{V_{cy}} = \frac{\sin \Psi_m}{\cos \psi_s} .$$
(6.97)

The *optimum tuning angle* is from (6.97)

$$\tan \Psi_m = \frac{V_{br}}{V_{cy}} \cos \psi_s .$$
(6.98)

This tuning is effected by a shift in the resonant frequency of the cavity with respect to the generator frequency by, for example, moving a tuner (Fig. 6.4) in or out. From (6.88) we get with (6.84,95,99) for the frequency shift or *frequency tuning*

$$\delta\omega = \omega - \omega_r = -\frac{\omega_r}{2Q_o} \frac{P_b}{P_{cy}} \cot \psi_s ,$$
(6.99)

where the cavity power is defined by

$$P_{cy} = \frac{V_{cy}^2}{2R_s}$$
(6.100)

and the beam power

$$P_b = I_b V_{cy} \sin \psi_s .$$
(6.101)

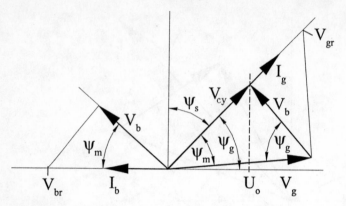

Fig. 6.9. Phasor diagram with optimum tuning angle

To determine the required generator power the components of the cavity voltage vector can be expressed by other quantities and we get from Fig. 6.9

$$V_{cy} \sin \psi_s = V_{gr} \cos \Psi_m \cos(\psi_g - \Psi_m) - V_{br} \cos^2 \Psi_m \qquad (6.102)$$

and

$$V_{cy} \cos \psi_s = V_{gr} \cos \Psi_m \sin(\psi_g - \Psi_m) + V_{br} \cos \Psi_m \sin \Psi_m . \qquad (6.103)$$

Combining both equations to eliminate the phase $(\psi_g - \Psi_m)$ we get

$$V_{gr}^2 = \left(V_{cy} \frac{\sin \psi_s}{\cos \Psi_m} + V_{br} \cos \Psi_m \right)^2 + \left(V_{cy} \frac{\cos \psi_s}{\cos \Psi_m} - V_{br} \sin \Psi_m \right)^2 \qquad (6.104)$$

and with (6.98,99) the required *generator power* is

$$
\begin{aligned}
P_g &= \frac{V_{cy}^2}{2R_s} \frac{(1+\beta)^2}{4\beta} \left[\left(\frac{\sin \psi_s}{\cos \Psi_m} + \frac{2I_b R_s}{V_{cy}(1+\beta)} \cos \Psi_m \right)^2 \right. \\
&\quad \left. + \left(\frac{\cos \psi_s}{\cos \Psi_m} - \frac{2I_b R_s}{V_{cy}(1+\beta)} \sin \Psi_m \right)^2 \right].
\end{aligned}
\qquad (6.105)
$$

This expression can be greatly simplified for *optimum matching* and becomes with (6.98,99)

$$P_{g,opt} = \frac{(1+\beta)^2}{8\beta R_s} \left(V_{cy} + \frac{2R_s I_b}{1+\beta} \sin \psi_s \right)^2 . \qquad (6.106)$$

Equation (6.106) represents a combination of beam current through I_b, rf power P_g, coupling coefficient β, and shunt impedance R_s to sustain a cavity voltage V_{cy}. Specifically, considering that the rf power P_g and coupling coefficient β is fixed by the hardware installed a maximum supportable beam current can be derived as a function of the desired or required cavity voltage. Solving for the *cavity voltage*, (6.106) becomes after some manipulation

$$V_{cy} = \frac{\sqrt{2\beta R_s}}{1+\beta} \left(\sqrt{P_{g,opt}} + \sqrt{P_{g,opt} - \frac{1+\beta}{\beta} P_b} \right).$$ (6.107)

This expression exhibits a limit for the beam current above which the second square root becomes imaginary. The condition for real solutions requires that

$$P_b \leq \frac{\beta}{1+\beta} P_g$$ (6.108)

leading to a limit of the *maximum sustainable beam current* of

$$I_b \leq \frac{\beta}{1+\beta} \frac{P_g}{V_a},$$ (6.109)

where

$$V_a = V_{cy} \sin \psi_s$$ (6.110)

is the desired effective *accelerating voltage* in the cavity.

Inspection of (6.106) shows that the required generator power can be further minimized by adjusting to the optimum *coupling coefficient* β. Optimum coupling can be derived from $\partial P_g / \partial \beta = 0$ with the solution

$$\beta_{opt} = 1 + \frac{2R_s I_b}{V_{cy}} \sin \psi_s = 1 + \frac{P_b}{P_{cy}}.$$ (6.111)

The minimum generator power required to produce an accelerating voltage $V_{cy} \sin \psi_s$ is therefore from (6.106) with (6.111)

$$P_{g,min} = \frac{V_{cy}^2}{2R_s} \beta_{opt} = \beta_{opt} P_{cy}$$ (6.112)

and the *optimum tuning angle* from (6.98)

$$\boxed{\tan \Psi_{opt} = \frac{\beta_{opt} - 1}{\beta_{opt} + 1} \cot \psi_s.}$$ (6.113)

In this operating condition all rf power from the generator is absorbed by the beam loaded cavity and no power reflection occurs. The maximum beam power is therefore $P_b = P_g - P_{cy}$ and the maximum beam current

$$\boxed{I_b \leq \frac{P_g}{V_{cy} \sin \psi_s} - \frac{V_{cy}}{R_s \sin \psi_s}.}$$ (6.114)

Conditions have been derived assuring most efficient power transfer to the beam by proper adjustment of the cavity power input coupler to obtain the optimum coupling coefficient. Of course this coupling coefficient is optimum only for a specific beam current which in most cases is chosen to be the maximum desired beam current.

We are now in a position to determine the total rf power flow. From conservation of energy we have

$$P_g = P_{cy} + P_b + P_r,$$
(6.115)

where P_r is the *reflected power* which vanishes for the case of optimum coupling.

Proper adjustment of the accelerating cavity can be tested with the help of a pulsed rf generator. Directional couplers in the transmission line from generator to cavity allow the monitoring of the forward and reflected power while a loop probe in the cavity detects the field strength in the cavity. Application of an rf pulse to the cavity results first in a large reflected power since the cavity is completely mismatched. Both, cavity wall losses and beam acceleration, require full fields to become optimally effective for matching. In terms of the circuit diagram the capacitance is empty and the load impedance zero at the very beginning of the pulse. As the capacitor charges up the field in the cavity reaches its operating value and the reflected power drops to zero for a properly adjusted coupling coefficient. Conversely, the coupling coefficient can be inferred from the measurement of the reflected power. At the moment the generator pulse comes to an end, the energy stored in the cavity starts to decay and from observation of the decay time the quality factor can be derived.

In Fig. 6.10 measured signals for the forward generator power, the particle beam, and the reflected power are shown for the pulsed operation of a 3 GHz *rf electron gun* [6.16]. In this case the coupling coefficient is adjusted to a significant beam loading and the initial mismatch is due to both the loading of the cavity and the development of the particle beam current from a hot cathode at the wall of the cavity. We observe a significant mismatch at the beginning and end of the generator pulse. As the beam current develops during the pulse the matching improves without reaching its optimum value due to a deliberate limitation of the beam current. Obviously, the effect of beam loading must be included to achieve maximum energy transfer from the generator to the particle beam.

6.2.3 Higher-Order Mode Losses in an rf Cavity

The importance of beam loading for accurate adjustments of the rf system has been discussed qualitatively but not yet quantitatively. In this paragraph, quantitative expressions will be derived for beam loading. Accelerating cavities constitute an impedance to a particle current and a bunch of particles with charge q passing through a cavity induces electromagnetic

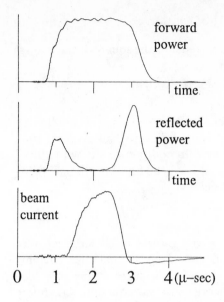

0   1   2   3   4 (μ–sec)

Fig. 6.10. Observation of beam loading in an accelerating cavity, generator power to the cavity P_g (top), beam current I_b (middle), and reflected power P_r (bottom)

fields into a broad frequency spectrum limited at the high frequency end by the bunch length. The magnitude of the excited frequencies in the cavity depends on the frequency dependence of the cavity impedance, which is a function of the particular cavity design and need not be known for this discussion. Fields induced within a cavity are called modes, oscillating at different frequencies with the lowest mode being the fundamental resonant frequency of the cavity. Although cavities are designed primarily for one resonant frequency, many *higher-order modes* or HOM's can be excited at higher frequencies. Such modes occur above the fundamental frequency first at distinct well-separated frequencies with increasing spectral densities at higher frequencies.

For a moment we consider here only the fundamental frequency and deal with higher-order modes later. Fields induced by the total bunch charge act back on individual particles modifying the overall accelerating voltage seen by the particle. To quantify this we use the *fundamental theorem of beam loading* formulated by *Wilson* [6.17] which states that each particle within a bunch sees one half of the induced field while passing through the cavity.

We prove this theorem by conducting a gedanken experiment proposed by *Wilson*. Consider a bunch of particles with charge q passing through a lossless cavity inducing a voltage V_{i1} in the fundamental mode. This induced field is opposed to the accelerating field since it describes a loss of energy. While the bunch passes through the cavity this field increases from zero reaching a maximum value at the moment the particle bunch leaves the cavity. Each particle will have interacted with this field and the energy loss corresponds to a fraction f of the induced voltage $V_{i,h}$ where the index h indicates that we consider only the fundamental mode. The total energy lost by the bunch of charge q is

$$\Delta E_1 = -q_1 f V_{i,h}. \tag{6.116}$$

This energy appears as field energy proportional to the square of the voltage

$$W_1 = c_1 V_{i,h}^2, \tag{6.117}$$

where c_1 is a constant.

Now consider another bunch with the same charge $q_2 = q_1 = q$ following behind the first bunch at a distance corresponding to half an oscillation period at the fundamental cavity frequency. In addition to its own induced voltage this second bunch will see the field from the first bunch, now being accelerating, and will therefore gain an energy

$$\Delta E_2 = q_1 V_{i,h} - q_2 f V_{i,h} = q V_{i,h} (1 - f). \tag{6.118}$$

After passage of the second charge, the cavity returns to the original state before the first charge arrived because the field from the first charge having changed sign exactly cancels the induced field from the second charge. The cavity has been assumed lossless and energy conservation requires therefore that $\Delta E_1 + \Delta E_2 = 0$ or $-qfV_{i,h} + qV_{i,h}(1 - f) = 0$ from which we get

$$f = \tfrac{1}{2} \tag{6.119}$$

proving the statement of the fundamental theorem of beam loading. The energy loss of a bunch of charge q due to its own induced field is therefore

$$\Delta E_1 = -\tfrac{1}{2} q V_{i,h}. \tag{6.120}$$

This theorem will be used to determine the energy transfer from cavity fields to a particle beam. To calculate the induced voltages in rf cavities, or in arbitrarily shaped vacuum chambers providing some impedance for the particle beam can become very complicated. For cylindrically symmetric cavities the induced voltages can be calculated numerically with programs like SUPERFISH[6.11], URMEL[6.12] or MAFIA[6.12].

For a more practical approach *Wilson* [6.17] introduced a *loss parameter* k which can be determined either by electronic measurements or by numerical calculations. This loss parameter for the fundamental mode loss of a bunch with charge q is defined by

$$\Delta E_h = k_h q^2 \tag{6.121}$$

and together with (6.120) we get the induced voltage

$$V_{i,h} = -2 k_h q \tag{6.122}$$

or after elimination of the charge from (6.121,122)

$$\Delta E_h = \frac{V_{i,h}^2}{4 k_h}, \tag{6.123}$$

where the index $_h$ indicates that the parameter should be taken at the fundamental frequency. The loss parameter can be expressed in terms of cavity parameters. From the definition of the cavity quality factor (6.63) and cavity losses from (6.60) we get

$$\frac{R_{\text{cy}}}{Q} = \frac{V^2}{\omega W},$$
(6.124)

where ω is the frequency and W the stored field energy in the cavity. Applying this to the induced field we note that ΔE_h is equal to the field energy W_h and combining (6.123,124) the loss parameter to the fundamental mode in a cavity with shunt impedance R_h and quality factor Q_h is

$$\boxed{k_h = \frac{\omega_h}{4} \frac{R_h}{Q_h}.}$$
(6.125)

The excitation of *higher-order mode fields* by the passing particle bunch leads to additional energy losses which are conveniently expressed in units of the energy loss to the fundamental mode

$$\Delta E_{\text{hom}} = (r_{\text{hom}} - 1)\,\Delta E_h,$$
(6.126)

where r_{hom} is the ratio of the total energy losses into all modes to the loss into the fundamental mode only. The induced higher order field energy in the cavity is therefore

$$W_{\text{hom}} = (r_{\text{hom}} - 1)\,W_h.$$
(6.127)

Again we may define a loss parameter k_n for an arbitrary nth mode and get analogous to (6.125)

$$k_n = \frac{\omega_n}{4} \frac{R_n}{Q_n},$$
(6.128)

where R_n and Q_n are the shunt impedance and quality factor for the nth mode, respectively. The total loss parameter due to all modes is by linear superposition

$$k = \sum_n k_n.$$
(6.129)

The task to determine the induced voltages has been reduced to the determination of the loss parameters for individual modes or if this is not possible or desirable we may use just the overall loss parameter k as may be determined experimentally. This is particularly convenient for cases where it is difficult to calculate the mode losses but much easier to measure the overall losses by electronic measurements.

The higher-order mode losses will become important for discussion of beam stability since these fields will act back on subsequent particles and bunches thus creating a coupling between different parts of one bunch or different bunches. These interactions will be discussed in more detail in Chap. 10.

Efficiency of energy transfer from cavity to beam: Higher-order mode losses affect the efficiency by which energy is transferred to the particle beam. Specifically, since the higher-order mode losses depend on the beam current we must expect some limitation in the current capability of the accelerator.

With these preparations we have now all information to calculate the transfer of energy from the cavity to the particle beam. Just before the arrival of a particle bunch let the cavity voltage as generated by the rf power source be

$$V_{\text{cy}} = V_{\text{g}}\, e^{i\Psi_{\text{g}}}\,, \tag{6.130}$$

where V_{g} is the generator voltage and Ψ_{g} the generator voltage phase with respect to the particle beam. To combine the generator voltage with the induced voltage we use phasor diagrams in the complex plane.

The generator voltage is shown in Fig. 6.11 as a vector rotated by the angle Ψ_{g} from the real axis representing the cavity state just before the beam passes. The beam induced voltage is parallel and opposite to the real axis. Both vectors add up to the voltage V just after the beam has left the cavity.

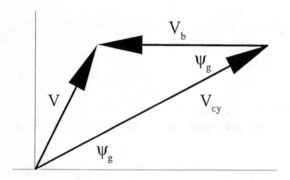

Fig. 6.11. Phasor diagram for cavity voltages with beam loading

The difference of the fundamental field energy before and after passage of the particle bunch is equal to the higher-order field energy and the energy transferred to the passing particle bunch and is from the phasor diagram 6.11

$$\begin{aligned}
\Delta E_{\text{hom}} &= \alpha\,(V_{\text{cy}}^2 - V^2) - W_{\text{hom}}\,, \\
&= \alpha\,(2\,V_{\text{cy}}\,V_{\text{b}}\cos\Psi_{\text{g}} - V_{\text{b}}^2) - W_{\text{hom}}\,,
\end{aligned} \tag{6.131}$$

where α is the proportionality factor between the energy gain ΔE and the square of the voltage defined by $\alpha = \Delta E/V^2$. With (6.121,122) we get from (6.131) the net energy transfer to a particle bunch [6.18]

$$\Delta E_1 = \alpha \left(2 V_{cy} V_b \cos \Psi_g - r_{hom} V_b \right)^2. \tag{6.132}$$

The energy stored in the cavity before arrival of the beam is $W_{cy} = \alpha V_{cy}^2$ and the *energy transfer efficiency* to the beam becomes

$$\eta = \frac{\Delta E_h}{W_{cy}} = 2 \frac{V_b}{V_{cy}} \cos \Psi_g - r_{hom} \frac{V_b^2}{V_{cy}^2}. \tag{6.133}$$

It is obvious from (6.133) that energy transfer is not guaranteed by the synchronicity condition or the power of the generator alone. Specifically, the second term in (6.133) becomes dominant for a large beam intensity and the efficiency may even become negative indicating reversed energy transfer from the beam to cavity fields. The energy transfer efficiency has a maximum for $V_b = \frac{\cos \Psi_g}{r_{hom}} V_{cy}$ and is

$$\eta_{max} = \frac{\cos^2 \Psi_g}{r_{hom}}, \tag{6.134}$$

a result first derived by *Keil, Schnell* and *Zotter* [6.19] and is frequently called the *Keil-Schnell-Zotter criterion*. The maximum energy transfer efficiency is limited by the phase of the generator voltage and the higher-order mode losses.

6.2.4 Beam Loading in Circular Accelerators

Only one passage of a bunch through a cavity has been considered in the previous section. In circular accelerators, however, particle bunches pass periodically through the accelerating cavities and we have to consider the cumulative build up of induced fields. Whenever a particle bunch is traversing a cavity the induced voltage from this passage is added to those still present from previous bunch traversals. For simplicity, we assume a number of equidistant bunches along the circumference of the ring, where adjacent bunches are separated by an integer number m_b of the fundamental rf wavelength. The induced voltage decays exponentially by a factor $e^{-\rho}$ between two consecutive bunches where

$$\rho = \frac{t_b}{t_d}, \tag{6.135}$$

t_b is the time between bunches and t_d the cavity voltage decay time for the fundamental mode. The phase of the induced voltage varies between the passage of two consecutive bunches by

$$\varphi = \omega_h t_b - 2\pi m_b. \tag{6.136}$$

At the time a bunch passes through the cavity the total induced voltages are then the superposition of all fields induced by previous bunches

$$V_i = V_{ih} \left(1 + e^{-\rho} e^{i\varphi} + e^{-2\rho} e^{i2\varphi} + \ldots \right) \tag{6.137}$$

shown in Fig. 6.12 as the superposition of all induced voltages in form of a phasor diagram together with the resultant induced voltage V_i.

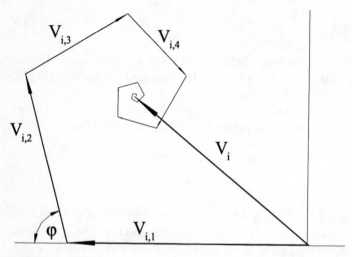

Fig. 6.12. Phasor diagram for the superposition of induced voltages in an accelerating cavity

The sum (6.137) can be evaluated for

$$V_i = V_{ih} \frac{1}{1 - e^{-\rho} e^{i\varphi}}, \tag{6.138}$$

which is the total induced voltage in the cavity just after the last bunch passes; however, the voltage seen by this last bunch is only half of the induced voltage and the total voltage V_b acting back on the bunch is therefore

$$V_b = V_{ih} \left(\frac{1}{1 - e^{-\rho} e^{i\varphi}} - \frac{1}{2} \right). \tag{6.139}$$

The voltage V_{ih} can be expressed in more practical units. Considering the damping time (6.73) for fields in a cavity we note that two damping times exist, one for the empty unloaded cavity t_{do} and a shorter damping time t_d when there is also a beam present. For the unloaded damping time we have from (6.73)

$$t_{do} = \frac{2 Q_{oh}}{\omega_h}, \tag{6.140}$$

where Q_{oh} is the unloaded quality factor. From (6.121,123) we get with $q = I_o\, t_b$ where I_o is the average beam current,

$$V_{ih} = \frac{\omega_h}{2\,Q_{oh}}\, R_h\, I_o\, t_b$$

and with (6.140)

$$V_{ih} = R_h\, I_o\, \frac{t_b}{t_{do}}. \qquad (6.141)$$

Introducing the *coupling coefficient* β we get from (6.84,140)

$$t_{do} = (1 + \beta)\, t_d. \qquad (6.142)$$

In analogy to (6.135) we define

$$\rho_o = \frac{\rho}{1 + \beta} = \frac{t_b}{t_{do}} \qquad (6.143)$$

and (6.141) becomes

$$V_{ih} = \rho_o\, R_h\, I_o. \qquad (6.144)$$

We are finally in a position to calculate from (6.139,144) the total beam induced cavity voltage V_b in the fundamental mode for circular accelerators.

Phase oscillation and stability: In the course of discussing phase oscillations we found it necessary to select carefully the synchronous phase depending on the particle energy being below or above the transition energy. Particularly, we found that phase stability for particles above transition energy requires the rf voltage to decrease with increasing phase, $dV_{cy}/d\psi < 0$. From the derivative of (6.102) with respect to ψ we find the condition $\sin(\psi - \Psi) > 0$ or

$$0 < \psi - \Psi < \pi. \qquad (6.145)$$

Since $V_{gr}\cos\Psi > 0$ we get from (6.103) and (6.145)

$$V_{cy}\cos\psi_s - V_{br}\sin\Psi > 0$$

or with (6.98)

$$\boxed{V_{br}\sin\psi_s < V_{cy}} \qquad (6.146)$$

which is *Robinson's phase-stability criterion* or the *Robinson condition* [6.20] for the tuning angle of the accelerator cavity. The maximum current that can be accelerated in a circular accelerator with stable phase oscillations is limited by the effective cavity voltage. In terms of rf power (6.146) is equivalent to

$$P_b = (1 + \beta) P_{cy} \tag{6.147}$$

and the stability condition for the *coupling coefficient* is

$$\beta > \beta_{opt} - 2. \tag{6.148}$$

The stability condition is always met for rf cavities with optimum coupling $\beta = \beta_{opt}$.

Robinson damping: Correct tuning of the rf system is a necessary but not a sufficient condition for stable phase oscillations. In [Ref. 6.3, Chap. 10] we found the occurrence of damping or antidamping due to forces that depend on the energy of the particle. Such a case occurs in the interaction of bunched particle beams with accelerating cavities or vacuum chamber components which act like narrow band resonant cavities. The revolution time of a particle bunch depends on the average energy of particles within a bunch and the Fourier spectrum of the bunch current being made up of harmonics of the revolution frequency is therefore energy dependent. On the other hand by virtue of the frequency dependence of the cavity impedance, the energy loss of a bunch in the cavity due to beam loading depends on the revolution frequency. We have therefore an energy dependent loss mechanism which can lead to damping or worse antidamping of *coherent phase oscillation* and we will therefore investigate this phenomenon in more detail. *Robinson* [6.20] studied first the dynamics of this effect generally referred to as *Robinson damping* or *Robinson instability*.

Above transition energy the revolution frequency is lower for higher bunch energies compared to the reference energy and vice versa. To obtain damping of coherent phase oscillations, we would therefore tune the cavity such that the bunch would loose more energy in the cavity while at higher energy during the course of coherent synchrotron oscillation and loose less energy at lower energies. In this situation, the impedance of the cavity should decrease with increasing frequency for damping to occur as demonstrated in Fig. 6.13.

Here the resonance curve or impedance spectrum is shown for the case of a resonant frequency above the beam frequency $h\omega_o$ in Fig. 6.13a and below the beam frequency in Fig 6.13b. Consistent with the arguments made above we would expect damping in case of Fig. 6.13b for a beam above transition and antidamping in case of Fig. 6.13a. Adjusting the resonance frequency of the cavity to a value below the beam frequency $h\omega_o$ where ω_o is the revolution frequency, is called *capacitive detuning*. Conversely, we would tune the cavity resonance frequency above the beam frequency ($\omega_r > h\omega_o$) or *inductively detune* the cavity for damping below transition energy (Fig. 6.13a).

In a more formal way we fold the beam-current spectrum with the impedance spectrum of the cavity and derive scaling laws for the damping as well as the shift in synchrotron frequency. During phase oscillations

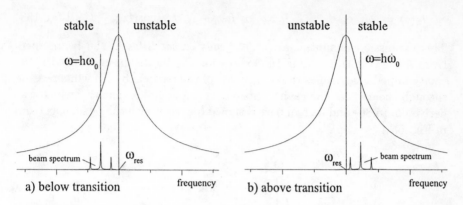

stable | unstable   unstable | stable

$\omega = h\omega_0$ $\omega = h\omega_0$

ω_{res} ω_{res}

beam spectrum   beam spectrum

a) below transition   b) above transition

frequency   frequency

Fig. 6.13. Cavity tuning for positive Robinson damping below and above transition energy

the revolution frequency is modulated and as a consequence the beam spectrum includes in addition to the fundamental frequency two side bands or *satellites*. The *beam-current spectrum* is composed of a series of harmonics of the revolution frequency up to frequencies with wavelength of the order of the bunchlength

$$I(t) = I_b + \sum_{n>0} I_n \cos(n\omega_0 t - \varphi), \tag{6.149}$$

where I_b is the average circulating beam current. The Fourier coefficient for bunches short compared to the wavelength of the harmonic is given by

$$I_n = 2I_b. \tag{6.150}$$

Here we restrict the discussion to the interaction between beam and cavity at the fundamental cavity frequency and the only harmonic of the beam spectrum is therefore the hth harmonic

$$I_h(t) = 2I_b \cos(h\omega_0 t - \varphi). \tag{6.151}$$

By virtue of coherent synchrotron oscillations the phase oscillates for each particle in a bunch like

$$\varphi(t) = \varphi_0 \sin \Omega_s t, \tag{6.152}$$

where φ_0 is the maximum amplitude and Ω_s the synchrotron oscillation frequency of the phase oscillation. We insert this into (6.151) and get after expanding the trigonometric functions for small oscillation amplitudes $\varphi_0 \ll 1$

$$I_h(t) = 2I_b \cos h\omega_o t - I_b \varphi_o \left[\cos(h\omega_o + \Omega_s)t - \cos(h\omega_o - \Omega_s)t\right]. \quad (6.153)$$

This expression exhibits clearly *sidebands* or *satellites* in the beam spectrum at $h\omega_o \pm \Omega_s$. Folding the expression for the beam current with the cavity impedance defines the energy loss of the particle bunch while passing through the cavity. The cavity impedance is a complex quantity which was derived in (6.86) and its real part is shown together with the beam spectrum in Fig. 6.14.

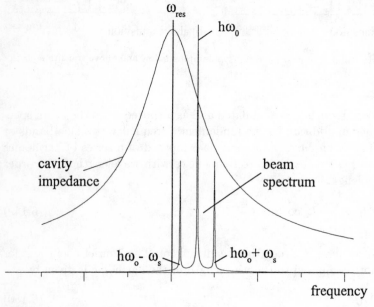

Fig. 6.14. Cavity impedance and beam spectrum in the vicinity of the fundamental rf frequency $\omega_{rf} = h\omega_o$

The induced voltage in the cavity by a beam $I_h(t) = I_h \cos h\omega_o t$ is

$$V_h = -Z\,I_h(t) = -Z_r\,I_h \cos h\omega_o t - Z_i\,I_h \sin h\omega_o t, \quad (6.154)$$

where we have split the impedance in its real and imaginary part and have expressed the imaginary part of the induced voltage by a $\pi/2$ phase shift. Applying (6.154) to all components of the beam current (6.153) we get the induced voltage in the cavity

$$
\begin{aligned}
V_h = &-Z_r^o\,2I_b \cos h\omega_o t - Z_i^o\,2I_b \sin h\omega_o t \\
&+ I_b\,Z_r^+\,\varphi_o \cos h\omega_o t \cos \Omega_s t - I_b\,Z_r^+\,\varphi_o \sin h\omega_o t \sin \Omega_s t \\
&+ I_b\,Z_i^+\,\varphi_o \sin h\omega_o t \cos \Omega_s t + I_b\,Z_i^+\,\varphi_o \cos h\omega_o t \sin \Omega_s t \\
&- I_b\,Z_r^-\,\varphi_o \cos h\omega_o t \cos \Omega_s t + I_b\,Z_r^-\,\varphi_o \sin h\omega_o t \sin \Omega_s t \\
&- I_b\,Z_i^-\,\varphi_o \sin h\omega_o t \cos \Omega_s t + I_b\,Z_i^-\,\varphi_o \cos h\omega_o t \sin \Omega_s t,
\end{aligned}
\quad (6.155)
$$

where Z°, Z^+, and Z^- are the real $_r$ and imaginary $_i$ cavity impedances at the frequencies $h\omega_o$, $h\omega_o + \Omega_s$, and $h\omega_o - \Omega_s$ respectively. We make use of the expression for the phase oscillation (6.152) and its derivative

$$\dot{\varphi}(t) = \varphi_o\,\Omega_s\,\cos\Omega_s t\,, \tag{6.156}$$

multiply the induced voltage spectrum (6.155) by the current spectrum (6.153) and get after averaging over fast oscillating terms at frequency $h\omega_o$

$$\langle V_h\,I_h\rangle = -2I_b^2\left\{Z_r^\circ - \left[Z_i^\circ - \tfrac{1}{2}(Z_i^+ + Z_i^-)\right]\varphi - \frac{Z_r^+ - Z_r^-}{2\,\Omega_s}\dot{\varphi}\right\}\,. \tag{6.157}$$

This is the rate of energy loss of the particle bunch into the impedance of the cavity. Dividing by the total circulating charge $T_o I_b$ we get the rate of relative energy loss per unit charge

$$\frac{\mathrm{d}\varepsilon}{\mathrm{d}t} = \frac{\langle eV_h\,I_h\rangle}{T_o I_b E_o} = -\frac{\ddot{\varphi}}{\beta c k_h \eta_c}\,, \tag{6.158}$$

where T_o is the revolution time and I_b the average beam current.

We made use of the relation between the energy deviation from the ideal energy and the rate of change of the phase [Ref. 6.3, Eq. (8.21)] on the r.h.s. of the equation. From (6.157,158) and making use of the definition of the *synchrotron frequency* in [Ref. 6.3, Eq. 8.33]

$$\Omega_{so}^2 = \frac{ck_h\eta_c}{E_o T_o}\,eV_{cy}\cos\psi_s$$

we get a differential equation of the form

$$\ddot{\varphi} + 2\alpha_R\,\dot{\varphi} + \omega_{cy}^2\,\varphi = 0 \tag{6.159}$$

with a *Robinson damping* decrement

$$\alpha_R = \frac{\beta\Omega_{so}}{2V_{cy}\cos\psi_s}\left(Z_r^+ - Z_r^-\right)I_b\,, \tag{6.160}$$

and a detuned cavity frequency

$$\omega_{cy}^2 = \frac{-2\beta\Omega_{so}^2}{V_{cy}\cos\psi_s}\left[Z_i^\circ - \tfrac{1}{2}\left(Z_i^+ + Z_i^-\right)\right]I_b\,. \tag{6.161}$$

The unperturbed phase equation [Ref. 6.3, Eq. (8.28)] is

$$\ddot{\varphi} + 2\alpha_{so}\,\dot{\varphi} + \Omega_{so}^2\,\varphi = 0\,.$$

Combining both, we derive a modification of both the damping and oscillation frequency. The combined *damping decrement* is

$$\alpha_s = \alpha_{so} + \frac{\beta \Omega_{so}}{V_{cy} \cos \psi_s} \left(Z_r^+ - Z_r^- \right) I_b > 0 \qquad (6.162)$$

where α_{so} is the radiation damping in electron accelerators. The total damping decrement must be positive for beam stability. The interaction of the beam with the accelerating cavity above transition where $\cos \psi_s < 0$ is stable for all values of the beam current if $Z_r^+ < Z_r^-$ or if the cavity resonant frequency is capacitively detuned. Due to the imaginary part of the impedance the interaction of beam and cavity leads to a *synchrotron oscillation frequency shift* given by

$$\Omega_s^2 = \Omega_{so}^2 - \frac{2\beta \Omega_{so}^2}{V_{cy} \cos \psi_s} \left[Z_i^o - \tfrac{1}{2} \left(Z_i^+ + Z_i^- \right) \right] I_b . \qquad (6.163)$$

This frequency shift has two components, the incoherent frequency shift due to the impedance Z_i^o at the fundamental beam frequency $h\omega_o$ and a frequency shift for *coherent bunch-phase oscillations* due to the imaginary part of the cavity impedances. For small frequency shifts $\Delta \Omega_s = \Omega_s - \Omega_{so}$ (6.163) can be linearized for

$$\frac{\Delta \Omega_s}{\Omega_{so}} = \frac{-I_b \beta}{V_{cy} \cos \psi_s} \left[Z_i^o - \tfrac{1}{2}(Z_i^+ + Z_i^-) \right] . \qquad (6.164)$$

The *cavity impedance* is from (6.85)

$$Z = R_s \frac{1 - iQ_o \frac{\omega^2 - \omega_r^2}{\omega_r \omega}}{1 + Q_o^2 \left(\frac{\omega^2 - \omega_r^2}{\omega_r \omega} \right)^2} . \qquad (6.165)$$

From the imaginary part of the cavity impedance and capacitive detuning we conclude that above transition energy the *incoherent synchrotron tune shift* is negative

$$\Delta \Omega_{s,\text{incoh}} < 0 \qquad (6.166)$$

while the *coherent synchrotron tune shift* is positive

$$\Delta \Omega_{s,\text{coh}} > 0 . \qquad (6.167)$$

This conclusion may in special circumstances be significantly different due to passive cavities in the accelerator. The shift in the synchrotron tune is proportional to the beam current and can be used as a diagnostic tool to determine the cavity impedance or its deviation from the ideal model (6.165).

In the preceding discussion it was assumed that only resonant cavities contribute to Robinson damping. This is correct to the extend that other cavity like structures of the vacuum enclosure in a circular accelerator have a low quality factor Q for the whole spectrum or at least at multiples of the revolution frequency and therefore do not contribute significantly to this effect through a persistent energy loss over many turns. Later we will see that such low-Q structures in the vacuum chamber may lead to other types of beam instability.

Potential-well distortion: The synchrotron frequency is determined by the slope of the rf voltage at the synchronous phase. In the last subsection the effect of beam loading at the cavity fundamental frequency was discussed demonstrating the need to include the induced voltages in the calculation of the synchrotron oscillation frequency. These induced voltages cause a perturbation of the *potential well* and as a consequence a change in the bunch length. In this subsection we will therefore also include higher-order interaction of the beam with its environment.

It is not possible to derive a general expression for the impedance of all components of a vacuum chamber in a circular accelerator. However, measurements [6.21] have shown that the *impedance spectrum* of circular accelerator vacuum chambers, while excluding accelerating cavities, has the form similar to that shown in [Ref. 6.3, Fig. 12.4] for the SPEAR storage ring which is reproduced for convenience in Fig. 6.15.

Up to the transition frequency f_t, which is determined by vacuum chamber dimensions, the impedance is predominantly inductive and becomes capacitive above the transition frequency. We are looking here only for fields with wavelength longer than the bunch length which may distort the rf voltage waveform such as to change the slope for the whole bunch. Later we will consider shorter wavelength which give rise to perturbations within the bunch. Because the bunch length is generally of the order of vacuum chamber dimensions we only need to consider the impedance spectrum below transition frequency which is predominantly inductive. To preserve generality, however, we assume a more general but still purely imaginary impedance defined by

$$Z(\omega)_\| = i\omega \mathcal{Z}_\| . \tag{6.168}$$

For mathematical simplicity in studying the modification of a finite bunch length due to potential-well distortions we use a parabolic particle distribution [6.22] in phase (Fig. 6.16) normalized to the bunch current $\int_{-\varphi_\ell}^{\varphi_\ell} I(\varphi)\,\mathrm{d}\varphi = I_\mathrm{b}$, i.e.

$$I(\varphi) = \frac{3\,I_\mathrm{b}}{4\,\varphi_\ell} \left(1 - \frac{\varphi^2}{\varphi_\ell^2}\right) , \tag{6.169}$$

where $2\varphi_\ell$ is the bunch length expressed in terms of a phase with respect to the fundamental rf wavelength.

The combined induced voltage in the whole vacuum chamber is

$$V_{\mathcal{Z}} = \mathcal{Z}_\| \frac{\mathrm{d}I}{\mathrm{d}t} = h\omega_\mathrm{o}\,\mathcal{Z}_\| \frac{\mathrm{d}I}{\mathrm{d}\varphi} = h\,\mathrm{Im}\left\{\frac{Z_\|}{n}\right\}\frac{\mathrm{d}I}{\mathrm{d}\varphi} , \tag{6.170}$$

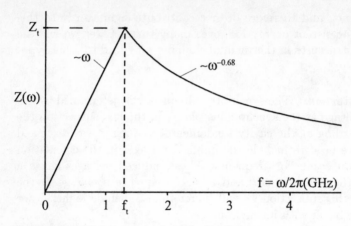

Fig. 6.15. SPEAR impedance spectrum [6.21]

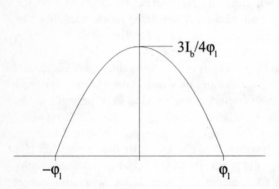

Fig. 6.16. Current distribution for potential-well distortion

where we have introduced the *normalized impedance*

$$\frac{Z_{\parallel}}{n} = i\omega_o Z_{\parallel} \tag{6.171}$$

which is the longitudinal impedance divided by the frequency in units of the revolution frequency or the *mode number* $n = \omega/\omega_o$. Inserting (6.169) into (6.170) we get the induced voltage

$$V_Z = -\frac{3h\,I_b}{2\,\varphi_\ell^3}\,\mathrm{Im}\left\{\frac{Z_{\parallel}}{n}\right\}\varphi \tag{6.172}$$

which must be added to the rf voltage $V_{\mathrm{rf}} = V_{\mathrm{cy}}\sin(\psi_s + \varphi)$. Forming an effective voltage we get

$$V_{\mathrm{eff}} = V_{\mathrm{cy}}\cos\psi_s\left(1 - \frac{3h\,I_b}{2\varphi_\ell^3 V_{\mathrm{cy}}\cos\psi_s}\,\mathrm{Im}\left\{\frac{Z_{\parallel}}{n}\right\}\right)\varphi + V_{\mathrm{cy}}\sin\psi_s\,. \tag{6.173}$$

This modification of the effective cavity voltage leads to an *incoherent shift* of the *synchrotron oscillation frequency*

$$
\boxed{\frac{\Omega_s^2}{\Omega_{so}^2} = 1 - \frac{3\eta_c e\, I_b}{4\pi\, \varphi_\ell^3\, E\nu_s^2}\, \mathrm{Im}\left\{\frac{Z_\parallel}{n}\right\},}
\qquad (6.174)
$$

where we used the definition of the synchrotron tune $\nu_s^2 = \frac{\eta_c eV_{cy}\cos\psi_s}{2\pi h E}$.

Above transition energy $\cos\psi_s < 0$ and therefore the frequency shift is positive for $\mathrm{Im}\{Z_\parallel/n\} < 0$ and negative for $\mathrm{Im}\{Z_\parallel/n\} > 0$. We note specifically that the shift depends strongly on the bunch length and increases with decreasing bunch length, a phenomenon we observe in all higher-order mode interactions.

Note that this shift of the synchrotron oscillation frequency does not appear for *coherent oscillations* since the induced voltage also moves with the bunch oscillation. The bunch center actually sees always the unaltered rf field and oscillates according to the slope of the unperturbed rf voltage. The coherent synchrotron oscillation frequency therefore need not be the same as the incoherent frequency. This has some ramification for the experimental determination of the synchrotron oscillation frequency.

The shift in incoherent synchrotron oscillation frequency reflects also a change in the equilibrium bunch length which is different for proton or ion beams compared to an electron bunch. The energy spread of radiating electron beams is determined only by quantum fluctuations due to the emission of synchrotron radiation and is independent of rf fields. The electron bunch length scales therefore inversely proportional to the synchrotron oscillation frequency and we get with $\Omega_s/\Omega_{so} = \sigma_{\ell o}/\sigma_\ell$ from (6.174) after solving for $\sigma_\ell/\sigma_{\ell o}$

$$
\frac{\sigma_\ell^3}{\sigma_{\ell o}^3} - \frac{\sigma_\ell}{\sigma_{\ell o}} - \frac{8\,\eta_c\, eI_b}{9\pi^2\,\sqrt{2\pi}\,\sigma_{\ell o}^3\, E\,\nu_s^2}\, \mathrm{Im}\left\{\frac{Z_\parallel}{n}\right\} = 0,
\qquad (6.175)
$$

where we replaced the parabolic current distribution by a gaussian distribution with equal total bunch current and equal intensity in the bunch center by setting $\varphi_\ell = 3\sqrt{2\pi}/4\,h\,\sigma_\ell/\bar R$ and where $\sigma_{\ell o}$ is the unperturbed bunch length.

Nonradiating particles, in contrast, must obey Liouville's theorem and the longitudinal beam emittance $\ell\,\Delta p$ will not change due to potential-well distortions. For proton or ion bunches we employ the same derivation for the bunch lengthening but note that the bunch length scales with the energy spread in such a way that the product of bunch length ℓ and momentum spread Δp remains constant. Therefore $\ell \propto 1/\sqrt{\Omega_s}$ and the perturbed bunch length is from (6.174) with $\ell = (\bar R/h)\,\varphi_\ell$

$$
\frac{\ell^4}{\ell_o^4} - \frac{3\,\eta_c\, eI_b}{4\pi\, E\nu_s^2}\,\frac{\bar R^3}{\ell_o^3}\, \mathrm{Im}\left\{\frac{Z_\parallel}{n}\right\}\frac{\ell}{\ell_o} - 1 = 0,
\qquad (6.176)
$$

where we have kept a finite value for the synchronous phase to cover the case of acceleration in a synchrotron. In a storage ring $\psi_s = 0$. Of course along with this perturbation of the proton or ion bunch length goes an opposite perturbation of the momentum spread.

6.3 Higher-Order Phase Focusing

The principle of phase focusing is fundamental for beam stability in circular accelerators and we find the *momentum compaction factor* to be a controlling quantity. Since the specific value of the momentum compaction determines critically the beam stability, it is interesting to investigate the consequences to beam stability as the momentum compaction factor varies. Specifically, we will discuss the situation where the linear momentum compaction factor is reduced to very small values and higher-order terms become significant. This is, for example, of interest in proton or ion accelerators going through *transition energy* during acceleration, or as we try to increase the quadrupole focusing in electron storage rings to obtain a small beam emittance, or when we intentionally reduce the momentum compaction to reduce the bunch length. In extreme cases, the momentum compaction factor becomes zero at transition energy or in an *isochronous storage ring* where the revolution time is made the same for all particles independent of the momentum. The linear theory of phase focusing would predict beam loss in such cases due to lack of phase stability. To accurately describe the beam stability, when the momentum compaction factor is small or vanishes, we obviously cannot completely ignore higher-order terms. Some of the higher-order effects on phase focusing will be discussed here. There are two main contributions to the higher-order momentum compaction factor, one from the dispersion function and the other from the momentum dependent path length. First, we derive the higher-order contributions to the dispersion function, and then apply the results to the principle of phase focusing to determine the perturbation on the beam stability.

6.3.1 Path Length in Higher Order

The *path length* together with the velocity of particles governs the time of arrival at the accelerating cavities from turn to turn and therefore defines the stability of a particle beam. Generally, only the linear dependence of the path length on particle momentum is considered. We find, however, higher-order chromatic contributions of the dispersion function to the path length as well as momentum independent contributions due to the finite angle of the trajectory with respect to the ideal orbit during betatron oscillations.

The path length for a particular trajectory from point $s = 0$ to point s in our curvilinear coordinate system can be derived from the integral

$L = \oint_0^s d\sigma$, where σ is the coordinate along the particular trajectory. This integral can be expressed by

$$L = \oint \sqrt{(1 + \kappa x)^2 + x'^2 + y'^2} \, ds, \qquad (6.177)$$

where the first term of the integrand represents the contribution to the path length due to curvature generated by bending magnets while the second and third term are contributions due to finite horizontal and vertical betatron oscillations. For simplicity, we ignore vertical bending magnets. Where this simplification cannot be made, it is straight forward to extend the derivation of the path length in higher order to include bending in the vertical plane as well. We expand (6.177) up to second order and get for the path length variation $\Delta L = L - L_o$

$$\Delta L = \oint \left(\kappa x + \tfrac{1}{2} \kappa^2 x^2 + \tfrac{1}{2} x'^2 + \tfrac{1}{2} y'^2 \right) ds + O(3). \qquad (6.178)$$

The particle amplitudes are composed of betatron oscillation, orbit distortions and off energy orbits

$$\begin{aligned} x &= x_\beta + x_o + \eta_o \, \delta + \eta_1 \, \delta^2 + \dots, \\ y &= y_\beta + y_o, \end{aligned} \qquad (6.179)$$

where (x_β, y_β) describes the betatron oscillations and $(\eta_o, \eta_1, \dots)$ are the linear and higher-order dispersion functions derived in Sec. 4.4. The quantities (x_o, y_o) describe a deviation of the actual orbit from the ideal orbit due to magnetic field and alignment errors.

Evaluating the integral (6.178), we note that the oscillatory character of (x_β, y_β) causes all terms linear in (x_β, y_β) to vanish while averaging over many turns. The orbit distortions (x_o, y_o) are statistical in nature since the correction in a real accelerator is done such that $\langle x_o \rangle = 0$ and $\langle x'_o \rangle = 0$. Betatron oscillations and orbit distortions are completely independent and therefore cross terms like $\langle x_\beta x_o \rangle$ vanish. The dispersion function η_o and the higher-order term η_1 are unique periodic solutions of the inhomogeneous equation of motion. For the betatron oscillations we assume a nonresonant tune which causes terms like $\langle x_\beta \eta_o \rangle$ to vanish as well. With these results the path length variation is

$$\begin{aligned} \Delta L &\approx \tfrac{1}{2} \oint \left(x_\beta'^2 + y_\beta'^2 + x_o'^2 + y_o'^2 + \kappa^2 \, x_\beta^2 + \kappa^2 \, x_o^2 \right) ds \\ &+ \delta \oint \kappa \eta_o \, ds + \delta^2 \oint \left(\kappa \eta_1 + \tfrac{1}{2} \kappa^2 \eta_o^2 + \tfrac{1}{2} \eta_o'^2 \right) ds. \end{aligned} \qquad (6.180)$$

There are three main contributions of which two are of chromatic nature. The finite transverse betatron oscillations as well as orbit distortions contribute to a second order increase in the path length of the beam transport system which is of nonchromatic nature. Equation (6.180) can be simplified by using the explicit expressions for the particle motion $x_\beta(s) = \sqrt{\epsilon_x \beta_x(s)} \sin \psi_x(s)$ and $x'_\beta = \sqrt{\epsilon_x/\beta_x} (\cos \psi - \alpha_x \sin \psi)$. Forming the square $x'^2_\beta = (\epsilon_x/\beta_x) (\cos^2 \psi - \alpha_x \sin 2\psi_x + \alpha_x^2 \sin^2 \psi_x)$ and averaging over all phases ψ_x

$$\oint x'^2_\beta \, ds = \epsilon_x \oint \frac{1}{\beta_x} (\cos^2 \psi + \alpha_x^2 \sin^2 \psi) \, ds ,$$

$$= \epsilon_x \oint \left(\frac{\cos 2\psi}{\beta_x} + \gamma_x \sin^2 \psi \right) \, ds \approx \tfrac{1}{2} \epsilon_x \oint \gamma_x \, ds , \tag{6.181}$$

where we used the simplifying expression $\sin^2 \psi \approx 1/2$. Similarly, we get

$$\oint y'^2_\beta \, ds \approx \tfrac{1}{2} \epsilon_y \oint \gamma_y \, ds \tag{6.182}$$

and

$$\oint \kappa^2 x^2_\beta \, ds \approx \tfrac{1}{2} \epsilon_x \oint \kappa^2 \beta_x \, ds . \tag{6.183}$$

The integrals are taken over the entire beam transport line of length L_o and using average values for the integrands the *path-length variation* is

$$\boxed{\begin{aligned} \frac{\Delta L}{L_o} &= + \tfrac{1}{4} \left(\epsilon_x \langle \gamma_x \rangle + \epsilon_y \langle \gamma_y \rangle + \epsilon_x \langle \kappa^2 \beta_x \rangle \right) \\ &\quad + \tfrac{1}{2} \langle x'^2_o \rangle + \tfrac{1}{2} \langle y'^2_o \rangle + \tfrac{1}{2} \langle x^2_o \rangle \\ &\quad + \alpha_c \delta + \left(\langle \kappa \eta_1 \rangle + \tfrac{1}{2} \langle \kappa^2 \eta^2_o \rangle + \tfrac{1}{2} \langle \eta'^2_o \rangle \right) \delta^2 . \end{aligned}} \tag{6.184}$$

In this expression for the path-length variation we find separate contributions due to betatron oscillations, orbit distortion and higher-order chromatic effects. We have used the emittance ϵ as the amplitude factor for betatron oscillation and get therefore a path length spread within the beam due to the finite beam emittance ϵ. Note specifically that for an electron beam this emittance scales by the factor n_σ^2 for gaussian tails where n_σ is the oscillation amplitude in units of the standard amplitude. For a stable machine condition the contribution of the orbit distortion is the same for all particles and can therefore be corrected by an adjustment of the rf frequency. We include these terms here, however, to allow the estimation of allowable tolerances for dynamic orbit changes.

6.3.2 Higher-Order Phase Space Motion

The longitudinal phase stability in a circular accelerator depends on the value of the momentum compaction η_c which actually regulates the phase focusing to obtain stable particle motion. This parameter is not a quantity that can be chosen freely in the design of a circular accelerator without jeopardizing other desirable design goals. If, for example, a small beam emittance is desired in an electron storage ring, or if for some reason it is desirable to have an isochronous ring where the revolution time for all particles is the same we find that the momentum compaction becomes very small. This in itself does not cause instability unless the momentum compaction approaches zero and higher-order chromatic terms modify phase focusing to the extent that the particle motion becomes unstable. To derive the conditions for the loss of phase stability, we evaluate the path length variation (6.184) with momentum in higher order

$$\frac{\Delta L}{L_o} = \alpha_c \delta + \alpha_1 \delta^2 + \xi + \mathcal{O}(3), \tag{6.185}$$

where ξ represents the momentum independent term

$$\xi = \frac{1}{4} \left(\varepsilon_x \langle \gamma_x \rangle + \varepsilon_y \langle \gamma_y \rangle + \varepsilon_x \langle \kappa^2 \beta_x \rangle \right) \tag{6.186}$$

and

$$\alpha_1 = \langle \kappa \eta_1 \rangle + \frac{1}{2} \langle \kappa^2 \eta_o^2 \rangle + \frac{1}{2} \langle \eta_o'^2 \rangle. \tag{6.187}$$

Following the derivation of the linear phase equation, we note that it is the variation of the revolution time with momentum rather than the path-length variation that affects the synchronicity condition. With

$$\eta_c = \frac{1}{\gamma^2} - \alpha_c \tag{6.188}$$

the differential equation for the phase oscillation to second order is

$$\frac{\partial \psi}{\partial t} = -\omega_{rf} \left(\eta_c \delta - \alpha_1 \delta^2 - \xi \right) \tag{6.189}$$

and for the momentum oscillation

$$\frac{\partial \delta}{\partial t} = \frac{e V_{rf}}{T_o \, cp_o} \left(\sin \psi - \sin \psi_s \right). \tag{6.190}$$

In linear approximation, where $\alpha_1 = 0$ and $\xi = 0$, a single pair of *fixed points* and *separatrices* exist in phase space. These fixed points can be found from the condition that $\dot{\psi} = 0$ and $\dot{\delta} = 0$ and lie on the abscissa for $\delta = 0$. The *stable fixed point* is located at $(\psi_{sf}, \delta_{sf}) = (\psi_s, 0)$ defining the center of the rf bucket where stable phase oscillations occur. The *unstable fixed point* at $(\psi_{uf}, \delta_{uf}) = (\pi - \psi_s, 0)$ defines the crossing point of the separatrices separating the trajectories of oscillations from those of librations.

Considering also higher-order terms in the theory of phase focusing leads to a more complicated pattern of phase space trajectories. Setting (6.190) equal to zero we note that the abscissae of the fixed points are at the same location as for the linear case

$$\psi_{1f} = \psi_s \quad \text{and} \quad \psi_{2f} = \pi - \psi_s. \tag{6.191}$$

The ordinates of the fixed points, however, are determined by solving the nonlinear equation (6.189) for $\dot{\psi} = 0$

$$\eta_c\,\delta_f - \alpha_1\,\delta_f^2 - \xi = 0 \qquad (6.192)$$

with the solutions

$$\boxed{\delta_f = +\frac{\eta_c}{2\,\alpha_1}\left(1 \pm \sqrt{1 - \Gamma}\right),} \qquad (6.193)$$

where

$$\Gamma = \frac{4\,\xi\,\alpha_1}{\eta_c^2}. \qquad (6.194)$$

Due to the quadratic character of equation (6.192) we get now two layers of fixed points with associated areas of oscillation and libration. In Figs. 6.17, 18a and 18b the phase diagrams are shown for increasing values of the perturbation α_1 while for now we set the momentum independent perturbation $\xi = 0$. Numerically the contour lines have been calculated from the Hamiltonian (6.198) with $\Delta/2\eta_c = 0.005$. The appearance of the second layer of stable islands and the increasing perturbation of the original rf buckets is obvious. There is actually a point (Fig. 6.18a) where the separatrices of both island layers merge. We will use this merging of the separatrices later to define a tolerance limit for the perturbation on the momentum acceptance.

The coordinates of the fixed points in the phase diagram are determined from (6.200,201) and are for the linear fixed points in the first layer

Point A: $\psi_A = \psi_s$, $\delta_A = \dfrac{\eta_c}{2\,\alpha_1}\left(1 - \sqrt{1 - \Gamma}\right)$,

Point B: $\psi_B = \pi - \psi_s$, $\delta_B = \dfrac{\eta_c}{2\,\alpha_1}\left(1 - \sqrt{1 - \Gamma}\right)$. $\qquad (6.195)$

The momenta of these fixed points are at $\delta = 0$ for $\Gamma = 0$ consistent with earlier discussions. As orbit distortions and betatron oscillations increase, however, we note a displacement of the equilibrium momentum as Γ increases.

The fixed points of the second layer of islands or rf buckets are displaced both in phase and in momentum with respect to the linear fixed points and are located at

Point C: $\psi_C = \psi_s$, $\delta_C = \dfrac{\eta_c}{2\,\alpha_1}\left(1 + \sqrt{1 - \Gamma}\right)$,

Point D: $\psi_D = \pi - \psi_s$, $\delta_D = \dfrac{\eta_c}{2\,\alpha_1}\left(1 + \sqrt{1 - \Gamma}\right)$. $\qquad (6.196)$

The dependence of the coordinates for the fixed points on orbit distortions and the amplitude of betatron oscillations becomes evident from (6.206,207). Specifically, we note a shift in the reference momentum of the beam by ξ/η_c

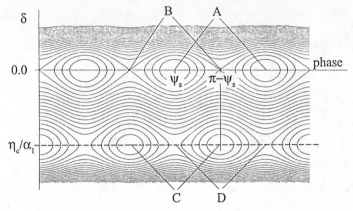

Fig. 6.17. Second-order longitudinal phase space for $\psi_s = 0$, $\xi = 0$ and weak perturbation $\alpha_1/\eta_c = -3.0$

Fig. 6.18. Higher-order longitudinal phase space diagrams for $\psi_s = 0$, $\xi = 0$ and strong perturbation $\alpha_1/\eta_c = -6.0$ **a** and $\alpha_1/\eta_c = -13.5$ **b**

as the orbit distortion increases as demonstrated in the examples shown in Figs. (6.19, 20, 21c, 21d). Betatron oscillations, on the other hand, cause a spread of the beam momentum in the vicinity of the fixed points. This readjustment of the beam momentum is a direct consequence of the principle of phase focusing whereby the particle follows a path such that the synchronicity condition is met. The phase space diagram of Fig. 6.17 is repeated in Fig. 6.19 with a parameter $2\xi/\eta_c = -0.125$ and in Fig. 6.20 with the further addition of a finite synchronous phase of $\psi_s = 0.7$ rad. In addition to the shift of the reference momentum a significant reduction in the momentum acceptance compared to the regular rf buckets is evident in both diagrams.

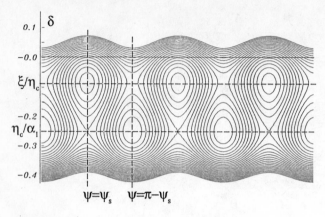

Fig. 6.19. Second-order longitudinal phase space for $\psi_s = 0$, $2\xi/\eta_c = -0.125$ and weak perturbation $\alpha_1/\eta_c = -3.0$

As long as the perturbation is small and $|\alpha_1| \ll |\eta_c|$ the new fixed points are located far away from the reference momentum and their effect on the particle dynamics can be ignored. The situation becomes very different whenever the linear momentum compaction becomes very small or even zero due to strong quadrupole focusing during momentum ramping through transition or in the case of the deliberate design of a *low α-lattice* for a *quasi isochronous* storage ring. In these cases higher order perturbations become significant and cannot be ignored. We cannot assume anymore that the perturbation term α_1 is negligibly small and the phase dynamics may very well become dominated by perturbations.

The perturbation α_1 of the momentum compaction factor depends on the perturbation of the dispersion function and is therefore also dependent on the sextupole distribution in the storage ring. Given sufficient sextupole families it is at least in principle possible to adjust the parameter α_1 to zero or a small value by a proper distribution of sextupoles.

6.3.3 Stability Criteria

Stability criteria for phase oscillations under the influence of higher order momentum compaction terms can be derived from the Hamiltonian. The nonlinear equations of motion (6.200,201) can be derived from the Hamiltonian

$$
\begin{aligned}
H = {} & \frac{e\,V_{cy}}{T_o\,cp_o} \left[\cos\psi - \cos\psi_s + (\psi - \psi_s)\sin\psi_s \right] \\
& - \omega_{rf} \left[\tfrac{1}{2}\,\eta_c\,\delta^2 - \tfrac{1}{3}\,\alpha_1\,\delta^3 - \xi\,\delta \right].
\end{aligned}
\tag{6.197}
$$

To eliminate inconsequential factors for the calculation of phase space trajectories we simplify (6.197) to

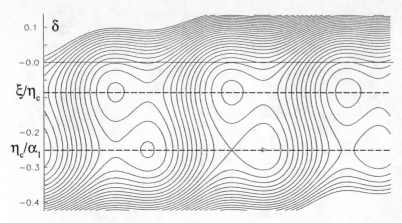

Fig. 6.20. Higher-order longitudinal phase space diagrams for $\psi_s = 0.7$, $2\xi/\eta_c = -0.125$ and a weak perturbation $\alpha_1/\eta_c = -3.0$

$$\tilde{H} = \Delta \left[\cos \psi - \cos \psi_s + (\psi - \psi_s) \sin \psi_s\right]$$
$$- 2\eta_c \delta^2 + \tfrac{4}{3}\alpha_1 \delta^3 + 4\xi \delta \,, \tag{6.198}$$

where

$$\Delta = \frac{2\,e\,V_{cy}}{T_o\,cp_o\,\omega_{rf}\,\eta_c}\,. \tag{6.199}$$

We may use (6.198) to calculate phase space trajectories and derive stability conditions for various combinations of the parameters Δ, the perturbation of the momentum compaction α_1, and the synchronous phase ψ_s (Figs. 6.17 – 20). In Fig. 6.21 the phase diagrams of Figs. 6.17 – 20 are displayed now as three-dimensional surfaces plots with the same parameters. Starting from the linear approximation where only regular rf buckets appear along the ψ-axis, we let the ratio α_1/η_c increase and find the second set of rf buckets to move in from large relative momentum errors δ_f toward the main rf buckets. A significant modification of the phase diagrams occurs when the perturbation reaches such values that the separatrices of both sets of buckets merge (Fig. 6.18a). A further increase of the perturbation quickly reduces the momentum acceptance of the rf system as can be noticed by comparing Figs. 6.18a and 18b or Figs. 6.21a and 21b. The effect of the momentum shift when $\xi \neq 0$ becomes obvious in diagrams 6.19,20 and 6.21c and d as well as the effect of a finite synchronous phase in 6.21d.

From these qualitative observations we derive a threshold of allowable perturbation α_1 above which the momentum acceptance of the system becomes significantly reduced. From Fig. 6.18a we take the condition for momentum stability when the separatrices of both sets of buckets merge which occurs when the Hamiltonian for both separatrices or for the fixed points (B) and (C) are equal and

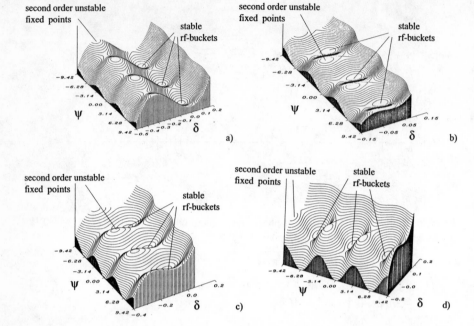

Fig. 6.21. Three dimensional rendition of Figs. 6.17 – 20

$$\tilde{H}(\pi - \psi_\mathrm{s}, \delta_\mathrm{B}) = \tilde{H}(\psi_\mathrm{s}, \delta_\mathrm{c}).$$ (6.200)

Equation (6.200) becomes in the form of (6.198)

$$\Delta \left(-2\cos\psi_\mathrm{s} + (\pi - 2\psi_\mathrm{s})\sin\psi_\mathrm{s}\right) - \delta_\mathrm{B}^2 + \frac{2}{3}\frac{\alpha_1}{\eta_\mathrm{c}}\delta_\mathrm{B}^3 + 2\frac{\xi}{\eta_\mathrm{c}}\delta_\mathrm{B}$$
$$= -\delta_\mathrm{c}^2 + \frac{2}{3}\frac{\alpha_1}{\eta_\mathrm{c}}\delta_\mathrm{c}^3 + 2\frac{\xi}{\eta_\mathrm{c}}\delta_\mathrm{c}\,.$$ (6.201)

Comparing (6.199) with the results of linear theory, we note that the maximum unperturbed momentum acceptance is related to the parameter Δ by

$$\Delta = \frac{1}{F(q)\sin\psi_\mathrm{s}}\left(\frac{\Delta p}{p_\mathrm{o}}\right)^2_\mathrm{max}\frac{|\eta_\mathrm{c}|}{\eta_\mathrm{c}},$$ (6.202)

where with $q = 1/\sin\psi_\mathrm{s}$

$$F(q) = [-2\cos\psi_\mathrm{s} + (\pi - 2\psi_\mathrm{s})\sin\psi_\mathrm{s}].$$ (6.203)

Equation (6.201) can be solved for the maximum momentum acceptance

$$\left(\frac{\Delta p}{p_o}\right)^2_{\max} = \frac{\eta_c}{|\eta_c|}\,(\delta_C^2 - \delta_B^2) + \frac{2}{3}\,\frac{\alpha_1}{|\eta_c|}\,(\delta_C^3 - \delta_B^3) + \frac{2\xi}{|\eta_c|}\,(\delta_C - \delta_B).$$ (6.204)

Using the expressions (6.193) for the coordinates of the fixed points, (6.204) eventually becomes with (6.194)

$$\left(\frac{\Delta p}{p_o}\right)^2_{\max} = \frac{\eta_c^2}{3\,\alpha_1^2}\,(1 - \Gamma)^{3/2},$$ (6.205)

and the *stability criterion* that the nonlinear perturbation not reduce the *momentum acceptance* is finally expressed by

$$\boxed{\; |\alpha_1| \leq \frac{|\eta_c|}{\sqrt{3}}\,\frac{(1 - \Gamma)^{3/4}}{\left(\frac{\Delta p}{p_o}\right)_{\text{desired}}}. \;}$$ (6.206)

From this criterion we note that the momentum independent perturbation Γ can further limit the momentum acceptance until there is for $\Gamma \geq 1$ no finite momentum acceptance left at all.

The *momentum shift* and the momentum acceptance as well as *stability limits* can be calculated analytically as a function of α_1 and the momentum independent term Γ. As long as the perturbation is small and (6.206) is fulfilled we calculate the momentum acceptance for the linear rf buckets from the value of the Hamiltonian (6.198). For stronger perturbations where the separatrices of both layers of rf buckets have merged and are actually exchanged (Fig. 6.17), a different value of the Hamiltonian must be chosen. The maximum stable synchrotron oscillation in this case is not anymore defined by the separatrix through fixed point B but rather by the separatrix through fixed point C. In the course of synchrotron oscillations a particle reaches maximum momentum deviations from the reference momentum at the phase $\psi = \psi_s$. We have two extreme momentum deviations, one at the fixed point (C), and the other half a synchrotron oscillation away. Both points have the same value of the Hamiltonian (6.198) and are related by

$$-\hat{\delta}^2 + \frac{2}{3}\,\frac{\alpha_1}{\eta_c}\,\hat{\delta}^3 + \frac{2c}{\eta_c}\,\hat{\delta} = -\delta_C^2 + \frac{2}{3}\,\frac{\alpha_1}{\eta_c}\,\delta_C^3 + \frac{2\xi}{\eta_c}\,\delta_C.$$ (6.207)

We replace δ_C from (6.196) and obtain a third-order equation for the maximum momentum acceptance $\hat{\delta}$

$$\hat{\delta}^2 + \frac{2}{3}\,\frac{\alpha_1}{\eta_c}\,\hat{\delta}^3 + \frac{2\xi}{\eta_c}\,\hat{\delta} = \frac{\eta_c}{6\,\alpha_1^2}\left[1 + (1 - \Gamma)^{\frac{3}{2}} - \frac{3}{2}\,\Gamma\right].$$ (6.208)

This third-order equation can be solved analytically and has the solutions

$$\hat{\delta}_1 = \frac{\eta_c}{\alpha_1}\,(1 - 2\sqrt{1 - \Gamma}),$$

$$\hat{\delta}_{2,3} = \frac{\eta_c}{\alpha_1}\,(1 + \sqrt{1 - \Gamma}).$$ (6.209)

Two of the three solutions are the same and define the momentum at the crossing of the separatrix at the fixed point (C) while the other solution determines the momentum deviation half a synchrotron oscillation away from the fixed point (C). We plot these solutions in Fig. 6.22 together with the momentum shift of the reference momentum at the fixed point (A). As long as there is no momentum independent perturbation $\Gamma = 0$ and the momentum acceptance is

$$-2 < \frac{\alpha_1}{-\eta_c} \delta < 1. \tag{6.210}$$

The asymmetry of the momentum acceptance obviously reflects the asymmetry of the separatrix. For $\alpha_1 \to 0$ the momentum acceptance in (6.210) diverges, which is a reminder that we consider here only the case where the perturbation α_1 exceeds the limit (6.206). In reality the momentum acceptance does not increase indefinitely but is limited by other criteria, for example, by the maximum rf voltage available. The momentum acceptance limits of (6.209) are further reduced by a finite beam emittance when $\Gamma \neq 0$ causing a spread in the revolution time. All beam stability is lost as Γ approaches unity and the *stability criterion* for stable synchrotron motion in the presence of betatron oscillations is defined by

$$\boxed{\frac{4\xi\alpha_1}{\eta_c^2} < 1,} \tag{6.211}$$

where the parameter ξ is defined by (6.186).

In evaluating the numerical value of ξ we must consider the emittances $\epsilon_{x,y}$ as amplitude factors. In case of a gaussian electron beam in a storage ring, for example, a long quantum lifetime can be obtained only if particles with betatron oscillation amplitudes up to at least seven standard values are stable. For such particles the emittance is $\epsilon = 7^2 \epsilon_\sigma$, where ϵ_σ is the beam emittance for one standard deviation. Similarly, the momentum acceptance must be large enough to include a momentum deviation of $\delta_{\max} \geq 7\sigma_E/E_0$.

In general, the stability criteria can be met easily if, by proper adjustment of sextupole magnets, the linear perturbation α_1 of the momentum compaction is set to zero. In this case, however, we must consider dynamic stability of the beam and storage ring to prevent α_1 to vary more than the stability criteria allow. Any dynamic variation of α_1 must meet the condition

$$\Delta\alpha_1 < \frac{\eta_c^2}{4\xi}. \tag{6.212}$$

Even if the quadratic term α_1 is made to approach zero we still must consider the momentum shift due to nonchromatic terms when $\xi \neq 0$. From (6.195) we have for the momentum shift of the stable fixed point A

$$\delta_A = \frac{-\eta_c}{2\alpha_1} \left(1 - \sqrt{1 - \Gamma}\right),$$

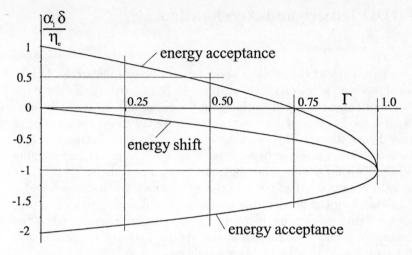

Fig. 6.22. Higher-order momentum acceptance

where Γ is small when $\alpha_1 \to 0$ and the square root can be expanded. In this limit the momentum shift becomes

$$\delta_A \to \frac{\xi}{\eta_c} \qquad \text{for} \qquad \alpha_1 \to 0. \tag{6.213}$$

To achieve low values of the momentum compaction, it is therefore also necessary to reduce the particle beam emittance. Case studies of isochronous lattices show, however, that this might be very difficult because the need to generate both positive and negative values for the dispersion function generates large values for the slopes of the dispersion leading to rather large beam emittances.

Adjusting the quadratic term α_1 to zero finally brings us back to the situation created when the linear momentum compaction was reduced to small values. One cannot ignore higher-order terms anymore. In this case we would expect that the quadratic and cubic perturbations of the momentum compaction will start to play a significant role since $\eta_c \approx 0$ and $\alpha_1 \approx 0$. The quadratic term α_3 will introduce a spread of the momentum compaction due to the momentum spread in the beam while the cubic term α_4 introduces a similar spread to the linear term α_1.

6.4 FODO Lattice and Acceleration

So far we have ignored the effect of acceleration in beam dynamics. In specific cases, however, acceleration effects must be considered specifically if the particle energy changes significantly along the beam line. In linear accelerators such a need occurs at low energies when we try to accelerate a large emittance beam through the small apertures of the accelerating sections. For example, when a positron beam is to be created the positrons emerging from a target within a wide solid angle are focused into the small aperture of a linear accelerator. After some initial acceleration in the presence of a solenoid field along the accelerating structure it is desirable to switch over to more economic quadrupole focusing. Even at higher energies when the beam diameter is much smaller than the aperture strong focusing is still desired to minimize beam break up instabilities.

Lattice structure: A common mode of focusing uses a FODO lattice in conjunction with the linac structure. We may, however, not apply the formalism developed for FODO lattices without modifications because the particle energy changes significantly along the lattice. A thin lens theory has been derived by Helm [6.23] based on a regular FODO channel in the particle reference system. Due to Lorentz contraction the constant quadrupole separation $\widetilde{L}$ in the particle system become increasing distances in the laboratory system as the beam energy increases. To show this quantitatively, we consider a FODO channel installed along a linear accelerator with a constant cell half length $\widetilde{L}$ in the particle system. In the laboratory system the cell length will vary according to the particle acceleration. The tick-marks along the scale in Fig. 6.23 indicate the locations of the quadrupoles and the distances between magnets in the laboratory system are designated by $L_1, L_2 \dots$.

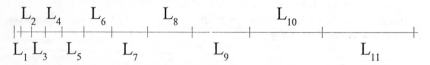

Fig. 6.23. FODO channel and acceleration

With α the acceleration in units of the rest energy per unit length and γ_0 the particle energy at the center of the first quadrupole the condition to have a FODO channel in the particle system is

$$\widetilde{L} = \int\limits_{s_o}^{s_1=s_o+L_1} \frac{ds}{1 + \frac{\alpha s}{\gamma_o}} = \frac{\gamma_o}{\alpha} \ln \left(1 + \frac{\alpha L_1}{\gamma_o} \right), \tag{6.214}$$

where s_o is the location of the first quadrupole center and s_1 that of the second. The quantity $2\widetilde{L}$ is the length of a FODO cell in the particle system and L_1 is the distance between the first and second quadrupole in the laboratory system. Solving for L_1 we get

$$L_1 = \widetilde{L} \frac{e^\kappa - 1}{\kappa}, \tag{6.215}$$

where

$$\kappa = \frac{\alpha}{\gamma_o} \widetilde{L}. \tag{6.216}$$

At the same time the beam energy has increased from γ_o to

$$\gamma_1 = \gamma_o + \alpha L_1. \tag{6.217}$$

Equation (6.214) can be applied to any of the downstream distances between quadrupoles. The nth distance L_n, for example, is determined by an integration from s_{n-1} to s_n or equivalently from 0 to L_n

$$\widetilde{L} = \int\limits_{o}^{L_n} \frac{ds}{1 + \frac{\alpha s}{\gamma_{n-1}}} = \frac{\gamma_{n-1}}{\alpha} \log \left(1 + \frac{\alpha L_n}{\gamma_{n-1}} \right). \tag{6.218}$$

While solving for L_n we express the energy γ_{n-1} by addition of the energy gains, $\gamma_{n-1} = \sum_i \Delta\gamma_i = \alpha \sum_i L_i$, taking the distances L_i from expressions (6.226,229) we get with $0 < i < n-1$

$$L_n = \widetilde{L} \frac{e^\kappa - 1}{\kappa} e^{(n-1)\kappa}. \tag{6.219}$$

In thin lens approximation we find thereby that the distances between successive quadrupoles in the laboratory system increase exponentially according to the relation (6.219) to resemble the focusing properties of a regular FODO channel with a cell length $2\widetilde{L}$ in the particle system under the influence of an accelerating field.

Such FODO channels are used to focus large emittance particle beams in linear accelerators as is the case for positron beams in positron linacs. For strong focusing as needed at low energies where the beam emittance is large, the thin lens approximation, however, is not accurate enough and a more exact formulation of the transformation matrices must be applied [6.24] which we will derive here in some detail.

6.4.1 Transverse Beam Dynamics and Acceleration

Transverse focusing can be significantly different along a linear accelerator due to the rapid changing particle energy compared to a fixed energy transport line and the proper beam dynamics must be formulated in the presence of longitudinal acceleration. To derive the correct equations of motion we consider the particle dynamics in the presence of the complete Lorentz force including electrical fields

$$\dot{\mathbf{p}} = e\,\mathbf{E} + [c]\,\frac{e}{c}\,[\dot{\mathbf{r}} \times \mathbf{B}]\,. \tag{6.220}$$

To solve this differential equation we consider a straight beam transport line with quadrupoles aligned along the s-coordinate as we would have in a linear accelerator. The accelerating fields are assumed to be uniform with a finite component only along the s-coordinate. At the location $\mathbf{r} = (x, y, s)$ the fields can be expressed by $\mathbf{E} = (0, 0, \alpha/e)$ and $\mathbf{B} = (gy, gx, 0)$, where the acceleration α is defined by

$$\alpha = e\,|\mathbf{E}|\,. \tag{6.221}$$

To evaluate (6.220) we express the time derivative of the momentum, $\mathbf{p} = \gamma\,m\,\dot{\mathbf{r}}$ by

$$\dot{\mathbf{p}} = \dot{\gamma}\,m\,\dot{\mathbf{r}} + \gamma\,m\,\ddot{\mathbf{r}}\,, \tag{6.222}$$

where $\gamma = E/(mc^2)$ is the particle energy in units of the rest energy. From $c\dot{p} = \dot{E}/\beta = c\alpha$ we find that $\dot{\gamma} = \alpha\beta/(mc)$ and (6.222) becomes for the x-component

$$\dot{p}_x = \frac{\alpha\beta}{c}\,m\,\dot{x} + \frac{E}{c^2}\,\ddot{x}\,. \tag{6.223}$$

In this subsection we make ample use of quantities α, β, γ being acceleration and relativistic parameters which should not be confused with the lattice functions which we will not need here. Bowing to convention we refrain from introducing new labels.

The variation of the momentum with time can be expressed also with the Lorentz equation (6.220) and with the specified fields we get

$$\dot{p}_x = -[c]\,e\,\beta\,g\,x\,. \tag{6.224}$$

We replace the time derivatives in (6.223) by derivatives with respect to the independent variable s

$$\dot{x} = \beta\,c\,x'\,,$$

$$\ddot{x} = \beta^2\,c^2\,x'' + \frac{\alpha}{\gamma^3\,m}\,x'\,,$$

and after insertion into (6.223) and equating with (6.224) the equation of motion becomes

$$\frac{d^2x}{ds^2} + \frac{\alpha/\beta}{\beta E}\frac{dx}{ds} + \frac{[c]\,e\,g}{\beta E}\,x = 0,\qquad(6.225)$$

where we used the relation $\beta^2 + 1/\gamma^2 = 1$. With $\frac{\alpha}{\beta} = \frac{dcp/ds}{cp_o}$ and defining the quantity

$$\eta_o = \frac{dcp/ds}{cp_o} = \frac{\alpha}{\beta cp_o}\qquad(6.226)$$

we get for the equation of motion in the horizontal plane, $u = x$

$$\frac{d^2u}{ds^2} + \frac{\eta_o}{1+\eta_o s}\frac{du}{ds} + \frac{k_o}{1+\eta_o s}\,u = 0,\qquad(6.227)$$

introducing the quadrupole strength $k_o = [c]\frac{eg}{cp_o}$. Equation (6.227) is valid also for the vertical plane $u = y$ if we only change the sign of the quadrupole strength k_o. Equation (6.227) is a *Bessel's differential equation* which becomes obvious by defining a new independent variable

$$\xi = \frac{2\beta}{\eta_o}\sqrt{k_o(1+\eta_o s)}\qquad(6.228)$$

transforming (6.227) into

$$\frac{d^2u}{d\xi^2} + \frac{1}{\xi}\frac{du}{d\xi} + u = 0\qquad(6.229)$$

which is the equation of motion in the presence of both transverse and longitudinal fields.

Analytical solutions: The solutions of the differential equation (6.229) are Bessel's functions of the first and second kind in zero order

$$u(s) = C_1\,I_0(\xi) + C_2\,Y_0(\xi).\qquad(6.230)$$

In terms of initial conditions (u_o, u_o') for $s = 0$ we can express the solutions in matrix formulation

$$\begin{pmatrix} u(s) \\ u'(s) \end{pmatrix} = \pi\frac{\sqrt{k_o}}{\eta_o}\begin{pmatrix} -I_o & Y_o \\ \frac{\sqrt{k_o}\,I_1}{\sqrt{1+\eta_o s}} & \frac{\sqrt{k_o}\,Y_1}{\sqrt{1+\eta_o s}} \end{pmatrix}\begin{pmatrix} Y_{1o} & \frac{Y_{oo}}{\sqrt{k_o}} \\ I_{1o} & \frac{I_{oo}}{\sqrt{k_o}} \end{pmatrix}\begin{pmatrix} u_o \\ u_o' \end{pmatrix}.\qquad(6.231)$$

Here we defined $Z_i = Z_i\left(\frac{2\beta}{\eta_o}\sqrt{k_o(1+\eta_o s)}\right)$ and $Z_{io} = Z_{io}\left(\frac{2\beta}{\eta_o}\sqrt{k_o}\right)$ where Z_i stands for either of the Bessel functions I_i or Y_i and $i = 0, 1$.

Transformation matrices: The *transformation matrix* for a drift space can be obtained from (6.231) by letting $k_o \to 0$ but it is much easier to just integrate (6.227) directly with $k_o = 0$. We get from (6.227) $\frac{u''}{u'} = -\frac{\eta_o}{1+\eta_o s}$, after logarithmic integration $u' = \frac{1}{1+\eta_o s}+$ const and after still another integration

$$u = u_o + \frac{u'_o}{\eta_o} \log(1 + \eta_o s) \tag{6.232}$$

or for a *drift space* of length L

$$\begin{pmatrix} u \\ u' \end{pmatrix} = \begin{pmatrix} 1 & \frac{1}{\eta_o} \log(1 + \eta_o L) \\ 0 & \frac{1}{1+\eta_o L} \end{pmatrix} \begin{pmatrix} u_o \\ u'_o \end{pmatrix}. \tag{6.233}$$

For most practical purposes we may assume that $2\frac{\sqrt{k_o}}{\eta_o} \gg 1$ and may, therefore, use asymptotic expressions for the Bessel functions. In this approximation the transformation matrix of a *focusing quadrupole* of length ℓ is

$$\begin{aligned} \mathcal{M}_f &= \begin{pmatrix} \sigma \cos \Delta\xi & \frac{\sigma}{\sqrt{k_o}} \sin \Delta\xi \\ -\sigma^3 \sqrt{k_o} \sin \Delta\xi & \sigma^3 \cos \Delta\xi \end{pmatrix} \\ &+ \begin{pmatrix} \frac{\sigma}{8}(\frac{3}{\xi_o} + \frac{1}{2}) \sin \Delta\xi & \frac{\sigma}{8\sqrt{k_o}} \frac{\Delta\xi}{\xi_o \xi_\ell} \cos \Delta\xi \\ \frac{3\sigma^3}{8} \frac{\Delta\xi}{\xi_o \xi_\ell} \sqrt{k_o} \cos \Delta\xi & -\frac{\sigma^3}{8}(\frac{1}{\xi_o} + \frac{3}{2}) \sin \Delta\xi \end{pmatrix}, \end{aligned} \tag{6.234}$$

where

$$\sigma^4 = \frac{1}{1 + \eta_o \ell} \tag{6.235}$$

and with $\Delta\xi = \xi_\ell - \xi_o$,

$$\begin{aligned} \xi_o &= \frac{2}{\eta_o} \sqrt{k_o} \quad \text{and} \\ \xi_\ell &= \frac{2}{\eta_o} \sqrt{k_o(1 + \eta_o \ell)}. \end{aligned} \tag{6.236}$$

Similarly we get for a *defocusing quadrupole*

$$\begin{aligned} \mathcal{M}_d &= \begin{pmatrix} \sigma \cosh \Delta\xi & \frac{\sigma}{\sqrt{k_o}} \sinh \Delta\xi \\ -\sigma^3 \sqrt{k_o} \sinh \Delta\xi & \sigma^3 \cosh \Delta\xi \end{pmatrix} \\ &+ \begin{pmatrix} \frac{\sigma}{8}(\frac{3}{\xi_o} + \frac{1}{2}) \sinh \Delta\xi & \frac{\sigma}{8\sqrt{k_o}} \frac{\Delta\xi}{\xi_o \xi_\ell} \cosh \Delta\xi \\ \frac{3\sigma^3}{8} \frac{\Delta\xi}{\xi_o \xi_\ell} \sqrt{k_o} \cosh \Delta\xi & -\frac{\sigma^3}{8}(\frac{1}{\xi_o} + \frac{3}{2}) \sinh \Delta\xi \end{pmatrix}. \end{aligned} \tag{6.237}$$

These transformation matrices can be further simplified for low accelerating fields noting that $\frac{\eta_o \ell}{4} \ll 1$. In this case $\xi_\ell - \xi_o \approx \sqrt{k_o}\ell = \psi$ and with

$$\Delta = \frac{1}{8}\left(\frac{3}{\xi_o} + \frac{1}{\xi_\ell}\right) \approx \frac{1}{8}\left(\frac{3}{\xi_\ell} + \frac{1}{\xi_o}\right) \tag{6.238}$$

we get for a *focusing quadrupole* the approximate transformation matrix

$$\mathcal{M}_f = \begin{pmatrix} \sigma & 0 \\ 0 & \sigma^3 \end{pmatrix} \left[\begin{pmatrix} \cos\psi & \frac{1}{\sqrt{k_o}}\sin\psi \\ -\sqrt{k_o}\sin\psi & \cos\psi \end{pmatrix} \right.$$
$$\left. + \begin{pmatrix} \Delta\sin\psi & 0 \\ 0 & -\Delta\sin\psi \end{pmatrix} \right] \tag{6.239}$$

and similar for a *defocusing quadrupole*

$$\mathcal{M}_d = \begin{pmatrix} \sigma & 0 \\ 0 & \sigma^3 \end{pmatrix} \left[\begin{pmatrix} \cosh\psi & \frac{1}{\sqrt{k_o}}\sinh\psi \\ -\sqrt{k_o}\sinh\psi & \cosh\psi \end{pmatrix} \right.$$
$$\left. + \begin{pmatrix} \Delta\sinh\psi & 0 \\ 0 & -\Delta\sinh\psi \end{pmatrix} \right]. \tag{6.240}$$

Finally, the transformation matrix for a *drift space* of length L in an accelerating system can be derived from either (6.239) or (6.240) by letting $k_o \to 0$ for

$$\mathcal{M}_o = \begin{pmatrix} 1 & -\frac{1}{\eta_o}\log\sigma^4 \\ 0 & \sigma^4 \end{pmatrix}, \tag{6.241}$$

where $\sigma^4 = 1/(1+\eta_o L)$ in agreement with (6.233). In the limit of vanishing accelerating fields $\eta_o \to 0$ and we obtain back the well-known transformation matrices for a drift space. Similarly, we may test (6.250,251) for consistency with regular transformation matrices.

In Eqs. (6.239) to (6.241) we have the transformation matrices for all elements to form a FODO channel in the presence of acceleration. We may now apply all formalisms used to derive periodic betatron, dispersion functions or beam envelopes as derived in [6.3] for regular FODO cells. Considering one half cell we note that the quadrupole strength k_o of the first half quadrupole is determined by the last half quadrupole of the previous FODO half cell. We have therefore two variables left, the half cell drift length L and the strength k_1 of the second half quadrupole of the FODO half cell, to fit the lattice functions to a symmetric solution by requiring that $\alpha_x = 0$ and $\alpha_y = 0$.

6.4.2 Adiabatic Damping

Transformation matrices derived in this section are not phase space conserving because their determinant is no more unity. The determinant for a drift space with acceleration is, for example,

$$\det\mathcal{M}_o = \sigma^4 = \frac{1}{1+\eta_o s} \tag{6.242}$$

which is different from unity if there is a finite acceleration. The two-dimensional (x, x') phase space, for example, is reduced according to (6.242) to

$$|\mathbf{x}, \mathbf{x}'| = \frac{1}{1 + \eta_o s} |\mathbf{x_o}, \mathbf{x_o'}| \qquad (6.243)$$

and the beam emittance defined by $\mathbf{x}$ and $\mathbf{x}'$ is therefore not preserved in the presence of accelerating fields. This phenomenon is known as *adiabatic damping* under which the beam emittance varies like

$$\boxed{\epsilon = \frac{1}{1 + \eta_o s} \epsilon_o = \frac{c p_o}{c p} \epsilon_o \, ,} \qquad (6.244)$$

where $\eta_o s = \Delta E / E_o$ is the relative energy gain along the length s of the accelerator. From this we see immediately that the *normalized phase space* area $c p \epsilon$ is conserved in full agreement with Liouville's theorem. In beam transport systems involving a linear accelerator it is therefore more convenient and dynamically correct to use the truly invariant normalized beam emittance defined by

$$\epsilon_n = \beta \gamma \epsilon , \qquad (6.245)$$

where $\gamma = E/mc^2$, E is the particle energy and $\beta = v/c$. This normalized emittance remains constant even when the particle energy is changing due to external electric fields. In the presence of dissipating processes like synchrotron radiation, scattering or damping, however, even the normalized beam emittance changes because Liouville's theorem of the conservation of phase space is not valid anymore.

From (6.243) we obtain formally the constancy of the normalized beam emittance by multiplying with the momentum p_o and factor $(1 + \eta_o s)$ for

$$|\mathbf{x}, (1 + \eta_o s) p_o \, \mathbf{x}'| = |\mathbf{x_o}, p_o \, \mathbf{x_o'}|$$

or with the transverse momenta $p_o \, \mathbf{x}' = \mathbf{p}_{ox}$ and $(1 + \eta_o s) p_o \mathbf{x}' = \mathbf{p}_x$

$$|\mathbf{x}, \mathbf{p}_x| = |\mathbf{x_o}, \mathbf{p}_{ox}| = \text{const.} \qquad (6.246)$$

This can be generalized to a five-dimensional phase space remembering that in this case $\det(\mathcal{M}_o) = \left(\frac{1}{1 + \eta_o s}\right)^3$ since the matrix has the form

$$\mathcal{M}_o = \begin{pmatrix} 1 & -\frac{4}{\eta_o} \log \sigma & 0 & 0 & 0 & 0 \\ 0 & \sigma^4 & 0 & 0 & 0 & 0 \\ 0 & 0 & 1 & -\frac{4}{\eta_o} \log \sigma & 0 & 0 \\ 0 & 0 & 0 & \sigma^4 & 0 & 0 \\ 0 & 0 & 0 & 0 & 1 & A \\ 0 & 0 & 0 & 0 & 0 & \sigma^4 \end{pmatrix} , \qquad (6.247)$$

where A is an rf related quantity irrelevant for our present arguments. For the five-dimensional phase space with coordinates $\mathbf{x}$, $\mathbf{p}_x$, $\mathbf{y}$, $\mathbf{p}_y$, τ, $\Delta\mathbf{E}$, where $\mathbf{p_x}$, $\mathbf{p_y}$ are the transverse momenta, τ the longitudinal position of the particle with respect to a reference particle and ΔE the energy deviation we get finally

$$|\mathbf{x},\ \mathbf{p}_x,\ \mathbf{y},\ \mathbf{p}_y,\ \tau,\ \Delta\mathbf{E}| \;=\; |\mathbf{x}_o,\ \mathbf{p}_{ox},\ \mathbf{y}_o,\ \mathbf{p}_{oy},\ \tau_o,\ \Delta\mathbf{E}_o| = \text{const}. \qquad (6.248)$$

These results do not change if we had included focusing in the transformation matrix. From (6.250,251) we see immediately that the determinants for both matrices are

$$\det(\boldsymbol{M}_{\rm f}) \approx \det(\boldsymbol{M}_{\rm d}) \approx \sigma^4 \qquad\qquad (6.249)$$

ignoring small terms proportional to Δ.

Problems

Problem 6.1. Derive expressions for the maximum electric field strength and the waveguide losses per unit length for the TE_{1o} mode in a rectangular waveguide. Use this result to design a waveguide for 3 GHz. Calculate the cut-off frequency, the phase and group velocities and the waveguide wavelength. What criteria did you use to choose the dimensions a and b? Sketch the electrical and magnetic fields.

Problem 6.2. Plot the electrical and magnetic field distribution for the three lowest order modes in a rectangular and cylindrical cavity. Calculate the shunt impedance and compare the results. Which type of cavity is more effective?

Problem 6.3. Derive a general expression of the shunt impedance for general TM-modes in a cylindrical cavity.

Problem 6.4. Assume the rectangular waveguides of problem 6.1 to be made of copper with a surface resistivity of $R_{\rm s} = 2.61\,10^{-7}\,\sqrt{f_{\rm rf}}\ (\Omega)$. What is the maximum rf power that can be carried in this waveguide if the electric fields are limited by breakdown to 1 MV/m?

Problem 6.5. Consider a 1.5 GeV electron storage ring composed of FODO cells and realistic bending magnets of length $L_c/4$, where L_c is the FODO-cell length. Further assume a 500 MHz rf system with 100kW of rf power available. How many copper cavities are required to sustain a circulating beam current of 400mA with a maximum energy spread of $\pm1\%$? Assume a smooth beam injection over many pulses. During injection the tuning angle must be adjusted. What is the required tuning range expressed in a frequency shift?

Problem 6.6. Consider an electron storage ring to be used as a damping ring for a linear collider. The energy is $E = 1.21 \text{GeV}$, circumference $C = 35.27 \text{m}$, bending radius $\rho = 2.037 \text{m}$, momentum compaction factor $\alpha_c = 0.01841$, rf harmonic number $h = 84$, cavity shunt impedance of $R_{\text{cy}} = 28$ $\text{M}\Omega/\text{m}$. An intense bunch of $N_e = 5 \cdot 10^{10}$ particles is injected in a single pulse and is stored for only a few msec to damp to a small beam emittance. Specify and optimize a suitable rf system and calculate the required rf, beam, and cavity power, cavity voltage, number of cavities, coupling factor first while ignoring beam loading and then with beam loading. Assume a quantum lifetime of 1 min. What tuning angle should be chosen before injection of the beam?

Problem 6.7. Determine the undisturbed synchrotron oscillation frequency for problem 6.5. For idealized copper pill box cavities choose a safe detuning to stabilize the Robinson instability and determine the change in the damping decrement and the coherent and incoherent synchrotron frequency for a beam current of 400mA evenly distributed over n bunches.

Problem 6.8. The ESRF synchrotron light source has a momentum compaction factor of $\eta_c = -3.1 \cdot 10^{-4}$, a first order perturbation $\alpha_1 = 5.5 \cdot 10^{-4}$ and $\langle \gamma_x \rangle \approx 1.8$ and $\langle \gamma_y \rangle \approx 0.8$. Calculate the maximum allowable beam emittance of the incoming beam to keep the energy shift of the rf bucket due to the largest betatron oscillations within 0.5%. The beam emittance of the stored beam is $6.9 \cdot 10^{-9}$ rad-m. What is the energy shift for particles oscillating with an amplitude of 10 sigma's? Compare with the rms energy spread of 0.1% ?

Problem 6.9. Derive (6.219) and show that the distances L_n between quadrupoles scale like the beam energy γ_n.

Problem 6.10. Sometimes it is desirable to lengthen the bunch of an electron beam which can be done by installing a second rf system tuned such that the slope of the rf voltage at the synchronous angle is reduced. Determine the frequency of such a second rf system and derive in linear approximation the bunch lengthening as a function of the second rf voltage.

7. Synchrotron Radiation

The phenomenon of synchrotron radiation has been introduced in a conceptual way in [7.1] and a number of basic relations have been derived. In this chapter we will approach the physics of synchrotron radiation in a more formal way to exhibit detailed characteristics. Specifically, we will derive expressions for the spatial and spectral distribution of photon emission in a way which is applicable later for special insertion devices.

7.1 Theory of Synchrotron Radiation

The theory of synchrotron radiation is intimately related to the electromagnetic fields generated by moving charged particles. Wave equations can be derived from Maxwell's equations and we will find that any charged particle under the influence of external forces can emit radiation. We will formulate the characteristics of this radiation and apply the results to highly relativistic particles.

7.1.1 Radiation Field

The electromagnetic fields for a single moving point charge will be derived first and then applied to a large number of particles. The fields are derived from *Maxwell's equations* for moving charges in vacuum where we set $\epsilon = 1$ and $\mu = 1$. With the *nabla operator* $\nabla = (\partial/\partial x, \partial/\partial y, \partial/\partial z)$ *Maxwell's equations* are of the form

$$\nabla \cdot \mathbf{E} = \frac{4\pi}{[4\pi\epsilon_{\mathrm{o}}]} \rho \,, \tag{7.1}$$

$$\nabla \cdot \mathbf{B} = 0 \,, \tag{7.2}$$

$$\nabla \times \mathbf{E} = -\frac{[c]}{c} \frac{\partial \mathbf{B}}{\partial t} \,, \tag{7.3}$$

$$[c]\,\nabla \times \mathbf{B} = \frac{4\pi}{[4\pi\epsilon_{\mathrm{o}}]} \rho\,\boldsymbol{\beta} + \frac{1}{c} \frac{\partial \mathbf{E}}{\partial t} \,, \tag{7.4}$$

where ρ is the charge density and $\boldsymbol{\beta} = \mathbf{v}/c$ the velocity vector of the particles. Since in vacuum $\nabla \cdot \mathbf{B} = 0$ we can derive the magnetic field from a *vector potential* $\mathbf{A}$ defined by

$$\mathbf{B} = \nabla \times \mathbf{A}.$$ (7.5)

Inserting the vector potential into the curl equation (7.3) we get
$\nabla \times (\mathbf{E} + \frac{[c]}{c} \frac{\partial}{\partial t} \mathbf{A}) = 0$, or after integration

$$\mathbf{E} = -\frac{[c]}{c} \frac{\partial}{\partial t} \mathbf{A} - \nabla \phi,$$ (7.6)

where ϕ is the *scalar potential*. Inserting (7.6) into (7.4) and remembering the vector relation $\nabla \times (\nabla \times \mathbf{A}) = \nabla(\nabla \mathbf{A}) - \nabla^2 \mathbf{A}$ we get with the *Lorentz gauge* $[c]\nabla \mathbf{A} + \frac{1}{c}\frac{d}{dt}\phi = 0$

$$\nabla^2 \mathbf{A} - \frac{1}{c^2} \frac{d^2}{dt^2} \mathbf{A} = -\frac{4\pi}{[4\pi\epsilon_0]} \rho\boldsymbol{\beta}.$$ (7.7)

Similarly, we derive the wave equation for the scalar potential

$$\nabla^2 \phi - \frac{1}{c^2} \frac{d^2}{dt^2} \phi = -\frac{4\pi}{[4\pi\epsilon_0]} \rho.$$ (7.8)

These are the well-known *wave equations* with the solutions

$$\mathbf{A}(t) = \frac{1}{[4\pi c\epsilon_0]} \frac{1}{c} \int \frac{\mathbf{v}\rho(x,y,z)}{R}\bigg|_{t_r} dx\,dy\,dz$$ (7.9)

and

$$\phi(t) = \frac{1}{[4\pi\epsilon_0]} \int \frac{\rho(x,y,z)}{R}\bigg|_{t_r} dx\,dy\,dz.$$ (7.10)

All quantities under the integrals must be evaluated at the *retarded time*

$$t_r = t - \frac{1}{c} R(t_r)$$ (7.11)

when the radiation was emitted by the moving charge. The quantity R is the distance between the observation point $P(x, y, z)$ and the location of the charge element $\rho(x_r, y_r, z_r)\,dx_r\,dy_r\,dz_r$ at the retarded time t_r. The vector

$$\mathbf{R} = (x_r - x, y_r - y, z_r - z)$$ (7.12)

points away from the observation point to the charge element at the retarded time t_r as shown in Fig. 7.1.

Special care must be exercised in performing the integrations. Although we consider only a point charge q, the integral in (7.10) cannot be replaced by q/R but must be integrated over a finite volume followed by a transition to a point charge. As we will see this is a consequence of the fact that the velocity of light is finite and therefore the movement of charge elements must be taken into account.

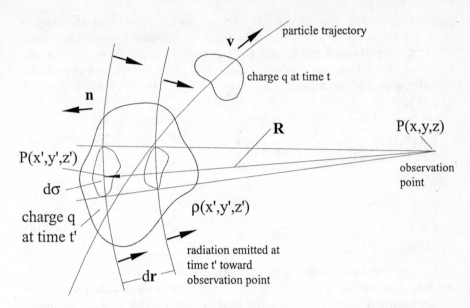

Fig. 7.1. Retarded position of a moving charge distribution

To define the quantities involved in the integration we use Fig. 7.1. The combined field at the observation point P at time t comes from all charges located at a distance R away from P. We consider the contribution from all charges contained within a spherical shell centered at P with a radius R and thickness dr to the radiation field at P and time t. Radiation emitted at time t_r will reach P at the time t. If $d\sigma$ is a surface element of the spherical shell, the volume element of charge is $dx\,dy\,dz = d\sigma\,dr$. The retarded time for the radiation from the outer surface of the shell is t_r and the retarded time for the radiation from the charge element on the inner surface of the shell is $t_r - dr/c$. From Fig. 7.1 we find the electromagnetic field observed at P at time t to originate from the fractional charges within the volume element $dr\,d\sigma$ or from the charge element $dq = \rho\,d\sigma\,dr$.

For electrical charges moving with the velocity $\mathbf{v}$, we have to include in the integration also those charge elements that move across the inner shell surface during the time dr/c. For a uniform charge distribution this additional charge is $dq = \rho\,\mathbf{vn}\,dt\,d\sigma$ where $\mathbf{n}$ is the vector normal to the surface of the shell and pointing away from the observer

$$\mathbf{n} = \frac{\mathbf{R}}{R}.$$ (7.13)

With $dt = dr/c$ and $\boldsymbol{\beta} = \mathbf{v}/c$, we get finally for both contributions to the charge element

$$dq = \rho\,(1 + \mathbf{n}\boldsymbol{\beta})\,dr\,d\sigma\,.$$ (7.14)

Depending on the direction of the velocity vector $\boldsymbol{\beta}$, we find an increase or a reduction in the radiation field from the moving charges. We solve (7.14) for $\rho \, dr \, d\sigma$ and insert into the integrals (7.9,10). Now we may use the assumption that the electrical charge is a point charge and get for the *retarded potentials* of a moving point charge q at the observation point P and at time t

$$\mathbf{A}(P,t) = \frac{1}{[4\pi c \epsilon_0]} \frac{q}{R} \frac{\boldsymbol{\beta}}{1+\mathbf{n}\boldsymbol{\beta}} \bigg|_{t_r} \tag{7.15}$$

and

$$\phi(P,t) = \frac{1}{[4\pi \epsilon_0]} \frac{q}{R} \frac{1}{1+\mathbf{n}\boldsymbol{\beta}} \bigg|_{t_r} . \tag{7.16}$$

These equations are known as the *Liénard-Wiechert potentials* and express the field potentials of a moving charge as functions of the charge parameters at the retarded time. To obtain the electric and magnetic fields we insert the retarded potentials into (7.5,6) noting that the differentiation must be performed with respect to the time t and location P of the observer while the potentials are expressed at the retarded time t_r.

In both equations for the vector and scalar potential we have the same denominator and we set for simplicity of writing

$$r = R(1+\mathbf{n}\boldsymbol{\beta}) . \tag{7.17}$$

It will become necessary to calculate the derivative of the retarded time with respect to the time t and since $t_r = t - R/c$ the time derivative of t_r is

$$\frac{dt_r}{dt} = 1 - \frac{1}{c} \frac{dR}{dt_r} \frac{dt_r}{dt} .$$

The variation of the distance R with the retarded time depends on the velocity $\mathbf{v}$ of the moving charge and is the projection of the vector $\mathbf{v} \, dt_r$ onto the unity vector $\mathbf{n}$. Therefore,

$$dR = \mathbf{vn} \, dt_r \tag{7.18}$$

and finally

$$\frac{dt_r}{dt} = \frac{1}{1+\mathbf{n}\boldsymbol{\beta}} = \frac{R}{r} . \tag{7.19}$$

From (7.6) the electrical field is described with (7.19) after a few manipulations by

$$[4\pi \epsilon_0] \frac{\mathbf{E}(t)}{q} = -\frac{1}{c} \frac{R}{r^2} \frac{\partial \boldsymbol{\beta}}{\partial t_r} + \frac{\boldsymbol{\beta} R}{c \, r^3} \frac{\partial r}{\partial t_r} + \frac{1}{r^2} \nabla_r r . \tag{7.20}$$

In evaluating the nabla operator and other differentials we remember that all parameters on the r.h.s. must be taken at the retarded time (7.11) which itself is also a function of the location of the observation point P. To distinguish between the ordinary nabla operator and the case where the dependence of the retarded time on the position $P(x, y, z)$ must be considered, we add to the nabla symbol the index $_r$ like ∇_r. The components of this operator are then $\frac{\partial}{\partial x}\big|_r = \frac{\partial}{\partial x} + \frac{\partial t_r}{\partial x}\frac{\partial}{\partial t_r}$, etc. We evaluate first

$$\nabla_r r = \nabla_r R + \nabla_r (\mathbf{R}\boldsymbol{\beta}) \tag{7.21}$$

and find for the first term with $\nabla R = -\mathbf{n}$ from (7.12)

$$\nabla_r R = -\mathbf{n} + \frac{\partial R}{\partial t_r}\nabla_r t_r . \tag{7.22}$$

For the gradient of the retarded time we get

$$\nabla t_r = \nabla\left[t - \frac{1}{c}R(t_r)\right] = -\frac{1}{c}\nabla_r R = -\frac{1}{c}\left(-\mathbf{n} + \frac{\partial R}{\partial t_r}\nabla t_r\right) \tag{7.23}$$

and performing the differentiation we get with $\frac{\partial x_r}{\partial t_r} = v_x$, etc.,

$$\frac{\partial R}{\partial t_r} = \left(\frac{\partial R}{\partial x_r}\frac{\partial x_r}{\partial t_r}, \frac{\partial R}{\partial y_r}\frac{\partial y_r}{\partial t_r}, \frac{\partial R}{\partial z_r}\frac{\partial z_r}{\partial t_r}\right) = \mathbf{n}\mathbf{v} .$$

Solving (7.23) for ∇t_r we get

$$\nabla t_r = \frac{\mathbf{R}}{cr} \tag{7.24}$$

and (7.22) becomes finally

$$\nabla_r R = -\mathbf{n} + \frac{\mathbf{R}}{r}(\boldsymbol{\beta}\mathbf{n}). \tag{7.25}$$

For the second term in (7.21) we note that the velocity $\mathbf{v}$ does not depend on the location of the observer and with $\nabla_r \mathbf{R} = -1$, (7.24) and

$$\frac{d\mathbf{R}}{dt_r} = \mathbf{v}, \tag{7.26}$$

we get for the second term in (7.21)

$$\nabla_r (\mathbf{R}\boldsymbol{\beta}) = -\boldsymbol{\beta} + \frac{\partial}{\partial t_r}(\mathbf{R}\boldsymbol{\beta})\nabla t_r = -\boldsymbol{\beta} + \left(\mathbf{R}\frac{\partial\boldsymbol{\beta}}{\partial t_r}\right)\frac{\mathbf{R}}{cr} + \beta^2\frac{\mathbf{R}}{r}. \tag{7.27}$$

For the complete evaluation of the electric field in (7.20) we finally need an expression for

$$\frac{\partial r}{\partial t_r} = \frac{\partial R}{\partial t_r} + \frac{\partial}{\partial t_r}\boldsymbol{\beta}\mathbf{R} = c\boldsymbol{\beta}\mathbf{n} + c\beta^2 + \mathbf{R}\frac{d\boldsymbol{\beta}}{dt_r}, \tag{7.28}$$

where we made use of (7.26). Collecting all differential expressions required in (7.20) we get with (7.21,25,27,28)

$$[4\pi\epsilon_o]\frac{\mathbf{E}(t)}{q} = \frac{1}{r^2}\left[-\mathbf{n}-\boldsymbol{\beta}+\frac{\mathbf{R}}{r}\left(\boldsymbol{\beta}\mathbf{n}+\beta^2+\frac{1}{c}\dot{\boldsymbol{\beta}}\mathbf{R}\right)\right]_{\mathrm{r}}$$
$$-\frac{R}{cr^2}\dot{\boldsymbol{\beta}}+\boldsymbol{\beta}\frac{R}{r^3}\left(\boldsymbol{\beta}\mathbf{n}+\beta^2+\frac{1}{c}\dot{\boldsymbol{\beta}}\mathbf{R}\right)\bigg|_{\mathrm{r}}, \tag{7.29}$$

where $\dot{\boldsymbol{\beta}} = \mathrm{d}\boldsymbol{\beta}/\mathrm{d}t_{\mathrm{r}}$. After some manipulation and using the vector relation

$$\mathbf{a}\times(\mathbf{b}\times\mathbf{c}) = \mathbf{b}(\mathbf{ac})-\mathbf{c}(\mathbf{ab}) \tag{7.30}$$

the equation for the electrical field of a charge q moving with the velocity $\mathbf{v}$ becomes

$$[4\pi\epsilon_o]\frac{\mathbf{E}}{q} = -\frac{1-\beta^2}{r^3}(\mathbf{R}+\boldsymbol{\beta}R)|_{\mathrm{r}}$$
$$+\frac{1}{cr^3}\left\{\mathbf{R}\times[(\mathbf{R}+R\boldsymbol{\beta})\times\dot{\boldsymbol{\beta}}]\right\}\bigg|_{\mathrm{r}}, \tag{7.31}$$

where we have added the index $_{\mathrm{r}}$ as a remainder that all quantities on the r.h.s. of (7.31) must be taken at the retarded time t_{r}.

This equation for the electric field of a moving charge has two distinct parts. The first part is inversely proportional to the square of the distance between radiation source and observer and depends only on the velocity of the charge. For a charge at rest $\beta = 0$ this term reduces to the *Coulomb field* of a point charge q. The area close to the radiating charge where this term is dominant is called the *Coulomb regime*. The field is directed toward the observer for a positive charge at rest and tilts into the direction of propagation as the velocity of the charge increases. For highly relativistic particles we note the Coulomb field becomes very small.

We will not further consider this regime since we are interested only in the radiation field far away from the moving charge. The second term in (7.31) is inversely proportional to the distance from the charge and depends on the velocity as well as on the acceleration of the charge. This term scales linear with the distance r falling off much slower than the Coulomb term and therefore reaches to large distances from the radiation source. We call this regime the *radiation regime* and the remainder of this chapter will focus on the discussion of the radiation from moving charges. The electrical field in the radiation regime is

$$[4\pi\epsilon_o]\frac{\mathbf{E}(t)}{q}\bigg|_{\mathrm{rad}} = \frac{1}{cr^3}\left\{\mathbf{R}\times[(\mathbf{R}+\boldsymbol{\beta}R)\times\dot{\boldsymbol{\beta}}]\right\}_{\mathrm{r}}. \tag{7.32}$$

The polarization of the electric field at the location of the observer is purely orthogonal to the direction of observation $\mathbf{R}$.

Similar to the derivation of the electric field, we can derive the expression for the magnetic field from (7.5) with (7.15) and get

$$\mathbf{B} = \nabla_r \times \mathbf{A} = q \left[\nabla_r \times \frac{\boldsymbol{\beta}}{r} \right] = \frac{q}{r} \left[\nabla_r \times \boldsymbol{\beta} \right] - \frac{q}{r^2} \left[\nabla_r r \times \boldsymbol{\beta} \right], \qquad (7.33)$$

where again all parameters on the r.h.s. must be evaluated for the retarded time t_r. The evaluation of the "retarded" curl operation $[\nabla_r \times \boldsymbol{\beta}]$ becomes obvious if we evaluate one component only, for example, the x component

$$\left(\frac{\partial}{\partial y} + \frac{\partial t_r}{\partial y} \frac{\partial}{\partial t_r} \right) \beta_z - \left(\frac{\partial}{\partial z} + \frac{\partial t_r}{\partial z} \frac{\partial}{\partial t_r} \right) \beta_y = [\nabla \times \boldsymbol{\beta}]_x + [\nabla t_r \times \dot{\boldsymbol{\beta}}]_x .$$

In a similar way, we get the other components and find

$$[\nabla_r \times \boldsymbol{\beta}] = [\nabla \times \boldsymbol{\beta}] + [\nabla t_r \times \dot{\boldsymbol{\beta}}] = \frac{1}{cr} [\mathbf{R} \times \dot{\boldsymbol{\beta}}], \qquad (7.34)$$

where we used (7.24) and the fact that the velocity $\boldsymbol{\beta}$ of the particle does not depend on the coordinates of the observation point, $[\nabla \times \boldsymbol{\beta}] = 0$. The gradient $\nabla_r r$ has been derived earlier in equations (7.21 – 27). Inserting these expressions into (7.33) we find the magnetic field of an electrical charge moving with the velocity $\mathbf{v}$

$$\begin{aligned}[4\pi c \epsilon_o] \frac{\mathbf{B}(t)}{q} = & -\frac{[\boldsymbol{\beta} \times \mathbf{n}]}{r^2} - \frac{R}{cr^2} [\dot{\boldsymbol{\beta}} \times \mathbf{n}] \Big|_r \\ & + \frac{R}{r^3} \left(\mathbf{n}\boldsymbol{\beta} + \beta^2 + \frac{\dot{\boldsymbol{\beta}}\mathbf{R}}{c} \right) [\boldsymbol{\beta} \times \mathbf{n}] \Big|_r .\end{aligned} \qquad (7.35)$$

Again, there are two distinct groups of terms for the radiation field. In case of the electrical field the terms that fall off like the square of the distance are the Coulomb fields. For magnetic fields such terms appear only if the charge is moving $\boldsymbol{\beta} \neq 0$ and are identical to the *Biot Savart fields*. Here we concentrate only on the far fields or radiation fields which decay inversely proportional to the distance from the source. The magnetic radiation field is then given by

$$[4\pi c \epsilon_o] \frac{\mathbf{B}(t)}{q} \Big|_{rad} = -\frac{R}{cr^2} [\dot{\boldsymbol{\beta}} \times \mathbf{n}] + \frac{R}{cr^3} (\dot{\boldsymbol{\beta}}\mathbf{R}) [\boldsymbol{\beta} \times \mathbf{n}] \Big|_r . \qquad (7.36)$$

Comparing the magnetic field (7.35) with the electrical field (7.29) reveals a very simple correlation between both fields. The magnetic field can be obtained from the electric field, and vice versa, by mere vector multiplication with the unit vector $\mathbf{n}$

$$\mathbf{B} = \frac{1}{[c]} [\mathbf{E} \times \mathbf{n}]_r . \qquad (7.37)$$

From this equation we can deduce special properties for the field directions by noting that the electric and magnetic fields are orthogonal to each other and both are orthogonal to the direction of observation $\mathbf{n}$. The existence of electric and magnetic fields can give rise to radiation for which the *Poynting vector* is

$$\mathbf{S}(t) = [4\pi c\epsilon_0]\frac{c}{4\pi}\,[\mathbf{E} \times \mathbf{B}]_\mathrm{r} = [4\pi c\epsilon_0]\frac{c}{4\pi}\,[\mathbf{E} \times (\mathbf{E} \times \mathbf{n})]_\mathrm{r}. \tag{7.38}$$

Using again the vector relation (7.30) and noting that the electric field is normal to $\mathbf{n}$, we get for the Poynting vector or the radiation flux in the direction to the observer

$$\mathbf{S}(t) = -[4\pi c\epsilon_0]\frac{c}{4\pi}\,E^2\,\mathbf{n}_\mathrm{r}. \tag{7.39}$$

Equation (7.39) defines the energy flux density measured at the observation point P and time t in form of synchrotron radiation per unit cross section and parallel to the direction of observation $\mathbf{n}$. All quantities expressing this energy flux are still to be taken at the retarded time. For practical reasons it becomes desirable to express the Poynting vector at the retarded time t_r as well. The energy flux at the observation point, in terms of the retarded time is then $\frac{dW}{dt_\mathrm{r}} = \frac{dW}{dt}\frac{dt}{dt_\mathrm{r}}$. Instead of (7.39) we express the Poynting vector with (7.19) like

$$\mathbf{S}_\mathrm{r} = \mathbf{S}\frac{dt}{dt_\mathrm{r}} = -[4\pi c\epsilon_0]\frac{c}{4\pi}\,E^2\,[(1+\boldsymbol{\beta}\,\mathbf{n})\,\mathbf{n}]_\mathrm{r}. \tag{7.40}$$

The Poynting vector in this form can be readily used for calculations like those determining the spatial distribution of the radiation power.

7.2 Synchrotron Radiation Power and Energy Loss

So far no particular choice of the reference system has been assumed, but a particularly simple reference frame F^* is the one moving with the charge assumed from now on to be a single particle with charge e. For an observer in this moving system the charge oscillates about a fixed point and the electric field in the radiation regime is from (7.31)

$$\mathbf{E}^*(t) = \frac{1}{[4\pi\epsilon_0]}\frac{e}{cR}\,[\mathbf{n} \times (\mathbf{n} \times \dot{\boldsymbol{\beta}}^*)]\Big|_\mathrm{r}. \tag{7.41}$$

The synchrotron radiation power per unit solid angle and at distance R from the source is from (7.40) with $\mathbf{v} = 0$

$$\frac{dP(t)}{d\Omega} = -\mathbf{n}\,\mathbf{S}\,R^2|_\mathrm{r} = [4\pi c\epsilon_0]\frac{c}{4\pi}\,\mathbf{E}^{*2}\,R^2|_\mathrm{r} \tag{7.42}$$

and with (7.41) we get the radiation power per unit solid angle at the retarded time t_r while introducing the classical particle radius $r_c \, mc^2 = e^2$ to obtain expressions which are independent of electromagnetic units

$$\frac{\mathrm{d}P}{\mathrm{d}\Omega} = \frac{r_c \, mc^2}{4\pi c} \left| (\mathbf{n} \times [\mathbf{n} \times \dot{\boldsymbol{\beta}}^*]) \right|_r^2 = \frac{r_c \, mc^2}{4\pi c} \left| \dot{\boldsymbol{\beta}}^* \right|_r^2 \sin^2 \Theta_r , \qquad (7.43)$$

where Θ_r is the retarded angle between the direction of acceleration and the direction of observation $\mathbf{n}$. Integration over all solid angle gives the total radiated power. With $\mathrm{d}\Omega = \sin \Theta_r \, \mathrm{d}\Theta_r \, \mathrm{d}\Phi$ where Φ is the azimuthal angle with respect to the direction of acceleration, the total radiation power is

$$P = \frac{2}{3} \frac{r_c \, mc^2}{c} \left| \dot{\boldsymbol{\beta}}^* \right|_r^2 , \qquad (7.44)$$

which has been derived first by *Larmor* [7.2] within the realm of classical electrodynamics. Emission of a quantized photon, however, exerts a recoil on the electron varying its energy slightly. *Schwinger* [7.3] investigated this effect and derived a correction to the radiation power by a factor

$$P = P_{\text{classical}} \left(1 - \frac{55}{16\sqrt{3}} \frac{\epsilon_c}{E} \right) , \qquad (7.45)$$

where ϵ_c is the classical photon energy. The correction is generally very small and we ignore therefore this quantum mechanical effect in our discussions.

The radiation power and spatial distribution is identical to that from a linear microwave antenna. The radiation is emitted normal to the direction of acceleration and has a $\sin^2$-distribution. Expression (7.44) can also be written in invariant form with

$$m \, \dot{\mathbf{v}}^* = \frac{\mathrm{d}\mathbf{p}}{\mathrm{d}\tau^*} , \qquad (7.46)$$

where $\tau^* = t/\gamma$ is the time in the reference system S^* of the charged particle, m is the mass of the charged particle and p the momentum. This is possible because we may construct a four momentum vector in the following way: The momentum of the radiation field is constant since momentum is emitted through radiation symmetrically and therefore $c \, \mathrm{d}\mathbf{p} = 0$. The energy component $\mathrm{d}U$ on the other hand is $\mathrm{d}U = P \, \mathrm{d}t$. Formulating also the r.h.s. in invariant form (7.44) becomes

$$P = \frac{2}{3} \frac{r_c \, mc^2}{c} \frac{1}{(mc)^2} \frac{\mathrm{d}p_\mu}{\mathrm{d}\tau^*} \frac{\mathrm{d}p^\mu}{\mathrm{d}\tau^*} . \qquad (7.47)$$

Expressing

$$\left(\frac{\mathrm{d}p_\mu}{\mathrm{d}\tau^*} \frac{\mathrm{d}p^\mu}{\mathrm{d}\tau^*} \right) = \left(\frac{\mathrm{d}p}{\mathrm{d}\tau^*} \right)^2 - \frac{1}{c^2} \left(\frac{\mathrm{d}E}{\mathrm{d}\tau^*} \right)^2 \qquad (7.48)$$

we get with $\mathbf{p} = \gamma m \mathbf{v}$ and $E = \gamma m c^2$ the radiation power in the laboratory system

$$P = \frac{2}{3} \frac{r_c m c^2}{c} \gamma^2 \left[\left(\frac{\mathrm{d}}{\mathrm{d}t} \gamma \boldsymbol{\beta} \right)^2 - \left(\frac{\mathrm{d}}{\mathrm{d}t} \gamma \right)^2 \right] . \tag{7.49}$$

and after evaluation of the derivatives and some manipulation this becomes

$$P = \frac{2}{3} \frac{r_c m c^2}{c} \gamma^6 \left[\dot{\boldsymbol{\beta}}^2 - (\boldsymbol{\beta} \times \dot{\boldsymbol{\beta}})^2 \right] , \tag{7.50}$$

where we have made use of the vector equation

$$\boldsymbol{\beta}^2 \dot{\boldsymbol{\beta}}^2 - (\boldsymbol{\beta} \dot{\boldsymbol{\beta}})^2 = (\boldsymbol{\beta} \times \dot{\boldsymbol{\beta}})^2 .$$

Equation (7.50) expresses the radiation power in a simple way and allows us to calculate other radiation characteristics based on beam parameters in the laboratory system. Specifically, we will distinguish between acceleration parallel $\dot{\boldsymbol{\beta}}_{\parallel}$ or perpendicular $\dot{\boldsymbol{\beta}}_{\perp}$ to the propagation $\boldsymbol{\beta}$ of the charge and set therefore

$$\dot{\boldsymbol{\beta}} = \dot{\boldsymbol{\beta}}_{\parallel} + \dot{\boldsymbol{\beta}}_{\perp} . \tag{7.51}$$

Insertion into (7.50) shows the total radiation power to consist of separate contributions from parallel and orthogonal acceleration. Separating both contributions we get the *synchrotron radiation power* for both parallel and transverse acceleration, respectively

$$P_{\parallel} = \frac{2}{3} \frac{r_c m c^2}{c} \gamma^6 \dot{\boldsymbol{\beta}}_{\parallel}^2 , \tag{7.52}$$

$$P_{\perp} = \frac{2}{3} \frac{r_c m c^2}{c} \gamma^4 \dot{\boldsymbol{\beta}}_{\perp}^2 . \tag{7.53}$$

Expressions have been derived that define the radiation power for parallel acceleration like in a linear accelerator or orthogonal acceleration found in circular accelerators or deflecting systems. We note a similarity for both contributions except for the energy dependence. At highly relativistic energies, the same acceleration force leads to much less radiation if the acceleration is parallel to the motion of the particle compared to orthogonal acceleration. Parallel acceleration is related to the accelerating force $\frac{\mathrm{d}\mathbf{p}_{\parallel}}{\mathrm{d}t}$ by $\dot{\mathbf{v}}_{\parallel} = \frac{1}{\gamma^3} \frac{\mathrm{d}\mathbf{p}_{\parallel}}{\mathrm{d}t}$ and after insertion into (7.52) the radiation power due to parallel acceleration becomes

$$P_{\parallel} = \frac{2}{3} \frac{r_c m c^2}{c} \frac{1}{(mc)^2} \left(\frac{\mathrm{d}\mathbf{p}_{\parallel}}{\mathrm{d}t} \right)^2 . \tag{7.54}$$

The radiation power for acceleration along the propagation of the charged particle is therefore independent of the energy of the particle and depends only on the accelerating force or with $(\mathrm{d}\mathbf{p}_{\parallel}/\mathrm{d}t) = \beta\,c\,(\mathrm{d}E/\mathrm{d}x)$ on the energy increase per unit length of accelerator.

In contrast, we find very different radiation characteristics for transverse acceleration as it happens, for example, during the transverse deflection of a charged particle in a magnetic field. The transverse acceleration $\dot{\mathbf{v}}_{\perp}$ is expressed by the Lorentz force

$$\frac{\mathrm{d}p_{\perp}}{\mathrm{d}t} = \gamma m \dot{\mathbf{v}}_{\perp} = \frac{[c]}{c}\,e\,[\mathbf{v} \times \mathbf{B}] \tag{7.55}$$

and after insertion into (7.53) the radiation power from transversely accelerated particles becomes

$$P_{\perp} = \tfrac{2}{3}\frac{r_c mc^2}{c}\frac{\gamma^2}{(mc)^2}\left(\frac{\mathrm{d}\mathbf{p}_{\perp}}{\mathrm{d}t}\right)^2. \tag{7.56}$$

Comparing (7.54) with (7.56) we find that the same accelerating force leads to a much higher radiation power by a factor γ^2 for transverse acceleration with respect to longitudinal acceleration. For all practical purposes technical limitations prevent the occurrence of sufficient longitudinal acceleration to generate noticeable radiation. We express the deflecting magnetic field B by the bending radius ρ and get the instantaneous synchrotron radiation power

$$P_{\gamma} = \tfrac{2}{3}\,r_c\,mc^2\,\frac{c\,\beta^4\,\gamma^4}{\rho^2} \tag{7.57}$$

or in more practical units

$$\boxed{P_{\gamma}(\mathrm{GeV/sec}) = \frac{c\,C_{\gamma}}{2\pi}\frac{E^4}{\rho^2}\,,} \tag{7.58}$$

where we use Sands' definition of the *radiation constant* [7.4] in

$$C_{\gamma} = \frac{4\pi}{3}\frac{r_c}{(mc^2)^3} = 8.8575\,10^{-5}\,\mathrm{m/GeV^3}\,, \tag{7.59}$$

with r_c the classical particle radius and mc^2 the rest energy of the particle. The numerical value is calculated for relativistic electrons and positrons.

From here on we will stop considering longitudinal acceleration unless specifically mentioned and replace therefore the index $\perp$ by setting $P_{\perp} = P_{\gamma}$. We also restrict from now on the discussion to singly charged particles and set $q = e$ ignoring extremely high energies where multiple charged ions may start to radiate.

The electromagnetic radiation of charged particles in transverse magnetic fields is proportional to the fourth power of the particle momentum $\beta\gamma$ and inversely proportional to the square of the bending radius ρ. The radiation emitted by charged particles being deflected in magnetic fields is called *synchrotron radiation*. The synchrotron radiation power increases very fast for high energy particles and provides the most severe limitation to the maximum energy achievable in circular accelerators. We note, however, also a strong dependence on the kind of particles involved in the process of radiation. Because of the much heavier mass of protons compared to the lighter electrons, we find appreciable synchrotron radiation only in circular electron accelerators. The radiation power of protons actually is smaller compared to that for electrons by the fourth power of the mass ratio or by the factor

$$\frac{P_e}{P_p} = 1836^4 = 1.36\,10^{13}.$$ (7.60)

In spite of this enormous difference measurable synchrotron radiation has been predicted by *Coisson* [7.5, 6] and was indeed detected at the 400 GeV proton synchrotron SPS at CERN [7.7,8]. Substantial synchrotron radiation is expected in circular proton accelerators at a beam energy of 10 TeV and more.

The knowledge of the synchrotron radiation power allows us now to calculate the energy loss of a particle per turn in a circular accelerator by integrating the radiation power along the circumference L_o of the circular accelerator

$$\Delta E = \oint P_\gamma \, \mathrm{dt} = \tfrac{2}{3}\, r_c mc^2 \, \beta^3 \, \gamma^4 \int_{L_o} \frac{1}{\rho^2} \, \mathrm{ds}.$$ (7.61)

If we assume an isomagnetic lattice where the bending radius is the same for all bending magnets $\rho = $ const, and integrate around a circular accelerator the energy loss per turn due to synchrotron radiation is given by

$$\Delta E = \oint P_\gamma \, \mathrm{dt} = P_\gamma \frac{2\pi\rho}{\beta c} = \frac{4\pi}{3}\, r_c mc^2 \, \beta^3 \, \frac{\gamma^4}{\rho}.$$ (7.62)

The integration obviously is to be performed only along those parts of the circular accelerator where synchrotron radiation occurs or along bending magnets only. In more practical units, the *energy loss* of relativistic electrons per revolution in a circular accelerator with an isomagnetic lattice and a bending radius ρ is given by

$$\boxed{\Delta E = C_\gamma \frac{E^4}{\rho}.}$$ (7.63)

From this energy loss per particle in each turn we calculate the total synchrotron radiation power for a beam of N_e particles. The total synchrotron radiation power for a single particle is its energy loss multiplied by the revolution frequency of the particle around the circular orbit. If L_o is the circumference of the orbit we have for the revolution frequency $f_{\mathrm{rev}} = \beta c / L_o$

and for the circulating particle current $I = e\,f_{\rm rev}\,N_{\rm e}$. The total *synchrotron radiation power* is then

$$P_\gamma({\rm MW}) \;=\; \frac{\Delta E}{e}\,I \;=\; C_\gamma\,\frac{E^4({\rm GeV}^4)}{\rho({\rm m})}\,I({\rm mA})\,. \tag{7.64}$$

The total synchrotron radiation power scales like the fourth power of energy and is inversely proportional to the bending radius. The strong dependence of the radiation on the particle energy causes severe practical limitations on the maximum achievable energy in a circular accelerator.

7.3 Spatial Distribution of Synchrotron Radiation

Expressions for the *radiation fields* and *Poynting vector* exhibit strong vectorial dependencies on the directions of motion and acceleration of the charged particles and on the direction of observation. These vectorial dependencies indicate that the radiation may not be emitted isotropically but rather into specific directions forming characteristic *radiation patterns*. In this section we will derive these spatial radiation characteristics and determine the direction of preferred radiation emission.

In (7.43) the radiation power per unit solid angle is expressed in the reference frame of the particle

$$\frac{{\rm d}P}{{\rm d}\Omega} \;=\; \frac{r_{\rm c}\,mc^2}{4\pi c}\,|\dot{\boldsymbol{\beta}}^*|^2\,\sin^2\Theta\,. \tag{7.65}$$

This equation shows a significant directionality of the radiation as shown in Fig. 7.2. The radiation power is mainly concentrated in the (y,x)-plane and is proportional to $\sin^2\Theta$ where Θ is the angle between the direction of acceleration, in this case the z-axis, and the direction of observation **n**. The radiation pattern in Fig. 7.2 is formed by the end points of vectors with the length $\frac{{\rm d}P}{{\rm d}\Omega}$ and angles Θ with respect to the z-axis. Because of symmetry, the radiation is isotropic with respect to the polar angle ϕ and therefore, the radiation pattern is rotation symmetric about the z-axis.

This pattern is the correct representation of the radiation for the reference frame of the radiating particle. We may, however, also consider this pattern as the radiation pattern from non relativistic particles like that from a linear radio antenna.

For relativistic particles the radiation pattern differs significantly from the non relativistic case. The Poynting vector in the form of (7.40) can be used to calculate the radiation power per unit solid angle in the direction to the observer $-\mathbf{n}$

$$\frac{{\rm d}P}{{\rm d}\Omega} \;=\; -\mathbf{n}\,\mathbf{S}_{\rm r}\,R^2 \;=\; [4\pi c\epsilon_0]\,\frac{c}{4\pi}\,E_{\rm r}^2\,(1+\boldsymbol{\beta}\mathbf{n})_{\rm r}\,R^2\,. \tag{7.66}$$

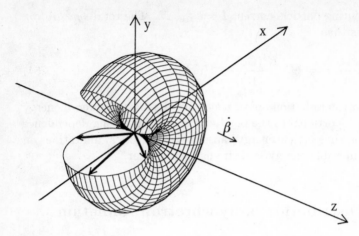

Fig. 7.2. Radiation pattern in the particle frame of reference or for nonrelativistic particles in the laboratory system

We calculate the spatial distribution of the synchrotron radiation for the case of acceleration orthogonal to the propagation of the particle as it happens in beam transport systems where the particles are deflected by a transverse magnetic fields. The particle is assumed to be located at the origin of a right-handed coordinate system as shown in Fig. 7.3 propagating in the z-direction and the orthogonal acceleration in this coordinate system occurs along the x-axis.

With the expression (7.32) for the electric fields in the *radiation regime* the spatial radiation power distribution (7.66) becomes

$$\frac{\mathrm{d}P}{\mathrm{d}\Omega} = \frac{c}{4\pi} r_c\, mc^2\, \frac{R^5}{c^3 r^5} \left\{ \mathbf{n} \times [(\mathbf{n} + \boldsymbol{\beta}) \times \dot{\boldsymbol{\beta}}] \right\}^2 \Big|_{\mathrm{r}}. \tag{7.67}$$

We will now replace all vectors by their components to obtain the directional dependency of the synchrotron radiation. The vector $\mathbf{n}$ pointing from the observation point to the source point of the radiation has from Fig. 7.3 the components

$$\mathbf{n} = (-\sin\theta\,\cos\phi, -\sin\theta\,\sin\phi, +\cos\theta), \tag{7.68}$$

where the angle θ is the angle between the direction of particle propagation and the direction of emission of the synchrotron light $-\mathbf{n}$. The x component of the acceleration can be derived from the Lorentz equation

$$\gamma m \dot{v}_{x\perp} = \frac{\mathrm{d}p_x}{\mathrm{d}t} = [c]\, e\, \beta_z B_y.$$

With $\frac{1}{\rho} = [c]\,\frac{eB_y}{cp} = [c]\,\frac{eB_y}{\gamma mcv}$ and $v_z \approx v$ we get for the acceleration vector

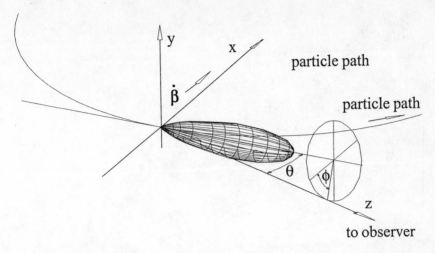

Fig. 7.3. Radiation geometry in the laboratory frame of reference for highly relativistic particles

$$\dot{\mathbf{v}}_\perp = (\dot{v}, 0, 0) = \left(\frac{v^2}{\rho}, 0, 0\right). \tag{7.69}$$

The velocity vector is

$$\mathbf{v} = (0, 0, v) \tag{7.70}$$

and after replacing the double vector product in (7.67) by a single vector sum

$$\mathbf{n} \times [(\mathbf{n} + \boldsymbol{\beta}) \times \dot{\boldsymbol{\beta}}] = (\mathbf{n} + \boldsymbol{\beta})(\mathbf{n}\dot{\boldsymbol{\beta}}) - \dot{\boldsymbol{\beta}}(1 + \mathbf{n}\boldsymbol{\beta}) \tag{7.71}$$

we may now square the r.h.s. of (7.67) and replace all vectors by their components. The denominator in (7.67) then becomes

$$r^5 = R^5 (1 + \mathbf{n}\boldsymbol{\beta})^5 = R^5 (1 - \beta \cos\theta)^5 \tag{7.72}$$

and the full expression for the radiation power exhibiting the spacial distribution is finally

$$\frac{\mathrm{d}P(t_\mathrm{r})}{\mathrm{d}\Omega} = \frac{r_\mathrm{c}\,c}{4\pi}\,mc^2\,\frac{\beta^4}{\rho^2}\,\frac{(1 - \beta \cos\theta)^2 - (1 - \beta^2)\sin^2\theta\,\cos^2\phi}{(1 - \beta \cos\theta)^5}. \tag{7.73}$$

This equation describes the instantaneous synchrotron radiation power per unit solid angle from charged particles moving with velocity v and being accelerated normal to the propagation by a magnetic field. The angle θ is the angle between the direction of observation $-\mathbf{n}$ and propagation $\mathbf{v}/v$. Integration over all angles results again in the total synchrotron radiation power (7.57).

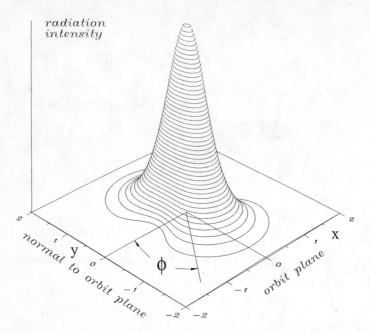

Fig. 7.4. Spatial distribution of synchrotron radiation ($x = \gamma\theta\cos\phi, y = \gamma\theta\sin\phi$)

In Fig. 7.4 the radiation power distribution is shown in real space as derived from (7.73). We note that the radiation is highly collimated in the forward direction along the z-axis which is also the direction of particle propagation. Synchrotron radiation in particle accelerators or beam lines is emitted whenever there is a deflecting electromagnetic field and emerges mostly tangentially from the particle trajectory. An estimate of the typical opening angle can be derived from (7.73). We set $\phi = 0$ and expand the cosine function for small angles $\cos\theta \approx 1 - \frac{1}{2}\theta^2$. With $\beta \approx 1 - \frac{1}{2}\gamma^{-2}$ we find the radiation power to scale like $(\gamma^{-2} + \theta^2)^{-3}$. The radiation power therefore is reduced to about one eighth the peak intensity at an emission angle of $\theta_\gamma = 1/\gamma$ or virtually all synchrotron radiation is emitted within an angle of

$$\theta_\gamma \approx \pm\frac{1}{\gamma} \tag{7.74}$$

with respect to the direction of the particle propagation.

From Fig. 7.4 we observe a slightly faster fall off for an azimuthal angle of $\phi = \frac{\pi}{2}$ which is normal to the plane of particle acceleration and propagation. Although the synchrotron radiation is emitted symmetrically within a small angle of the order of $\pm\frac{1}{\gamma}$ with respect to the direction of particle propagation, the radiation pattern from a relativistic particle as observed in the laboratory is very different in the deflecting plane from that in the

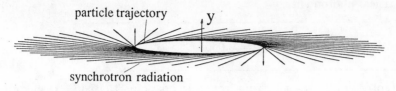

synchrotron radiation

Fig. 7.5. Synchrotron radiation from a particle accelerator

nondeflecting plane. While the particle radiates from every point along its path, the direction of this path changes in the deflecting plane but does not in the nondeflecting plane. The synchrotron radiation pattern from a bending magnet therefore resembles the form of a swath where the radiation is emitted evenly and tangentially from every point of the particle trajectory as shown in Fig. 7.5.

The extreme collimation of the synchrotron radiation and its high intensity in high energy electron accelerators can cause significant heating problems as well as desorption of gas molecules from the surface of the vacuum chamber. In addition, the high density of thermal energy deposition on the vacuum chamber walls can cause significant mechanical stresses causing cracks in the material. A careful design of the radiation absorbing surfaces to avoid damage to the integrity of the material is required. On the other hand, this same radiation is a valuable source of photons for a wide variety of research applications where, specifically, the collimation of the radiation together with the small source dimensions are highly desired features of the radiation.

7.4 Synchrotron Radiation Spectrum

Synchrotron radiation from relativistic charged particles is emitted over a wide spectrum of photon energies. The basic characteristics of this spectrum can be derived from simple principles as suggested by *Jackson* [7.9]. In [Ref. 7.1, Sect. 9.1.5] we derived a characteristic parameter, the *critical photon energy*, defined by

$$\hbar\omega_c = \tfrac{2}{3}\,\hbar c\,\frac{\gamma^3}{\rho}\,. \tag{7.75}$$

The significance of the critical photon energy is its definition of the upper bound for the synchrotron radiation spectrum. The spectral intensity falls off rapidly for frequencies above the *critical photon frequency*. In practical units, the critical photon frequency is

$$\omega_c = C_\omega\,\frac{E^3}{\rho} \quad \text{with} \quad C_\omega = 3.37\,10^{18}\,\frac{\text{m}}{\text{sec GeV}^3} \tag{7.76}$$

and the critical photon energy

$$\epsilon_c(\text{keV}) = 2.218 \frac{E^3(\text{GeV}^3)}{\rho(\text{m})} = 0.0665\, E^2(\text{GeV}^2)\, B(\text{kG}). \qquad (7.77)$$

The synchrotron radiation spectrum from relativistic particles in a circular accelerator is made up of harmonics of the particle revolution frequency ω_o and extends to values up to and beyond the critical frequency (7.75). Generally, a real synchrotron radiation beam from say a storage ring will not display this harmonic structure. The distance between the harmonics is extremely small compared to the extracted photon frequencies in the VUV and x-ray regime while the line width is finite due to the energy spread in a beam of many particles and the spectrum becomes therefore continuous. For a single pass of particles through a bending magnet in a beam transport line, we observe the same spectrum, although now genuinely continuous as can be derived with the use of Fourier transforms of a single light pulse. Specifically, the maximum frequency is the same assuming similar parameters.

7.4.1 Radiation Field in the Frequency Domain

Synchrotron radiation is emitted within a wide range of frequencies. As we have seen in the previous paragraph, a particle orbiting in a circular accelerator emits light flashes with a repetition equal to the revolution frequency. We expect therefore in the radiation frequency spectrum all harmonics of the revolution frequency up to very high frequencies limited only by the very short duration of the radiation being sent into a particular direction toward the observer. The number of harmonics increases with beam energy and reaches at the critical frequency the order of γ^4.

The frequency spectrum of synchrotron radiation has been derived by many authors. In this text we will stay closer to the derivation by *Jackson*[7.9] than others. The general method to derive the frequency spectrum is to transform the electric field from the time domain to the frequency domain by the use of Fourier transforms. Applying this method, we will determine the radiation characteristics of the light emitted by a single pass of a particle in a circular accelerator at the location of the observer. The electric field at the observation point has a strong time dependence and is given by (7.32) while the total radiation energy for one pass is from (7.43)

$$\frac{dW}{d\Omega} = -\int_{-\infty}^{\infty} \frac{dP}{d\Omega}\, dt = \int_{-\infty}^{\infty} \mathbf{S_r n} R^2\, dt = [4\pi c \epsilon_o] \frac{cR^2}{4\pi} \int_{-\infty}^{\infty} \mathbf{E}_r^2(t)\, dt. \quad (7.78)$$

The transformation from the time domain to the frequency domain is performed by a Fourier transform or an expansion into Fourier harmonics. This is the point where the particular characteristics of the transverse acceleration depend on the magnetic field distribution and are, for example, different in a single bending magnet as compared to an oscillatory wiggler magnet. We use here the method of Fourier transforms to describe the electric field of a single particle passing only once through a homogeneous bending magnet. In case of a circular accelerator the particle will appear periodically with the period of the revolution time and we expect a correlation of the frequency spectrum with the revolution frequency. This is indeed the case and we will later discuss the nature of this correlation. Expressing the electrical field $\mathbf{E}_{\mathrm{r}}(t)$ by its Fourier transform, we set

$$\mathbf{E}_{\mathrm{r}}(\omega) \,=\, \int\limits_{-\infty}^{\infty} \mathbf{E}_{\mathrm{r}}(t)\,\mathrm{e}^{-\mathrm{i}\omega t}\,\mathrm{d}t\,, \tag{7.79}$$

where $-\infty \leq \omega \leq \infty$. Applying *Parseval's theorem* we have

$$\int\limits_{-\infty}^{\infty} |\,\mathbf{E}_{\mathrm{r}}(\omega)\,|^2\,d\omega \,=\, 2\pi \int\limits_{-\infty}^{\infty} |\,\mathbf{E}_{\mathrm{r}}(t)\,|^2\,\mathrm{d}t \tag{7.80}$$

and the total absorbed radiation energy from a single pass of a particle is therefore

$$\frac{\mathrm{d}W}{\mathrm{d}\Omega} \,=\, [4\pi c\epsilon_{\mathrm{o}}]\,\frac{c}{8\pi^2}\,R^2 \int\limits_{-\infty}^{\infty} |\,\mathbf{E}_{\mathrm{r}}(\omega)\,|^2\,\mathrm{d}\omega\,. \tag{7.81}$$

Evaluating the electrical field by its Fourier components we derive an expression for the spectral distribution of the radiation energy

$$\frac{\mathrm{d}^2W}{\mathrm{d}\omega\,\mathrm{d}\Omega} \,=\, [4\pi c\epsilon_{\mathrm{o}}]\,\frac{c}{4\pi^2}\,\mathbf{E}_{\mathrm{r}}^2(\omega)\,R^2\,, \tag{7.82}$$

where we have implicitly used the fact that $\mathbf{E}_{\mathrm{r}}(\omega) = \mathbf{E}_{\mathrm{r}}(-\omega)$ since $\mathbf{E}_{\mathrm{r}}(t)$ is real. To calculate the Fourier transform we use (7.32) and note that the electrical field is expressed in terms of quantities at the retarded time t_{r}. The calculation is simplified if we express the whole integrand in (7.79) at the retarded time and get with $t_{\mathrm{r}} = t - \frac{1}{c}R(t_{\mathrm{r}})$ and $\mathrm{d}t_{\mathrm{r}} = \frac{R(t_{\mathrm{r}})}{r}\,\mathrm{d}t$ instead of (7.79)

$$\mathbf{E}_{\mathrm{r}}(\omega) \,=\, \frac{1}{[4\pi\epsilon_{\mathrm{o}}]}\,\frac{e}{c} \int\limits_{-\infty}^{\infty} \frac{\mathbf{R} \times [(\mathbf{R} + \boldsymbol{\beta}\,R) \times \dot{\boldsymbol{\beta}}]}{r^2\,R}\Big|_{\mathrm{r}}\,\mathrm{e}^{-\mathrm{i}\omega\,(t_{\mathrm{r}} + \frac{R}{c})}\,\mathrm{d}t_{\mathrm{r}}\,. \tag{7.83}$$

We require now that the radiation be observed at a point sufficiently far away from the source that during the time of emission the vector $\mathbf{R}(t_{\mathrm{r}})$ does not change appreciably in direction. This assumption is generally justified since the duration of the photon emission is of the order of $1/(\omega_{\mathrm{L}}\gamma)$ where $\omega_{\mathrm{L}} = c/\rho$ is the *Larmor frequency*. The observer therefore should be at a distance from the source large compared to ρ/γ. Equation (7.83) together with (7.17) may then be written like

$$\mathbf{E}_r(\omega) \;=\; \frac{1}{[4\pi\epsilon_0]}\,\frac{e}{cR}\int\limits_{-\infty}^{\infty}\frac{\mathbf{n}\times[(\mathbf{n}+\boldsymbol{\beta})\times\dot{\boldsymbol{\beta}}]}{(1+\mathbf{n}\boldsymbol{\beta})}\Big|_r\,\mathrm{e}^{-i\omega\,(t_r+\frac{R}{c})}\,\mathrm{d}t_r\,. \tag{7.84}$$

With

$$\frac{\mathbf{n}\times[(\mathbf{n}+\boldsymbol{\beta})\times\dot{\boldsymbol{\beta}}]}{(1+\mathbf{n}\boldsymbol{\beta})^2}\;=\;\frac{\mathrm{d}}{\mathrm{d}t_r}\,\frac{\mathbf{n}\times(\mathbf{n}\times\boldsymbol{\beta})}{1+\mathbf{n}\boldsymbol{\beta}}\,, \tag{7.85}$$

we integrate (7.84) by parts while noting that the integrals vanish at the boundaries and get

$$\mathbf{E}_r(\omega) \;=\; \frac{1}{[4\pi\epsilon_0]}\,\frac{-i\,e\,\omega}{cR}\int\limits_{-\infty}^{\infty}[\mathbf{n}\times[\mathbf{n}\times\boldsymbol{\beta}]]_r\;\mathrm{e}^{-i\omega\,(t_r+\frac{R}{c})}\,\mathrm{d}t_r\,. \tag{7.86}$$

After insertion into (7.82) the spectral and spatial intensity distribution is

$$\boxed{\;\frac{\mathrm{d}^2 W}{\mathrm{d}\omega\,\mathrm{d}\Omega} \;=\; \frac{r_c\,mc^2}{4\pi^2\,c}\,\omega^2\left|\int\limits_{-\infty}^{\infty}[\mathbf{n}\times[\mathbf{n}\times\boldsymbol{\beta}]]\,\mathrm{e}^{-i\omega\,(t_r+\frac{R}{c})}\,\mathrm{d}t_r\right|_r^2.\;} \tag{7.87}$$

The spectral and spatial radiation distribution depends on the Fourier transform of the particle trajectory which itself is a function of the magnetic field. The trajectory in a uniform dipole field is different from say the step function of real lumped bending magnets or oscillating deflecting fields from wiggler magnets and the radiation characteristics may therefore be different. In this chapter, we will concentrate only on a uniform dipole field and postpone the discussion of specific radiation characteristics for insertion devices to Chap. 11.

The integrand in (7.87) can be expressed in component form to simplify integration. For that we consider a fixed coordinate system (x, y, z) as shown in Fig. 7.6. The observation point is far away from the source point and we focus on the radiation that is centered about the tangent to the orbit at the source point. The observation point P and the vectors $\mathbf{R}$ and $\mathbf{n}$ are therefore within the (y, z)-plane and radiation is emitted at angles θ with respect to the z-axis.

The vector from the origin of the coordinate system P_o to the observation point P is $\mathbf{r}$, the vector $\mathbf{R}$ is the vector from P to the particle at P_p and $\mathbf{r}_p$ is the vector from the origin to P_p. With this we have

$$\mathbf{r} \;=\; \mathbf{r}_p - \mathbf{R}(t_r)\,, \tag{7.88}$$

where $\mathbf{r}_p$ and $\mathbf{R}$ are taken at the retarded time. The exponent in (7.87) is then

$$\omega \left[t_\mathrm{r} + \frac{1}{c} R(t_\mathrm{r}) \right] = \omega \left[t_\mathrm{r} + \frac{1}{c} \mathbf{n}\, \mathbf{R}(t) \right] = \omega \left(t_\mathrm{r} + \frac{1}{c} \mathbf{n} \mathbf{r}_\mathrm{p} - \frac{1}{c} \mathbf{n} \mathbf{r} \right) \quad (7.89)$$

and the term $-\omega \frac{\mathbf{n}\mathbf{r}}{c}$ is independent of the time generating only a constant phase factor which is completely irrelevant for the spectral distribution and may therefore be ignored.

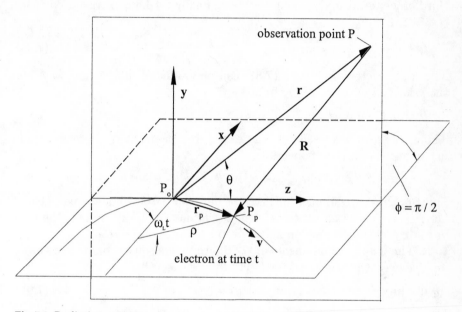

Fig. 7.6. Radiation geometry

In determining the vector components, we note from Fig. 7.6 that now the coordinate system is fixed in space. Following the above discussion the azimuthal angle is constant and set to $\phi = \frac{\pi}{2}$. With these assumptions, we get the vector components for the vector $\mathbf{n}$ from (7.67)

$$\mathbf{n} = (0, -\sin\theta, -\cos\theta). \quad (7.90)$$

The vector $\mathbf{r}_\mathrm{p}$ is defined by Fig. 7.6 and depends on the exact variation of the deflecting magnetic field along the path of the particles. Here we assume a constant bending radius ρ and have

$$\mathbf{r}_\mathrm{p} = [-\rho\,(1 - \cos\omega_\mathrm{L} t_\mathrm{r}),\, 0,\, \rho \sin\omega_\mathrm{L} t_\mathrm{r}], \quad (7.91)$$

where $\omega_\mathrm{L} = \frac{\beta c}{\rho}$ is the Larmor frequency. From these component representations we get for the vector product

$$\frac{\mathbf{n}\mathbf{r}_\mathrm{p}}{c} = -\frac{\rho}{c} \sin\omega_\mathrm{L} t_\mathrm{r} \cos\theta. \quad (7.92)$$

Noting that both arguments of the trigonometric functions in (7.92) are very small, the factor $t_r + \frac{1}{c}\mathbf{n}\mathbf{r}_p$ in (7.89) after expansion of the trigonometric functions up to third order in t_r becomes

$$t_r + \frac{\mathbf{n}\mathbf{r}_p}{c} = t_r - \frac{\rho}{c}\left(\omega_L t_r - \tfrac{1}{6}\omega_L^3 t_r^3\right)\left(1 - \tfrac{1}{2}\theta^2\right). \tag{7.93}$$

With $\omega_L = \frac{\beta c}{\rho}$ we get $t_r(1 - \frac{\rho}{c}\omega_L) = (1-\beta)t_r \approx t_r/(2\gamma^2)$. Keeping only up to third order terms in $\omega_L t_r$ and θ we have finally for high energetic particles $\beta \approx 1$

$$t_r + \frac{\mathbf{n}\mathbf{r}_p}{c} = \tfrac{1}{2}\left(\gamma^{-2} + \theta^2\right)t_r + \tfrac{1}{6}\omega_L^2 t_r^3. \tag{7.94}$$

The triple vector product in (7.87) can be evaluated in a similar way. For the velocity vector we derive from Fig. 7.6

$$\boldsymbol{\beta} = \beta\left(-\sin\omega_L t_r,\; 0,\; \cos\omega_L t_r\right). \tag{7.95}$$

The vector relations (7.30,92) can be used to express the triple vector product in terms of its components

$$\mathbf{n}\times[\mathbf{n}\times\boldsymbol{\beta}] = \beta\left(\sin\omega_L t_r, +\tfrac{1}{2}\sin 2\theta\,\cos\omega_L t_r + \sin^2\theta\,\cos\omega_L t_r\right). \tag{7.96}$$

Splitting this three-dimensional vector into two parts will allow us to characterize the *polarization states* of the radiation. To do this we take the unit vector $\mathbf{u}_\perp$ in the x-direction and $\mathbf{u}_\parallel$ a unit vector normal to $\mathbf{u}_\perp$ and normal to $\mathbf{R}$. The y and z components of (7.96) are then also the components of $\mathbf{u}_\parallel$ and we may express the vector (7.96) by

$$\mathbf{n}\times[\mathbf{n}\times\boldsymbol{\beta}] = \beta\,\sin\omega_L t_r\,\mathbf{u}_\perp + \sin\theta\,\cos\omega_L t_r\,\mathbf{u}_\parallel. \tag{7.97}$$

Inserting (7.94) and (7.97) into the integrand (7.86) we get with $\beta \approx 1$

$$\mathbf{E}_r(\omega) = -\frac{1}{[4\pi\epsilon_0]}\frac{we}{cR}\int_{-\infty}^{\infty}\left(-\sin\omega_L t_r\,\mathbf{u}_\perp + \sin\theta\,\cos\omega_L t_r\,\mathbf{u}_\parallel\right)$$
$$\exp\left\{-\mathrm{i}\,\tfrac{1}{2}\frac{\omega}{\gamma^2}\left[(1+\gamma^2\theta^2)\,t_r + \tfrac{1}{3}\gamma^2\omega_L^2 t_r^3\right]\right\}\mathrm{d}t_r. \tag{7.98}$$

Two polarization directions have been defined for the electric radiation field. One of which, $\mathbf{u}_\perp$ is in the plane of the particle path being perpendicular to the particle velocity and to the deflecting magnetic field. Following *Sokolov* and *Ternov* [7.10] we call this the σ-*mode* $\mathbf{u}_\perp \equiv \mathbf{u}_\sigma$. The other polarization direction in the plane containing the deflecting magnetic field and the observation point is perpendicular to $\mathbf{R}$ and is called the π-*mode* $\mathbf{u}_\parallel = \mathbf{u}_\pi$. Since the emission angle θ is very small, we find this polarization direction to be mostly parallel to the magnetic field. Noting that most accelerators or beam lines are constructed in the horizontal plane the polarizations are also often referred to as the horizontal polarization for the σ-mode and as the vertical polarization for the π-mode.

7.4.2 Spectral Distribution in Space and Polarization

As was pointed out by *Jackson*[7.9], the integration over infinite times does not invalidate our expansion of the trigonometric functions where we assumed the argument $\omega_{\mathrm{L}} t_{\mathrm{r}}$ to be small. Although the integral (7.98) extends over all past and future times, the integrand oscillates rapidly for all but the lowest frequencies and therefore only times of the order $ct_{\mathrm{r}} = \pm\rho/\gamma$ centered about $t_{\mathrm{r}} = 0$ contribute to the integral. This is a direct consequence of the fact that the radiation is emitted in the forward direction and therefore only photons from a very small segment of the particle trajectory reach the observation point. For very small frequencies of the order of the Larmor frequency, however, we must expect considerable deviations from our results. In practical circumstances such low harmonics will, however, not propagate in the vacuum chamber [7.11] and the observed photon spectrum therefore is described accurately for all practical purposes.

The integral in (7.98) can be expressed by *modified Bessel's functions* in the form of Airy's integrals as has been pointed out by *Schwinger* [7.12]. Since the deflection angle $\omega_{\mathrm{L}} t_{\mathrm{r}}$ is very small, we may use the replacements $\sin\omega_{\mathrm{L}} t_{\mathrm{r}} \approx \omega_{\mathrm{L}} t_{\mathrm{r}}$ and $\cos\omega_{\mathrm{L}} t_{\mathrm{r}} \approx 1$. Inserting the expression for the electric field (7.98) into (7.81) we note that cross terms of both polarizations vanish $\mathbf{u}_{\perp}\mathbf{u}_{\parallel} = 0$ and the radiation intensity can therefore be expressed by two separate orthogonal polarization components. Introducing in (7.98) the substitutions

$$\omega_{\mathrm{L}} t_{\mathrm{r}} = \sqrt{\frac{1}{\gamma^2} + \theta^2}\; x\,,$$

$$\xi = \tfrac{1}{3}\frac{\omega}{\omega_{\mathrm{L}}}\frac{1}{\gamma^3}\left(1 + \gamma^2\theta^2\right)^{3/2} = \tfrac{1}{2}\frac{\omega}{\omega_{\mathrm{c}}}\left(1 + \gamma^2\theta^2\right)^{3/2}\,, \qquad (7.99)$$

where $\hbar\omega_{\mathrm{c}}$ is the critical photon energy, the argument in the exponential factor of (7.98) becomes

$$-\mathrm{i}\tfrac{1}{2}\frac{\omega}{\gamma^2}\left[\left(1 + \gamma^2\theta^2\right)t_{\mathrm{r}} + \tfrac{1}{3}\gamma^2\omega_{\mathrm{L}}^2 t_{\mathrm{r}}^3\right] = -\mathrm{i}\tfrac{1}{2}\xi\left(3x + x^3\right)\,.$$

With these substitutions, (7.98) can be evaluated noting that only even terms contribute to the integral. With $\omega_{\mathrm{L}} t_{\mathrm{r}}$ and θ being small quantities we get integrals of the form [7.13]

$$\int_0^\infty \cos\left[\tfrac{1}{2}\xi\left(3x + x^3\right)\right]\mathrm{d}x = \frac{1}{\sqrt{3}}K_{1/3}(\xi)\,,$$

$$\int_0^\infty x\sin\left[\tfrac{1}{2}\xi\left(3x + x^3\right)\right]\mathrm{d}x = \frac{1}{\sqrt{3}}K_{2/3}(\xi)\,, \qquad (7.100)$$

where the functions K_ν are modified Bessel's functions of the second kind. These functions assume finite values for small arguments but vanish exponentially for large arguments as shown in Fig. 7.7. Fast converging series for these modified Bessel's functions with fractional index has been derived in [7.14]. The Fourier transform of the electrical field (7.98) finally becomes

$$
\mathbf{E_r}(\omega) = \frac{1}{[4\pi\epsilon_\mathrm{o}]} \frac{\sqrt{3}e}{cR} \frac{\omega}{\omega_\mathrm{c}} \gamma\left(1 + \gamma^2\theta^2\right)
$$
$$
\times \left[K_{2/3}(\xi)\,\mathbf{u}_\sigma - \mathrm{i}\,\frac{\gamma\theta K_{1/3}(\xi)}{\sqrt{1 + \gamma^2\theta^2}}\,\mathbf{u}_\pi \right]
$$

(7.101)

describing the *radiation field* far from the source for particles traveling through a uniform magnetic dipole field. Later we will modify this expression to make it suitable for particle motion in undulators or other nonuniform fields.

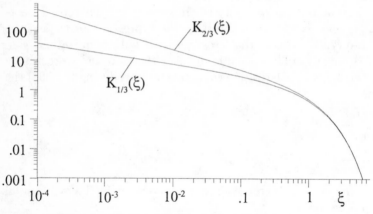

Fig. 7.7. Modified Bessel's functions $K_{1/3}(x)$ and $K_{2/3}(x)$

Radiation distribution: The spectral distribution of the synchrotron radiation is proportional to the square of the electrical field components and is from (7.82,101)

$$
\frac{\mathrm{d}^2W}{\mathrm{d}\omega\,\mathrm{d}\Omega} = \frac{3r_\mathrm{c}\,mc^2\gamma^2}{4\pi^2 c}\frac{\omega^2}{\omega_\mathrm{c}^2}\left(1 + \gamma^2\theta^2\right)^2
$$
$$
\times \left[K_{2/3}^2(\xi) + \frac{\gamma^2\theta^2}{1 + \gamma^2\theta^2}K_{1/3}^2(\xi) \right].
$$

(7.102)

The *radiation spectrum* has two components of orthogonal *polarization*, one in the plane of the particle trajectory and the other almost parallel to the deflecting magnetic field. In (7.101) both polarizations appear explicitly through the orthogonal unit vectors. Forming the square of the electrical field to get the radiation intensity, cross terms disappear because of the orthogonality of the unit vectors $\mathbf{u}_\sigma$ and $\mathbf{u}_\pi$. The expression for the radiation intensity therefore preserves separately the two polarization modes in the square brackets of (7.102) representing the σ-mode and π-mode of polarization, respectively.

It is interesting to study the spatial distribution for the two polarization modes in more detail. Not only are the intensities very different but the spatial distribution is different too. The spatial distribution of the σ-mode is directed mainly in the forward direction while the π-mode, in contrast, has zero intensity in the direction of the orbit $\theta = 0$ but emits radiation into two lobes at finite angles. In Figs. 7.8 and 7.9 the instantaneous radiation lobes are shown for both the σ- and the π-mode at a frequency of $\omega/\omega_c = 1$ being emitted tangentially from the orbit at the origin of the coordinate system.

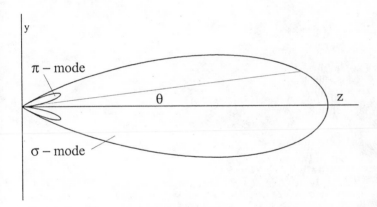

Fig. 7.8. Radiation lobes for σ- and π-polarization modes

The complete radiation pattern along the orbit is obtained by moving this pattern and those for other frequencies along the orbit resulting in a swath of radiation in the deflecting plane and little divergence in the nondeflecting plane.

The spatial radiation pattern varies, however, with the frequency of the radiation. Specifically, the angular distribution concentrates more and more in the forward direction as the radiation frequency increases. The radiation distribution in frequency and angular space is shown for both the σ- (Fig. 7.10) and the π-mode (Fig. 7.11) at the fundamental frequency. The high collimation of synchrotron radiation in the forward direction makes

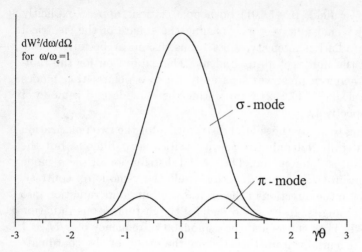

Fig. 7.9. Angular radiation distribution for σ- and π-polarization modes

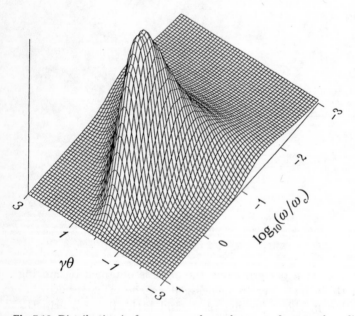

Fig. 7.10. Distribution in frequency and angular space for σ-mode radiation

it a prime research tool to probe materials and its atomic and molecular properties.

The radiation intensity W from a single electron and for a single pass may not always be the most useful parameter. A more useful parameter is the photon flux per unit solid angle into a frequency bin $\Delta\omega/\omega$ and for a circulating beam current I

$$\frac{\mathrm{d}N_{\mathrm{ph}}}{\mathrm{d}\theta\mathrm{d}\psi} = \frac{\mathrm{d}W}{\mathrm{d}\Omega}\frac{1}{\hbar}\frac{I}{e}\frac{\Delta\omega}{\omega}. \tag{7.103}$$

Here we have replaced the solid angle by its components, the vertical angle θ and the bending angle ψ. For numerical calculations of photon fluxes, we may use the graphic representation of the modified Bessel's function in Fig. 7.7 together with (7.102) and get in more practical units the differential photon flux

$$\boxed{\frac{\mathrm{d}N_{\mathrm{ph}}}{\mathrm{d}\theta\mathrm{d}\psi} = C_\Omega\,E^2\,I\,\frac{\Delta\omega}{\omega}\frac{\omega^2}{\omega_{\mathrm{c}}^2}\,K_{2/3}^2(\xi)\,F(\xi,\theta)\,,} \tag{7.104}$$

where

$$C_\Omega = \frac{3\,\alpha}{4\pi^2\,e\,(mc^2)^2} = 1.3273\,10^{16}\,\frac{\text{photons}}{\text{sec mrad}^2\,\text{GeV}^2\,\text{A}}\,, \tag{7.105}$$

I the circulating particle beam current, α the fine structure constant, and

$$F(\xi,\theta) = (1+\gamma^2\,\theta^2)^2\left[1+\frac{\gamma^2\,\theta^2}{1+\gamma^2\,\theta^2}\frac{K_{1/3}^2(\xi)}{K_{2/3}^2(\xi)}\right]. \tag{7.106}$$

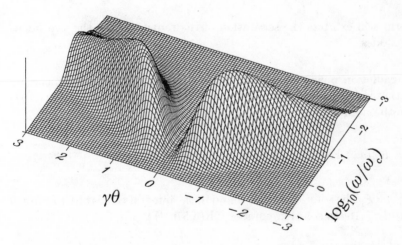

Fig. 7.11. Distribution in frequency and angular space for π-mode radiation

Equation (7.104) gives the photon flux for both the σ-mode radiation in the forward direction within an angle of about $\pm1/\gamma$ and for the π-mode off axis.

Harmonic representation: Expression (7.102) can be transformed into a different formulation emphasizing the harmonic structure of the radiation

spectrum. The equivalence between both formulations has been shown by
Sokolov and Ternov [7.10] expressing the modified Bessel's functions $K_{1/3}$
and $K_{2/3}$ by regular Bessel's functions of high order. With $\nu = \omega / \omega_L$ the
asymptotic formulas for $\nu \gg 1$ are [7.10]

$$K_{1/3}(\xi) = \frac{\pi\sqrt{3}}{\sqrt{1-\beta^2\cos^2\theta}} J_\nu(\nu\beta\cos\theta),$$

$$K_{2/3}(\xi) = \frac{\pi\sqrt{3}}{1-\beta^2\cos^2\theta} J'_\nu(\nu\beta\cos\theta),$$

(7.107)

where $\xi = \frac{\nu}{3}(1-\beta^2\cos^2\theta)^{3/2} \approx \frac{\nu}{3}(\frac{1}{\gamma^2}+\beta^2\theta^2)^{3/2}$ for small angles.

These approximations are justified since we are only interested in very
large harmonics of the revolution frequency. The harmonic number ν for
the critical photon frequency, for example, is given by ω_c/ω_L and is $\nu_c = \frac{3}{2}\gamma^3$ which for practical cases is generally a very large number. Inserting
these approximations into (7.102) gives the the formulation that has been
derived first by *Schott* [7.15 – 17] in 1907 long before synchrotron radiation
was discovered in an attempt to calculate the radiation intensity of atomic
spectral lines

$$\frac{d^2W}{d\omega\,d\Omega} = r_c mc^2 \frac{c}{2\pi\rho^2}\nu^2\left[J'^2_\nu(\nu\cos\theta)+\theta^2 J^2_\nu(\nu\cos\theta)\right].$$

(7.108)

This form still exhibits the separation of the radiation into the two polarization modes.

Spatial radiation power distribution: Integrating over all frequencies we
obtain the *angular distribution* of the synchrotron radiation. From (7.102)
we note the need to perform integrals of the form

$$\int_0^\infty \omega^2 K_\mu^2(a\omega)\,d\omega$$

where $a\omega = \xi$. The solution can be found in the integral tables of *Gradshteyn*
and *Ryzhik* [7.18] as solution number GR(6.576.4) [1]

$$\int_0^\infty \omega^2 K_\mu^2(a\omega)\,d\omega = \frac{\pi^2}{32\,a^3}\frac{1-4\mu^2}{\cos\pi\mu}.$$

(7.109)

Applying this solution to (7.102) and integrating over all frequencies we get
for the angular distribution of the synchrotron radiation

[1] In this paragraph we will need repeatedly results from mathematical tables. We abbreviate such solutions with the first letters of the authors names and the formula number.

$$\frac{dW}{d\Omega} = \frac{21}{32} \frac{r_c \, mc^2}{\rho} \frac{\gamma^5}{(1+\gamma^2\,\theta^2)^{5/2}} \left[1 + \frac{5}{7} \frac{\gamma^2\,\theta^2}{1+\gamma^2\,\theta^2} \right]. \tag{7.110}$$

This result is consistent with the angular radiation power distribution (7.74) where we found that the radiation is collimated very much in the forward direction with most of the radiation energy being emitted within an angle of $\pm 1/\gamma$. There are two contributions to the total radiation intensity, the σ- and the π-mode. The σ-mode has a maximum intensity in the forward direction, while the maximum intensity for the π-mode occurs at an angle of $\theta_\pi = 1/(\sqrt{2.5}\gamma)$.

The quantity W is the radiation energy per unit solid angle from a single electron and a single pass and the average radiation power is therefore $P_\gamma = W/T_{\text{rev}}$ or (7.110) becomes

$$\boxed{\frac{dP_\gamma}{d\Omega} = \frac{2}{3} \frac{r_c \, mc^2}{2\pi} \frac{c}{\rho^2} \frac{\gamma^5}{(1+\gamma^2\,\theta^2)^{5/2}} \left[1 + \frac{5}{7} \frac{\gamma^2\,\theta^2}{1+\gamma^2\,\theta^2} \right].} \tag{7.111}$$

Integrating (7.111) over all angles θ we find the synchrotron radiation power into both polarization modes, the σ-mode perpendicular to the magnetic field and the π-mode parallel to the magnetic field. In doing so, we note first that (7.111) can be simplified with (7.57) and $\beta = 1$

$$\frac{dP_\gamma}{d\Omega} = \frac{21}{32} \frac{P_\gamma}{2\pi} \frac{\gamma}{(1+\gamma^2\,\theta^2)^{5/2}} \left[1 + \frac{5}{7} \frac{\gamma^2\,\theta^2}{1+\gamma^2\,\theta^2} \right]. \tag{7.112}$$

This result is consistent with the result in Sect. 7.3 although we have now the radiation separated into the two orthogonal polarization components. The total power in each polarization mode can be derived from (7.112) by integration over all angles. First, we integrate over all points along the circular orbit and get a factor 2π since the observed radiation power does not depend on the location along the orbit. Continuing the integration over all angles of θ, we find the contributions to the integral to become quickly negligible for angles larger than $1/\gamma$. If it were not so, we could not have used (7.112) where the trigonometric functions have been replaced by their small arguments. Both terms in (7.112) can be integrated readily and the first term becomes with GR(2.271.6) [7.18]

$$\int\limits_{\theta_{\max}\gamma \ll -1}^{\theta_{\max}\gamma \gg 1} \frac{\gamma \, d\theta}{(1+\gamma^2\,\theta^2)^{5/2}} = \frac{4}{3}.$$

The second term is with GR[2.272.7] [7.18]

$$\int\limits_{\theta_{\max}\gamma \ll -1}^{\theta_{\max}\gamma \gg 1} \frac{\gamma^3\,\theta^2 \, d\theta}{(1+\gamma^2\,\theta^2)^{7/2}} = \frac{4}{15}.$$

With these integrals and (7.112) the radiation power into the σ and π-mode is finally with P_γ from (7.57)

$$P_\sigma = \frac{7}{8}\, P_\gamma \qquad\qquad\qquad\qquad\qquad\qquad (7.113)$$

and

$$P_\pi = \frac{1}{8}\, P_\gamma \qquad\qquad\qquad\qquad\qquad\qquad (7.114)$$

with

$$P_\sigma + P_\pi = P_\gamma\,. \qquad\qquad\qquad\qquad\qquad (7.115)$$

The horizontally polarized component of synchrotron radiation greatly dominates the photon beam characteristics and only 12.5 % of the total intensity is polarized in the vertical plane. In the forward direction the σ-polarization reaches even almost 100%. This high polarization of the radiation provides a valuable characteristic for experimentation with synchrotron light and gives in addition cause for a polarization process of the particle beam orbiting in a circular accelerator like in a storage ring [7.19]. Obviously the sum of both components is equal to the total radiation power.

Asymptotic solutions: Expressions for the radiation distribution can be greatly simplified if we restrict the discussion to very small or very large arguments of the modified Bessel's functions for which approximate expressions exist [7.20]. Knowledge of the radiation distribution at very low photon frequencies becomes important for experiments using such radiation or for beam diagnostics where the beam cross section is being imaged to a TV camera using the visible part of the radiation spectrum. To describe this visible part of the spectrum, we may in most cases assume that the photon frequency is much lower than the critical photon frequency.

For very small arguments or low frequencies and small angles, we find the approximate identities AS(9.6.9) [7.20]

$$K^2_{2/3}(\xi) \approx 2^{2/3}\, \Gamma^2(2/3) \left(\frac{\omega}{\omega_c}\right)^{-4/3} \frac{1}{(1+\gamma^2\theta^2)^2} \qquad \text{for} \quad \xi \to 0 \qquad (7.116)$$

and

$$K^2_{1/3}(\xi) \approx \frac{\Gamma^2(1/3)}{2^{2/3}} \left(\frac{\omega}{\omega_c}\right)^{-2/3} \frac{1}{1+\gamma^2\theta^2} \qquad \text{for} \quad \xi \to 0, \qquad (7.117)$$

where $\xi = 1/2\,(\omega/\omega_c)\,(1+\gamma^2\theta^2)^{3/2}$ from (7.99). Inserting this into (7.99) the photon flux spectrum in the forward direction becomes for $\theta = 0$ and $\omega/\omega_c \ll 1$

$$\frac{\mathrm{d}\mathcal{N}_{\mathrm{ph}}}{\mathrm{d}\theta\mathrm{d}\psi} \approx C_{\varOmega}\, E^2\, I\, \varGamma^2(\sqrt[2]{3}) \left(\frac{2\omega}{\omega_{\mathrm{c}}}\right)^{2/3} \frac{\Delta\omega}{\omega}, \tag{7.118}$$

where the gamma functions $\varGamma(\sqrt[1]{3}) = 2.6789385$ and $\varGamma(\sqrt[2]{3}) = 1.3541179$. The photon spectrum at very low frequencies is independent of the particle energy since $\omega_{\mathrm{c}} \propto E^3$. Clearly, in this approximation there is no angular dependence for the σ-mode radiation and the intensity increases with frequency. The π-mode radiation on the other hand is zero for $\theta = 0$ and increases in intensity with the square of θ as long as the approximation is valid.

To calculate the rms angular distribution of the photon flux we must evaluate the integral

$$\langle\theta^2\rangle = \frac{\int \theta^2 \frac{\mathrm{d}N_{\mathrm{ph}}}{\mathrm{d}\theta}\, \mathrm{d}\theta}{\int \frac{\mathrm{d}N_{\mathrm{ph}}}{\mathrm{d}\theta}\, \mathrm{d}\theta}. \tag{7.119}$$

These integrals have been evaluated, for example, in [7.21] and we will give only the results here. For very low photon frequencies the angular spread is given for the σ-mode radiation by

$$\sqrt{\langle\theta^2\rangle} = \sqrt{\frac{5}{12}\,\frac{\mathcal{A}\mathrm{i}(0)}{-\mathcal{A}'\mathrm{i}(0)}} \left(\frac{\omega_{\mathrm{L}}}{\omega}\right)^{1/3} = \frac{0.7560}{\nu^{1/3}} \tag{7.120}$$

and for the π-mode radiation by

$$\sqrt{\langle\theta^2\rangle} = \sqrt{\frac{9}{12}\,\frac{\mathcal{A}\mathrm{i}(0)}{-\mathcal{A}'\mathrm{i}(0)}} \left(\frac{\omega_{\mathrm{L}}}{\omega}\right)^{1/3} = \frac{1.0143}{\nu^{1/3}}, \tag{7.121}$$

where $\nu = \frac{\omega}{\omega_{\mathrm{L}}}$ is the harmonic number of the synchrotron radiation. In either case, we find the angular distribution to open up inversely proportional to the third root of the photon frequency.

For large arguments of the modified Bessel's functions or for high frequencies and large angles of emission different approximations hold. In this case the approximate expressions are actually the same for both Bessel's functions indicating the same exponential drop off for high energetic photons AS(9.7.2) [7.20]

$$K_{2/3}^2(\xi) \approx \frac{\pi}{2}\frac{\mathrm{e}^{-2\xi}}{\xi} \quad \text{and} \quad K_{1/3}^2(\xi) \approx \frac{\pi}{2}\frac{\mathrm{e}^{-2\xi}}{\xi} \quad \text{for} \quad \xi \to \infty. \tag{7.122}$$

The radiation distribution in this approximation becomes from (7.102)

$$\frac{\mathrm{d}^2 W}{\mathrm{d}\omega\,\mathrm{d}\theta\mathrm{d}\psi} = \frac{3\, r_{\mathrm{c}}\, mc^2}{4\pi\, c}\, \gamma^2\, \frac{\omega}{\omega_{\mathrm{c}}}\, \mathrm{e}^{-2\xi} \sqrt{1 + \gamma^2\theta^2}, \tag{7.123}$$

where the spatial radiation distribution is greatly determined by the exponential factor. The relative amplitude with respect to the forward direction scales therefore like

$$\exp\left\{-\frac{\omega}{\omega_c}\left[\left(1+\gamma^2\theta^2\right)^{3/2}-1\right]\right\}.$$

We look for that angle for which the intensity has fallen to $1/e$. Since $\omega \gg \omega_c$ this angle is very small $\gamma\theta \ll 1$ For this reason we can ignore other θ-dependent factors and the exponential factor becomes $1/e$ for

$$\frac{\omega}{\omega_c}\frac{3}{2}\gamma^2\theta^2_{1/e}=1.$$

Solving for the angle where the intensity is reduced to $1/e$ of the forward flux, we get finally

$$\theta_{1/e} \approx \sqrt{\frac{2}{3}\frac{\omega_c}{\omega}}\frac{1}{\gamma} \qquad \text{for } \omega \gg \omega_c. \tag{7.124}$$

The high energy end of the synchrotron radiation spectrum is more and more collimated into the forward direction. The angular distribution is graphically illustrated for both polarization modes in Figs. 7.10,11.

7.4.3 Angle-Integrated Spectrum

Synchrotron radiation is emitted over a wide range of frequencies and it is of great interest to know the exact frequency distribution of the radiation. Since the radiation is very much collimated in the forward direction it is useful to integrate over all angles of emission to obtain the total spectral photon flux that might be accepted by a beam line with proper aperture. To that goal, we integrate (7.102) with respect to the emission angles and obtain the frequency spectrum of the radiation. The emission angle θ appears in (7.102) in a rather complicated way which makes it difficult to perform the integration directly. We replace therefore the modified Bessel's functions by *Airy's functions* defined by AS(10.4.14) and AS(10.4.31) [7.20]

$$\begin{aligned}\mathrm{Ai}(z) &= \frac{\sqrt{z}}{\sqrt{3}\,\pi}K_{1/3}(\xi),\\ \mathrm{Ai}'(z) &= -\frac{z}{\sqrt{3}\,\pi}K_{2/3}(\xi),\end{aligned} \tag{7.125}$$

where the modified Bessel's functions are defined by (7.100) and with

$$\eta = \frac{3}{4}\frac{\omega}{\omega_c} \tag{7.126}$$

we get from (7.99)

$$z = \left(\frac{3}{2}\xi\right)^{2/3} = \eta^{2/3}\left(1+\gamma^2\theta^2\right). \tag{7.127}$$

We apply this derivation for a periodic motion of the particles orbiting in a circular accelerator. The spectral distribution of the radiation power can be obtained by noting that the differential radiation energy (7.102) is emitted every time the particle passes by the source point. A pulse of radiation is sent towards the observation point at a time interval equal to the revolution time $T_{\mathrm{rev}} = \frac{c}{2\pi\rho}$. The spectral power distribution can now be expressed by Airy functions and becomes

$$\frac{\mathrm{d}^2 P}{\mathrm{d}\omega\,\mathrm{d}\Omega} = \frac{9}{2\pi}\frac{P_\gamma\,\gamma}{\omega_{\mathrm{c}}}\left[\eta^{2/3}\,\mathcal{A}\mathrm{i'}^2(z) + \eta^{4/3}\,\gamma^2\,\theta^2\,\mathcal{A}\mathrm{i}^2(z)\right]. \qquad (7.128)$$

To obtain the photon frequency spectrum, we integrate over all angles of emission which is accomplished by integrating along the orbit contributing a mere factor of 2π and over the angle θ. Although this latter integration is to be performed between $-\pi$ and $+\pi$, we choose the mathematically easier integration from $-\infty$ to $+\infty$ because the Airy functions fall off very fast for large arguments. In fact, we have seen already that most of the radiation is emitted within a very small angle of $\pm 1/\gamma$. The integrals to be solved are of the form

$$\int\limits_{-\infty}^{\infty}\theta^n\,\mathcal{A}\mathrm{i}^2\left[\eta^{2/3}\left(1 + \gamma^2\,\theta^2\right)\right]\mathrm{d}\theta\,,$$

where the functions $\mathcal{A}\mathrm{i}$ are general Airy's functions and n an integer equal to 0 or 2. We concentrate first on the second term in (7.128), set $n = 2$ and form with (7.100,124) the square of the Airy function

$$\theta^2\,\mathcal{A}\mathrm{i}^2(z) = \frac{1}{\pi^2}\int\limits_0^{\infty}\theta^2\,\cos\left[\tfrac{1}{3}\,x^3 + z\,x\right]\mathrm{d}x\int\limits_0^{\infty}\cos\left[\tfrac{1}{3}\,y^3 + z\,y\right]\mathrm{d}y\,.$$

We solve these integrals by making use of the trigonometric relation

$$2\cos\frac{\alpha + \beta}{2}\cos\frac{\alpha - \beta}{2} = \cos\alpha + \cos\beta\,.$$

After introducing the substitutions $x + y = s$ and $x - y = t$, we obtain integrals over two terms which are symmetric in s and t and therefore can be set equal to get

$$\theta^2\,\mathcal{A}\mathrm{i}^2(z) = \frac{1}{2\,\pi^2}\int\limits_0^{\infty}\!\!\int\limits_0^{\infty}\theta^2\,\cos\left[\tfrac{1}{12}\left(s^3 + 3st^2\right) + zs\right]\mathrm{d}s\,\mathrm{d}t\,, \qquad (7.129)$$

where the factor $1/2$ comes from the transformation of the area element $\mathrm{d}x\,\mathrm{d}y = \frac{\mathrm{d}s}{\sqrt{2}}\frac{\mathrm{d}t}{\sqrt{2}}$. In our problem we replace the argument z by the expression $z = \eta^{2/3}\left(1 + \gamma^2\,\theta^2\right)$ and integrate over the angle θ

$$\int\limits_{-\infty}^{\infty} \theta^2 \, \mathcal{A}\mathrm{i}^2(z) \, \mathrm{d}\theta \; = \tag{7.130}$$

$$\frac{1}{\pi^2} \int\limits_{0}^{\infty} \int\limits_{0}^{\infty} \int\limits_{0}^{\infty} \theta^2 \, \cos\left[\tfrac{1}{12}\left(s^3 + 3st^2\right) + s\eta^{2/3}\left(1 + \gamma^2\,\theta^2\right)\right] \mathrm{d}s \, \mathrm{d}t \, \mathrm{d}\theta \, .$$

The integrand is symmetric with respect to θ and the integration therefore needs to be performed only from 0 to ∞ with the result being doubled. We also note that the integration is taken over only one quadrant of the (s, t)-space. Simplifying further the integration the number of variables in the argument of the cosine function can be reduced in the following way. We note the coefficient $\tfrac{1}{4}\,t^2 + \eta^{2/3}\gamma^2\,\theta^2$ which is the sum of squares. Setting $\tfrac{1}{2}t = r\cos\varphi$ and $\eta^{1/3}\gamma\theta = r\sin\varphi$ this term becomes simply r^2. The area element transforms like $\mathrm{d}\theta \, \mathrm{d}t = \dfrac{2}{\eta^{1/3}\gamma}\, r \, \mathrm{d}r \, \mathrm{d}\varphi$ and integrating over φ from 0 to $\pi/2$ since we need integrate only over one quarter plane, (7.130) becomes finally

$$\int\limits_{-\infty}^{\infty} \theta^2 \mathcal{A}\mathrm{i}^2(z)\mathrm{d}\theta = \frac{1}{2\eta\gamma^3\pi} \int\limits_{0}^{\infty}\int\limits_{0}^{\infty} r^2 \cos\left[\tfrac{1}{12}\,s^3 + s(\eta^{2/3} + r^2)\right] r \, \mathrm{d}r \, \mathrm{d}s \, . \tag{7.131}$$

The integrand of (7.131) has now a form close to that of an Airy integral and we will try to complete that similarity. With $q = \left(\tfrac{3}{2}\,\xi\right)^{1/3} x$ the definition of the the Airy function AS(10.4.31) [7.20] is consistent with (7.100) and (7.125)

$$\mathcal{A}\mathrm{i}(z) = \frac{1}{\pi} \int\limits_{0}^{\infty} \cos\left(\tfrac{1}{3}\,q^3 + zq\right) \mathrm{d}q \, . \tag{7.132}$$

Equation (7.131) can be modified into a similar form by setting

$$w^3 = \tfrac{1}{4}\,s^3 \qquad \text{and} \qquad s\left(\eta^{2/3} + r^2\right) = y\,w.$$

Solving for w we get

$$w = \frac{s}{2^{2/3}}$$

and with $y = 2^{2/3}\left(\eta^{2/3} + r^2\right)$, $\mathrm{d}s = 2^{2/3}\,\mathrm{d}w$ and $\mathrm{d}y = 2^{5/3}\,r\,\mathrm{d}r$ (7.131) becomes

$$\int\limits_{-\infty}^{\infty} \theta^2 \, \mathcal{A}\mathrm{i}^2(z) \, \mathrm{d}\theta \; = \; \frac{1}{4\,\eta\,\gamma^3} \int\limits_{y_0}^{\infty} \left(\frac{y}{2^{2/3}} - \eta^{2/3}\right) \mathcal{A}\mathrm{i}(y) \, \mathrm{d}y \, , \tag{7.133}$$

where we have used the definition of Airy's function and where the integration starts at

$$y_{\mathrm{o}} = 2^{2/3}\,\eta^{2/3} = \left(\frac{3}{2}\frac{\omega}{\omega_{\mathrm{c}}}\right)^{2/3}$$

(7.134)

corresponding to $r = 0$.

We may separate this integral into two parts and get a term $y\,\mathcal{A}i(y)$ under one of the integrals. This term is by definition of the Airy's functions AS(10.4.1) [7.20] equal to $\mathcal{A}i''$. Integration of this second derivative gives

$$\int_{y_{\mathrm{o}}}^{\infty} \mathcal{A}i''(y)\,\mathrm{d}y = -\,\mathcal{A}i'(y_{\mathrm{o}})$$

and collecting all terms in (7.133) we have finally

$$\int_{-\infty}^{\infty} \theta^2\,\mathcal{A}i^2(z)\,\mathrm{d}\theta = -\frac{1}{4\,\eta^{1/3}\,\gamma^3}\left[\frac{\mathcal{A}i'(y_{\mathrm{o}})}{y_{\mathrm{o}}} + \int_{y_{\mathrm{o}}}^{\infty} \mathcal{A}i(y)\,\mathrm{d}y\right].$$

(7.135)

The derivation of the complete spectral radiation power distribution (7.128) requires also the evaluation of the integral $\int \mathcal{A}i'^2(z)\,\mathrm{d}\theta$. This can be done with the help of the integral $\int \mathcal{A}i^2(z)\,\mathrm{d}\theta$ and the integral we have just derived. We follow a similar derivation that lead us just from (7.130) to (7.131) and get instead of (7.135)

$$\int_{-\infty}^{\infty} \mathcal{A}i^2(z)\,\mathrm{d}\theta = \frac{1}{2\,\eta^{1/3}\,\gamma} \int_{y_{\mathrm{o}}}^{\infty} \mathcal{A}i(y)\,\mathrm{d}y\,.$$

(7.136)

Recalling the definition of the argument $z = \eta^{2/3}\,(1+\gamma^2\,\theta^2)$, we differentiate (7.136) twice with respect to $\eta^{2/3}$ to get

$$2\int_{-\infty}^{\infty} [\mathcal{A}i(z)\,\mathcal{A}i''(z) + \mathcal{A}i'^2]\,\mathrm{d}\theta = -\frac{2^{1/3}}{\eta^{1/3}\,\gamma}\,\mathcal{A}i'(y_{\mathrm{o}})\,.$$

(7.137)

Using the relation $\mathcal{A}i''(z) = z\,\mathcal{A}i(z)$ and the results (7.134,135) in (7.137) we get

$$\int_{-\infty}^{\infty} \mathcal{A}i'^2(z)\,\mathrm{d}\theta = -\frac{\eta^{1/3}}{4\,\gamma}\left[3\frac{\mathcal{A}i'(y_{\mathrm{o}})}{y_{\mathrm{o}}} + \int_{y_{\mathrm{o}}}^{\infty} \mathcal{A}i(y)\,\mathrm{d}y\right].$$

(7.138)

At this point all integrals have been derived that are needed to describe the spectral radiation power separately in both polarization modes and the spectral radiation power from (7.128) becomes

$$\frac{dP}{d\omega} = \frac{27}{16} \frac{P_\gamma}{\omega_c} \frac{\omega}{\omega_c} \left\{ \left[-3 \frac{Ai'(y_0)}{y_0} - \int\limits_{y_0}^{\infty} Ai(y)\, dy \right] \right.$$

$$\left. - \left[\frac{Ai'(y_0)}{y_0} + \int\limits_{y_0}^{\infty} Ai(y)\, dy \right] \right\} . \tag{7.139}$$

The first term describes the σ-mode of polarization and the second term the π-mode. Combining both polarization modes, we may derive a comparatively simple expression for the spectral radiation power. To this goal, we replace the Airy's functions by modified Bessel's functions

$$\frac{Ai'(y_0)}{y_0} = -\frac{1}{\sqrt{3}\,\pi} K_{2/3}(x_0), \tag{7.140}$$

where from (7,125,126,133) $x_0 = \omega/\omega_c$. With $\sqrt{y}\, dy = dx$, the recurrence formula $2 K'_{2/3} = -K_{1/3} - K_{5/3}$ and (7.125) the Airy integral is finally

$$\int\limits_{y_0}^{\infty} Ai(y)\, dy = -\frac{2}{\sqrt{3}\,\pi} \int\limits_{\omega/\omega_c}^{\infty} K'_{2/3}\, x\, dx - \frac{1}{\sqrt{3}\pi} \int\limits_{\omega/\omega_c}^{\infty} K_{5/3}(\xi)\, d\xi,$$

$$= \frac{2}{\sqrt{3}\,\pi} K_{2/3}(\xi) - \frac{1}{\sqrt{3}\pi} \int\limits_{\omega/\omega_c}^{\infty} K_{5/3}(\xi) d\xi. \tag{7.141}$$

We use (7.139,140) in (7.139) and get the simple expression for the synchrotron radiation spectrum

$$\frac{dP}{d\omega} = \frac{P_\gamma}{\omega_c} \frac{9\sqrt{3}}{8\pi} \frac{\omega}{\omega_c} \int\limits_{\omega/\omega_c}^{\infty} K_{5/3}(x)\, dx = \frac{P_\gamma}{\omega_c} S(\omega/\omega_c), \tag{7.142}$$

where we defined the universal function

$$S(\omega/\omega_c) = \frac{9\sqrt{3}}{8\pi} \frac{\omega}{\omega_c} \int\limits_{\omega/\omega_c}^{\infty} K_{5/3}(x)\, dx = \frac{P_\gamma}{\omega_c}. \tag{7.143}$$

The spectral distribution depends only on the critical frequency ω_c, the total radiation power and a purely mathematical function. This result has been derived originally by *Ivanenko* and *Sokolov* [7.22] and independently by *Schwinger* [7.12]. Specifically, it should be noted that the *synchrotron radiation spectrum*, if normalized to the critical frequency, does not depend on the particle energy and is represented by the universal function shown in Fig. 7.12. The energy dependence is contained in the cubic dependence of the critical frequency acting as a scaling factor for the real spectral distribution.

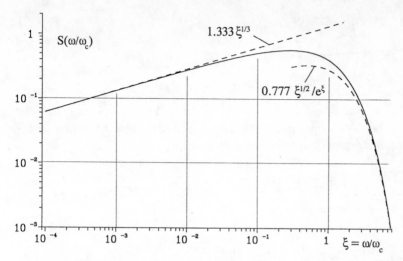

Fig. 7.12. Universal function: $S(\omega/\omega_c) = \frac{9\sqrt{3}}{8\pi} \frac{\omega}{\omega_c} \int\limits_{\omega/\omega_c}^{\infty} K_{5/3}(x)\,\mathrm{d}x$

The mathematical function is properly normalized as we can see by integrating over all frequencies from $x_o = \omega/\omega_c$ to infinity.

$$\int\limits_0^{\infty} \frac{\mathrm{d}P}{\mathrm{d}\omega}\,\mathrm{d}\omega = \frac{9\sqrt{3}}{8\pi} P_\gamma \int\limits_0^{\infty} x_o \left[\int\limits_{x_o}^{\infty} K_{5/3}(x)\,\mathrm{d}x\right] \mathrm{d}x_o. \tag{7.144}$$

After integration by parts the result can be derived from GR[6.561.16] [7.18]

$$\int\limits_0^{\infty} \frac{\mathrm{d}P}{\mathrm{d}\omega}\,\mathrm{d}\omega = \frac{9\sqrt{3}}{16\pi} P_\gamma \int\limits_0^{\infty} x_o^2\, K_{5/3}(x_o)\,\mathrm{d}x_o = \Gamma\left(2+\tfrac{1}{3}\right)\Gamma\left(\tfrac{2}{3}\right). \tag{7.145}$$

Using the triplication formula AS(6.1.19) [7.20] the product of the gamma functions becomes

$$\Gamma\left(2+\tfrac{1}{3}\right)\Gamma\left(\tfrac{2}{3}\right) = \frac{4}{9}\frac{2\pi}{\sqrt{3}}. \tag{7.146}$$

With this equation the proper normalization of (7.144) is demonstrated

$$\int\limits_0^{\infty} \frac{\mathrm{d}P}{\mathrm{d}\omega}\,\mathrm{d}\omega = P_\gamma. \tag{7.147}$$

Of more practical use is the *spectral photon flux* per unit angle of deflection in the bending magnet. With the photon flux $\mathrm{d}N_{\mathrm{ph}} = \mathrm{d}P/\hbar\omega$ we get from (7.142)

$$\frac{dN_{\text{ph}}}{d\psi} = \frac{P_\gamma}{2\pi \hbar \omega_c} \frac{\Delta\omega}{\omega} S(\omega/\omega_c) \tag{7.148}$$

and with (7.57) and (7.75)

$$\frac{dN_{\text{ph}}}{d\psi} = \frac{4\alpha}{9} \gamma \frac{\Delta\omega}{\omega} \frac{I}{e} S(\omega/\omega_c), \tag{7.149}$$

where ψ is the deflection angle in the bending magnet and α the fine structure constant. In practical units this becomes

$$\boxed{\frac{dN_{\text{ph}}}{d\psi} = C_\psi I E \frac{\Delta\omega}{\omega} S(\omega/\omega_c)} \tag{7.150}$$

with

$$C_\psi = \frac{4\alpha}{9 \, mc^2 \, e} = 3.967 \, 10^{16} \frac{\text{photons}}{\text{sec mrad A GeV}}. \tag{7.151}$$

The synchrotron radiation spectrum in Fig. 7.12 is rather uniform up to the critical frequency beyond which the intensity falls off rapidly. Equation (7.142) is not well suited for quick calculation of the radiation intensity at a particular frequency. We may, however, express (7.142) in much simpler form for very low and very large frequencies making use of limiting expressions of Bessel's functions for large and small arguments.

For small arguments we find with AS(9.6.9) [7.20]

$$K_{5/3}(x \to 0) \approx \Gamma\left(5/3\right) \frac{2^{2/3}}{x^{5/3}}$$

which allows us to integrate (7.145) readily. For small arguments or long wavelength we get instead of (7.142)

$$\frac{dP}{d\omega} \approx \frac{P_\gamma}{\omega_c} \frac{9\sqrt{3}}{8\pi} 2^{2/3} \, \Gamma(2/3) \left(\frac{\omega}{\omega_c}\right)^{1/3} \approx 1.333 \left(\frac{\omega}{\omega_c}\right)^{1/3} \frac{P_\gamma}{\omega_c} \tag{7.152}$$

for $\omega/\omega_c \ll 1$.

For high photon frequencies $\omega \gg \omega_c$ the modified Bessel's function becomes from AS(9.7.2) [7.20]

$$K_{5/3}(x) \approx \sqrt{\frac{\pi}{2x_o}} \, e^{-x}$$

and after integration with GR(3.361.1) and GR(3.361.2) [7.18] (7.142) becomes for $\omega/\omega_c \gg 1$

$$\frac{dP}{d\omega} \approx \frac{9\sqrt{3}}{8\sqrt{2\pi}} \sqrt{\frac{\omega}{\omega_c}} \, e^{\omega/\omega_c} \frac{P_\gamma}{\omega_c} \approx 0.77736 \sqrt{\frac{\omega}{\omega_c}} \, e^{\omega/\omega_c} \frac{P_\gamma}{\omega_c}. \tag{7.153}$$

Both approximations are included in Fig. 7.12 and display actually a rather good representation of the real spectral radiation distribution. Specifically, we note the slow increase in the radiation intensity at low frequencies and the exponential drop off above the critical frequency.

Problems

Problem 7.1. Derive (7.31) from (7.29) and show from (7.31) that the electrical field in the radiation regime is purely orthogonal to the direction of propagation.

Problem 7.2. Integrate the radiation power distribution (7.73) over all solid angles and prove that the total radiation power is equal to (7.57).

Problem 7.3. Calculate and plot the radiated power within an opening angle $1/\gamma$ as a fraction of the total radiation power as a function of photon energy. Similarly calculate and plot the opening angle in units of $1/\gamma$ which includes 68% (corresponding to one sigma if the distribution were gaussian) of the total radiation power as a function of photon energy. In which photon energy region are we correct to say that most radiation is emitted within an angle of $1/\gamma$?

Problem 7.4. The design of the European *Large Hadron Collider* [7.23] calls for a circular proton accelerator for energies up to 10 TeV. The circumference is 26.7km and the bending radius $\rho = 2668$ m. Calculate the energy loss per turn due to synchrotron radiation and the critical photon energy. What is the synchrotron radiation power for a circulating beam of 164 ma? Calculate the spatial radiation distribution for photons with $\epsilon_{ph} = \epsilon_c$ in the vertical plane and determine the typical opening angle in units of $1/\gamma$.

Problem 7.5. Design a synchrotron radiation source for a photon energy of your choice. Use a simple FODO lattice and specify the minimum beam energy, beam current, and bending radius which will produce from a bending magnet a photon flux of 10^{14} photons/sec/mrad at the desired photon energy and into a band width of $\delta\omega/\omega = 1\%$. What is the width of the spectrum for which the photon flux is at least 10^{11} photons/sec/mrad. How big is your ring assuming a 30% fill factor for bending magnets?

Problem 7.6. Determine the vertical angle at which the radiation intensity has fallen off to 10% as a function of photon frequency ω/ω_c. Plot the result and derive approximate expressions for very low or very large photon energies.

Problem 7.7. Synchrotron radiation is very intense and well collimated. Imagine a table top synchrotron radiation source which could be used as a flood lamp for visible light. Derive some main ring specifications which would be required to produce a radiation intensity in the visible spectrum of say 1kW. What is the efficiency from rf power th useful radiation power?

Problem 7.8. In the presently most intense synchrotron radiation source, the *European Synchrotron Radiation Facility*, ESRF, in Grenoble (France) an electron beam of 100 mA circulates in a storage ring at an energy of 6 GeV. The bending magnet field is 10 kGauss. Calculate and plot the radiation power into a deflection angle of 10mrad as a function of photon frequency. Normalize the radiation power to the rf power delivered to the beam.

Problem 7.9. Assume a proton storage ring in space surrounding the earth at an average radius of 22,000 km. Further assume that 100 MW of RF power is available to compensate for synchrotron radiation. What is the maximum achievable proton energy for a circulating beam of 10 μA that can be reached with permanent magnets producing a maximum field of 5 kGauss? Is the energy limited by the maximum magnetic field or the available Rf power? Calculate energy loss per turn, critical photon energy, and total radiation power. What are the answers in case of electrons?

Problem 7.10. Consider the storage ring of problem 7.9 and store a beam of muons at the maximum possible energy as limited by RF power or magnet field. What is the muon beam lifetime if we consider an energy acceptance of the storage ring of 3% ?

Problem 7.11. Take an electron beam with a gaussian particle distribution in the storage ring of problem 7.5 and estimate the rms energy loss per turn into synchrotron radiation from all quadrupoles. What fraction of synchrotron radiation is produced in quadrupoles?

8. Hamiltonian Many-Particle Systems

Mathematical tools have been derived in previous chapters to describe the dynamics of singly charged particles in electromagnetic fields. While the knowledge of single-particle dynamics is essential for the development of particle beam transport systems, we are still missing a formal treatment of the behavior of *multiparticle beams*. In principle a multiparticle beam can be described simply by calculating the trajectories of every single particle within this beam, a procedure that is obviously too inefficient to be useful for the description of any real beam involving a very large number of particles.

In this paragraph we will derive concepts to describe the collective dynamics of a beam composed of a large number of particles and its evolution along a transport line utilizing statistical methods that lead to well defined descriptions of the total beam parameters. Mathematical problems arise only when we have a particle beam with neither few particles nor very many particles. Numerical methods must be employed if the number of particles are of importance and where statistical methods would lead to incorrect results.

The evolution of a particle beam has been derived based on *Liouville's theorem* assuring the constancy of the particle density in phase space. However, this concept has not allowed us to determine modifications of particle distributions due to external forces. Particle distributions are greatly determined by particle source parameters, quantum effects due to synchrotron radiation, nonlinear magnetic fields, collisions with other particles in the same beam, with particles in another beam or with atoms of the residual gases in the beam environment to name only a few phenomena that could influence that distribution. In this chapter, we will derive mathematical methods that allow the determination of particle distributions under the influence of various external electromagnetic forces.

8.1 The Vlasov Equation

To study the development of a particle beam along a transport line, we will concentrate on the evolution of a particle density *distribution function* $\Psi(\mathbf{r}, \mathbf{p}, t)$ in six-dimensional phase space where every particle is represented by a single point. We consider a volume element of the phase space that is small enough that we may assume the particle density to be constant within that element and determine its evolution in time. In doing so, we will further assume a large, statistically significant number of particles in each volume element and only a slow variation of the particle density from one volume element to any adjacent volume element. To simplify the equations we restrict the following discussion to the two-dimensional phase space (w, p_w) and use exclusively normalized coordinates $w = x/\sqrt{\beta}$.

The dynamics of a collection of particles can be studied by observing the evolution of their phase space. Specifically, we may choose a particular phase space element and follow it along its path taking into account the forces acting on it. To do this we select a phase space element in form of a rectangular box defined by the four corner points P_i in Fig. 8.1.

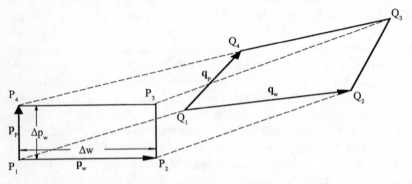

Fig. 8.1. Two-dimensional motion of a rectangle in phase space

At the time t these corners have the coordinates

$$
\begin{aligned}
&P_1(w, p_w), \\
&P_2(w + \Delta w, p_w), \\
&P_3(w + \Delta w, p_w + \Delta p_w), \\
&P_4(w, p_w + \Delta p_w).
\end{aligned}
\tag{8.1}
$$

A short time Δt later, this rectangular box will have moved and may be deformed into a new form more like a parallelogram (Q_1, Q_2, Q_3, Q_4) as shown in Fig. 8.1. In determining the volume of the new box at time $t + \Delta t$ we will assume the conservation of particles allowing no particles to be generated or getting lost. To keep the derivation general the rate of change in the conjugate variables is set to

$$
\begin{aligned}
\dot{w} &= f_w(w, p_w, t), \\
\dot{p}_w &= g_w(w, p_w, t),
\end{aligned}
\tag{8.2}
$$

where $\dot{w} = dw/dt$ and $\dot{p}_w = dp_w/dt$ and the time interval Δt is small enough to allow linear expansion of the particle motion. In other words, the time interval shall be chosen such that no physical parameters of the dynamical system change significantly and a Taylor's expansion can be applied. The new corners of the volume element are then given by

$$Q_1\left[w + f_w(w, p_w, t)\,\Delta t,\; p_w + g_w(w, p_w, t)\,\Delta t\right],$$
$$Q_2\left[w + \Delta w + f_w(w + \Delta w, p_w, t)\,\Delta t,\right.$$
$$\left. p_w + g_w(w + \Delta w, p_w, t)\,\Delta t\right],$$
$$Q_3\left[w + \Delta w + f_w(w + \Delta w, p_w + \Delta p_w, t)\,\Delta t,\right. \qquad (8.3)$$
$$\left. p_w + \Delta p_w + g_w(w + \Delta w, p_w + \Delta p_w, t)\,\Delta t\right],$$
$$Q_4\left[w + f_w(w, p_w + \Delta p_w, t)\,\Delta t,\right.$$
$$\left. p_w + \Delta p_w + g_w(w, p_w + \Delta p_w, t)\,\Delta t\right].$$

The goal of our discussion is now to derive an expression for the particle density $\Psi(w, p_w, t)$ after a time Δt. Because of the conservation of particles we have

$$\Psi(w + f_w\,\Delta t, p_w + g_w\,\Delta t, t + \Delta t)\,\Delta A_Q = \Psi(w, p_w, t)\,\Delta A_P, \qquad (8.4)$$

where ΔA_P and ΔA_Q are the areas in phase space as defined by the corner points P_i and Q_i, respectively. From Fig. 8.1 and (8.1) we derive an expression for the phase space areas which are at the starting time t

$$\Delta A_P = \Delta w\,\Delta p_w \qquad (8.5)$$

and at the time $t + \Delta t$ from (8.3)

$$\Delta A_Q = \Delta w\,\Delta p_w\left[1 + \left(\frac{\partial f_w}{\partial w} + \frac{\partial g_w}{\partial p_w}\right)\Delta t\right], \qquad (8.6)$$

where Taylor's expansions have been used for the functions f_w and g_w retaining only linear terms. To prove (8.6) we note that the area ΔA_P has the form of a rhombus with its sides determined by two vectors and the area, therefore, is equal to the determinant formed by these two vectors. In our case these vectors are $\mathbf{P}_1 = (\Delta w, 0)$ pointing from P_1 to P_2 and $\mathbf{P}_2 = (0, \Delta p_w)$ pointing from P_1 to P_4. The area therefore is

$$|\mathbf{P}_1, \mathbf{P}_2| = \begin{vmatrix} \Delta w & 0 \\ 0 & \Delta p_w \end{vmatrix} = \Delta w\,\Delta p_w = \Delta A_P \qquad (8.7)$$

in agreement with (8.5). A time interval Δt later these vectors will have changed as determined by (8.2). Each of the corner points P_i is moving although with different speed thus distorting the rectangle P_i into the shape Q_i of Fig. 8.1. To calculate the new vectors defining the distorted area we expand the functions f_w and g_w in a Taylor's series at the point (w, p_w). While, for example, the w-component of the movement of point P_1 along the w coordinate is given by $f_w\,\Delta t$ the same component for P_2 changes by $f_w\,\Delta t + \frac{\partial f_w}{\partial w}\,\Delta w\,\Delta t$. The w-component of the vector $\mathbf{Q}_1$ therefore becomes

$\Delta w + \frac{\partial f_w}{\partial w} \Delta w \Delta t$. In a similar way, we can calculate the p-component of this vector as well as both components for the vector $\mathbf{Q}_2$. The phase space area of the distorted rectangle (Q_1, Q_2, Q_3, Q_4) at time $t + \Delta t$ with these vector components is then given by

$$|\mathbf{Q}_1, \mathbf{Q}_2| = \begin{vmatrix} \Delta w + \frac{\partial f_w}{\partial w} \Delta w \Delta t & \frac{\partial f_w}{\partial p_w} \Delta p_w \Delta t \\ \frac{\partial g_w}{\partial w} \Delta w \Delta t & \Delta p_w + \frac{\partial g_w}{\partial p_w} \Delta p_w \Delta t \end{vmatrix} = \Delta A_Q. \qquad (8.8)$$

Dropping second-order terms in Δt we get indeed the expression (8.6). Obviously, the phase space volume does not change if

$$\frac{\partial f_w}{\partial w} + \frac{\partial g_w}{\partial p_w} = 0$$

in agreement with the result obtained in [Ref. 8.1, Sect. 5.4.2] where we have assumed that the Lorentz force is the only force acting on the particle. In this paragraph, however, we have made no such restrictions and it is this generality that allows us to derive, at least in principle, the particle distribution under the influence of any forces. The factor

$$\left[1 + \left(\frac{\partial f_w}{\partial w} + \frac{\partial g_w}{\partial p_w} \right) \Delta t \right] \qquad (8.9)$$

in (8.6) is the general *Wronskian* of the transformation and is not necessarily equal to unity. We have found such an example in [Ref. 8.1, Sect. 3.3] in the form of *adiabatic damping*. Indeed we have damping or antidamping whenever the Wronskian is different from unity.

To illustrate this we use the example of a *damped harmonic oscillator* which is described by the second-order differential equation

$$\ddot{w} + 2\alpha_w \dot{w} + \omega_o^2 w = 0, \qquad (8.10)$$

or in form of a set of two linear differential equations

$$\begin{aligned} \dot{w} &= \omega_o p_w = f_w(w, p_w, t), \\ \dot{p}_w &= -\omega_o w - 2\alpha_w p_w = g_w(w, p_w, t). \end{aligned} \qquad (8.11)$$

From this we find indeed the relation

$$\frac{\partial f_w}{\partial w} + \frac{\partial g_w}{\partial p_w} = -2\alpha_w, \qquad (8.12)$$

where α_w is the *damping decrement* of the oscillator. We have obtained on a general basis that the phase space density for harmonic oscillators will vary only if damping forces are present. Here we use the term damping in a very general way including excitation depending on the sign of the damping decrement α_w. The designation α_w for the damping decrement may potentially lead to some confusion with the same use for the betatron function $\alpha = \frac{1}{2}\beta'$. However, we choose here to rather require some care than introduce against common use new designations for the damping decrement or the betatron functions. We also note that for all cases where the damping time is long compared to the oscillation time, and we consider here only such cases, the damping occurs for both conjugate trajectories.

The derivation in two-dimensional phase space can easily be generalized to six-dimensional phase space with the generalized volume element

$$\Delta V_P = \Delta\mathbf{r}\,\Delta\mathbf{p} \tag{8.13}$$

at time t and a time interval Δt later

$$\Delta V_Q = \Delta\mathbf{r}\,\Delta\mathbf{p}\left[1 + \nabla_r\mathbf{f}\,\Delta t + \nabla_p\mathbf{g}\,\Delta t\right]. \tag{8.14}$$

Here the *Nabla operators* are defined by

$$\nabla_r = \left(\frac{\partial}{\partial w},\frac{\partial}{\partial v},\frac{\partial}{\partial u}\right) \quad \text{and} \quad \nabla_p = \left(\frac{\partial}{\partial p_w},\frac{\partial}{\partial p_v},\frac{\partial}{\partial p_u}\right), \tag{8.15}$$

where (w,v,u) are normalized variables and the vector functions $\mathbf{f}$ and $\mathbf{g}$ are defined by the components $\mathbf{f} = (f_w, f_v, f_u)$ and $\mathbf{g} = (g_w, g_v, g_u)$.

Equation (8.4) can now be reduced further after applying a Taylor's expansion to the density function Ψ. With (8.5,6) and keeping only linear terms

$$\frac{\partial\Psi}{\partial t} + f_w\frac{\partial\Psi}{\partial w} + g_w\frac{\partial\Psi}{\partial p_w} = -\left(\frac{\partial f_w}{\partial w} + \frac{\partial g_w}{\partial p_w}\right)\Psi. \tag{8.16}$$

It is straightforward to generalize this result again to six-dimensional phase space

$$\boxed{\frac{\partial\Psi}{\partial t} + \mathbf{f}\,\nabla_r\Psi + \mathbf{g}\,\nabla_p\Psi = -\left(\nabla_r\mathbf{f} + \nabla_p\mathbf{g}\right)\Psi,} \tag{8.17}$$

which is called the *Vlasov equation*. If there is no damping the r.h.s. of the Vlasov Equation vanishes and we have

$$\frac{\partial\Psi}{\partial t} + \mathbf{f}\,\nabla_r\Psi + \mathbf{g}\,\nabla_p\Psi = 0. \tag{8.18}$$

This is simply the total time derivative of the phase space density Ψ telling us that in the absence of damping it remains a constant of motion. The preservation of the phase space density is *Liouville's theorem* and we have derived in this paragraph on a very general basis the validity of this theorem for a Hamiltonian system with vanishing dissipating forces $(\nabla_r\mathbf{f} + \nabla_p\mathbf{g}) = 0$.

Equation (8.18) describes the evolution of a multiparticle system in phase space where the physics of the particular particle dynamics is introduced through the functions $\mathbf{f}(\mathbf{r},\mathbf{p},t)$ and $\mathbf{g}(\mathbf{r},\mathbf{p},t)$. The definition of these

functions in (8.2) appears similar to that for the Hamiltonian equations of
motion. If the variables $\mathbf{r}$ and $\mathbf{p}$ are canonical variables we may indeed derive
these functions from the Hamiltonian

$$
\begin{aligned}
\dot{\mathbf{r}} &= \nabla_p \mathcal{H} = \mathbf{f}, \\
\dot{\mathbf{p}} &= -\nabla_r \mathcal{H} = \mathbf{g},
\end{aligned}
\tag{8.19}
$$

where $\mathcal{H}$ is the Hamiltonian of the system. We are, therefore, at least in
principle, able to solve the evolution of a multiparticle system in phase
space if its Hamiltonian is known. It should be emphasized, however, that
the variables (w, p) need not be canonical to be used in the Vlasov equation.

It is interesting to apply the Vlasov equation to a simple one-dimensional
harmonic oscillator (1.51) with vanishing perturbation. Introducing the
canonical variable p through $\dot{w} = \nu p$ the Hamiltonian becomes

$$
\mathcal{H}_o = \tfrac{1}{2}\nu p^2 + \tfrac{1}{2}\nu w^2
\tag{8.20}
$$

and the equations of motion are

$$
\begin{aligned}
\dot{w} &= +\frac{\partial \mathcal{H}_o}{\partial p} = \nu p = f, \\
\dot{p} &= -\frac{\partial \mathcal{H}_o}{\partial w} = -\nu w = g.
\end{aligned}
\tag{8.21}
$$

It is customary for harmonic oscillators and similarly for particle beam
dynamics to use the oscillation phase as the independent or "time" variable.
Since we have not made any specific use of the real time in the derivation of
the Vlasov equation, we choose here the phase as the "time" variable. For
the simple case of an undamped harmonic oscillator $\frac{\partial f}{\partial w} = 0$ and $\frac{\partial g}{\partial p} = 0$
and consequently, the Vlasov equation becomes from (8.16) with (8.21)

$$
\frac{\partial \Psi}{\partial \varphi} + \nu p \frac{\partial \Psi}{\partial w} - \nu w \frac{\partial \Psi}{\partial p} = 0.
\tag{8.22}
$$

In cylindrical phase space coordinates $(w = r\cos\theta,\ p = r\sin\theta,\ \varphi)$ this reduces to the simple equation

$$
\frac{\partial \Psi}{\partial \varphi} - \nu \frac{\partial \Psi}{\partial \theta} = 0.
\tag{8.23}
$$

Any differentiable function with the argument $(r, \theta + \nu\varphi)$ can be a solution
of (8.23) describing the evolution of the particle density Ψ with time

$$
\Psi(w, p_w, \varphi) = F(r, \theta + \nu\varphi),
\tag{8.24}
$$

The particle distribution in (w, p_w)-phase space merely rotates about
the center with the frequency ν and remains otherwise unchanged as shown
in Fig. 8.2. This is just another way of saying that an ensemble of many

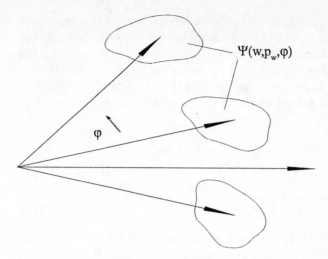

$\Psi(\mathrm{w},\mathrm{p}_{\mathrm{w}},\varphi)$

φ

Fig. 8.2. Beam motion in phase space

particles behaves like the sum of all individual particles since any interaction between particles as well as damping forces have been ignored. In
(x, x')-phase space this rotation is deformed into a "rotation" along elliptical trajectories. The equation of motion in (w, p_w)-phase space is solved
by $r = $ const indicating that the amplitude r is a constant of motion. In
(x, x')-phase space we set $w = x/\sqrt{\beta}$ and $p = \sqrt{\beta}\, x' + \frac{\alpha}{\sqrt{\beta}}\, x$ and get from
$r^2 = w^2 + p_w^2$ for this constant of motion

$$\beta\, x'^2 + 2\alpha\, xx' + \gamma\, x^2 \;=\; \text{const} \tag{8.25}$$

which is the *Courant-Snyder invariant* [Ref. 8.1, Eq. (5.125)]. The Vlasov
equation allows us to generalize this result collectively to all particles in a
beam. Any particular particle distribution a beam may have at the beginning of the beam transport line or circular accelerator will be preserved as
long as damping or similar effects are absent.

8.1.1 Betatron Oscillations and Perturbations

The Vlasov equation has been derived on very general arguments and will
prove to be a useful tool to derive particle beam parameters. Specifically,
it allows us to study the influence of arbitrary macroscopic field on the
particle density in phase space and on the characteristic frequency of the
particle motion. To demonstrate this we expand the example of the harmonic
oscillator to include also *perturbation* terms. For such a perturbed system
the equation of motion is

$$\ddot{w} + \nu_o^2\, w \;=\; \nu_o^2\, \beta^{\frac{3}{2}} \sum_{n>0} p_n\, \beta^{\frac{n}{2}}\, w^n , \tag{8.26}$$

where the coefficients p_n are the strength parameters for the nth order perturbation term defined in (5.1) and the amplitude w is the normalized betatron oscillation amplitude. The Vlasov equation allows us to calculate the impact of these perturbation terms on the betatron frequency. We demonstrate this first with a linear perturbation term $(n = 1)$ caused by a gradient field error $p_1 = -\delta k$ in a quadrupole. In this case the equation of motion is from (8.26)

$$\ddot{w} + \nu_{\mathrm o}^2 w = -\nu_{\mathrm o}^2 \beta^2 \, \delta k \, w \tag{8.27}$$

or

$$\ddot{w} + \nu_{\mathrm o}^2 \left(1 + \beta^2 \delta k\right) w = 0 . \tag{8.28}$$

This second-order differential equation can be replaced by two first-order differential equations which is in general the most straight forward way to obtain the functions (8.2)

$$
\begin{aligned}
\dot{w} &= \nu_{\mathrm o} \sqrt{1 + \beta^2 \, \delta k} \, p , \\
\dot{p} &= -\nu_{\mathrm o} \sqrt{1 + \beta^2 \, \delta k} \, w .
\end{aligned}
\tag{8.29}
$$

Here we have assumed that the betatron function β and the quadrupole field error δk are uniformly distributed along the beam line and therefore can be treated as constants. This approach is justified here since we are interested only in the average oscillation frequency of the particles and not in fast oscillating terms. The desired result can be derived directly from (8.29) without any further mathematical manipulation by comparison with (8.21). From there the oscillating frequency for the perturbed system is given by

$$\nu = \nu_{\mathrm o} \sqrt{1 + \beta^2 \, \delta k} \approx \nu_{\mathrm o} \left(1 + \tfrac{1}{2} \beta^2 \delta k\right) , \tag{8.30}$$

for small perturbations. The betatron frequency shift can be expressed by the lowest order harmonic of the Fourier expansion for the periodic perturbation function $\nu_{\mathrm o} \beta^2 \, \delta k$ to give

$$2\pi\nu_{\mathrm o} \left(\beta^2 \, \delta k\right)_{\mathrm o} = \oint \nu_{\mathrm o} \beta^2 \, \delta k \, \mathrm{d}\varphi = \oint \beta \, \delta k \, \mathrm{d}s$$

making use of the definition for the betatron phase. The *tune shift* $\delta\nu$ due to quadrupole field errors is therefore

$$\delta\nu = \nu - \nu_{\mathrm o} = \frac{1}{4\pi} \oint \beta \, \delta k \, \mathrm{d}s \tag{8.31}$$

in agreement with [Ref. 8.1, Eq. (7.49)]. Again, the Vlasov equation confirms this result for all particles irrespective of the distribution in phase space.

This procedure can be expanded to any order of perturbation. From the differential equation (8.26) one gets in analogy to the equations of motion (8.29)

$$\dot{w} = \nu_{\mathrm{o}} \sqrt{1 - \beta^{3/2} \sum_{n>0} p_n \beta^{n/2} w^{n-1} \, p}\,,$$

$$\dot{p} = -\nu_{\mathrm{o}} \sqrt{1 - \beta^{3/2} \sum_{n>0} p_n \beta^{n/2} w^{n-1} \, w}\,. \tag{8.32}$$

For small perturbations the solution for the unperturbed harmonic oscillator $w(\varphi) = w_{\mathrm{o}} \sin(\nu\varphi + \delta)$ may be used where δ is an arbitrary phase constant. The *tune shift* $\Delta\nu = \nu - \nu_{\mathrm{o}}$ is thus

$$\Delta\nu = -\sum_{n>0} \frac{1}{4\pi} \oint p_n \, \beta^{\frac{n+1}{2}} \, w_{\mathrm{o}}^{n-1} \, \sin^{n-1}[\nu_{\mathrm{o}} \, \varphi(s) + \delta] \, \mathrm{d}s\,. \tag{8.33}$$

Not all perturbation terms contribute to a tune variation. All even terms $n = 2m$ where m is an integer, integrate, for example, to zero in this approximation and a sextupole field therefore does not contribute to a tune shift or tune spread. This conclusion must be modified, however, due to higher-order approximations which become necessary when perturbations cannot be considered small anymore. Further, we find from (8.33) that the tune shift is independent of the particle oscillation amplitude only for quadrupole field errors $n = 1$. For higher-order multipoles the tune shift becomes amplitude dependent resulting in a *tune spread* within the particle beam rather than a coherent tune shift for all particles of the beam.

In a particular example the tune spread caused by a single octupole, $n = 3$, in a circular accelerator is given by

$$\Delta\nu_3 = -\frac{\epsilon_w}{8\pi} \oint p_3 \beta^2 \, \mathrm{d}s\,, \tag{8.34}$$

where $w_{\mathrm{o}}^2 = \epsilon_w$ is the emittance of the beam. Similar results can be found for higher-order multipoles.

8.1.2 Damping

At the beginning of this section we have decided to ignore damping and have used the undamped Vlasov equation (8.18). Damping or antidamping effects do, however, occur in real systems and it is interesting to investigate if the Vlasov equation can be used to derive some general insight into damped systems as well. For a damped oscillator we use (8.11,12) to form the Vlasov equation in the form of (8.16). Instead of the phase we now use the real time as the independent variable to allow the intuitive definition of the *damping decrement* as the relative decay of the oscillation amplitude with time

$$\frac{\partial \Psi}{\partial t} + \omega_{\rm o}\, p_w\, \frac{\partial \Psi}{\partial w} - (\omega_{\rm o}\, w + 2\alpha_w\, p_w)\, \frac{\partial \Psi}{\partial p_w} = +2\alpha_w\, \Psi \,. \tag{8.35}$$

This partial differential equation can be solved analytically in a way similar to the solution of the undamped harmonic oscillator by using cylindrical coordinates. For very weak damping we expect a solution close to (8.24) where the amplitude r in phase space was a constant of motion. For a damped oscillator we try to form a similar *invariant* from the solution of a damped harmonic oscillator

$$w = w_{\rm o}\, {\rm e}^{-\alpha_w t}\, \cos \sqrt{\omega_{\rm o}^2 - \alpha_w^2}\, t = r\, {\rm e}^{-\alpha_w t}\, \cos \theta \,. \tag{8.36}$$

With the conjugate component $\omega_{\rm o}\, p_w = \dot{w}$, we form the expression

$$\frac{\omega_{\rm o} p_w + \alpha_w w}{\sqrt{\omega_{\rm o}^2 - \alpha_w^2}} = -w_{\rm o}\, {\rm e}^{-\alpha_w t}\, \sin \sqrt{\omega_{\rm o}^2 - \alpha_w^2}\, t = -r\, {\rm e}^{-\alpha_w t}\, \sin \theta \tag{8.37}$$

and eliminate the phase θ from (8.36,37) keeping only terms linear in the damping decrement α_w we obtain the "invariant"

$$r^2\, {\rm e}^{-2\alpha_w t} = w^2 + p_w^2 + 2\, \frac{\alpha_w}{\omega_{\rm o}}\, w\, p_w \,. \tag{8.38}$$

Obviously if we set $\alpha_w = 0$ we have the invariant of the harmonic oscillator. The time dependent factor due to finite damping modifies this "invariant". However, for cases where the damping time is long compared to the oscillation period we may still consider (8.38) a quasi invariant. The phase coordinate θ can be derived from (8.36,37) as a function of w and p_w as may be verified by insertion into the differential equation (8.35). The solution for the phase space density of a damped oscillator is of the form

$$\Psi(w, p_w, t) = {\rm e}^{2\alpha_w t}\, F(r, \Phi) \,, \tag{8.39}$$

where $F(r, \Phi)$ is any arbitrary but differentiable function of r and Φ and the phase Φ is defined by

$$\begin{aligned} \Phi &= \theta + \sqrt{\omega_{\rm o}^2 - \alpha_w^2}\, t \\ &= \arctan\left(-\frac{\omega_{\rm o}\, p_w + \alpha_w w}{\sqrt{\omega_{\rm o}^2 - \alpha_w^2}\, w} \right) + \sqrt{\omega_{\rm o}^2 - \alpha_w^2}\, t \,. \end{aligned} \tag{8.40}$$

For very weak damping $\alpha_w \to 0$ and the solution (8.39) approaches (8.24) where $\alpha_w = 0$ and $\nu\varphi = \omega_{\rm o} t$ as expected. Therefore even for finite damping a particle distribution rotates in phase space although with a somewhat reduced rotation frequency due to damping. The particle density Ψ, however, changes exponentially with time due to the factor ${\rm e}^{2\alpha_w t}$. For damping $\alpha_w > 0$, we get an increase in the phase space density at the distance R from the beam center. At the same time the real particle oscillation amplitudes (w, p_w) are being reduced proportional to $e^{-\alpha_w t}$ and the increase in the phase space density at R reflects the concentration of particles in the beam center from larger amplitudes due to damping.

In conclusion we found that in systems where velocity dependent forces exist, we have damping $\alpha_w > 0$ or antidamping $\alpha_w < 0$ of the oscillation amplitudes. As has been discussed in the previous chapter such forces do exist in accelerators leading to damping effects. Mostly, however, the Vlasov equation is applied to situations where particles interact with self or external fields that can lead to instabilities. It is the task of particle beam dynamics to determine the nature of such interactions and to derive the circumstances under which the damping coefficient α_w, if not zero, is positive for damping or negative leading to beam instability.

8.2 Damping of Oscillations in Electron Accelerators

In electron accelerators we are concerned mainly with damping effects caused by the emission of synchrotron radiation. All six degrees of freedom for particle motion are damped. Damping of energy oscillations occurs simply from the fact that the synchrotron radiation power is energy dependent. Therefore a particle with a higher energy than the reference particle radiates more and a particle with less energy radiates less. The overall effect is that the energy deviation is reduced or damped. Damping of the transverse motion is principally a geometric effect. The photons of synchrotron radiation are emitted into the direction of the particle motion. Therefore part of the energy loss is correlated to a loss in transverse momentum. On the other hand, the lost energy is restored through accelerating fields with longitudinal components only. The overall effect of an energy loss during the course of betatron oscillations is therefore a loss of transverse momentum which leads to a reduction in the transverse oscillation amplitude, an effect we call damping. In the next section, we will discuss the physics leading to damping and derive the appropriate damping decrement for different modes of oscillations.

8.2.1 Damping of Synchrotron Oscillations

In a real beam particles are spread over a finite distribution of energies close to the reference energy. The magnitude of this energy spread is an important parameter to be considered for both beam transport systems as well as for experimental applications of particle beams. In general an energy spread as small as possible is desired to minimize chromatic aberrations and for improved accuracy of experimental observation. We will therefore derive the parametric dependence of damping and discuss methods to reduce the energy spread within a particle beam.

To do this, we consider a beam of electrons being injected with an arbitrary energy distribution into a storage ring ignoring incidental beam losses during the injection process due to a finite energy acceptance. From [Ref. 8.1, Chap. 8] it is known that particles in a storage ring undergo synchrotron oscillations which are oscillations about the ideal momentum and the ideal longitudinal position. Since energy and time or equivalently energy and longitudinal position are conjugate phase space variables, we will investigate both the evolution of the energy spread as well as the longitudinal distribution or bunch length of the particle beam.

The evolution of energy spread or bunch length of the particle beam will depend very much on the nature of particles and their energy. For heavy particles like protons or ions there is no synchrotron radiation damping and therefore the phase space for such beams remains constant. As a consequence, the energy spread or bunch length also stays a constant. A similar situation occurs for electrons or positrons at very low energies since synchrotron radiation is negligible. Highly relativistic electrons, however, produce intense synchrotron radiation leading to a strong damping effect which is discussed below in more detail.

The *damping decrement* α_w is defined in the Vlasov equation by

$$\frac{\partial f}{\partial w} + \frac{\partial g}{\partial p} = -2\alpha_w \tag{8.41}$$

and can be calculated with the knowledge of the functions f and g. For the conjugate variables (w, p_w) we use the time deviation of a particle with respect to the synchronous particle $w = \tau$ as shown in Fig. 8.3 and the difference of the particle's energy E from the *synchronous* or *reference energy* E_s and set $p_w = \epsilon = E - E_s$.

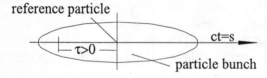

reference particle

$\tau > 0$

$ct = s$

particle bunch

Fig. 8.3. Longitudinal particle position

Since $f = \frac{d\tau}{dt} = \dot{\tau}$ and $g = \frac{d\epsilon}{dt} = \dot{\epsilon}$ we have to determine the rate of change for the conjugate variables. The rate of change of τ can be derived from [Ref. 8.1, Eq. (8.18)] with $cp_0 \approx E_s$ and is

$$\frac{d\tau}{dt} = -\eta_c \beta^2 \frac{\epsilon}{E_s}, \tag{8.42}$$

where we have replaced the phase by the time $\dot{\psi} = c\beta k_0 \dot{\tau}$ and the relative momentum error by the relative energy error since we consider here only highly relativistic particles. The latter replacement is a matter of convenience since we will be using the energy gain in accelerating fields.

The energy rate of change $\dot\epsilon$ is the balance of the energy gained in accelerating fields and the energy lost due to synchrotron radiation or other losses

$$\dot\epsilon = \frac{1}{T}\left[eV_{\mathrm{rf}}(\tau_s + \tau) - U(E_s + \epsilon)\right].\tag{8.43}$$

Here T is the time it takes the particles to travel the distance L. The energy gain within the distance L for a particle traveling a time τ behind the reference or synchronous particle is $eV_{\mathrm{rf}}(\tau_s + \tau)$ and U is the energy loss to synchrotron radiation along the same distance of travel.

We apply these expressions to the simple situation of a *linear accelerator* of length L where the momentum compaction factor vanishes $\alpha_c = 0$ and where there is no energy loss due to synchrotron radiation $U \equiv 0$. Further we ignore for now other energy losses and have

$$\boxed{\begin{aligned} f &= \dot\tau = \frac{1}{\beta^2\,\gamma^2}\,\frac{\epsilon}{E_{\mathrm{s}}}\,, \\ g &= \dot\epsilon = \frac{1}{T}\,eV_{\mathrm{rf}}(\tau_s + \tau). \end{aligned}}\tag{8.44}$$

Inserted into (8.41) we find the damping decrement to vanish which is consistent with observation and with the phenomenon of *adiabatic damping*. This name is unfortunate in the sense that it does not actually describe a damping effect in phase space as we just found out but rather describes the variation of the relative energy spread with energy which is merely a consequence of the constant phase space density or *Liouville's theorem*. From the Vlasov equation we learn that in the absence of damping the energy spread ϵ stays constant as the particle beam gets accelerated. Consequently, the relative energy spread $\frac{\epsilon}{E_s}$ decreases as we would expect for adiabatic damping. In [Ref. 8.1, Sect. 3.3] the same phenomenon was discussed for the transverse phase space. The Vlasov equation still can be used to describe adiabatic damping but we need to use the relative energy spread as one of the variables. Instead of the second equation (8.43) we have then

$$g = \frac{\mathrm{d}}{\mathrm{d}t}\frac{\epsilon}{E} = \frac{\frac{\epsilon}{E_s} - \frac{\epsilon}{E_o}}{\Delta t}\,,\tag{8.45}$$

where E_o is the particle energy at time t_o, E_s is the energy at time $t = t_o + \mathrm{d}t$ and $E_s = E_o + a\,\mathrm{d}t$ with $a = \frac{\mathrm{d}E_s}{\mathrm{d}t} = \dot E_s$ the acceleration per unit time. We choose the time interval $\mathrm{d}t$ small enough so that $a\,\mathrm{d}t \ll E_o$ and get

$$g = -\frac{\epsilon}{E_s}\frac{a}{E_o} = -\frac{\epsilon}{E_o}\frac{\dot E}{E_o}.\tag{8.46}$$

The damping decrement becomes from (8.41) with $\delta = \frac{\epsilon}{E_o}$ and $\partial f/\partial\tau = 0$

$$\frac{\partial g}{\partial \delta} = -\frac{\dot{E}}{E_s} = -2\,\alpha_w = 2\frac{1}{\delta}\frac{d\delta}{dt} \qquad (8.47)$$

and after integration

$$\int \frac{d\delta}{\delta} = \ln\frac{\delta}{\delta_o} = -\frac{1}{2}\int \frac{\dot{E}}{E_s}\,dt = -\frac{1}{2}\ln\frac{E_s}{E_o}$$

or

$$\delta = \frac{\epsilon}{E_o}\sqrt{\frac{E_o}{E_s}}. \qquad (8.48)$$

The relative energy spread in the beam is reduced during acceleration inversely proportional to the square root of the energy. This reduction of the relative energy spread is called *adiabatic damping*.

Returning to the general case we apply a Taylor's expansion to the rf voltage in (8.44) and get for terms on the r.h.s. of (8.43) keeping only linear terms

$$e\,V_{\text{rf}}(\tau_s + \tau) = e\,V_{\text{rf}}(\tau_s) + e\frac{V_{\text{rf}}}{\partial \tau}\,\tau,$$

$$-U(E_s + \epsilon) = -U(E_s) - \left.\frac{\partial U}{\partial E}\right|_{E_s}\epsilon. \qquad (8.49)$$

Since the energy gain from the rf field $e V_{\text{rf}}(\tau_s)$ for the synchronous particle just compensates its energy loss $U(E_s)$, we have instead of (8.43) now

$$\dot{\epsilon} = \frac{1}{T}\left[e\,\dot{V}_{\text{rf}}(\tau_s)\,\tau - \left.\frac{\partial U}{\partial E}\right|_{E_s}\epsilon\right], \qquad (8.50)$$

where we have set $\dot{V}_{\text{rf}} = \frac{\partial V_{\text{rf}}}{\partial \tau}$. The *synchrotron oscillation damping decrement* can now be derived from (8.41) with (8.44,51) to give

$$\alpha_\epsilon = +\frac{1}{2}\frac{1}{T}\left.\frac{\partial U}{\partial E}\right|_{E_s}. \qquad (8.51)$$

We will now derive the damping decrement for the case that the energy loss is only due to synchrotron radiation. The energy loss along the transport line L is given by

$$U = \frac{1}{c}\int_0^L P_\gamma\,d\sigma, \qquad (8.52)$$

where P_γ is the synchrotron radiation power (7.57) and the integration is taken along the actual particle trajectory σ. If $\rho(s)$ is the bending radius along s, we have

$$\frac{\mathrm{d}\sigma}{\mathrm{d}s} = 1 + \frac{x}{\rho}.$$

With $x = x_\beta + \eta \frac{\epsilon}{E_\mathrm{s}}$ and averaging over many betatron oscillations, we get $\langle x_\beta \rangle = 0$ and

$$\frac{\mathrm{d}\sigma}{\mathrm{d}s} = 1 + \frac{\eta}{\rho} \frac{\epsilon}{E}. \tag{8.53}$$

This asymmetric averaging of the betatron oscillation only is permissible if the synchrotron oscillation frequency is much lower than the betatron oscillation frequency as is the case in circular accelerators. With $\mathrm{d}\sigma = [1 + (\eta/\rho)(\epsilon/E_\mathrm{s})]\,\mathrm{d}s$ in (8.52), the energy loss for a particle of energy $E_\mathrm{s} + \epsilon$ is

$$U(E_\mathrm{s} + \epsilon) = \frac{1}{c} \int_L P_\gamma \left(1 + \frac{\eta}{\rho} \frac{\epsilon}{E_\mathrm{s}}\right) \mathrm{d}s \tag{8.54}$$

or after differentiation with respect to the energy

$$\left.\frac{\partial U}{\partial E}\right|_{E_\mathrm{s}} = \frac{1}{c} \int_L \left[\frac{\mathrm{d}P_\gamma}{\mathrm{d}E} + P_\gamma \frac{\eta}{\rho} \frac{1}{E_\mathrm{s}}\right]_{E_\mathrm{s}} \mathrm{d}s. \tag{8.55}$$

The synchrotron radiation power is proportional to the square of the energy and the magnetic field $P_\gamma \sim E_\mathrm{s}^2 B_\mathrm{o}^2$ which we use in the expansion

$$\frac{\mathrm{d}P_\gamma}{\mathrm{d}E} = \frac{\partial P_\gamma}{\partial E} + \frac{\partial P_\gamma}{\partial B_\mathrm{o}} \frac{\partial B}{\partial E} = 2\frac{P_\gamma}{E_\mathrm{s}} + 2\frac{P_\gamma}{B} \frac{\partial B}{\partial x} \frac{\partial x}{\partial E}. \tag{8.56}$$

The variation of the synchrotron radiation power with energy depends directly on the energy but also on the magnetic field if there is a field gradient $\frac{\partial B}{\partial x}$ and a finite dispersion function $\eta = E_\mathrm{s} \frac{\partial x}{\partial E}$. The magnetic field as well as the field gradient is to be taken at the reference orbit. Collecting all these terms and setting $\frac{1}{B_\mathrm{o}} \frac{\partial B}{\partial x} = \rho k$ we get for (8.55)

$$\begin{aligned}
\left.\frac{\partial U}{\partial E}\right|_{E_\mathrm{s}} &= \frac{1}{c} \int_L \left(2\frac{P_r}{E_\mathrm{s}} + 2\frac{P_\gamma}{E_\mathrm{s}} \rho k \eta + \frac{P_\gamma}{E_\mathrm{s}} \frac{\eta}{\rho}\right)\Bigg|_{E_\mathrm{s}} \mathrm{d}s, \\
&= \frac{U_s}{E_\mathrm{s}} \left[2 + \frac{1}{cU_s} \int_L P_\gamma \eta \left(\frac{1}{\rho} + 2\rho k\right)\Bigg|_{E_\mathrm{s}} \mathrm{d}s\right],
\end{aligned} \tag{8.57}$$

where we have made use of $U_s = \frac{1}{c} \int_L P_\gamma(E_\mathrm{s})\,\mathrm{d}s$. Recalling the expressions for the synchrotron radiation power and energy loss $P_\gamma = C_\gamma E_\mathrm{s}^4/\rho^2$ and $U_s = C_\gamma E_\mathrm{s}^4 \int \mathrm{d}s/\rho^2$, we may simplify (8.57) for

$$\left.\frac{\partial U}{\partial E}\right|_{E_\mathrm{s}} = \frac{U_s}{E_\mathrm{s}} (2 + \vartheta). \tag{8.58}$$

The ϑ-parameter depends on the particular lattice and is defined by

$$\vartheta = \frac{\int_L \eta \left(\frac{1}{\rho^3} + 2\frac{k}{\rho} \right) ds}{\int_L \frac{1}{\rho^2} ds}. \tag{8.59}$$

We note in (8.59) two terms. The first term η/ρ^3 is purely geometric and is a consequence of a longer path length (8.53) in curved systems like for sector magnets. Since $\rho > 0$ and $\eta > 0$ for most circular accelerators this term increases the synchrotron oscillation damping. This is to be expected because a particle with a higher energy travels a longer path in the sector bending magnet for positive dispersion and therefore, radiates a longer time loosing an extra amount of energy. Conversely, a particle with a lower energy than the ideal energy travels a shorter path through the sector magnet and looses less energy compared to the reference particle. For rectangular magnets this term vanishes because the total path length is independent of the beam energy or the displacement of the equilibrium orbit. The correction for arbitrary wedge magnets is somewhere in between and can be easily derived for a particular magnet.

The second term in (8.59) is due to a change in the magnetic field with the transverse position of the beam in the magnet. Such a variation occurs only in synchrotron magnets. This field variation may be positive or negative and consequently, the correction to the damping is also positive or negative. For a FODO lattice, we have additional damping since the dispersion is typically largest in the focusing quadrupoles where $k > 0$ and therefore $(\eta k)_{QF} - (\eta k)_{QD} > 0$. While it seems a positive effect to have additional damping of the synchrotron oscillations, we will find later in this chapter that the total damping in 6-dimensional phase space is a constant. Therefore, any additional damping caused by a special lattice design in one degree of freedom will show up as antidamping in another degree of freedom for the particle motion.

Obviously, this additional damping or antidamping term does not occur for lattices utilizing pure dipole and quadrupole magnets. Such a lattice where bending and focusing is performed in separate magnets is called a *separated-function lattice*. Alternatively, lattices where bending and focusing is combined in the same magnets are called *combined-function lattices*.

With (8.58) and (8.59) we finally get from (8.51) the *damping decrement for synchrotron oscillations*

$$\boxed{\alpha_\epsilon = \frac{U_s}{2\,T\,E_s}\,(2 + \vartheta) = \frac{U_s}{2\,T\,E_s}\,J_\epsilon = \frac{\langle P_\gamma \rangle}{2\,E_s}\,J_\epsilon,} \tag{8.60}$$

where we have introduced the *partition number*

$$J_\epsilon = 2 + \vartheta. \tag{8.61}$$

Since all parameters except ϑ are positive we have shown that the synchrotron oscillations for radiating particles are damped. A potential situation for antidamping can be created if $\vartheta < -2$.

8.2.2 Damping of Vertical Betatron Oscillations

Particles orbiting in a circular accelerator undergo transverse betatron oscillations. These oscillations are damped in electron rings due to the emission of synchrotron radiation. To calculate the damping decrement, we assume accelerating fields evenly distributed around the ring to restore the lost energy. In practice this is not true since only few rf cavities in a ring are located at one or more places around the ring. As long as the revolution time around the ring is small compared to the damping time, however, we need not consider the exact location of the accelerating cavities and may assume an even and uniform distribution around the ring.

First we will derive the damping decrement for the vertical betatron oscillation. In a plane accelerator with negligible coupling this motion is independent of the other oscillations. This is not the case for the horizontal betatron motion which is coupled to the synchrotron oscillation due to the presence of a finite dispersion function. We will therefore derive the vertical damping decrement first and then discuss a very general theorem applicable for the damping in circular accelerators. This theorem together with the damping decrement for the synchrotron and vertical betatron oscillations will enable us to derive the horizontal damping in a much simpler way than would be possible in a more direct way.

In normalized coordinates the functions f and g are for the vertical plane

$$
\begin{aligned}
\frac{dw}{d\varphi} &= +\nu p = f(w, p, \varphi), \\
\frac{dp}{d\varphi} &= -\nu w = g(w, p, \varphi),
\end{aligned}
\tag{8.62}
$$

where $\nu = \nu_y$, $w = y/\sqrt{\beta_y}$, $\frac{1}{\nu_y}\frac{dw}{d\varphi} = \sqrt{\beta_y}\,y' - \frac{1}{2}\frac{\beta_y'}{\sqrt{\beta_y}}\,y$ and $\nu_y\,\varphi = \psi_y$ is the vertical betatron phase.

Due to the emission of a synchrotron radiation photon alone the particle does not change its position y nor its direction of propagation y'. With this we derive now the damping within a path element Δs which includes the emission of photons as well as the appropriate acceleration to compensate for that energy loss. Just after the emission of the photon but before the particle interacts with accelerating fields let the transverse momentum and total energy be $p_\perp$ and E_s, respectively. The slope of the particle trajectory is therefore (Fig. 8.4)

$$
y_o' = \frac{cp_\perp}{\beta E_s}.
\tag{8.63}
$$

Energy is transferred from the accelerating cavity to the particle at the rate of the synchrotron radiation power P_γ and the particle energy increases in the cavity of length Δs from E_s to $E_s + P_\gamma\frac{\Delta s}{\beta c}$ and the slope of the particle trajectory becomes at the exit of the cavity of length Δs due to this acceleration

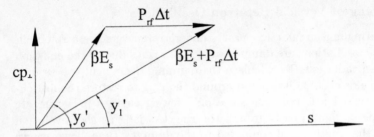

Fig. 8.4. Acceleration and damping

$$y_1' = \frac{cp_\perp}{\beta E_s + P_\gamma \frac{\Delta s}{c}} \approx \frac{cp_\perp}{\beta E_s} \left(1 - \frac{P_\gamma}{\beta E_s} \frac{\Delta s}{c} \right).$$
(8.64)

We are now in a position to express the functions f and g in terms of physical parameters. The function f is the variation of the particle position w per unit phase advance and is expressed by

$$f = \frac{\Delta w}{\Delta \varphi} = \frac{y_1 - y_o}{\sqrt{\beta_y} \, \Delta \varphi} = \frac{y_o'}{\sqrt{\beta_y}} \frac{\Delta s}{\Delta \varphi} = \nu \sqrt{\beta_y} \, y_o',$$
(8.65)

where we made use of $\Delta \varphi = \Delta s / (\nu \beta)$. The damping decrement will depend on the derivation $\frac{df}{dw}$ which can be seen from (8.65) to vanish since f does not depend on w

$$\frac{\partial f}{\partial w} = 0.$$
(8.66)

The variation of the conjugate variable p with phase is from the first of equations (8.62)

$$\frac{\Delta p}{\Delta \varphi} = \frac{\frac{dw_1}{d\varphi} - \frac{dw_o}{d\varphi}}{\nu \, \Delta \varphi}.$$
(8.67)

From linear beam dynamics, we find

$$\frac{dw_1}{d\varphi} - \frac{dw_o}{d\varphi} = \sqrt{\beta_y} \, (y_1' - y_o') - \tfrac{1}{2} \frac{\beta_y'}{\sqrt{\beta_y}} \, (y_1 - y_o)$$

and get with (8.64,65)

$$g(w, p, \varphi) = \frac{\Delta p}{\Delta \varphi} = \frac{-\sqrt{\beta_y} \frac{P_\gamma}{\beta c E_s} \Delta s \, y_o' + F(y)}{\nu \, \Delta \varphi}.$$
(8.68)

The function $F(y)$ is a collection of y-dependent terms that become irrelevant for our goal. The degree of damping will be determined by the value of the derivative $\frac{\partial g}{\partial p}$ which with $y_o' = \frac{1}{\sqrt{\beta_y}} \frac{dw}{d\varphi} + \tfrac{1}{2} \beta_y' \frac{1}{\beta_y} y_o$ becomes

$$\frac{\partial g}{\partial p} = \nu \frac{\partial g}{\partial \frac{dw}{d\varphi}} = \frac{P_\gamma}{\beta c E_{\mathrm{s}}} \frac{\Delta s}{\Delta \varphi}. \tag{8.69}$$

In the derivation of (8.69) we have used the betatron phase as the "time" and get therefore the damping per unit betatron phase advance. Transforming to the real time we have with $\frac{\Delta s}{\beta c \Delta \varphi} = \frac{T_{\mathrm{rev}}}{2\pi}$ and (8.41)

$$\frac{\partial g}{\partial p} = \frac{P_\gamma}{E_{\mathrm{s}}} \frac{T_{\mathrm{rev}}}{2\pi} = -2\alpha_y \frac{T_{\mathrm{rev}}}{2\pi}$$

and solving for the *vertical damping decrement*[1]

$$\boxed{\alpha_y = -\frac{\langle P_\gamma \rangle}{2E_{\mathrm{s}}}.} \tag{8.70}$$

In this last equation we have used the average synchrotron radiation power which is the appropriate quantity in case of a nonisomagnetic ring.

The damping of the vertical betatron function is proportional to the synchrotron radiation power. This fact can be used to increase damping when so desired by increasing the synchrotron radiation power from special magnets in the lattice structure.

8.2.3 Robinson's Damping Criterion

The general motion of charged particles extends over all six degrees of freedom in phase space and therefore the particle motion is described in six-dimensional phase space as indicated in the general Vlasov equation (8.17). It is, however, a fortunate circumstance that it is technically possible to construct accelerator components in such a fashion that there is only little or no coupling between different pairs of conjugate coordinates. As a consequence, we can generally treat horizontal betatron oscillations separate from the vertical betatron oscillations and both of them separate from synchrotron oscillations. Coupling effects that do occur will be treated as perturbations. There is some direct coupling via the dispersion function between synchrotron and particularly the horizontal betatron oscillations but the frequencies are very different with the synchrotron oscillation frequency being in general much smaller than the betatron oscillation frequency. Therefore in most cases the synchrotron oscillation can be ignored while discussing transverse oscillations and we may average over many betatron oscillations when we discuss synchrotron motion.

---

[1] Generally, the letter α_y is used for the vertical damping decrement. Since in beam dynamics α_y is also used to identify a lattice function, a mixup of the quantities could occur. We have chosen not to use a different nomenclature, however, since this choice is too deeply entrenched in the community. With some care, confusion can be avoided.

A special property of particle motion in six-dimensional phase space must be introduced allowing us to make general statements about the overall damping effects in a particle beam. We start from the Vlasov equation (8.17)

$$\frac{\partial \Psi}{\partial t} + \mathbf{f}\,\nabla_r \Psi + \mathbf{g}\,\nabla_p \Psi = -\left(\nabla_r \mathbf{f} + \nabla_p \mathbf{g}\right)\Psi$$

and define a *total damping decrement* α_t by setting

$$\nabla_r \mathbf{f} + \nabla_p \mathbf{g} = -2\alpha_t . \tag{8.71}$$

The total damping decrement is related to the individual damping decrements of the transverse and longitudinal oscillations but the precise dependencies are not yet obvious. In the derivation of (8.17), we have expanded the functions f and g in a Taylor series neglecting all terms of second or higher order in time and got as a result the simple expression (8.71) for the overall damping. Upon writing (8.71) in component form, we find from the components of the l.h.s. that the overall damping decrement α_t is just the sum of all three individual damping decrements and we may therefore set

$$\nabla_r \mathbf{f} + \nabla_p \mathbf{g} = -2\alpha_t = -2(\alpha_x + \alpha_y + \alpha_\epsilon). \tag{8.72}$$

From this equation and the linearity of the functions $\mathbf{f}$ and $\mathbf{g}$ describing the physics of the dynamical system general characteristics of the damping process can be derived. The damping decrement does not depend on the dynamic variables of the particles and coupling terms do not contribute to damping. The damping rate is therefore the same for all particles within a beam. In the following paragraphs, we will discuss in more detail the general characteristics of synchrotron radiation damping. Specifically, we will determine the functions $\mathbf{f}$ and $\mathbf{g}$ and derive an expression for the total damping.

We consider a small section of a beam transport line or circular accelerator including all basic processes governing the particle dynamics. These processes are focusing, emission of photons and acceleration. All three processes are assumed to occur evenly along the beam line. The six-dimensional phase space to be considered is

$$(x, x', y, y', \tau, \epsilon). \tag{8.73}$$

During the short time Δt some of the transverse coordinates change and it is those changes that determine eventually the damping rate. Neither the emission of a synchrotron radiation photon nor the absorption of energy in the accelerating cavities causes any change in the particle positions $x, y,$ and τ. Indicating the initial coordinates by the index $_o$ and setting $\Delta t = \frac{\Delta s}{\beta c}$ we get for the evolution of the particle positions within the length element Δs in the three space dimensions

$$x = x_{\text{o}} + x'_{\text{o}} \, \Delta s \,,$$
$$y = y_{\text{o}} + y'_{\text{o}} \, \Delta s \,,$$
$$\tau = \tau_{\text{o}} + \eta_{\text{c}} \frac{\epsilon_{\text{o}}}{E_{\text{s}}} \frac{\Delta s}{\beta c} \,. \tag{8.74}$$

The conjugate coordinates vary in a somewhat more complicated way. First we note that the Vlasov equation does not require the conjugate coordinates to be canonical variables. Indeed this derivation will become simplified if we do not use canonical variables but use the slopes of the particle trajectories with the reference path and the energy deviation. The change of the slopes due to focusing is proportional to the oscillation amplitude and vanishes on average. Emission of a synchrotron radiation photon occurs typically within an angle of $\pm\frac{1}{\gamma}$ causing a small transverse kick to the particle trajectory. In general, however, this transverse kick will be very small and we may assume for all practical purposes the slope of the transverse trajectory not to be altered by photon emission. Forces parallel to the direction of propagation of the particles can be created, however, through the emission of synchrotron radiation photons. In this case, the energy or energy deviation of the particle will be changed like

$$\epsilon = \epsilon_{\text{o}} - P_{\gamma} \frac{\Delta s}{\beta c} + P_{\text{rf}} \frac{\Delta s}{\beta c} \,. \tag{8.75}$$

Here we use the power P_{γ} to describe the synchrotron radiation energy loss rate a particle may suffer during the time $\Delta t = \frac{\Delta s}{\beta c}$. No particular assumption has been made about the nature of the energy loss except that during the time Δt it be small compared to the particle energy. To compensate this energy loss the particles become accelerated in rf cavities. The power P_{rf} is the energy flow from the cavity to the particle beam, not to be confused with the total power the rf source delivers to the cavity.

The transverse slopes x' and y' are determined by the ratio of the transverse to the longitudinal momentum $u' = p_u/p_s$ where u stands for x or y, respectively. During the acceleration in the rf cavity the transverse momentum does not change but the total kinetic energy increases from E_{s} to $E_{\text{s}} + P_{\text{rf}} \frac{\Delta s}{\beta c}$. As a consequence, the transverse slope of the trajectory is reduced and is after a distance Δs

$$u' = \frac{c p_u}{c p_s + P_{\text{rf}} \beta \frac{\Delta s}{\beta c}} \approx u'_{\text{o}} - \frac{P_{\text{rf}}}{E_{\text{s}}} \frac{\Delta s}{\beta c} u'_{\text{o}} \,.$$

Explicitly the transverse slopes vary now like

$$x' = x'_{\text{o}} - \frac{P_{\text{rf}}}{E_{\text{s}}} \frac{\Delta s}{\beta c} x'_{\text{o}} \,,$$
$$y' = y'_{\text{o}} - \frac{P_{\text{rf}}}{E_{\text{s}}} \frac{\Delta s}{\beta c} y'_{\text{o}} \,. \tag{8.76}$$

All ingredients are available now to formulate with (8.75,76) expressions for the functions $\mathbf{f}$ and $\mathbf{g}$ in component form

$$
\begin{aligned}
\mathbf{f} &= \left(x'_{\mathrm{o}},\, y'_{\mathrm{o}},\, \eta_c\, \frac{\epsilon}{E_{\mathrm{s}}} \right), \\
\mathbf{g} &= \left(-\frac{P_{\mathrm{rf}}}{E_{\mathrm{s}}}\, x'_{\mathrm{o}},\, -\frac{P_{\mathrm{rf}}}{E_{\mathrm{s}}}\, y'_{\mathrm{o}},\, -P_\gamma + P_{\mathrm{rf}} \right)
\end{aligned}
\tag{8.77}
$$

With these expressions we evaluate (8.72) and find that $\nabla_r \mathbf{f} = 0$. For the determination of $\nabla_p \mathbf{g}$ we note that the cavity power P_{rf} does not depend on the particle energy and the derivative of the radiation power with respect to the particle energy is

$$
-\frac{\partial P_\gamma}{\partial \epsilon} = -2\frac{P_\gamma}{E_{\mathrm{s}}}.
\tag{8.78}
$$

Finally we note that the rf power P_{rf} is just equal to the radiation power P_γ and get finally from (8.72)

$$
\boxed{\; \alpha_x + \alpha_y + \alpha_\epsilon = 2\,\frac{P_\gamma}{E_{\mathrm{s}}}. \;}
\tag{8.79}
$$

The sum of all damping decrements is a constant, a result which has been derived first by *Robinson* [8.2] and is known as the *Robinson criterion*. The total damping depends only on the synchrotron radiation power and the particle energy and variations of magnetic field distribution in the ring keeping the radiation power constant will not affect the total damping rate but may only shift damping from one degree of freedom to another.

8.2.4 Damping of Horizontal Betatron Oscillations

With the help of the Robinson criterion, the damping decrement for the horizontal betatron oscillation can be derived by simple subtraction. Inserting (8.61,71) into (8.79) and solving for the *horizontal damping decrement* we get

$$
\boxed{\; \alpha_x = \frac{P_\gamma}{2\,E_{\mathrm{s}}}\,(1-\vartheta). \;}
\tag{8.80}
$$

All damping rates differ only by a factor involving the ϑ-parameter and it is therefore customary to define *damping-partition numbers*

$$
\begin{aligned}
J_\epsilon &= 2+\vartheta, \\
J_y &= 1, \\
J_x &= 1-\vartheta.
\end{aligned}
\tag{8.81}
$$

Applying the *Robinson criterion* to the damping-partition numbers, we finally find the simple result

$$\sum J_i = 4.$$ (8.82)

No matter what type of magnet lattice we use, the total damping depends only on the synchrotron radiation power and the particle energy. We may, however, vary the distribution of the damping rates through the ϑ-parameter to different oscillation modes by proper design of the focusing and bending lattice in such a way that one damping rate is modified in the desired way limited only by the onset of antidamping in another mode. Specifically, this is done by introducing gradient bending magnets with a field gradient such as to produce the desired sign of the ϑ parameter.

8.3 The Fokker-Planck Equation

From the discussions of the previous section it became clear that the Vlasov equation is a useful tool to determine the evolution of a multiparticle system under the influence of forces depending on the physical parameters of the system through differentiable functions. If, however, the trajectory of a system in phase space depends only on its instantaneous physical parameters where the physics of the particle dynamics cannot be expressed by differentiable functions, the Vlasov equation will not be sufficient to describe the full particle dynamics. A process which depends only on the state of the system at the time t and not on its history is called a *Markoff process*.

In particle beam dynamics we have frequently the appearance of such processes where forces are of purely statistical nature like those caused, for example, by the quantized emission of synchrotron radiation photons or by collisions with other particles within the same bunch or residual gas atoms. To describe such a situation we still have variations of the coordinates per unit time similar to those in (8.2) but we must add a term describing the statistical process and we set therefore

$$\begin{aligned}
\dot{w} &= f_w(w, p_w, t) + \sum \xi_i \, \delta\,(t - t_i)\,, \\
\dot{p}_w &= g_w(w, p_w, t) + \sum \pi_i \, \delta\,(t - t_i)\,,
\end{aligned}$$ (8.83)

where ξ_i and π_i are instantaneous statistical changes in the variables w and p_w with a statistical distribution in time t_i and where $\delta(t - t_i)$ is the Dirac delta function. The probabilities $P_w(\xi)$ and $P_p(\pi)$ for statistical occurrences with the amplitudes ξ and π and require that these probability functions are normalized and centered

$$\int P_w(\xi)\,\mathrm{d}\xi = 1, \qquad \int P_p(\pi)\,\mathrm{d}\pi = 1,$$

$$\int P_w(\xi)\,\xi\,\mathrm{d}\xi = 0, \qquad \int P_p(\pi)\,\pi\,\mathrm{d}\pi = 0. \tag{8.84}$$

The first equations normalize the probability amplitudes and the second equations are true for symmetric statistical processes. The sudden change in the amplitude Δw_i or momentum Δp_{wi} due to one such process is given by

$$\Delta w_i = \int \xi_i\,\delta(t - t_i)\,\mathrm{d}t = \xi_i,$$

$$\Delta p_{wi} = \int \pi_i\,\delta(t - t_i)\,\mathrm{d}t = \pi_i. \tag{8.85}$$

Analogous to the discussion of the evolution of phase space in the previous section, we will now formulate a similar evolution including statistical processes. At the time $t + \Delta t$, the particle density in phase space is taken to be $\Psi(w, p_w, t + \Delta t)$ and we intend to relate this to the values of the particle density at time t. During the time interval Δt there are finite probabilities $P_w(\xi)$, $P_p(\pi)$ that the amplitude $(w - \xi)$ or the momentum $(p_w - \pi)$ be changed by a statistical process to become w or p_w. This definition of the probability function also covers the cases where particles during the time Δt either jump out of the phase space area ΔA_P or appear in the phase space area ΔA_Q.

To determine the number of particles ending up within the area ΔA_Q, we look at all area elements ΔA_P which at time t are a distance $\Delta w = \xi$ and $\Delta p_w, = \pi$ away from the final area element ΔA_Q at time $t + \Delta t$. As a consequence of our assumption that the particle density is only slowly varying in phase space, we may assume that the density Ψ is uniform within the area elements ΔA_P eliminating the need for a local integration. We may now write down the expression for the phase space element and the particle density at time $t + \Delta t$ by integrating over all values of ξ and π

$$\Psi(w + f_w\,\Delta t, p_w + g_w\Delta t, t + \Delta t)\,\Delta A_Q$$

$$= \Delta A_P \int\limits_{-\infty}^{+\infty}\int\limits_{-\infty}^{+\infty} \Psi(w - \xi, p_w - \pi, t)\,P_w(\xi)\,P_p(\pi)\,\mathrm{d}\xi\,\mathrm{d}\pi. \tag{8.86}$$

Here the volume elements ΔA_P and ΔA_Q are given by (8.5,6), respectively. The statistical fluctuations may in general be of any magnitude. In particle beam dynamics, however, we find that the fluctuations with reasonable probabilities are small compared to the values of the variables w and p_w. The phase space density can therefore be expanded into a Taylor series where we retain linear as well as quadratic terms in ξ and π. Thus, we get

$$\Psi\left(w-\xi, p_w-\pi, t\right) = \tag{8.87}$$

$$\Psi_{\mathrm{o}} - \xi\frac{\partial\Psi_{\mathrm{o}}}{\partial w} - \pi\frac{\partial\Psi_{\mathrm{o}}}{\partial p_w} + \tfrac{1}{2}\xi^2\frac{\partial^2\Psi_{\mathrm{o}}}{\partial w^2} + \tfrac{1}{2}\pi^2\frac{\partial^2\Psi_{\mathrm{o}}}{\partial p_w^2} + \xi\,\pi\,\frac{\partial^2\Psi_{\mathrm{o}}}{\partial w\partial p_w}\,,$$

where $\Psi_{\mathrm{o}} = \Psi(w, p_w, t)$. With (8.84) the integrals in (8.86) become

$$\int\int \Psi(w-\xi, p_w-\pi, t)\,P_w(\xi)\,P_p(\pi)\,\mathrm{d}\xi\,\mathrm{d}\pi = \tag{8.88}$$

$$\Psi_{\mathrm{o}} + \tfrac{1}{2}\frac{\partial^2\Psi_{\mathrm{o}}}{\partial w^2}\int \xi^2\,P_w(\xi)\,\mathrm{d}\xi + \tfrac{1}{2}\frac{\partial^2\Psi_{\mathrm{o}}}{\partial p_w^2}\int \pi^2 P_p(\pi)\,\mathrm{d}\pi\,.$$

For simplicity, we leave off the integration limits which are still from $-\infty$ to $+\infty$. If we now set $\mathcal{N}$ to be the number of statistical occurrences per unit time we may simplify the quadratic terms on the r.h.s. of (8.88) by setting

$$\tfrac{1}{2}\int \xi^2\,P_w(\xi)\,\mathrm{d}\xi = \tfrac{1}{2}\left\langle \mathcal{N}_\xi\,\xi^2 \right\rangle\Delta t\,,$$

$$\tfrac{1}{2}\int \pi^2\,P_p(\pi)\,\mathrm{d}\pi = \tfrac{1}{2}\left\langle \mathcal{N}_\pi\,\pi^2 \right\rangle\Delta t\,, \tag{8.89}$$

and get similarly to the derivation of the Vlasov equation in Sect. 8.1 from (8.87)

$$\frac{\partial\Psi_{\mathrm{o}}}{\partial t} + f_w\frac{\partial\Psi_{\mathrm{o}}}{\partial w} + g_w\frac{\partial\Psi_{\mathrm{o}}}{\partial p_w} = \tag{8.90}$$

$$-\left(\frac{\partial f_w}{\partial w} + \frac{\partial g_w}{\partial p_w}\right)\Psi_{\mathrm{o}} + \tfrac{1}{2}\left\langle \mathcal{N}_\xi\,\xi^2 \right\rangle\frac{\partial^2\Psi_{\mathrm{o}}}{\partial w^2} + \tfrac{1}{2}\left\langle \mathcal{N}_\pi\,\pi^2 \right\rangle\frac{\partial^2\Psi_{\mathrm{o}}}{\partial p_w^2}\,.$$

This partial differential equation is identical to the Vlasov equation except for the statistical excitation terms and is called the *Fokker-Planck equation*. We define *diffusion coefficients* describing the flow in ξ and π space by

$$D_\xi = \tfrac{1}{2}\left\langle \mathcal{N}_\xi\,\xi^2 \right\rangle,$$

$$D_\pi = \tfrac{1}{2}\left\langle \mathcal{N}_\pi\,\pi^2 \right\rangle, \tag{8.91}$$

and the Fokker-Planck equation becomes finally

$$\boxed{\frac{\partial\Psi}{\partial t} + f_w\frac{\partial\Psi}{\partial w} + g_w\frac{\partial\Psi}{\partial p_w} = 2\,\alpha_w\,\Psi + D_\xi\frac{\partial^2\Psi}{\partial w^2} + D_\pi\frac{\partial^2\Psi}{\partial p_w^2}\,.} \tag{8.92}$$

For the case of a damped oscillator the Fokker-Planck equation can be derived similar to (8.35) and is

$$\frac{\partial\Psi}{\partial t} + \omega_{\mathrm{o}}\,p_w\frac{\partial\Psi}{\partial w} - (\omega_{\mathrm{o}}w + 2\,\alpha_w\,p_w)\frac{\partial\Psi}{\partial p_w} =$$

$$2\,\alpha_w\,\Psi + D_\xi\frac{\partial^2\Psi}{\partial w^2} + D_\pi\frac{\partial^2\Psi}{\partial p_w^2}\,. \tag{8.93}$$

This form of the Fokker-Planck equation will be very useful to describe a particle beam under the influence of diffusion processes. In the following section, we will derive general solutions which will be applicable to specific situations in accelerator physics.

8.3.1 Stationary Solution of the Fokker-Planck Equation

A unique stationary solution exists for the particle density distribution described by the partial differential equation (8.93). To derive this solution we transform (8.93) to cylindrical coordinates $(w, p_w) \to (r, \theta)$ with $w = r \cos \theta$ and $p_w = r \sin \theta$ and we note terms proportional to derivatives of the phase space density with respect to the angle θ. One of these terms $\omega_0 \Psi_\theta$ exists even in the absence of diffusion and damping and describes merely the betatron motion in phase space while the other terms depend on damping and diffusion. The diffusion terms will introduce a statistical mixing of the phases θ and after some damping times any initial azimuthal variation of the phase space density will be washed out. We are here only interested in the stationary solution and therefore set all derivatives of the phase space density with respect to the phase θ to zero. In addition we find it necessary to average square terms of $\cos \theta$ and $\sin \theta$. With these assumptions the Fokker-Planck equation (8.93) becomes after some manipulations in the new coordinates

$$\frac{\partial \Psi}{\partial t} = 2\alpha_w \Psi + \left(\alpha_w r + \frac{D}{r} \right) \frac{\partial \Psi}{\partial r} + D \frac{\partial^2 \Psi}{\partial r^2}, \tag{8.94}$$

where we have defined a total diffusion coefficient

$$D = \tfrac{1}{2} \left(D_\xi + D_\pi \right). \tag{8.95}$$

Equation (8.94) has some similarity with, for example, wave equations in quantum mechanics which are solved by the method of separation of variables and we expect the stationary solution for the phase space density to be of the form $\Psi(r, t) = \sum_n F_n(t) \, G_n(r)$. The solution $G_n(r)$ must meet some particular boundary conditions. Specifically, at time $t = 0$, we may have any arbitrary distribution of the phase space density $G_{no}(r)$. Furthermore, we specify that there be a wall at $r = R$ beyond which the phase space density drops to zero and consequently, the boundary conditions are

$$\begin{aligned} G_n(r < R) &= G_{no}(r), \\ G_n(r > R) &= 0. \end{aligned} \tag{8.96}$$

By the method of separation of the constants we find for the functions $F_n(t)$

$$F_n(t) = \text{const.} \, e^{-\alpha_n t}, \tag{8.97}$$

where the quantity $-\alpha_n$ is the separation constant. The general form of the solution for (8.94) may now be expressed by a series of orthogonal functions or eigenmodes of the distribution $G_n(r)$ which fulfill the boundary conditions (8.96)

$$\Psi(r,t) = \sum_{n\geq 0} c_n\, G_n(r)\, e^{-\alpha_n t}. \qquad (8.98)$$

The amplitudes c_n in (8.98) will be determined to fit the initial density distribution

$$\Psi_o(r, t=0) = \sum_{n\geq 0} c_n\, G_{no}(r). \qquad (8.99)$$

With the ansatz (8.98) we get from (8.94) for each of the eigenmodes the following second-order differential equation:

$$\frac{\partial^2 G_n}{\partial r^2} + \left(\frac{1}{r} + \frac{\alpha_w}{D}r\right)\frac{\partial G_n}{\partial r} + \frac{\alpha_w}{D}\left(2 + \frac{\alpha_n}{\alpha_w}\right)G_n = 0. \qquad (8.100)$$

All terms with a coefficient $\alpha_n > 0$ vanish after some time due to damping (8.97). Negative values for the damping decrements $\alpha_n < 0$ define instabilities which we will not consider here. Stationary solutions, therefore require the separation constants to be zero $\alpha_n = 0$. Furthermore, all solutions G_n must vanish at the boundary $r = R$ where R may be any value including infinity if there are no physical boundaries at all to limit the maximum particle oscillation amplitude. In the latter case where there are no walls, the differential equation (8.100) can be solved by the stationary distribution

$$\Psi(r,t) = \sum_{\substack{n\geq 0 \\ \alpha_n=0}} c_n\, G_n(r) \propto \exp\left(-\frac{\alpha_w}{2D}r^2\right) \qquad (8.101)$$

which can easily be verified by backinsertion into (8.100). The solution for the particle distribution in phase space under the influence of damping α_w and statistical fluctuations D is a *gaussian distribution* with the standard width

$$\sigma_r = \sqrt{\frac{D}{\alpha_w}}. \qquad (8.102)$$

Normalizing the phase space density the *stationary solution* of the Fokker-Planck equation for a particle beam under the influence of damping and statistical fluctuations is

$$\boxed{\Psi(r) = \frac{1}{\sqrt{2\pi}\,\sigma_r}\, e^{-r^2/2\sigma_r^2}.} \qquad (8.103)$$

Eigenfunctions for which the eigenvalues α_n are not zero, are needed to describe an arbitrary particle distribution, e.g., a rectangular distribution at time $t = 0$. The Fokker-Planck equation, however, tells us that after some damping times these eigensolutions have vanished and the gaussian distribution is the only stationary solution left. The gaussian distribution is not restricted to the r-space alone. The particle distribution in equilibrium between damping and fluctuations is also gaussian in the normalized phase space (w, p_w) as well as in real space. With $r^2 = w^2 + p_w^2$ we get immediately for the density distribution in (w, p_w)-space

$$\Psi(w, p_w) = \frac{1}{2\pi \sigma_w \sigma_{p_w}} \, e^{-w^2/2\sigma_w^2} \, e^{-p_w^2/2\sigma_{p_w}^2} \,, \qquad (8.104)$$

where we have set $\sigma_w = \sigma_{p_w} = \sqrt{\frac{D}{\alpha_w}}$. The standard deviation in w and p_w is the same as for r which is to be expected since all three quantities have the same dimension and are linearly related.

In real space we have for $u = x$ or $u = y$ by definition $u = \sqrt{\beta_u}\, w$ and $p = \frac{\dot{w}}{\nu}$ where $\dot{w} = \frac{dw}{d\varphi}$. On the other hand, $p = \sqrt{\beta_x}\, x' - \frac{\beta'}{\sqrt{\beta}}\, x$ and inserted into (8.101) the density distribution in real space is

$$\boxed{\Psi(u, u') \propto \exp\left(-\frac{\gamma_u u^2 - \beta'_u uu' + \beta_u u'^2}{2\sigma_w^2}\right).} \qquad (8.105)$$

This distribution describes the particle distribution in real phase space where particle trajectories follow tilted ellipses. Note that we carefully avoid replacing the derivative of the betatron function with $\beta' = -2\alpha$ because this would lead to a definite confusion between the damping decrement and the betatron function. To further reduce confusion we also use the damping times $\tau_i = \alpha_i^{-1}$. Integrating the distribution (8.105) for all values of the angles u', for example, gives the particle distribution in the horizontal or vertical midplane. Using the mathematical relation $\int_\infty e^{-p^2 x^2 \pm qx}dx = \frac{\sqrt{\pi}}{p}\, e^{q^2/(4p^2)}$ [8.3] we get

$$\boxed{\Psi(u) = \frac{1}{\sqrt{2\pi}\,\sqrt{\beta_u}\,\sigma_w}\, e^{-u^2/2\sigma_u^2}\,,} \qquad (8.106)$$

where the standard width of the horizontal gaussian particle distribution is

$$\boxed{\sigma_u = \sqrt{\beta}\,\sigma_w = \sqrt{\beta}\,\sqrt{\tau_u D_u}\,.} \qquad (8.107)$$

The index $_u$ has been added to the diffusion and damping terms to indicate that these quantities are in general different in the horizontal and vertical plane. The damping time depends on all bending magnets, vertical and horizontal, but only on the damping-partition number for the plane under consideration. Similar distinction applies to the diffusion term.

In a similar way, we get the distribution for the angles by integrating (8.105) with respect to u

$$\Psi(u') = \frac{\sqrt{\beta}}{\sqrt{2\pi}\sqrt{1 + \frac{1}{4}\beta'^2}\,\sigma_w}\,\exp\left[-\frac{\beta\,u'^2}{2\left(1 + \frac{1}{4}\beta'^2\right)\sigma_w^2}\right], \qquad (8.108)$$

where the standard width of the angular distribution is

$$\sigma'_u = \sqrt{\frac{4 + \beta'^2}{4\,\beta}}\,\sigma_w = \sqrt{\frac{4 + \beta'^2}{4\,\beta}}\,\sqrt{\tau_u\,D_u}\,. \qquad (8.109)$$

We have not made any special assumption as to the horizontal or vertical plane and find in (8.107 – 110) the solutions for the particle distribution in both planes.

In the *longitudinal phase space* the equations of motion are mathematically equal to equations (8.10,11). First we define new variables

$$\dot{w} = -\frac{\Omega_{\mathrm{so}}}{\eta_{\mathrm{c}}}\,\dot{\tau}\,, \qquad (8.110)$$

where Ω_{so} is the *synchrotron oscillation frequency*, η_{c} the momentum compaction and τ the longitudinal deviation of a particle from the reference particle. The conjugate variable we define by

$$p = -\frac{\dot{\epsilon}}{E_{\mathrm{o}}}\,, \qquad (8.111)$$

where ϵ is the energy deviation from the reference energy E_{o}. After differentiation of (8.50) and making use of (8.51) and the definition of the synchrotron oscillation frequency [Ref. 8.1, Eq. (8.36)], we use these new variables and obtain the two first-order differential equations

$$\dot{w} = \Omega_{\mathrm{s}}\,p\,,$$
$$\dot{p} = -\Omega_{\mathrm{s}}\frac{\epsilon}{E_{\mathrm{o}}} - 2\alpha_\epsilon\,p\,. \qquad (8.112)$$

These two equations are of the same form as (8.10,11) and the solution of the longitudinal Fokker-Planck equation is therefore similar to (8.105 – 109). The energy distribution within a particle beam under the influence of damping and statistical fluctuations becomes with $p = \delta = \epsilon/E_{\mathrm{o}}$

$$\Psi(\delta) = \frac{1}{\sqrt{2\pi}\,\sigma_\delta}\,e^{-\delta^2/2\sigma_\delta^2}\,, \qquad (8.113)$$

where the standard value for the *energy spread* in the particle beam is defined by

$$\frac{\sigma_\epsilon}{E_o} = \sqrt{\tau_\epsilon D_\epsilon}. \tag{8.114}$$

In a similar way, we get for the conjugate coordinate τ by setting $w = \frac{\Omega_s}{\eta_c}\tau$ the distribution

$$\Psi(\tau) = \frac{1}{\sqrt{2\pi}\,\sigma_\tau}\,e^{-\tau^2/2\sigma_\tau^2}. \tag{8.115}$$

The standard width of the *longitudinal particle distribution* is finally

$$\sigma_\tau = \frac{|\eta_c|}{\Omega_s}\sqrt{\tau_\epsilon D_\epsilon}. \tag{8.116}$$

The deviation in time τ of a particle from the synchronous particle is equivalent to the distance of these two particles. Since $s = c\beta\tau$ we may define the standard value for the *bunch length* by

$$\sigma_\ell = c\beta \frac{|\eta_c|}{\Omega_s}\sqrt{\tau_\epsilon D_\epsilon}. \tag{8.117}$$

By application of the Fokker-Planck equation to systems of particles under the influence of damping and statistical fluctuations, we were able to derive expressions for the particle distribution within the beam. In fact, we were able to determine that the particle distribution is gaussian in all six degrees of freedom. Since such a distribution does not exhibit any definite boundary for the beam, it becomes necessary to define the size of the distributions in all six degrees of freedom by the standard value of the gaussian distribution. Specific knowledge of the nature for the statistical fluctuations are required to determine the numerical values of the beam sizes.

In Chap. 9 we will apply these results to determine the equilibrium beam emittance in an electron positron storage ring where the statistical fluctuations are generated by quantized emission of synchrotron radiation photons.

8.3.2 Particle Distribution Within a Finite Aperture

The particle distribution in an electron beam circulating in a storage ring assumes the form of a gaussian if we ignore the presence of walls containing the beam. All other modes of particle distribution are associated with a finite damping time and vanish therefore after a short time. In a real storage ring we must, however, consider the presence of vacuum chamber walls which cut off the gaussian tails of the particle distribution. Although the particle intensity is very small in the far tails of a gaussian distribution, we cannot cut off those tails too tight without reducing significantly the beam

lifetime. Due to quantum excitation, we observe a continuous flow of particles from the beam core into the tails and back by damping toward the core.
A reduction of the aperture into the gaussian distribution absorbs therefore
not only those particles which populate these tails at a particular moment
but also all particles which reach occasionally large oscillation amplitudes
due to the emission of a high energy photon. The absorption of particles
due to this effect causes a reduction in the beam lifetime which we call the
quantum lifetime.

The presence of a wall modifies the particle distribution especially close
to the wall. This modification is described by normal mode solutions with a
finite damping time which is acceptable now because any aperture less than
an infinite aperture absorbs beam particles thus introducing a finite beam
lifetime. Cutting off gaussian tails at large amplitudes will not affect the
gaussian distribution in the core and we look therefore for small variations
of the gaussian distribution which become significant only quite close to the
wall. Instead of (8.101) we try the ansatz

$$\Psi(r,t) = e^{-\frac{\alpha_w}{2D}r^2} g(r) e^{-\alpha t}, \tag{8.118}$$

where $1/\alpha$ is the time constant for the distribution, with the boundary
condition that the particle density be zero at the aperture or acceptance
defining wall $r = A$ or

$$\Psi(A,t) = 0. \tag{8.119}$$

Equation (8.118) must be a solution of (8.94) and back insertion of (8.118)
into (8.94) gives the condition on the function $g(r)$

$$g'' + \left(\frac{1}{r} - \frac{r}{\sigma^2}\right) g' + \frac{\alpha}{\alpha_w \sigma^2} g = 0. \tag{8.120}$$

Since $g(r) = 1$ in case there is no wall, we expand into a power series

$$g(r) = 1 + \sum_{k \geq 1} C_k x^k, \quad \text{where} \quad x = \frac{r^2}{2\sigma^2}. \tag{8.121}$$

Inserting (8.121) into (8.120) and collecting terms of equal powers in r we
derive the coefficients

$$C_k = \frac{1}{(k!)^2} \prod_{p=1}^{p=k} (p - 1 - X) \approx -\frac{(k-1)!}{(k!)^2} X, \tag{8.122}$$

where $X = \frac{\alpha}{2\alpha_w} \ll 1$. The approximation $X \ll 1$ is justified since we expect
the vacuum chamber wall to be far away from the beam center such that
the expected quantum lifetime $1/\alpha$ is long compared to the damping time

$1/\alpha_w$ of the oscillation under consideration. With these coefficients (8.121) becomes

$$g(r) = 1 - \frac{\alpha}{2\alpha_w} \sum_{k \geq 1} \frac{1}{k\,k!}\, x^k \,. \tag{8.123}$$

For $x = A^2/(2\sigma^2) \gg 1$ where A is the amplitude or amplitude limit for the oscillation w, the sum in (8.123) can be replaced by an exponential function

$$\sum_{k \geq 1} \frac{1}{k\,k!}\, x^k \approx \frac{e^x}{x} \,. \tag{8.124}$$

From the condition $g(A) = 0$ we finally get for the *quantum lifetime* $\tau_q = 1/\alpha$

$$\boxed{\tau_q = \tfrac{1}{2}\,\tau_w\, \frac{e^x}{x} \,,} \tag{8.125}$$

where

$$x = \frac{A^2}{2\sigma^2} \,. \tag{8.126}$$

The quantum lifetime is related to the damping time. To make the quantum life time very large of the order of 50 or more hours, the aperture must be at least about $7\,\sigma_w$ in which case $x = 24.5$ and $e^x/x = 1.8 \cdot 10^9$.

The aperture A is equal to the transverse acceptance of a storage ring for a one-dimensional oscillation like the vertical betatron oscillation while longitudinal or energy oscillations are limited through the maximum energy acceptance allowed by the rf voltage. Upon closer look, however, we note a complication for horizontal betatron oscillations and synchrotron oscillations because of the coupling from energy oscillation into transverse position due to a finite dispersion function. We also have assumed that $\alpha/(2\alpha_w) \ll 1$ which is not true for tight apertures of less than one sigma. Both of these situations have been investigated in detail [8.4, 5] and the interested reader is referred to those references.

Specifically, if the acceptance A of a storage ring is defined at a location where there is also a finite dispersion function, *Chao* [8.4] derives a combined quantum lifetime of

$$\tau = \frac{e^{n^2/2}}{\sqrt{2\pi}\,\alpha_x\, n^3} \frac{1}{(1+r)\,\sqrt{r(1-r)}} \,, \tag{8.127}$$

where $n = A/\sigma_{\mathrm{T}}$, $\sigma_{\mathrm{T}}^2 = \sigma_x^2 + \eta^2\sigma_\delta^2$, $r = \eta^2\sigma_\delta^2/\sigma_{\mathrm{T}}^2$, A the transverse aperture, η the dispersion function at the same location as A, σ_x the transverse beam size and $\sigma_\delta = \sigma_\epsilon/E$ the standard relative energy width in the beam.

8.3.3 Particle Distribution in the Absence of Damping

To obtain a stationary solution for the particle distribution it was essential that there were eigensolutions with vanishing eigenvalues $\alpha_n = 0$. As a result, we obtained an equilibrium solution where the statistical fluctuations are compensated by damping. In cases where there is no damping, we would expect a different solution with particles spreading out due to the effect of diffusion alone. This case can become important in very high energy electron positron *linear colliders* where an extremely small beam emittance must be preserved along a long beam transport line. The differential equation (8.100) becomes in this case

$$\frac{\partial^2 G_n}{\partial r^2} + \frac{1}{r}\frac{\partial G_n}{\partial r} + \frac{\alpha_n}{D}\,G_n = 0\,. \tag{8.128}$$

We will assume that a beam with a gaussian particle distribution is injected into a damping free transport line and we therefore look for solutions of the form

$$\Psi_n(r,t) = c_n\,G_n(r)\,\mathrm{e}^{-\alpha_n t}\,, \tag{8.129}$$

where

$$G_n(r) = \mathrm{e}^{-r^2/2\sigma_{\mathrm{o}}^2} \tag{8.130}$$

with σ_{o} being the beam size at $t = 0$. We insert (8.130) into (8.128) and obtain an expression for the eigenvalues α_n

$$\alpha_n = \frac{2D}{\sigma_{\mathrm{o}}^2} - \frac{D}{\sigma_{\mathrm{o}}^4}\,r^2\,. \tag{8.131}$$

The time dependent solution for the particle distribution now becomes

$$\Psi(r,t) = A\exp\left(-\frac{2D}{\sigma_{\mathrm{o}}^2}t\right)\exp\left[\left(-\frac{r^2}{2\sigma_{\mathrm{o}}^2}\right)\left(1 - \frac{2D}{\sigma_{\mathrm{o}}^2}t\right)\right]\,. \tag{8.132}$$

Since nowhere a particular mode is used we have omitted the index n.

The solution (8.132) exhibits clearly the effect of the diffusion in two, respects. The particle density decays exponentially with the decrement $2D/\sigma_{\mathrm{o}}^2$. At the same time the distribution remains to be gaussian although being broadened by diffusion. The time dependent beam size σ is given by

$$\sigma^2(t) = \frac{\sigma_{\mathrm{o}}^2}{1 - \frac{2D}{\sigma_{\mathrm{o}}^2}t} \approx \sigma_{\mathrm{o}}^2\left(1 + \frac{2D}{\sigma_{\mathrm{o}}^2}t\right)\,, \tag{8.133}$$

where we have assumed that the diffusion term is small $(2D/\sigma_{\mathrm{o}}^2)\,t \ll 1$. Setting $\sigma^2 = \sigma_u^2 = \epsilon_u\,\beta_u$ for the plane u where β_u is the betatron function at the observation point of the beam size σ_u and get for the time dependent beam emittance

$$\epsilon_u = \epsilon_{uo} + \frac{2D}{\beta_u} t \qquad (8.134)$$

or for the rate of change

$$\frac{d\epsilon}{dt} = \frac{2D}{\beta} = \frac{1}{\beta} (D_\xi + D_\pi). \qquad (8.135)$$

Due to the diffusion coefficient D we obtain a continuous increase of the beam emittance in cases where no damping is available.

The Fokker-Planck diffusion equation provides a tool to describe the evolution of a particle beam under the influence of conservative forces as well as statistical processes. Specifically, we found that such a system has a stationary solution in cases where there is damping. The stationary solution for the particle density is a gaussian distribution with the standard width of the distribution σ given by the diffusion constant and the damping decrement.

In particular, the emission of photons due to synchrotron radiation has the properties of a Markoff process and we find therefore the particle distribution to be gaussian. Indeed we will see that this is true in all six dimensions of phase space.

Obviously not every particle beam is characterized by the stationary solution of the Fokker-Planck equation. Many modes contribute to the particle distribution and specifically at time $t = 0$ the distribution may have any arbitrary form. However, it has been shown that after a time long compared to the damping times only one nontrivial stationary solution is left, the gaussian distribution.

In particle beam transport systems such "long" times are available only in circular accelerators for light particles like electrons and positrons. Only for these particles is the synchrotron radiation strong enough to lead to damping times which are shorter than the lifetimes of beams in synchrotrons or storage rings. For extremely high beam energies in the order of some tens of TeV, the synchrotron radiation of protons and antiprotons becomes significant as well and damping and quantum effects can be observed. In this case the distinction of electron beams and proton beams vanishes and the physics becomes the same.

Problems

Problem 8.1. An arbitrary particle distribution of beam injected into a storage ring damps out while a gaussian distribution evolves with a standard width specific to the ring design. What happens if a beam from another storage ring with a gaussian distribution is injected? Explain in some detail why this beam changes its distribution to the ring specific gaussian distribution.

Problem 8.2.   Derive from the Vlasov equation an expression for the synchrotron frequency while ignoring damping. A second rf system with different frequency can be used to change the synchrotron tune. Determine a system that would reduce the synchrotron tune for the reference particle to zero while providing the required rf voltage at the synchronous phase constant. What is the relationship between both voltages and phases? Is the tune shift the same for all particles?

Problem 8.3.   Formulate an expression for the equilibrium bunch length in a storage ring with two rf systems of different frequencies to control bunch length.

Problem 8.4.   To reduce coupling instabilities between bunches of a multibunch beam it is desirable to give each bunch a different synchrotron tune. This can be done, for example, by employing two rf systems operating at harmonic numbers h and $h + 1$. Determine the ratio or required rf voltages to split the tunes between successive bunches by $\delta\nu/\nu_s$.

Problem 8.5.   Attempt to damp out the energy spread of a storage ring beam in the following way. At a location where the dispersion function is finite one could insert a TM_{110}-mode cavity. Such a cavity produces accelerating fields which vary linear with the transverse distance of a particle from the reference path. This together with a linear change in particle energy due to the dispersion would allow the correction of the energy spread in the beam. Derive the complete Vlasov equation for such an arrangement and discuss the six-dimensional dynamics. Show that it is impossible to achieve a monochromatic stable beam.

Problem 8.6.   Energy loss of a particle beam due to synchrotron radiation provides damping. Show that energy loss due to interaction with an external electromagnetic field does not provide damping.

Problem 8.7.   Consider a 1.5 GeV electron storage ring with a bending field of 15 kGauss. Let the bremsstrahlung lifetime be 100 hr, the Coulomb scattering lifetime 50 hr and the Touschek lifetime 60 hr. Calculate the total beam lifetime including quantum excitation as a function of aperture. How many "sigma's" (A/σ) must the aperture be in order not to reduce the beam lifetime by more than 10% due to quantum excitation?

Problem 8.8.   Derive an expression for the diffusion due to elastic scattering of beam particles on residual gas atoms. How does the equilibrium beam emittance of an electron beam scale with gas pressure and beam energy? Determine an expression for the required gas pressure to limit the emittance growth of a proton or ion beam to no more than 1% per hour and evaluate numerical for a proton emittance of 10^{-9} rad-m at an energy of 300 GeV. Is this a problem if the achievable vacuum pressure is 1 nTorr? Concentrating the allowable scattering to one location of 10 cm length (gas jet as a target) in a ring of 4 km circumference, calculate the tolerable pressure of the gas jet.

Problem 8.9.   Consider a long beam transport line made up of FODO cells for a 500 GeV electron beam with an emittance of 10^{-11} rad-m. For a straight, 1 km long beam line determine the FODO cell parameters and tolerance on quadrupole alignment to keep the emittance growth along the beam line to less than 10%.

Problem 8.10.   For future linear electron colliders it may be desirable to provide a switching of the beams from one experimental detector to another. Imagine a linear collider system with two experimental stations separated transversely by 50m. To guide the beams from the linear accelerators to the experimental stations use translating FODO cells and determine the parameters required to keep the emittance growth of a beam to less than 10% (beam emittance 10^{-11} rad-m at 500 GeV).

Problem 8.11.   Use the Fokker-Planck equation and derive an expression for the equilibrium beam emittance of a coupled beam.

9. Particle Beam Parameters

Different applications of particle beams require different and often very specific beam characteristics like cross section, divergence, energy spread or pulse structure. To a large extend such parameters can be adjusted by particular application of focusing and other forces. In this chapter, we will discuss some of these methods of beam optimization and manipulation.

9.1 Particle Distribution in Phase Space

The beam emittance for particle beams is primarily defined by characteristic source parameters and source energy. Given perfect matching between different accelerators and beam lines during subsequent acceleration this source emittance is reduced inversely proportional to the particle momentum by *adiabatic damping* and stays constant in terms of normalized emittance. This describes accurately the situation for proton and ion beams, for nonrelativistic electrons and electrons in linear accelerators.

The beam emittance for relativistic electrons, however, evolves fundamentally different in circular accelerators. Relativistic electron and positron beams passing through bending magnets emit synchrotron radiation, a process that leads to quantum excitation and damping. As a result the original beam emittance at the source is completely replaced by an equilibrium emittance that is unrelated to the source characteristics.

9.1.1 Diffusion Coefficient and Synchrotron Radiation

Emission of a photon causes primarily a change of the particle energy but as a consequence, the characteristics of the particle motion is changed as well. Neither position nor the direction of the particle trajectory is changed during the forward emission of photons along the direction of the particle propagation, ignoring for now the small transverse perturbation due to the finite opening angle of the radiation by $\pm 1/\gamma$. From beam dynamics, however, we know that different reference trajectories exist for particles with different energies. Two particles with energies cp_0 and $cp_1 > cp_0$ follow two different reference trajectories separated at the position s along the beam transport line by a distance

$$\Delta x = \eta(s) \frac{cp_1 - cp_o}{cp_o}, \tag{9.1}$$

where $\eta(s)$ is the dispersion function and cp_o the reference energy. Although particles in general do not exactly follow these reference trajectories they do perform betatron oscillations about them. The sudden change of the particle energy during the emission of a photon consequently leads to a sudden change in the reference path and thereby to a sudden change in the betatron oscillation amplitude.

Following the discussion of the *Fokker-Planck equation* in Chap. 8, we may derive a diffusion coefficient from these sudden changes in the coordinates. Using normalized coordinates $w = x/\sqrt{\beta}$, the change in the betatron amplitude at the moment a photon of energy ϵ_γ is emitted becomes

$$\Delta w = \xi = -\frac{\eta(s)}{\sqrt{\beta_x}} \frac{\epsilon_\gamma}{E_o}. \tag{9.2}$$

Similarly, the conjugate coordinate $\dot{w} = \sqrt{\beta_x}\, x'_\beta + \alpha_x\, x_\beta$ changes by

$$\Delta \dot{w} = \pi = -\sqrt{\beta_x}\, \eta'(s) \frac{\epsilon_\gamma}{E_o} - \frac{\alpha_x}{\sqrt{\beta_x}} \eta(s) \frac{\epsilon_\gamma}{E_o}. \tag{9.3}$$

The frequency at which these statistical variations occur are the same both for ξ and π and is equal to the number of photons emitted per unit time

$$\mathcal{N}_\xi = \mathcal{N}_\pi = \mathcal{N}. \tag{9.4}$$

From (9.2,3) we get

$$\xi^2 + \pi^2 = \left(\frac{\epsilon_\gamma}{E_o}\right)^2 \left[\frac{\eta^2}{\beta_x} + (\sqrt{\beta_x}\, \eta' + \frac{\alpha_x}{\sqrt{\beta_x}} \eta)^2\right] = \left(\frac{\epsilon_\gamma}{E_o}\right)^2 \mathcal{H}, \tag{9.5}$$

where we have defined a special lattice function

$$\boxed{\mathcal{H} = \beta_x \eta'^2 + 2\alpha_x \eta\, \eta' + \gamma_x \eta^2.} \tag{9.6}$$

We are interested in the average value of the total diffusion coefficient (8.95)

$$D = \tfrac{1}{2} \langle \mathcal{N}\,(\xi^2 + \pi^2)\rangle_s = \frac{1}{2\,E_o^2} \langle \mathcal{N}\,\langle \epsilon_\gamma^2 \rangle\, \mathcal{H}\rangle_s, \tag{9.7}$$

where the average $\langle \cdots \rangle_s$ is to be taken along the whole transport line or the whole circumference of a circular accelerator. Since photon emission does not occur outside of bending magnets, the average is taken only along the length of the bending magnets. To account for the large variations in the photon energies, we use the rms value of the photon energies $\langle \epsilon_\gamma^2 \rangle$.

The number of photons emitted per unit time with frequencies between ω and $\omega + d\omega$ is simply the spectral radiation power at this frequency divided by the photon energy $\hbar\omega$ and is from (7.142)

$$\frac{\mathrm{d}n(\omega)}{\mathrm{d}\omega} = \frac{1}{\hbar\omega}\frac{\mathrm{d}P(\omega)}{\mathrm{d}\omega} = \frac{P_\gamma}{\hbar\omega_c^2}\frac{9\sqrt{3}}{8\pi}\int\limits_\zeta^\infty K_{5/3}(x)\,\mathrm{d}x\,,\tag{9.8}$$

where $\zeta = \omega/\omega_c$. The total photon flux is by integration over all frequencies

$$\mathcal{N} = \frac{P_\gamma}{\hbar\omega_c}\frac{9\sqrt{3}}{8\pi}\int\limits_0^\infty\int\limits_\zeta^\infty K_{5/3}(x)\,\mathrm{d}x\,\mathrm{d}\zeta\tag{9.9}$$

which becomes with GR(6.561.16) and $\Gamma(\tfrac{11}{6})\,\Gamma(\tfrac{1}{6}) = 5\pi/3$ after integration by parts from AS(6.1.17)

$$\mathcal{N} = \frac{P_\gamma}{\hbar\omega_c}\frac{9\sqrt{3}}{8\pi}\int\limits_0^\infty K_{5/3}(\zeta)\,\mathrm{d}\zeta = \frac{15\sqrt{3}}{8}\frac{P_\gamma}{\hbar\omega_c}.\tag{9.10}$$

The rms value of the photon energy $\langle\epsilon_\gamma^2\rangle$ can be derived in the usual way from the spectral distribution

$$\langle\epsilon_\gamma^2\rangle = \frac{\hbar^2}{\mathcal{N}}\int\limits_0^\infty\omega^2\,n(\omega)\,\mathrm{d}\omega = \frac{9\sqrt{3}\,P_\gamma\,\hbar\omega_c}{8\pi\,\mathcal{N}}\int\limits_0^\infty\zeta^2\int\limits_\zeta^\infty K_{5/3}(x)\,\mathrm{d}x\,\mathrm{d}\zeta\tag{9.11}$$

and is after integration by parts

$$\langle\epsilon_\gamma^2\rangle = \frac{P_\gamma\,\hbar\omega_c}{\mathcal{N}}\frac{9\sqrt{3}}{8\pi}\frac{1}{3}\int\limits_0^\infty\zeta^3\,K_{5/3}(\zeta)\,\mathrm{d}\zeta\,.\tag{9.12}$$

The integral of the modified Bessel's function in (9.12) is from GR[6.561.16] $4\,\Gamma(2+\tfrac{5}{6})\,\Gamma(2-\tfrac{5}{6})$ where we use again AS(6.1.17) for $\Gamma(\tfrac{5}{6})\,\Gamma(\tfrac{1}{6}) = 2\pi$. Collecting all terms

$$\mathcal{N}\langle\epsilon_\gamma^2\rangle = \frac{55}{24\sqrt{3}}P_\gamma\,\hbar\omega_c\tag{9.13}$$

and the diffusion coefficient (9.7) becomes

$$D = \tfrac{1}{2}\langle\mathcal{N}(\xi^2+\pi^2)\rangle_s = \frac{55}{48\sqrt{3}}\frac{\langle P_\gamma\,\hbar\omega_c\,\mathcal{H}\rangle_s}{E_o^2}.\tag{9.14}$$

The stationary solution for the *Fokker-Planck equation* has been derived describing the equilibrium particle distribution in phase space under the influence of quantum excitation and damping. In all six dynamical degrees of freedom the equilibrium distribution is a gaussian distribution and the standard value of the distribution width is determined by the damping time and the respective diffusion coefficient. In this chapter, we will be able to calculate quantitatively the diffusion coefficients and from that the beam parameters.

9.1.2 Quantum Excitation of Beam Emittance

High energy electron or positron beams passing through a curved beam transport system suffer from quantum excitation which is not compensated by damping since there is no acceleration. In Sect. 8.3.3 we have discussed this effect and found the transverse beam emittance to increase linear with time (8.135) and we get with (9.14)

$$\frac{d\epsilon_x}{c\,dt} = \frac{d\epsilon_x}{ds} = \frac{55}{24\sqrt{3}}\frac{r_e\,\hbar c}{mc^2}\gamma^5\left\langle\frac{\mathcal{H}}{\rho^3}\right\rangle_s. \tag{9.15}$$

There is a strong energy dependence of the emittance increase along the beam transport line and therefore the effect becomes only significant for very high beam energies as proposed for linear collider systems. Since the emittance blow up depends on the lattice function $\mathcal{H}$, we would choose a very strong focusing lattice to minimize the dilution of the beam emittance. For this reason the beam transport system for the linear collider at the *Stanford Linear Accelerator Center* [9.1] is composed of very strongly focusing combined bending magnets.

Particle distributions become modified each time we inject a beam into a circular accelerator with significant synchrotron radiation. Arbitrary particle distributions can be expected from injection systems before injection into a circular accelerator. If the energy in the circular accelerator is too small to produce significant synchrotron radiation the particular particle distribution is preserved according to Liouville's theorem while all particles merely rotate in phase space as discussed in Sect. 8.1. As the beam energy is increased or if the energy is sufficiently high at injection to generate significant synchrotron radiation, all modes in the representation of the initial particle distribution vanish within a few damping times while only one mode survives or builds up which is the gaussian distribution with a standard width given by the diffusion constant and the damping time. In general, any deviation from this unique equilibrium solution and be it only a mismatch to the correct orientation of the beam in phase space will persist for a time not longer than a few damping times.

9.1.3 Horizontal Equilibrium Beam Emittance

In circular accelerators like storage rings damping will counteract quantum excitation leading to an equilibrium. The beam size is related to damping and the diffusion coefficient from (8.107) like

$$\frac{\sigma_x^2}{\beta_x} = \tau_x\,D_x. \tag{9.16}$$

Damping times have been derived in Sect. 8.2 and with (9.7) the horizontal beam size σ_x at a location where the value of the betatron function is β_x becomes

$$\frac{\sigma_x^2}{\beta_x} = \frac{\langle \mathcal{N} \langle \epsilon_\gamma^2 \rangle \mathcal{H} \rangle_s}{2 E_\mathrm{o} J_x \langle P_\gamma \rangle_s}. \tag{9.17}$$

The ratio σ_x^2/β_x is consistent with our earlier definition of the beam emittance. For a particle beam which is in equilibrium between quantum excitation and damping, this ratio is defined as the *equilibrium beam emittance* being equivalent to the beam emittance for all particles within one standard value of the gaussian distribution. For further simplification, we make use of the expression (9.13) and get with the radiation power (7.57) and the critical frequency (7.75) the *horizontal beam emittance*

$$\epsilon_x = C_\mathrm{q} \gamma^2 \frac{\langle \mathcal{H}/| \rho^3 | \rangle_s}{J_x \langle 1/\rho^2 \rangle_s}, \tag{9.18}$$

where we adopted *Sands'* [9.2] definition of a quantum excitation constant

$$C_\mathrm{q} = \frac{55}{32\sqrt{3}} \frac{\hbar c}{mc^2} = 3.84 \cdot 10^{-13}\,\mathrm{m}. \tag{9.19}$$

The equilibrium beam emittance scales like the square of the beam energy and depends further only on the bending radius and the lattice function $\mathcal{H}$. From the definition of $\mathcal{H}$ the horizontal equilibrium beam emittance depends on the magnitude of the dispersion function and can therefore be adjusted to small or large values depending on the strength of the focusing for the dispersion function.

9.1.4 Vertical Equilibrium Beam Emittance

The vertical beam emittance follows from (9.18) considering that the dispersion function and therefore $\mathcal{H}$ vanishes. Consequently, the equilibrium vertical beam emittance seems to be zero because there is only damping but no quantum excitation as in the horizontal plane. In this case, however, we can no longer ignore the fact that the photons are emitted into a finite although small angle about the forward direction of particle propagation. Each such emission causes in addition to a loss in the particle energy also a transverse recoil deflecting the particle trajectory. The photons are emitted typically within an angle $1/\gamma$ generating a transverse kick without changing the betatron oscillation amplitude. With $\delta y = 0$ and $\delta y' = \frac{1}{\gamma} \frac{\epsilon_\gamma}{E_\mathrm{o}}$, we get for the statistical variations

$$\xi^2 = 0,$$
$$\pi^2 = \beta_y \frac{1}{\gamma^2} \left(\frac{\epsilon_\gamma}{E_\mathrm{o}} \right)^2. \tag{9.20}$$

Following a derivation similar to that for the horizontal beam emittance, we get for the *vertical beam emittance*

$$\epsilon_y = C_\mathrm{q} \frac{\langle \beta_y/|\rho^3| \rangle_s}{J_y \langle 1/\rho^2 \rangle_s}. \tag{9.21}$$

This is the fundamentally lower limit of the equilibrium beam emittance due to the finite emission angle of synchrotron radiation. For an isomagnetic ring the *vertical beam emittance*

$$\epsilon_y = C_\mathrm{q} \frac{\langle \beta_y \rangle_s}{J_y |\rho|} \tag{9.22}$$

does not depend on the particle energy but only on the bending radius and the average value of the betatron function. In most practical circular accelerator designs, both the bending radius and the betatron function are of similar magnitude and the fundamental emittance limit therefore is of the order of $C_\mathrm{q} = 10^{-13}$ radian meter, indeed very small compared to actually achieved beam emittances.

The assumption that the vertical dispersion function vanishes in a flat circular accelerator is true only for an ideal ring. Dipole field errors, quadrupole misalignments and any other source of undesired dipole fields create a vertical closed orbit distortion and an associated vertical dispersion function. This vertical dispersion function, often called *spurious dispersion function*, is further modified by orbit correction magnets but it is not possible to completely eliminate it because the location of dipole errors are not known.

Since the diffusion coefficient D is quadratic in the dispersion function (9.7) we get a contribution to the vertical beam emittance from quantum excitation similar to that in the horizontal plane. Indeed, this effect on the vertical beam emittance is much larger than that due to the finite emission angle of photons discussed above and is therefore together with coupling the dominant effect in the definition of the vertical beam emittance.

The contribution to the vertical beam emittance is in analogy to the derivation leading to (9.18)

$$\Delta \epsilon_y = C_\mathrm{q} \gamma^2 \frac{\langle \mathcal{H}_y/|\rho^3| \rangle_s}{J_y \langle 1/\rho^2 \rangle_s}, \tag{9.23}$$

where $\mathcal{H}_y$ is the average value along the ring circumference, i.e.

$$\mathcal{H}_y = \langle \beta_y \eta_y'^2 + 2 \alpha_y \eta_y \eta_y' + \gamma_y \eta_y^2 \rangle_s. \tag{9.24}$$

To minimize this effect orbit correction schemes must be employed which not only correct the equilibrium orbit but also the perturbation to the dispersion function. Of course the same effect with similar magnitude occurs also in the horizontal plane but is in general negligible compared to ordinary quantum excitation.

9.2 Equilibrium Energy Spread and Bunch Length

The statistical processes caused by the emission of synchrotron radiation photons affect not only the four transverse dimensions of phase space but also the energy-time phase space. Particles orbiting in a circular accelerator emit photons with a statistical distribution of energies while only the average energy loss is replaced in the accelerating cavities. This leaves a residual statistical distribution of the individual particle energies which we have derived in Sect. 8.3 to be gaussian just like the transverse particle distribution with a standard width given by (8.114). The conjugate coordinate is the "time" $w = \frac{\Omega}{\eta_c}\tau$ where τ is the deviation in time of a particle from the synchronous particle, and ϵ the energy deviation of a particle from the reference energy E_o.

The emission of a photon will not change the position of the particle in time and therefore $\xi = 0$. The conjugate coordinate being the particle energy will change due to this event by the magnitude of the photon energy and we have $\pi = \epsilon_\gamma/E_o$. Comparing with (9.5), we note that we get the desired result analogous to the transverse phase space by setting $\mathcal{H} = 1$ and using the correct damping time for longitudinal motion. The *equilibrium energy spread* becomes then from (8.114) in analogy to (9.18)

$$\frac{\sigma_\epsilon^2}{E_o^2} = C_q \gamma^2 \frac{\langle |1/\rho^3|\rangle_s}{J_\epsilon \langle 1/\rho^2\rangle_s} \tag{9.25}$$

which in a separated function lattice depends only on the particle energy and the bending radius. In a fully or partially combined function lattice, the partition number J_ϵ can be modified providing a way to vary the energy spread.

There is also a related equilibrium distribution in the longitudinal dimension which defines the length of the particle bunch. This distribution is also gaussian and the standard *bunch length* is from (8.116,117)

$$\sigma_s = c\beta \frac{|\eta_c|}{\Omega_s} \frac{\sigma_\epsilon}{E_o}. \tag{9.26}$$

The equilibrium bunch length not only depends on the particle energy and the bending radius but also on the focusing lattice through the momentum compaction factor and the partition number as well as on rf parameters included in the synchrotron oscillation frequency Ω_s. To exhibit the scaling, we introduce lattice and rf parameters into (9.26) to get with (9.25) and the definition of the synchrotron frequency

$$\Omega_{so}^2 = \omega_{rev}^2 \frac{h\,\eta_c\,e\widehat{V}_o \cos\psi_s}{2\pi\,\beta\,cp_o}$$

an expression for the equilibrium bunch length

$$\sigma_s^2 = \frac{2\pi\,C_\mathrm{q}}{(mc^2)^2}\,\frac{\eta_c\,E_\mathrm{o}^3\,R^2}{J_\epsilon\,h\,e\widehat{V}_\mathrm{o}\,\cos\psi_s}\,\frac{\langle\,|\,1/\rho^3\,|\,\rangle_s}{\langle\,1/\rho^2\,\rangle_s}\,, \tag{9.27}$$

where R is the average radius of the ring. The bunch length can be modified through more parameters than any other characteristic beam parameter in the six-dimensional phase space. Lattice design affects the resulting bunch length through the momentum compaction factor and the partition number. Strong focusing results in a small value for the momentum compaction factor and a small bunch length. Independent of the strength of the focusing, the momentum compaction factor can in principle be adjusted to any value including zero and negative values by allowing the dispersion function to change sign along a circular accelerator because the momentum compaction factor is basically the average of the dispersion function $\alpha_c = \langle\eta/\rho\rangle$. In this degree of approximation the bunch length could therefore be reduced to arbitrarily small values by reducing the momentum compaction factor. However, close to the transition energy phase focusing to stabilize synchrotron oscillations is lost.

Introduction of gradient magnets into the lattice modifies the partition numbers as we have discussed in Sect. 8.2.1. As a consequence, both the energy spread and bunch length increase or decrease at the expense of the opposite effect on the horizontal beam emittance. The freedom to adjust any of these three beam parameters in this way is therefore limited but nonetheless an important means to make adjustments if necessary. Obviously, the rf frequency as well as the rf voltage have a great influence on the bunch length. The bunch length scales inversely proportional to the square root of the rf frequency and is shortest for high frequencies. Generally, no strong arguments exist to choose a particular rf frequency but might become more important if control of the bunch length is important for the desired use of the accelerator. The bunch length is also determined by the rate of change of the rf voltage in the accelerating cavities at the synchronous phase

$$\dot{V}(\psi_s) = \frac{\mathrm{d}}{\mathrm{d}\psi}\,\widehat{V}\,\sin\psi\,\big|_{\psi=\psi_s} = \widehat{V}\,\cos\psi_s\,.$$

For a single frequency rf system the bunch length can be shortened when the rf voltage is increased. To lengthen the bunch the rf voltage can be reduced up to a point where the rf voltage would fail to provide a sufficient energy acceptance.

9.3 Phase-Space Manipulation

The distribution of particles in phase space is given either by the injector characteristics and injection process or in the case of electron beams by the equilibrium of quantum excitation due to synchrotron radiation and damping. The result of these processes are not always what is desired and it is therefore useful to discuss some method to modify the particle distribution in phase space within the validity of Liouville's theorem. Modification of proton and ion distributions in phase space must be done adiabatically and have been discussed in [Ref. 9.3, Chap. 10]. Here we concentrate mainly on the phase-space manipulation of electron beams.

9.3.1 Exchange of Transverse Phase-Space Parameters

In beam dynamics we are often faced with the desire to change the beam size in one of the six phase-space dimensions. Liouville's theorem tells us that this is not possible with macroscopic fields unless we let another dimension vary as well so as not to change the total volume in six-dimensional phase space.

A very simple way of exchanging phase-space dimensions is the increase or decrease of one transverse dimension at the expense of its conjugate coordinate. A very wide and almost parallel beam, for example, can be focused to a small spot size. At the focal point, however, the beam divergence has become very large. Obviously, this process can be reversed too and we describe such a process as the rotation of a beam in phase space or as *phase-space rotation*.

A more complicated but often very desirable exchange of parameters is the reduction of beam emittance in one plane at the expense of the emittance in the other plane. Is it, for example, possible to reduce say the vertical beam emittance to zero at the expense of the horizontal emittance? Although Liouville's theorem would allow such an exchange other conditions in Hamiltonian theory will not allow this kind of exchange in multidimensional phase space. The condition of symplecticity is synonymous with Liouvilles theorem only in one dimension. For n dimensions the symplecticity condition imposes a total of $n(2n-1)$ conditions on the dynamics of particles in phase space [9.4]. These conditions impose an important practical limitation on the exchange of phase space between different degrees of freedom. Specifically, it is not possible to reduce the smaller of two phase-space dimensions further at the expense of the larger emittance, or if the phase space is the same in two dimensions neither can be reduced at the expense of the other.

9.3.2 Exchange of Longitudinal Phase-Space Parameters

So far we have discussed only the exchange of transverse phase-space parameters. Longitudinal phase space can be exchanged also by special application of magnetic and rf fields. Specifically, we often face the problem to compress the bunch to a very short length at the expense of energy spread.

Bunch compression: For linear colliders the following problem exists. The required very small transverse beam emittances can be obtained only in storage rings specially designed for low equilibrium beam emittances. Therefore, an electron beam is injected from a conventional source into a *damping ring* specially designed for low equilibrium beam emittance. After storage for a few damping times the beam is ejected from the damping ring again and transferred to the linear accelerator to be further accelerated. During the damping process in the damping ring, however, the bunch length will also reach its equilibrium value which in practical storage rings is significantly longer than could be accommodated in, for example, an S-band or X-band linear accelerator. The bunch length must be shortened.

This is done in a specially designed beam transport line between the damping ring and linear accelerator consisting of a *nonisochronous transport line* and an accelerating section installed at the beginning of this line (Fig. 9.1).

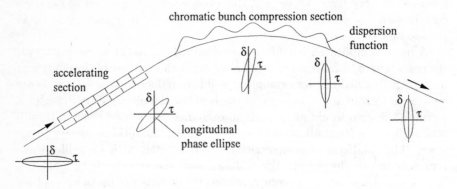

Fig. 9.1. Bunch-compressor system (schematic)

The accelerating section is phased such that the center of the bunch does not see any field while the particles ahead of the bunch center are accelerated and the particles behind the bunch center are decelerated. Following this accelerating section, the particles travel through a curved beam transport system with a finite momentum compaction factor $\alpha_{\mathrm{c}} = \frac{1}{L_{\mathrm{o}}} \int_{\mathrm{o}}^{L_{\mathrm{o}}} \frac{\eta}{\rho}\,\mathrm{d}s$ where L_{o} is the length of the beam transport line. Early particles within a bunch, having been accelerated, follow a longer path than the reference particles in the center of the bunch while the decelerated particles being late with respect to the bunch center follow a short cut. All particles are considered highly relativistic and the early particles fall back toward the bunch center while late particles catch up with the bunch center. If the parameters of the beam transport system are chosen correctly the bunch length reaches its minimum value at the desired location at, for example, the entrance of the linear accelerator. From that point on the phase-space rotation is halted because of lack of momentum compaction in a straight line. Liouville's theorem is not violated because the energy spread in the beam has been increased through the phase dependent acceleration in the *bunch-compression* system.

Formulating this bunch compression in more mathematical terms, we start from a particle distribution in longitudinal phase space described by the phase ellipse

$$\hat{\tau}_o^2 \, \epsilon^2 + \hat{\epsilon}_o^2 \, \tau^2 \; = \; \hat{\tau}_o^2 \, \hat{\epsilon}_o^2 \; = \; a^2 \,, \tag{9.28}$$

where a is the longitudinal emittance and τ is the particle location along the bunch measured from the bunch center. If $\tau > 0$ the particle trails the bunch center.

In the first step of bunch compression, we apply an acceleration

$$\Delta\epsilon \; = \; -eV_o \sin\omega_{\rm rf}\tau \; \approx \; -eV_o \, \omega_{\rm rf}\tau \,. \tag{9.29}$$

The particle energy is changed according to its position along the bunch. Replacing ϵ in (9.28) by $\epsilon + \Delta\epsilon$ and sorting we get

$$\hat{\tau}_o^2 \, \epsilon^2 - 2\hat{\tau}_o^2 \, eV_o \, \omega_{\rm rf} \, \epsilon\tau + (\hat{\tau}_o^2 \, e^2 V_o^2 \, \omega_{\rm rf}^2 + \hat{\epsilon}_o^2) \, \tau^2 \; = \; a^2 \,, \tag{9.30}$$

where the appearance of the cross term indicates the rotation of the ellipse. The second step is the actual bunch compression in the nonisochronous transport line of length L and momentum compaction $\Delta s/L = \eta_c \, \epsilon/(cp_o)$. Traveling though this beam line a particle experiences a shift in time of

$$\Delta\tau \; = \; \frac{\Delta s}{\beta c} \; = \; \frac{L \, \eta_c}{\beta c} \frac{\epsilon}{cp_o} \,. \tag{9.31}$$

Again, the time τ in (9.30) is replaced by $\tau + \Delta\tau$ to obtain the phase ellipse at the end of the bunch compressor of length L. The shortest bunch length occurs when the phase ellipse becomes upright. The coefficient for the cross term must therefore be zero giving a condition for minimum bunch length

$$eV_o \; = \; -\frac{cp_o \, \beta c}{L \, \eta_c \, \omega_{\rm rf}} \,. \tag{9.32}$$

From the remaining coefficient of ϵ^2 and τ^2, we get the bunch length after compression

$$\hat{\tau} = \frac{\hat{\epsilon}_o}{eV_{rf}\,\omega_{rf}} \tag{9.33}$$

and the energy spread

$$\hat{\epsilon} = \hat{\tau}_o\,\omega_{rf}\,eV_{rf}\,, \tag{9.34}$$

where we used the approximation $\hat{\tau}_o\,eV_o\,\omega_{rf} \gg \hat{\epsilon}_o$. This is justified because we must accelerate particles at the tip of the bunch by much more than the original energy spread for efficient bunch compression. Liouville's theorem is obviously kept intact since

$$\hat{\epsilon}\,\hat{\tau} = \hat{\epsilon}_o\,\hat{\tau}_o\,.$$

For tight bunch compression one needs a particle beam with small energy spread and an accelerating section with a high rf voltage and frequency. Of course, the same parameters contribute to the increase of the energy spread which can become the limiting factor in bunch compression. If this is the case, one could compress the bunch as much as is acceptable, then accelerate the beam to higher energies to reduce the energy spread by adiabatic damping, and then go through a bunch compression again.

Alpha magnet: Bunch compression requires two steps: first, an accelerating system must create a correlation between particle energy and position, then we utilize a nonisochronous magnetic transport line to rotate the particle distribution in phase space until the desired bunch length is reached.

The first step can be eliminated in the case of an electron beam generated in an rf gun. Here the electrons emerge from a cathode which is inserted into an rf cavity [9.5]. The electrons are accelerated immediately where the acceleration is a strong function of time because of the rapidly oscillating field. In Fig. 9.2 the result from computer simulations of the particle distribution in phase space [9.6] is shown for an electron beam from a 3 GHz rf gun [9.7, 8] (Fig. 9.3).

For bunch compression we use an *alpha magnet* which got its name from the alpha like shape of the particle trajectories. This magnet is basically a quadrupole split in half where the other half is simulated by a magnetic mirror plate at the vertical midplane. While ordinarily a particle beam would pass through a quadrupole along the axis or parallel to this axis this is not the case in an alpha magnet. The particle trajectories in an alpha magnet have very unique properties which were first recognized by *Enge* [9.9]. Most obvious is the fact that the entrance and exit point is the same for all particles independent of energy. The same is true also for the total deflection angle. *Borland* [9.10] has analyzed the particle dynamics in an alpha magnet in detail and we follow his derivation here. Particles entering the alpha magnet fall under the influence of the Lorentz force

$$\mathbf{F} = -e\mathbf{E} + \frac{e}{c}[\mathbf{v} \times \mathbf{B}]\,, \tag{9.35}$$

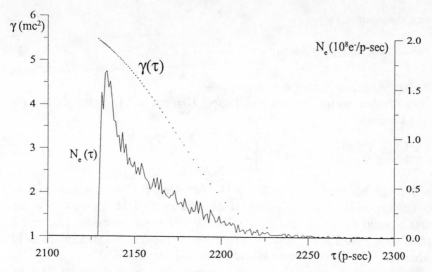

Fig. 9.2. Particle distribution in phase space for an electron beam from an rf gun with thermionic cathode

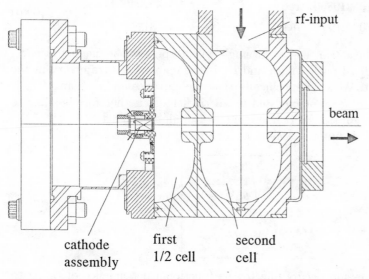

Fig. 9.3. Cross section of an rf gun

where we ignore the electrical field. Replacing the magnetic field by its gradient $\mathbf{B} = (g\,u_3, 0, g\,u_1)$ we get in the coordinate system of Fig. 9.4 the equation of motion,

$$\frac{\mathrm{d}^2\mathbf{u}}{\mathrm{d}s^2} = -\sigma^2 \left[\frac{\mathrm{d}\mathbf{u}}{\mathrm{d}s} \times \mathbf{u} \right], \tag{9.36}$$

where the scaling factor

$$\sigma^2 = \frac{e\,g}{mc^2\,\beta\gamma} = 5.86674 \cdot 10^{-4}\,\frac{g\,(\mathrm{G/cm})}{\beta\gamma}\,\mathrm{cm}^{-2}\,, \tag{9.37}$$

and the coordinate vector $\mathbf{u} = (u_1, u_2, u_3)$.

By introducing normalized coordinates $\mathbf{U} = \sigma\,\mathbf{u}$ and path length $S = \sigma s$ (9.36) becomes

$$\frac{\mathrm{d}^2\mathbf{U}}{\mathrm{d}S^2} = -\left[\frac{\mathrm{d}\mathbf{U}}{\mathrm{d}S} \times (U_3, 0, U_1)\right]\,. \tag{9.38}$$

The remarkable feature of (9.38) is the fact that it does not exhibit any dependence on the particle energy or the magnetic field. One solution for (9.38) is valid for all operating conditions and beam energies. The alpha shaped trajectories are similar to each other and scale with energy and field gradient according to the normalization introduced above.

Equation (9.38) can be integrated numerically and in doing so *Borland* obtains for the characteristic parameters of the normalized trajectory in an alpha magnet[9.10]

$$
\begin{aligned}
\theta_\alpha &= 0.7105219800 \quad \mathrm{rad}\,, & S_\alpha &= 4.6420994651\,, \\
& 40.709910710 \quad \mathrm{deg}\,, & \widehat{U}_1 &= 1.8178171151\,,
\end{aligned} \tag{9.39}
$$

where θ_α is the entrance and exit angle with respect to the magnet face, S_α is the normalized path length and $\widehat{U}_1$ is the apex of the trajectory in the alpha magnet. We note specifically that the entrance and exit angle θ_α is independent of beam energy and magnetic field. It is therefore possible to construct a beam transport line including an alpha magnet.

Upon introducing the scaling factor (9.37), (9.39) becomes

$$
\begin{aligned}
s_\alpha(\mathrm{cm}) &= \frac{S_\alpha}{\sigma} = 191.655\,\sqrt{\frac{\beta\gamma}{g\,(\mathrm{G/cm})}}\,, \\[2mm]
\hat{u}_1(\mathrm{cm}) &= \frac{\widehat{U}_1}{\sigma} = 75.0513\,\sqrt{\frac{\beta\gamma}{g\,(\mathrm{G/cm})}}\,.
\end{aligned} \tag{9.40}
$$

Bunch compression occurs due to the functional dependence of the path length on the particle energy. Taking the derivative of (9.40) with respect to the particle momentum $\tilde{p}_\mathrm{o} = \beta\gamma$, one gets the compression

$$\frac{\mathrm{d}s_\alpha(\mathrm{cm})}{\mathrm{d}\tilde{p}_\mathrm{o}} = \frac{75.0513}{\sqrt{2\,g\,(\mathrm{G/cm})\,\tilde{p}_\mathrm{o}}}\,. \tag{9.41}$$

For bunch compression $\mathrm{d}s_\alpha < 0$, higher momentum particles must arrive first because they follow a longer path and therefore fall back with respect

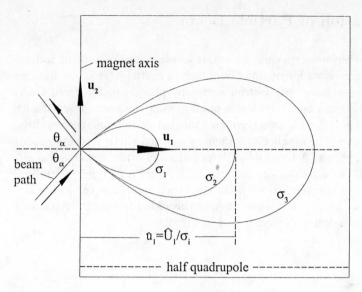

Fig. 9.4. Alpha magnet and particle trajectories

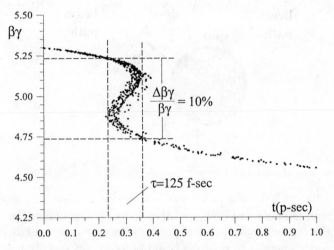

Fig. 9.5. Particle distribution in longitudinal phase space after compression in an alpha magnet

to later particles. For example, an electron beam with the phase-space distribution from Fig. 9.2 becomes compressed as shown in Fig. 9.5.

Because of the small longitudinal emittance of the beam it is possible to generate very short electron bunches of some 50 f-sec (rms) duration which can be used to produce intense coherent far infrared radiation [9.11].

9.4 Polarization of Particle Beam

For high energy physics experimentation, it is sometimes important to have beams of transversely or longitudinally polarized particles. It is possible, for example, to create *polarized electron beams* by photoemission from GaAs cathodes [9.12] From a beam dynamics point of view, we are concerned with the transport of polarized beams through a magnet system and the resulting polarization status. The magnetic moment vector of a particle rotates about a magnetic field vector. A longitudinally polarized electron traversing a vertical dipole field would therefore experience a rotation of the longitudinal polarization about the vertical axis. On the other hand, the vertical polarization would not be affected while passing through a horizontally bending magnet. This situation is demonstrated in Fig. 9.6.

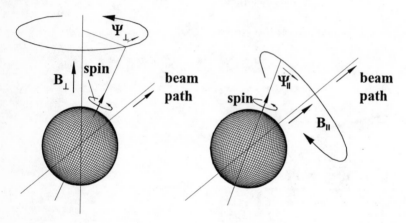

Fig. 9.6. Precession of the particle spin in a transverse or longitudinal magnetic field

Similarly, a longitudinal polarization is not affected by a solenoid field. In linear collider facilities, specific *spin rotators* are introduced to manipulate the electron spin in such a way as to preserve beam polarization and obtain the desired spin direction at the collision point. For the preservation of beam polarization, it is important to understand and formulate *spin dynamics*.

Electron and positron beams circulating for a long time in a storage ring can become polarized due to the reaction of continuous emission of transversely polarized synchrotron radiation. The evolution of the polarization has been studied in detail by several researchers [9.13 – 16] and the *polarization time* is given by [9.16]

$$\frac{1}{\tau_{\text{pol}}} = \frac{5\sqrt{3}}{8} \frac{e^2 \hbar \gamma^5}{m^2 c^2 \rho^3} \qquad (9.42)$$

and the theoretically maximum achievable polarization is 92.38 %. The polarization time is a strong function of beam energy and is very long for low energies. At energies of several GeV, however, this time becomes short compared to the storage time of an electron beam in a storage ring.

This build up of polarization is counteracted by nonlinear magnetic field errors which cause precession of the spin depending on the betatron amplitude and energy of the particle thus destroying polarization. Again, we must understand spin dynamics to minimize this depolarization. Simple relations exist for the rotation of the spin while the particle passes through a magnetic field. To rotate the spin by a magnetic field, there must be a finite angle between the spin direction and that of the magnetic field. The *spin rotation angle* about the axis of a transverse field depends on the angle between the spin direction $\boldsymbol{\sigma}_s$ with $|\boldsymbol{\sigma}_s| = 1$ and magnetic field $\mathbf{B}_\perp$ and is given by [9.15]

$$\Psi_\perp = C_\perp \left(1 + \frac{1}{\gamma^2}\right) |[\boldsymbol{\sigma}_s \times \mathbf{B}_\perp]| \ell, \qquad (9.43)$$

where

$$\eta_{\text{g}} = \frac{g-2}{2} = 0.00115965, \qquad (9.44)$$

$$C_\perp = \frac{e\,\eta_{\text{g}}}{mc^2} = 0.068033 \,(\text{kG}^{-1}\,\text{m}^{-1}), \qquad (9.45)$$

g the *gyromagnetic constant* and $B_\perp \ell$ the integrated transverse magnetic field strength. Apart from a small term $1/\gamma$, the spin rotation is independent of the energy. In other words, a spin component normal to the field direction can be rotated by 90° while passing though a magnetic field of 23.09 kG m. It is therefore not important at what energy the spin is rotated.

Equation (9.43) describes the situation in a flat storage ring with horizontal bending magnets only unless the polarization of the incoming beam is strictly vertical. Any horizontal or longitudinal polarization component would precess while the beam circulates in the storage ring. As long as this *spin precession* is the same for all particles the polarization would be preserved. Unfortunately, the small energy dependence of the precession angle and the finite energy spread in the beam would wash out the polarization. On the other hand the vertical polarization of a particle beam is preserved in an ideal storage ring. Field errors, however, may introduce a depolarization effect. Horizontal field errors from misalignments of magnets, for example, would rotate the vertical spin. Fortunately, the integral of all horizontal field components in a storage ring is always zero along the closed orbit and the net effect on the vertical polarization is zero. Nonlinear fields, however, do not cancel and must be minimized to preserve the polarization.

A transverse spin can also be rotated about the longitudinal axis of a solenoid field and the rotation angle is

$$
\boxed{
\begin{aligned}
\Psi_\parallel &= \frac{e}{E}\left(1 + \eta_{\mathrm g}\frac{\gamma}{1+\gamma}\right)\,|[\boldsymbol\sigma_s \times \mathbf B_\parallel]|\ell \\
&\approx \frac{C_\parallel}{E}\left(1 + \eta_{\mathrm g}\frac{\gamma}{1+\gamma}\right)\,|[\boldsymbol\sigma_s \times \mathbf B_\parallel]|\ell,
\end{aligned}}
\tag{9.46}
$$

where

$$
C_\parallel = e = 0.02997925 \left(\frac{\mathrm{GeV}}{\mathrm{kG\,m}}\right).
\tag{9.47}
$$

In a solenoid field it is therefore possible to rotate a horizontal polarization into a vertical polarization, or vice versa. Spin rotation in a longitudinal field is energy dependent and desired spin rotations should therefore be done at low energies if possible.

The interplay between rotations about a transverse axis and the longitudinal axis is responsible for a *spin resonance* which destroys whatever beam polarization exists. To show this we assume a situation where the polarization vector precesses just by 2π, or an integer multiple n thereof, while the particle circulates once around the storage ring. In this case $\psi_\perp = n \cdot 2\pi$, $eB_\perp\ell/E = 2\pi$, and we get from (9.43)

$$
n = \eta_{\mathrm g}\,(1+\gamma).
\tag{9.48}
$$

For $n = 1$, resonance occurs at a beam energy of $E = 440.14$ MeV. At this energy any small longitudinal field interacts with the polarization vector at the same phase, eventually destroying any transverse polarization. This resonance occurs at equal energy intervals of

$$
E_n(\mathrm{MeV}) = 440.14 + 440.65\,(n-1)
\tag{9.49}
$$

and can be used in storage rings as a precise energy calibration.

In Fig. 9.7 spin dynamics is shown for the case of a linear collider where a longitudinally polarized beam is desired at the collision point. A longitudinally polarized beam is generated from a source and accelerated in a linear accelerator. No rotation of the polarization direction occurs because no magnetic field are involved yet. At some medium energy the beam is transferred to a damping ring to reduce the beam emittance. To preserve polarization in the damping ring the polarization must be vertical. In Fig. 9.7, we assume a longitudinal polarized beam coming out of the linear accelerator. A combination of transverse fields, to rotate the longitudinal into a horizontal spin, followed by a solenoid field which rotates the horizontal

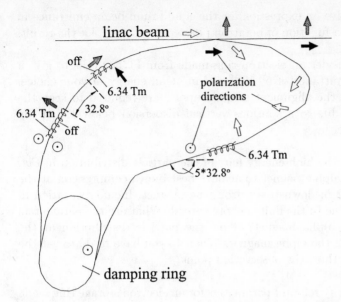

Fig. 9.7. Spin manipulation during beam transfer from linear accelerator to damping ring and back

into a vertical spin, is used in the transport line to the damping ring to obtain the desired vertical spin orientation. This orientation is in line with all magnets in the damping ring and the polarization can be preserved.

To obtain the desired rotation in the beam transport magnets at a given energy, the beam must be deflected by a specific deflection angle which is from (9.43)

$$\theta = \frac{e}{\beta E} B_\perp \ell = \frac{\Psi_\perp}{\eta_g} \frac{1}{1+\gamma}. \tag{9.50}$$

Coming out of the damping ring the beam passes through a combination of two solenoids and two transverse field sections. Depending on which solenoid is turned on, we end up with a longitudinal or transverse polarization at the entrance of the linac. By the use of both solenoids any polarization direction can be realized.

Problems

Problem 9.1. Show that the horizontal damping partition number is negative in a fully combined function FODO lattice as employed in older synchrotron accelerators. Why, if there is horizontal antidamping in such synchrotrons, is it possible to retain beam stability during acceleration? What happens if we accelerate a beam and keep it orbiting in the synchrotron at some higher energy?

Problem 9.2. Derive an expression for the equilibrium beam emittance in a FODO lattice as a function of betatron phase per cell and plot the result.

Problem 9.3. Consider an electron ring made from FODO cells to give a horizontal beam emittance of $5 \cdot 10^{-9}$ rad-m at an energy of your choice. What tolerance on the alignment of quadrupoles are required to keep the vertical emittance due to a spurious vertical dispersion below 1% of the horizontal emittance.

Problem 9.4. Use the high energy part of the particle distribution in Fig. 9.2 and specify an alpha magnet to achieve best bunch compression at the observation point 2 m downstream from the magnet. Include variation of particle velocities due to the finite energy spread. What are the parameters of the loop in the alpha magnet? Show the particle distribution at the entrance and exit of the alpha magnet. What do you have to do to get the shortest bunch length at the observation point?

Problem 9.5. Specify relevant parameters for an electron storage ring made of FODO cells with the goal to produce a very short equilibrium bunch length of $\sigma_\ell = 1$ mm. Use superconducting cavities for bunch compression with a maximum field of 10 MV/m and a total length of not more than 10% of the ring circumference.

Problem 9.6. Describe spin rotations in matrix formulation.

Problem 9.7. Consider an electron storage ring for an energy of 30 GeV and a bending radius of $\rho = 500$ m and calculate the polarization time. The vertical polarization will be perturbed by spurious dipole fields. Use statistical methods to calculate the rms change of polarization direction per unit time and compare with the polarization time. Determine the alignment tolerances to get a polarized beam.

10. Collective Phenomena

Transverse and longitudinal beam dynamics as discussed in earlier chapters is governed by purely single-particle effects where the results do not depend on the presence of other particles or any interactive environment. Space-charge effects were specifically excluded to allow the detailed discussion of *single-particle dynamics*. This restriction is sometimes too extreme and collective effects must be taken into account where significant beam intensities are desired. In most applications high beam intensities are desired and it is therefore prudent in particular cases to test for the appearance of space charge and intensity effects.

Collective effects can be divided into two distinct groups according to the physics involved. The compression of a large number of charged particles into a small volume increases the probability for collisions of particles within the same beam. Because particles perform synchrotron (or move at different velocities for coasting beams) and betatron oscillations statistical collisions occur in longitudinal, as well as transverse phase space; the other group of collective effects includes effects which are associated with electromagnetic fields generated by the collection of all particles in a beam.

The study and detailed understanding of the cause and nature of *collective effects* or *collective instabilities* is important for a successful design of the accelerator including corrective measures. Most accelerator design and developments are conducted to eliminate collective effects as much as possible through self-imposed limitation on the performance or installation of feedback systems and other stabilizing control mechanisms. Beyond that, we also must accept limitations in beam performance imposed by nature or lack of understanding and technological limits. Pursuit of accelerator physics is the attempt to explore and push such limits as far as nature allows.

10.1 Statistical Effects

Coupling of individual particles to the presence of other particles may occur through very short range forces in collisions with each other or through longer range forces caused by electromagnetic fields originating from the electric charge of particle beams. In this section, we will discuss two important statistical effects originating from collision processes within a particle bunch.

10.1.1 Schottky Noise

Electrical current is established by moving charged particles. The finite electrical charge of elementary particles gives rise to statistical variations of the electrical current. This phenomenon has been observed and analyzed by *Schottky* [10.1] and we will discuss this *Schottky noise* in the realm of particle dynamics in circular accelerators. The information included in the Schottky noise is of great diagnostic importance for the nondestructive determination of particle beam parameters, a technique which has been developed at the ISR [10.2] and has become a standard tool of beam diagnostics.

We consider a particle k with charge q orbiting in an accelerator with the angular revolution frequency ω_k and define a particle line density by $2\pi\bar{R}\lambda(t) = 1$ where $2\pi\bar{R}$ is the circumference of the ring. On the other hand, we may describe the orbiting particle by delta functions

$$q = q \int_0^{2\pi} \sum_{m=-\infty}^{+\infty} \delta(\omega_k t + \theta_k - 2\pi m)\,\mathrm{d}\theta\,,$$

where ω_k is the angular revolution frequency of the particle k and θ_k its phase at time $t = 0$. The delta function can be expressed by a Fourier series and the line-charge density at time t becomes

$$q\lambda_k(t) = \frac{q}{2\pi\bar{R}}\left[1 + 2\sum_{n=0}^{\infty}\cos(n\omega_k t + n\theta_k)\right]\,. \tag{10.1}$$

From a pick up electrode close to the circulating particle, we would obtain a signal with a frequency line spectrum $\omega = n\omega_k$ where n is an integer. In a real particle beam there are many particles with a finite spread of revolution frequencies ω_k and therefore the harmonic lines $n\omega_k$ spread out proportionally to n. For not too high harmonic numbers the frequency spreads do not yet overlap and we are able to measure the distribution of revolution frequencies. Tuning the spectrum analyzer to ω, we observe a signal with an amplitude proportional to $N(\omega/n)\frac{\delta\omega}{n}$ where $N(\omega/n)$ is the particle distribution in frequency space and $\delta\omega$ the frequency resolution of the spectrum analyzer. The signal from the pick up electrode is proportional to the line-charge density which is at the frequency ω from (10.1)

$$q\lambda_{\mathrm{rms}}(\omega) = \frac{\sqrt{2}\,q}{2\pi\bar{R}}\sqrt{N(\omega/n)\frac{\delta\omega}{n}} \tag{10.2}$$

and has been derived first by Schottky for a variety of current sources [10.1]. The spread in the revolution frequency originates from a momentum spread in the beam and measuring the *Schottky spectrum* allows its nondestructive determination.

Individual particles orbiting in an accelerator perform transverse betatron oscillations which we describe, for example, in the vertical plane by

$$y_k(t) = a_k \cos(\nu_k \omega_k t + \psi_k),$$ (10.3)

where a_k is the amplitude and ψ_k the phase of the betatron oscillation for the particle k at time $t = 0$. The difference signal from two pick up electrodes above and below the particle beam is, in linear approximation, proportional to the product of the betatron amplitude (10.3) and the line-charge density (10.1) and of the form

$$
\begin{aligned}
D_k(t) &= A_k \sum_{n=0}^{\infty} \cos[(n - \nu_k)(\omega_k t + \phi_k)] \\
&+ A_k \sum_{n=0}^{\infty} \cos[(n + \nu_k)(\omega_k t + \varphi_k)],
\end{aligned}
$$ (10.4)

where we have ignored terms at frequencies $n\omega_k$. The *transverse Schottky signal* is composed of two side bands for each harmonic at frequencies

$$\omega = (n \pm \nu_k)\,\omega_k.$$ (10.5)

which are also called the *slow wave* for $\omega = (n - \nu_k)\omega_k$ and the *fast wave* for $\omega = (n + \nu_k)\omega_k$.

The longitudinal Schottky noise depends on the rms contribution of all particles which are spread over a range of revolution frequencies due to a momentum spread and over betatron frequencies by virtue of the chromaticity. For $\Delta\omega_{\mathrm{rms}} = \eta_c \omega_o \delta_{\mathrm{rms}}$ and $\Delta\nu_{\mathrm{rms}} = \xi_y \delta_{\mathrm{rms}}$ where ω_o is the revolution frequency of the bunch center, $\delta_{\mathrm{rms}} = \Delta p_{\mathrm{rms}}/p_o$ the rms relative momentum error, η_c the momentum compaction and ξ_y the vertical chromaticity, the frequency distribution of the signal from the pick up is

$$
\begin{aligned}
\omega &= [n \pm (\nu_{yo} + \xi_y \delta_k)] \,(\omega_o + \eta_c \omega_o \delta_k) \\
&= (n \pm \nu_{yo})\,\omega_o + [(n \pm \nu_{yo})\,\eta_c \pm \xi_y]\,\omega_o\,\delta_k + \mathcal{O}(\delta^2).
\end{aligned}
$$ (10.6)

The momentum spread δ_k causes a frequency spread which is different for the slow and fast wave. In case of a positive chromaticity and above transition, for example, $\eta_c < 0$ and the frequency spreads add up for the slow wave and cancel partially for the fast wave. This has been verified experimentally for a coasting proton beam in the ISR [10.2].

A transverse Schottky scan may exhibit the existence of weak resonances which can dilute the particle density, specifically in a coasting proton or ion beam. To control coasting beam instabilities which we will discuss later in this chapter, it is desirable to make use of *Landau damping* by introducing a large momentum and tune spread. This tune spread, however, can be sufficiently large to spread over high order resonances and blow up that part of the beam which oscillates at those resonance frequencies. A Schottky scan can clearly identify such a situation as reported in [10.2].

In this text we are able to touch only the very basics of Schottky noise and the interested reader is referred to references [10.2 – 6] for more detailed discussions on the theory and experimental techniques to obtain Schottky scans and how to interpret the signals.

10.1.2 Stochastic Cooling

The "noise" signal from a circulating particle beam includes information which can be used to drive a feedback system in such a way as to reduce the beam emittance, longitudinal as well as transverse. Due to the finite number of particles in a realistic particle beam, the instantaneous center of a beam at the location of a pick up electrode exhibits statistical variations. This statistical displacement of a slice of beam converts to a statistical slope a quarter betatron wavelength downstream. The signal from the small statistical displacement of the beam at the pick up electrode can be amplified and fed back to the beam through a kicker magnet located an odd number of quarter wavelength downstream, assuming that the statistical variations do not smear out between pick up electrode and kicker. *Van der Meer* [10.7] proposed this approach to reduce the transverse proton beam emittance in ISR for increased luminosity and the process is now known as *stochastic cooling*.

This procedure is not a statistical process and we must ask ourselves if this is an attempt to circumvent Liouville's theorem. It is not. Due to the finite number of particles in the beam, the phase space is not uniformly covered by particles but rather exhibits many holes. The method of stochastic cooling detects the moment one of these holes appears on one or the other side of the beam in phase space. At the same moment, the whole emittance is slightly shifted with respect to the center of the phase space and this shift can be both detected and corrected. The whole process of *stochastic cooling* therefore only squeezes the "air" out of the particle distribution in phase space. The most prominent application of this method occurs in the cooling of an antiproton beam to reach a manageable beam emittance for injection into high energy proton antiproton colliders. To discuss this process in more detail, theoretically as well as technically, would exceed the scope of this text and the interested reader is referred to a series of articles published in [10.8].

10.1.3 Touschek Effect

The concentration of many particles into small bunches increases the probability for elastic collisions between particles. This probability is further enhanced considering that particles perform transverse betatron as well as longitudinal synchrotron oscillations. In each degree of freedom, we have limits to the oscillation amplitudes and if the amplitude of particles exceeds such limits due, for example, to a collision with another particle one or both particles can get lost. In this section, we discuss the process of single collisions where the momentum transfer is large enough to lead to the loss of both particles involved in the collision and postpone the discussion of multiple collisions with small momentum transfer to the next section.

We may consider two collision processes which could lead to beam loss. First, we observe two particles performing synchrotron oscillations and colliding head-on in such a way that they transfer their longitudinal momentum into transverse momentum. This collision process is insignificant in particle accelerators because the longitudinal motion includes not enough momentum to increase the betatron oscillations amplitude significantly for particle loss. On the other hand, transverse oscillations of particles include large momenta and a transfer into longitudinal momenta can lead to the loss of both particles. This effect was discovered on the first electron storage ring ever constructed [10.9, 10] and we therefore call this the *Touschek effect*.

In this text, we will not pursue a detailed derivation of the collision process and refer the interested reader to, for example, [10.11 – 13]. Of particular interest is the expression for the beam lifetime as a result of particle losses due to a momentum transfer into the longitudinal phase space exceeding the rf bucket acceptance of $\Delta p/p_0|_{\mathrm{rf}}$. Everytime such a transfer occurs both particles involved in the collision are lost. The beam decay rate is proportional to the number of particles in the bunch and the beam current therefore decays exponentially. Last, but not least, a loss occurs only if there is sufficient momentum in the transverse motion to exceed the rf momentum acceptance. We assume the momentum acceptance to be limited by the rf voltage although it could be limited by other reasons like a small aperture where the dispersion is large. Combining these parameters in a collision theory results in a beam lifetime for a beam with gaussian particle distribution given by

$$\boxed{\frac{1}{\tau} = -\frac{1}{N_{\mathrm{b}}}\frac{\mathrm{d}N_{\mathrm{b}}}{\mathrm{d}t} = \frac{r_{\mathrm{c}}^2 \, c \, N_{\mathrm{b}}}{8\pi \, \sigma_x \, \sigma_y \, \sigma_\ell} \frac{\lambda^3}{\gamma^2} D(\epsilon),} \tag{10.7}$$

where r_{c} is the classical particle radius, $\sigma_x, \sigma_y, \sigma_\ell$ are the standard values of the gaussian bunch width, height and length, respectively, and $\lambda^{-1} = \Delta p/p_0|_{\mathrm{rf}}$ the momentum acceptance parameter. The function $D(\epsilon)$ (Fig. 10.1) is defined by [10.13]

$$D(\epsilon) = \sqrt{\epsilon}\left[-\tfrac{3}{2}\mathrm{e}^{-\epsilon} + \frac{\epsilon}{2}\int_\epsilon^\infty \frac{\ln u}{u}\mathrm{e}^{-u}\,\mathrm{d}u \right.$$
$$\left. + \tfrac{1}{2}(3\epsilon - \epsilon\ln\epsilon + 2)\int_\epsilon^\infty \frac{\mathrm{e}^{-u}}{u}\,\mathrm{d}u\right], \tag{10.8}$$

where the argument is

$$\epsilon = \left(\frac{\Delta p_{\mathrm{rf}}}{\gamma \sigma_p}\right)^2 \quad \text{with} \quad \sigma_{\mathrm{p}} = \frac{mc\gamma \, \sigma_x}{\beta_x}. \tag{10.9}$$

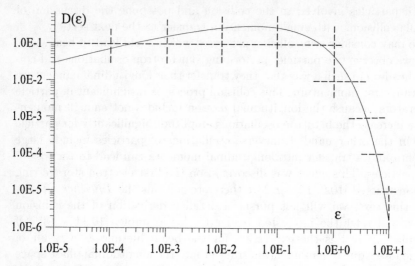

Fig. 10.1. Touschek lifetime function $D(\epsilon)$

Particle losses due to the Touschek effect is particularly effective at low energies and where the rf acceptance is small. For high particle densities $N_b/(\sigma_x\,\sigma_y\,\sigma_\ell)$ the rf acceptance should therefore be maximized. This seems to be the wrong thing to do because the bunch length is reduced at the same time and the particle density becomes even higher but a closer look at (10.7) shows us that the Touschek lifetime increases faster with rf acceptance than it decreases with bunch length.

10.1.4 Intra-Beam Scattering

The Touschek effect describes collision processes which lead to immediate loss of both colliding particles. In reality, however, there are many other collisions with only small exchanges of momentum. While these collisions do not lead to immediate particle loss, there might be sufficiently many during a damping time in electron storage rings or during the storage time for proton and ion beams to cause a significant increase in the bunch volume, or in the case of a coasting beam an increase in beam cross section. During the discussion of the Touschek effect we neglected the transfer from the longitudinal momentum space into transverse momentum space because the transverse momentum acceptance is larger than the longitudinal acceptance and particles are not lost during such an exchange. This is not appropriate any more for the *multiple Touschek effect* or *intra-beam scattering* where we are interested in all collisions.

The multiple Touschek effect was observed in the first ever constructed storage ring AdA (Anello di Accumulatione) in Frascati, Italy. The Touschek effect had been expected and analyzed before but did give too pessimistic beam lifetimes compared to those observed in AdA. A longer beam lifetime had been obtained because of multiple elastic scattering between particles increasing the bunch volume and thereby reducing the Touschek effect [10.10].

During the exchange of momentum as a consequence of collisions between particles within the same bunch or beam each degree of freedom can increase its energy or temperature because the beam is able to absorb any amount of energy from the rf system. We are particularly interested in the growth times of transverse and longitudinal emittances to asses the longterm integrity of the particle beam. The *multiple Touschek effect* or *intra-beam scattering* has been studied extensively [10.14 – 16] and we will not repeat here the derivations but merely recount the results.

The growth time of the beam emittances for gaussian particle distributions are [10.14,15] for the longitudinal phase space or momentum and bunch distribution

$$\tau_p^{-1} = \frac{1}{2\sigma_p^2} \frac{\mathrm{d}\sigma_p^2}{\mathrm{d}t} = A \frac{\sigma_h^2}{\sigma_p^2} f(a,b,c),\tag{10.10}$$

where the particle bunch density is expressed by

$$A = \frac{r_c^2 \, c \, N_b}{64\pi^2 \sigma_s \, \sigma_p \, \sigma_{x_\beta} \, \sigma_{y_\beta} \, \sigma_{x'} \, \sigma_{y'} \, \beta^3 \, \gamma^4}\tag{10.11}$$

with the standard dimensions of a gaussian distribution for the bunch length σ_s, the relative momentum spread σ_p, horizontal and vertical betatron amplitudes σ_{x_β} and σ_{y_β} as well as horizontal and vertical divergences $\sigma_{x'}$ and $\sigma_{y'}$. N_b is the number of particles per bunch, which must be replaced by $2\sqrt{\pi}\sigma_p/C$ for a coasting beam in an accelerator of circumference C. The constants r_c and $\beta = v/c$, finally, are the classical particle radius and particle velocity in units of the velocity of light.

The function

$$f(a,b,c) = 8\pi \int\limits_0^1 \left\{ \ln\left[\frac{c^2}{2} \left(\frac{1}{\sqrt{p}} + \frac{1}{\sqrt{q}} \right) \right] - 0.577.. \right\} \frac{1 - 3x^2}{\sqrt{pq}} \, \mathrm{d}x\tag{10.12}$$

where

$$p = a^2 + x^2(1 - a^2), \qquad q = b^2 + x^2(1 - b^2),$$

$$a = \frac{\sigma_h}{\gamma\sigma_{x'}}, \qquad b = \frac{\sigma_h}{\gamma\sigma_{y'}},$$

$$\sigma_h^2 = \frac{\sigma_p^2 \, \sigma_{x_\beta}^2}{\sigma_{x_\beta}^2 + \eta^2 \, \sigma_p^2}, \qquad c^2 = \beta^2 \sigma_h^2 \frac{\sqrt{2\pi}\sigma_{y_\beta}}{r_c}.$$

The transverse emittance growth times are similarly given by

$$\tau_x^{-1} = \frac{1}{2\sigma_{x_\beta}^2} \frac{\mathrm{d}\sigma_{x_\beta}^2}{\mathrm{d}t} = A \left[f(1/a, b/a, c/a) + \frac{\eta^2 \sigma_p^2}{\sigma_{x_\beta}} f(a,b,c) \right],\tag{10.13}$$

and

$$\tau_y^{-1} = \frac{1}{2\sigma_{y\beta}^2} \frac{\mathrm{d}\sigma_{y\beta}^2}{\mathrm{d}t} = A f(\frac{1}{b}, \frac{a}{b}, \frac{c}{b}).$$ (10.14)

These expressions allow the calculations of the emittance growth rate, which for electron accelerators are in most cases negligible compared to radiation damping but become significant in proton and ion storage rings where high particle densities and long storage times are desired. From the density factor A it is apparent that high particle density in six-dimensional phase space increases the growth rates while this effect is greatly reduced at higher beam energies.

10.2 Collective Self Fields

The electric charge of a particle beam can become a major contribution to the forces encountered by individual particles while travelling along a beam transport line or orbiting in a circular accelerator. These forces may act directly from beam to particle or may originate from electromagnetic fields excited by the particle bunch interacting with its surrounding vacuum chamber. In this section, we will derive expressions for the fields from a collection of particles and determine the force due to these fields on an individual test particle. We use the particle charge q rather than the elementary charge e to cover particles with multiple charges like ions for which we have $q = eZ$. For all cases to be correct, we should distinguish between the electrical charge of particles in the beam and that of the individual test particle. This, however, would significantly complicate the expressions and we use therefore the same charge for both the beam and test particles. In a particular situation whenever particles of different charges are considered, the sign and value of the charge factors in the formulas must be reconsidered.

10.2.1 Transverse Self Fields

Expressions for *space-charge fields* originating from a beam of charged particles have been derived in [Ref. 10.17, Sect. 12.1.1] and we obtained for a gaussian transverse distribution of particles with charge q the electric fields

$$
\begin{aligned}
E_x &= \left[\frac{1}{4\pi\epsilon_o} \right] \frac{2q\lambda}{\sigma_x(\sigma_x + \sigma_y)} x \\
E_y &= \left[\frac{1}{4\pi\epsilon_o} \right] \frac{2q\lambda}{\sigma_y(\sigma_x + \sigma_y)} y,
\end{aligned}
$$ (10.15)

and the magnetic fields

$$B_x = - \left[\frac{c\,\mu_\mathrm{o}}{\sqrt{4\pi}} \right] \frac{2\,q\,\lambda\,\beta}{\sigma_y\,(\sigma_x + \sigma_y)}\,y$$

$$B_y = + \left[\frac{c\,\mu_\mathrm{o}}{\sqrt{4\pi}} \right] \frac{2\,q\,\lambda\,\beta}{\sigma_x\,(\sigma_x + \sigma_y)}\,x\,. \qquad (10.16)$$

The local linear particle density λ is defined by

$$\lambda(z) = \int\!\!\int \rho(x,y,z)\mathrm{d}x\mathrm{d}y\,, \qquad (10.17)$$

where $\rho(x,y,z)$ is the local particle density normalized to the total number of particles in the beam $\int_{-\infty}^{\infty} \lambda(z)\mathrm{d}z = N_\mathrm{p}$. With these fields and the Lorentz equation, we formulate the transverse force acting on a single particle within the same particle beam. Since both expressions for the electrical and magnetic field differ only by the factor β we may, for example, derive from the Lorentz equation the vertical force on a particle with charge q

$$F_y = q\,(1 - \beta^2)\,E_y = \left[\frac{1}{4\pi\epsilon_\mathrm{o}} \right] \frac{2q^2\,\lambda}{\gamma^2\,\sigma_y\,(\sigma_x + \sigma_y)}\,y\,, \qquad (10.18)$$

where $\gamma = 1/\sqrt{1 - \beta^2}$.

The space-charge force appears at its strongest for nonrelativistic particles and diminishes quickly like $1/\gamma^2$ for relativistic particles. In accelerator physics, however, particle beams are carried from low to high energies and therefore space-charge effects may become important during some or all phases of acceleration. This is specifically true for heavy particles like protons and ions for which the relativistic parameter γ is rather low for most any practically achievable particle energies.

The radial fields for a round beam with radius r_o and uniform transverse particle density are from (10.15,16)

for $\quad r \le r_\mathrm{o} \qquad \begin{cases} E_r(z) = \left[\frac{1}{4\pi\epsilon_\mathrm{o}} \right] 2\,q\,\lambda(z)\,\frac{r}{r_\mathrm{o}^2} \\[2mm] B_\varphi(z) = - \left[\frac{c\,\mu_\mathrm{o}}{\sqrt{4\pi}} \right] 2\,q\,\lambda(z)\,\beta\,\frac{r}{r_\mathrm{o}^2}\,. \end{cases} \qquad (10.19)$

Similarly

for $\quad r \ge r_\mathrm{o} \qquad \begin{cases} E_r(z) = \left[\frac{1}{4\pi\epsilon_\mathrm{o}} \right] 2\,q\,\lambda(z)\,\frac{1}{r} \\[2mm] B_\varphi(z) = - \left[\frac{c\,\mu_\mathrm{o}}{\sqrt{4\pi}} \right] 2\,q\,\lambda(z)\,\beta\,\frac{1}{r}\,. \end{cases} \qquad (10.20)$

For $r \gg r_\mathrm{o}$ (10.20) is also true for arbitrary beam cross sections and transverse particle distributions.

10.2.2 Fields from Image Charges

In the discussion of space charges, we ignored so far the effect of metallic and magnetic surfaces close to the beam. The electromagnetic self fields of the beam circulating in a metallic vacuum chamber and between ferromagnetic poles of lattice magnets must meet certain boundary conditions on such surfaces. Appropriate corrections to the free space electromagnetic fields have been derived by *Laslett* [10.18] by adding the electromagnetic fields from all image charges to the fields of the particle beam itself.

Following Laslett, we consider a particle beam with metallic and ferromagnetic boundaries as shown in Fig. 10.2. For full generality, let the elliptical particle beam be displaced in the vertical plane by $\bar{y}$ from the midplane, the metallic vacuum chamber and magnet pole are simulated as pairs of infinitely wide parallel surfaces at $\pm b$ and $\pm g$, respectively, and the observation point of the fields be at y.

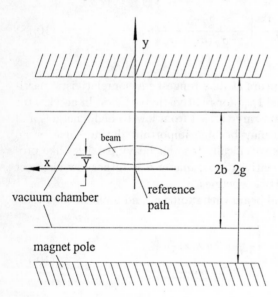

Fig. 10.2. Particle beam with metallic and ferromagnetic boundaries

The linear particle density is

$$\lambda = \frac{N_{\mathrm{p}}}{n_{\mathrm{b}} \ell_{\mathrm{b}}} = \frac{N_{\mathrm{p}}}{n_{\mathrm{b}} \sqrt{2\pi}\, \sigma_\ell}, \tag{10.21}$$

where N_{p} is the total number of particle in the circulating beam, n_{b} the number of bunches, $\ell_{\mathrm{b}} = \sqrt{2\pi}\sigma_\ell$ the effective bunch length and σ_ℓ the standard bunch length for a gaussian distribution.

The locations and strength of the electrical images of a line current in the configuration of Fig. 10.2 are shown in Fig. 10.3. The boundary condition for electric fields is $E_z(b) = 0$ on the surface of the metallic vacuum chamber and is satisfied if the image charges change sign from image to image. To calculate the electrical field $E_y(y)$, we add the contributions from all image fields in the infinite series

$$E_{y,\text{image}}(y) = \left[\frac{1}{4\pi\epsilon_\circ}\right] 2q\lambda$$

$$\left(\frac{1}{2b - \bar{y} - y} - \frac{1}{2b + \bar{y} + y} - \frac{1}{4b + \bar{y} - y} + \frac{1}{4b - \bar{y} + y}\right.$$

$$+\frac{1}{6b - \bar{y} - y} - \frac{1}{6b + \bar{y} + y} - \frac{1}{8b + \bar{y} - y} + \frac{1}{8b - \bar{y} + y}$$

$$\left.+\frac{1}{10b - \bar{y} - y} - \frac{1}{10b + \bar{y} + y} - \cdots\right). \qquad (10.22)$$

These image fields must be added to the direct field of the line charge to meet the boundary condition that the electric field enter metallic surfaces perpendicular. Equation (10.22) can be split into two series with factors $(\bar{y} + y)$ and $(\bar{y} - y)$ in the numerator; we get after some manipulations with $\bar{y} + y \ll b$ and $\bar{y} - y \ll b$

$$E_{y,\text{image}}(y) = \left[\frac{1}{4\pi\epsilon_\circ}\right] \frac{q\lambda}{b^2} \left[\sum_{m=1}^{\infty} \frac{\bar{y} + y}{(2m-1)^2} + \sum_{m=1}^{\infty} \frac{\bar{y} - y}{4m^2}\right],$$

$$= \left[\frac{1}{4\pi\epsilon_\circ}\right] \frac{q\lambda}{b^2} \left[(\bar{y} + y)\frac{\pi^2}{8} + (\bar{y} - y)\frac{\pi^2}{24}\right], \qquad (10.23)$$

$$= \left[\frac{1}{4\pi\epsilon_\circ}\right] \frac{q\lambda}{b^2} \frac{\pi^2}{12} (2\bar{y} + y) = \left[\frac{1}{4\pi\epsilon_\circ}\right] \frac{4q\lambda}{b^2} \epsilon_1 (2\bar{y} + y).$$

The electric image fields depend linearly on the deviations $\bar{y}$ and y of bunch center and particle, respectively, from the axis and act therefore like a quadrupole causing a tune shift.

A similar derivation is used to get the magnetic image fields due to ferromagnetic surfaces at $\pm g$ above and below the midplane. The magnetic field lines must enter the magnetic pole faces perpendicular and the image currents therefore flow in the same direction as the line current causing a magnetic force on the test particle which is opposed to that by the magnetic field of the beam itself.

Bunched beams generate high frequency electromagnetic fields which do not reach ferromagnetic surfaces because of eddy current shielding by the metallic vacuum chamber. For magnetic image fields we distinguish therefore between dc and ac image fields. The dc Fourier component of a bunched beam current is equal to twice the average beam current $qc\beta\lambda B$, where the *Laslett bunching factor B* is the bunch occupation along the ring circumference defined by

$$B = \frac{\bar{\lambda}}{\lambda} = \frac{n_b \ell_b}{2\pi R}. \qquad (10.24)$$

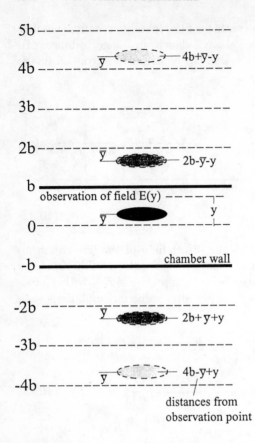

distances from
observation point

Fig. 10.3. Location and source of image fields

The dc magnetic image fields are derived similar to electric image fields with $B_\varphi = -2\lambda\beta/r$ from (10.19,20) and are with (10.24)

$$B_{x,\text{image,dc}}(y) = \left[\frac{c\mu_o}{\sqrt{4\pi}}\right] \frac{2q\lambda\beta}{g^2} B \left[\sum_{m=1}^{\infty} \frac{\bar{y}+y}{(2m-1)^2} + \sum_{m=1}^{\infty} \frac{\bar{y}-y}{4m^2}\right],$$

$$= \left[\frac{c\mu_o}{\sqrt{4\pi}}\right] \frac{q\lambda\beta}{g^2} B \left[(\bar{y}+y)\frac{\pi^2}{8} + (\bar{y}-y)\frac{\pi^2}{24}\right],$$

$$= \left[\frac{c\mu_o}{\sqrt{4\pi}}\right] \frac{4q\lambda\beta}{g^2} B \epsilon_2 (2\bar{y}+y). \qquad (10.25)$$

The magnetic image fields must penetrate the metallic vacuum chamber to reach ferromagnetic poles. This is no problem for dc or low frequency field components but in case of bunched beams relevant frequencies are rather high and eddy current shielding of the vacuum chamber for ac magnetic fields must be taken into account. In most cases, these ac fields are of rather high frequency and we may assume that they do not penetrate the thick metallic vacuum chamber. Consequently, we ignore here the effect of ferromagnetic poles and consider only the contribution of magnetic ac image fields due to eddy currents in vacuum chamber walls. Similar to electric image fields, the magnetic image fields are in analogy to (10.23)

$$
\begin{aligned}
B_{x,\text{image,ac}}(y) &= -\left[\frac{c\,\mu_o}{\sqrt{4\pi}}\right]\frac{q\,\lambda\,\beta}{b^2}\,(1-B)\,\frac{\pi^2}{12}\,(2\bar{y}+y)\,, \\
&= -\left[\frac{c\,\mu_o}{\sqrt{4\pi}}\right]\frac{4\,q\,\lambda\,\beta}{b^2}\,(1-B)\,\epsilon_1\,(2\bar{y}+y)\,,
\end{aligned}
\tag{10.26}
$$

where the factor $(1-B)$ accounts for the subtraction of the dc component $\beta\lambda B$.

Similar to the electric image fields, the magnetic image fields must be added to the direct magnet fields (10.16) from the beam current to meet the boundary condition of normal field components at ferromagnetic surfaces.

The coefficients ϵ_1 and ϵ_2 are the *Laslett form factors* which are for infinite parallel plate vacuum chambers and magnetic poles

$$
\epsilon_1 = \frac{\pi^2}{48} \quad \text{and} \quad \epsilon_2 = \frac{\pi^2}{24}.
\tag{10.27}
$$

The vacuum chamber and ferromagnetic poles are similar to infinitely wide surfaces. While this is a sufficiently accurate approximation for the magnet poles, corrections must be applied for circular or elliptical vacuum chambers. *Laslett* [10.18] has derived what we call now Laslett form factors for vacuum chambers with elliptical cross sections and variable aspect ratios which are compiled in Table 10.1.

Table 10.1. Laslett incoherent tune shift form factors for elliptical vacuum chambers [10.18]

a/b:[1]	1	5/4	4/3	3/2	2/1	∞
ϵ_1:	0	0.090	0.107	0.134	0.172	0.206

[1] a is the horizontal and b the vertical half-axis of an elliptical vacuum chamber cross section

All relevant field components have been identified and we collect these fields first for $\bar{y}=0$ and obtain from (10.15,23) for the electric field in the vertical mid plane

$$
E_y(y) = \left[\frac{1}{4\pi\epsilon_o}\right]\frac{2\,q\,\lambda}{\sigma_y\,(\sigma_x+\sigma_y)}\left[1+\frac{2\sigma_y\,(\sigma_x+\sigma_y)}{b^2}\,\epsilon_1\right]y.
\tag{10.28}
$$

From (10.16,25) the dc magnetic field is

$$
B_{x,\text{dc}} = -\left[\frac{c\,\mu_o}{\sqrt{4\pi}}\right]\frac{2\,q\,\lambda\,\beta\,B}{\sigma_y\,(\sigma_x+\sigma_y)}\left[1-\frac{2\sigma_y\,(\sigma_x+\sigma_y)}{g^2}\,\epsilon_2\right]y
\tag{10.29}
$$

and from (10.26) the ac magnetic field

$$
B_{x,\text{ac}} = -\left[\frac{c\mu_o}{\sqrt{4\pi}}\right]\frac{2\,q\,\lambda\,\beta}{\sigma_y(\sigma_x+\sigma_y)}\left[1+\frac{2\sigma_y(\sigma_x+\sigma_y)}{b^2}\epsilon_1\right](1-B)y.
\tag{10.30}
$$

Tacitly, we have assumed that the transverse particle distribution is gaussian which is a true representation of an electron beam but may not be correct for proton or ion beams. The standard deviations σ of a gaussian distribution are very well defined and can therefore be replaced by other quantities like the full-width half maximum or as the particle distribution may require.

The electromagnetic force due to space charge on individual particles in a beam has been derived and it became obvious that image field effects can play a significant role in the perturbation of the beam. The fields scale linear with amplitude for very small amplitudes and act therefore like focusing quadrupoles. At larger amplitudes, however, the fields turn over and evanesce like $1/r$. Consequently, the field gradient is negative decaying quickly with amplitude.

A complete set of direct and image fields have been derived which must be considered to account for space-charge effects. Similar derivations lead to other field components necessary to determine horizontal space-charge forces. In most accelerators, however, the beam cross section is flat and so is the vacuum chamber and the magnet pole aperture. As a consequence, we expect the space-charge forces to be larger in the vertical plane than in the horizontal plane.

10.2.3 Space-Charge Effects

The Lorentz force on individual particles can be calculated from the space-charge fields and we get

$$F_y = \left[\frac{1}{4\pi\epsilon_o}\right] \frac{2\,q^2\,f_p\lambda\,(1-\beta^2 f_v)}{\sigma_y\,(\sigma_x + \sigma_y)}\,f_{corr}\,y = \left[\frac{1}{4\pi\epsilon_o}\right] q^2\,\mathcal{F}\,y\,, \qquad (10.31)$$

where the correction factor due to image fields is with $\beta^2\gamma^2 = \gamma^2 - 1$

$$f_{corr} = 1 + \frac{2\sigma_y\,(\sigma_x + \sigma_y)}{b^2}\left\{\epsilon_1[1 + (\gamma^2 - 1)B] + \epsilon_2(\gamma^2 - 1)\frac{b^2}{g^2}B\right\} \quad (10.32)$$

and

$$\mathcal{F} = \left[\frac{1}{4\pi\epsilon_o}\right] \frac{2\,q^2\,f_p\lambda\,(1-\beta^2 f_v)}{\sigma_y\,(\sigma_x + \sigma_y)}\,f_{corr}\,, \qquad (10.33)$$

The factors f_p and f_v determine signs depending on the kind of particles interacting and the direction of travel with respect to each other. Specifically $f_p = \text{sign}(q\,q_b)$ where q is the charge of a test particle and q_b the charge of the field creating particles, e.g., the charge of a bunch. Similarly, $f_v = \text{sign}(\mathbf{v}\,\mathbf{v_b})$ where $\mathbf{v}$ is the direction of travel for the test particle and $\mathbf{v_b}$ the direction of travel of the bunch. To calculate the space-charge force of head-on colliding proton and antiproton beams, for example, we would set $f_p = -1$ and $f_v = -1$.

There is a significant cancellation of two strong terms, the repulsive electrical field and the focusing magnetic field, expressed by the factor $1 - \beta^2$

for space-charge forces within a highly relativistic beam. This cancellation can be greatly upset if particle beams become partially neutralized by collecting other particles of opposite charge within the beams potential well. For example, proton beams can trap electrons in the positive potential well as can electron beams trap positive ions in the negative potential well. To avoid such partial neutralization and appearance of unnecessarily strong space-charge effects, *clearing electrodes* must be installed over much of the ring circumference to extract with electrostatic fields low energy electrons or ions from the particle beam.

The electromagnetic space-charge force on an individual particle within a particle beam increases linearly with its distance from the axis. A similar force occurs for the horizontal plane and both fields therefore act like a quadrupole causing a tune shift. This has been recognized and analyzed early by *Kerst* [10.19] and *Blewett* [10.20]. A complete treatment of space charge dominated beams can be found in [10.21]. The equation of motion under the influence of *space charge forces* can be written in the form

$$m\gamma\ddot{u} + Du = \frac{\partial F_u}{\partial u}u \qquad \text{with} \quad u = (x, y). \tag{10.34}$$

We get the regular form $u'' + (k_{\mathrm{o}} + \Delta k)\,u = 0$ with $\ddot{u} = u''\,c^2\,\beta^2$ and $f_v = 1$, where k_{o} describes the quadrupole strength and the space-charge strength is expressed by

$$\Delta k = \frac{1}{mc^2\gamma\beta^2}\frac{\partial F_u}{\partial u} = -\frac{2\,r_{\mathrm{c}}}{\beta^2\,\gamma^3}\frac{\lambda}{\sigma_y\,(\sigma_x + \sigma_y)}\,f_{\mathrm{corr}} \tag{10.35}$$

where $r_{\mathrm{c}} = [4\pi\epsilon_{\mathrm{o}}]^{-1}\,q^2/(mc^2)$ is the *classical particle radius*

$$
\begin{aligned}
r_{\mathrm{e}} &= 2.817938\ 10^{-15}\,\text{m} \qquad \text{for electrons and} \\
r_{\mathrm{p}} &= 1.534698\ 10^{-18}\,\text{m} \qquad \text{for protons}.
\end{aligned}
\tag{10.36}
$$

For ions with charge multiplicity Z and atomic number A the classical particle radius is $r_{\mathrm{ion}} = r_{\mathrm{p}}\,Z^2/A$.

Space-charge dominated beams: So far *space-charge effects* or *space-charge focusing* has been consistently neglected in the discussions on transverse beam dynamics. In cases of low beam energy and high particle densities, it might become necessary to include space-charge effects. Space-charge forces are defocusing in both planes and compensation therefore requires additional focusing in both planes. However, it should be noted that particles closer to the beam surface will not experience the same linear space-charge defocusing as those near the axis and therefore a compensation of space-charge focusing works only for part of the beam. Here, we will not get involved with the dynamics of heavily space charge dominated particle beams but try to derive a criterion by which we can decide whether or not space-charge forces are significant in transverse particle beam optics.

This distinction becomes obvious from the equation of motion including space charges. From (10.34,35) we get the equation of motion

$$u'' + \left[k_o - \frac{2\,r_c}{\beta^2\,\gamma^3}\,\frac{\lambda}{\sigma_y\,(\sigma_x + \sigma_y)}\,f_{\text{corr}} \right] u = 0\,, \tag{10.37}$$

where we ignored the image current corrections. Space-charge forces can be neglected if the integral of the space-charge force over a length L which is characteristic for the average distance between quadrupoles in the beam line is small compared to the typical integrated quadrupole length $k_o \ell_q$

$$\frac{2\,r_c}{\beta^2\,\gamma^3} \int_L \frac{\lambda\,f_{\text{corr}}}{\sigma_y\,(\sigma_x + \sigma_y)}\,\mathrm{d}z \ll k_o \ell_q\,. \tag{10.38}$$

The effect of space-charge focusing is most severe where the beam cross section is smallest and (10.38) should therefore be applied specifically to such sections of the beam transport line. Obviously, the application of this formula requires some subjective judgement as to how much smaller space-charge effects should be. To aid this judgement, one might also calculate the average betatron phase shift caused by space-charge forces and compare with the total phase advance along the beam line under investigation. In this case we look for

$$\frac{2\,r_c}{\beta^2\,\gamma^3} \int_L \frac{\beta_u \lambda\,f_{\text{corr}}}{\sigma_y\,(\sigma_x + \sigma_y)}\,\mathrm{d}z \ll \psi_o(L) \tag{10.39}$$

to determine the severity of space-charge effects. The nominal phase advance $\psi_{o,u}(L)$ is defined such that $\psi_{o,u}(0) = 0$ at the beginning of the beam line.

Space-charge tune shift: Space-charge focusing may not significantly perturb the lattice functions but may cause a big enough tune shift in a circular accelerator moving the beam onto a resonance. The beam current is therefore limited by the maximum allowable tune shift in the accelerator which is for a linear focusing force $F(s)$ given by

$$\Delta\nu_u = -\frac{1}{4\pi}\,\frac{r_c}{\beta^2\,\gamma} \int_0^{L_{\text{int}}} F(s)\,\beta_u\,\mathrm{d}s\,. \tag{10.40}$$

The integration in (10.40) is taken over that part of the path in each revolution where the force is effective. For the effect on particles within the same beam this is the circumference and for the beam-beam effect it is the total length of all head on collisions per turn.

The tune shifts are not the same for all particles due to the nonuniform charge distribution within a beam. Only particles in the very center of the beam suffer the largest tune shift while particles with increasing betatron oscillation amplitudes are less affected. The effect of space charge therefore introduces a tune spread rather than a coherent tune shift and we refer to this effect as the *incoherent space-charge tune shift*.

As a particular case, we consider here the space-charge tune shift of a particle within a beam of equal species particles. Applying the Lorentz force (10.31) with (10.33) the space-charge tune shift becomes from (10.40)

$$\Delta\nu_{u,\mathrm{sc}} = -\frac{r_{\mathrm{c}}\lambda}{2\pi}\frac{f_{\mathrm{p}}\left(1-\beta^2 f_{\mathrm{v}}\right)}{\beta^2\gamma}\int\frac{\beta_u}{\sigma_u\left(\sigma_x+\sigma_y\right)}f_{\mathrm{corr}}\,\mathrm{d}s\,, \tag{10.41}$$

where the local linear particle density λ is defined by (10.21) and r_{c} is the classical particle radius.

The maximum *incoherent space-charge tune shift* is from (10.41) with $f_{\mathrm{p}}=1$, $f_{\mathrm{v}}=1$ and $(1-\beta^2)=1/\gamma^2$ and (10.33)

$$\Delta\nu_{u,\mathrm{sc},\mathrm{incoh}} = -\frac{r_{\mathrm{c}}\lambda}{2\pi\,\beta^2\,\gamma^3}\left[\int\limits_0^{2\pi\bar{R}}\frac{\beta_u}{\sigma_u\left(\sigma_x+\sigma_y\right)}\,\mathrm{d}s\right. \tag{10.42}$$

$$\left.+\,2(1+\beta^2\gamma^2 B)\int\limits_0^{L_{\mathrm{vac}}}\frac{\beta_u\,\epsilon_1}{b^2}\,\mathrm{d}s+2\beta^2\gamma^2 B\int\limits_0^{L_{\mathrm{mag}}}\frac{\beta_u\,\epsilon_2}{g^2}\,\mathrm{d}s\right],$$

where the integration length L_{vac} is equal to the total length of the vacuum chamber and L_{mag} is the total length of magnets along the ring circumference. Note, however, that this last term appears only at low frequencies because of eddy-current shielding in the vacuum chamber at high frequencies. A tune shift observed on a betatron side band of a high harmonic of the revolution frequency may not exhibit a tune shift due to this term while one might have a contribution at low frequencies.

A *coherent space-charge tune shift* can be identified by setting $y=\bar{y}$ in the field expressions (10.23,25,26) to determine the fields at the bunch center. The calculation is similar to that for the incoherent space-charge tune shift except that we define for this case new *Laslett form factors*

$$\xi_2 = \frac{\pi^2}{16} \tag{10.43}$$

for the image fields from the magnetic pole and form factors ξ_1 which depend on the aspect ratio of an elliptical vacuum chamber (Table 10.2).

Table 10.2. Laslett coherent tune shift form factors for elliptical vacuum chambers [10.18]

a/b:[1]	1	5/4	4/3	3/2	2/1	∞
ξ_1:	0.500	0.533	0.559	0.575	0.599	0.617

[1] a is the horizontal and b the vertical half-axis of an elliptical vacuum chamber cross section

The coherent space-charge tune shift is analogous to (10.42)

$$\Delta\nu_{u,sc,coh} = -\frac{r_c\lambda}{2\pi\,\beta^2\,\gamma^3}\left[\int_0^{2\pi\bar{R}}\frac{\beta_u}{\sigma_u\,(\sigma_x+\sigma_y)}\,ds\right.$$

$$\left.+2(1+\beta^2\gamma^2 B)\int_0^{L_{vac}}\frac{\beta_u\,\xi_1}{b^2}\,ds+2\beta^2\gamma^2 B\int_0^{L_{mag}}\frac{\beta_u\,b^2\,\xi_2}{g^2}\,ds\right]. \tag{10.44}$$

In both cases, we may simplify the expressions significantly for an approximate calculation by applying *smooth approximation* $\overline{\beta_u}\approx\bar{R}/\nu_{ou}$ and assuming a uniform vacuum chamber and magnet pole gaps. With these approximations, (10.41) becomes

$$\Delta\nu_{u,sc} = -\frac{r_c\,N_{tot}\,\bar{R}}{2\pi\,\nu_{ou}\,B}\frac{f_p\,(1-\beta^2 f_v)}{\beta^2\,\gamma}\frac{\langle f_{corr}\rangle}{\bar{\sigma}_u\,(\bar{\sigma}_x+\bar{\sigma}_y)}, \tag{10.45}$$

where

$$\langle f_{corr}\rangle = 1+\frac{\bar{\sigma}_u\,(\bar{\sigma}_x+\bar{\sigma}_y)}{\bar{b}^2}\left[\epsilon_1(1+\beta^2\gamma^2 B)+\epsilon_2\beta^2\gamma^2\frac{\bar{b}^2}{\bar{g}^2}B\right]. \tag{10.46}$$

Symbols with an overbar are the values of quantities averaged over the circumference of the ring and ν_{ou} is the unperturbed tune in the plane (x,y). The *incoherent tune shift* (10.42) becomes then

$$\boxed{\begin{aligned}\Delta\nu_{u,sc,incoh} &\approx -\frac{r_c\,N_{tot}\,\bar{R}}{2\pi\,\nu_{ou}\,B\,\beta^2\gamma^3}\left[\frac{1}{\bar{\sigma}_u\,(\bar{\sigma}_x+\bar{\sigma}_y)}\right.\\ &\left.+2(1+\beta^2\gamma^2 B)\frac{\epsilon_1}{\bar{b}^2}+2\beta^2\gamma^2 B\frac{\epsilon_2}{\bar{g}^2}\,\eta_b\right],\end{aligned}} \tag{10.47}$$

where $\eta_b = L_{mag}/(2\pi\bar{R})$ is the magnet fill factor and the coherent tune shift (10.44) becomes

$$\boxed{\begin{aligned}\Delta\nu_{u,sc,coh} &\approx -\frac{r_c\,N_{tot}\,\bar{R}}{2\pi\,\nu_{ou}\,B\,\beta^2\gamma^3}\left[\frac{1}{\bar{\sigma}_u\,(\bar{\sigma}_x+\bar{\sigma}_y)}\right.\\ &\left.+\frac{2(1+\beta^2\gamma^2 B)}{\bar{b}^2}\xi_1+\frac{2\beta^2\gamma^2 B}{\bar{g}^2}\xi_2\,\eta_b\right].\end{aligned}} \tag{(10.48)}$$

The tune shift diminishes proportional to the third power of the particle energy. As a matter of fact in electron machines of the order of 1 GeV or more, space-charge tune shifts are in most cases negligible. For low energy protons and ions, however, this tune shift is of great importance and must be closely controlled to avoid beam loss due to nearby resonances. While a maximum allowable tune shift of $0.15-0.25$ seems reasonable to avoid crossing a strong third order or half-integer resonance, practically achieved tune shifts can be significantly larger of the order $0.5 - 0.6$ [10.22 – 24]. However, independent of the maximum tune shift actually achieved in a particular ring, space charge forces ultimately lead to a limitation of the beam current.

Beam-beam tune shift: In case of counterrotating beams colliding at particular interaction points in a colliding-beam facility, we always have $f_v = -1$ but the colliding particles still may be of equal or opposite charge. In addition, there is no contribution from magnetic image fields since collisions do not occur within magnets. Even image fields from vacuum chambers are neglected because the *beam-beam interaction* happens only over a very short distance. A particle in one beam will feel the field from the other beam only during the time it travels through the other beam which is equal to the time it takes the particle to travel half the effective length of the oncoming bunch. With these considerations in mind, we obtain for the *beam-beam tune shift* in the vertical plane from (10.41) with $f_{\text{corr}} = 1$ and assuming head on collisions of particle-antiparticle beams ($f_e = -1$)

$$\boxed{\Delta\nu_{y,\text{bb}} = \frac{r_c\,N_{\text{tot}}}{2\pi\,B\,\gamma}\frac{\beta_y^*}{\sigma_y^*\,(\sigma_x^* + \sigma_y^*)}} \tag{10.49}$$

and in the horizontal plane

$$\boxed{\Delta\nu_{x,\text{bb}} = \frac{r_c\,N_{\text{tot}}}{2\pi\,B\,\gamma}\frac{\beta_x^*}{\sigma_x^*\,(\sigma_x^* + \sigma_y^*)}\,,} \tag{10.50}$$

where * indicates that the quantities be taken at the interaction point. In cases where other particle combinations are brought into collision or when both beams cross under an angle these equations must be appropriately modified to accurately describe the actual situation.

10.2.4 Longitudinal Space-Charge Field

Within a continuous particle beam travelling along a uniform vacuum chamber we do not expect longitudinal fields to arise. We must, however, consider what happens if the longitudinal charge density is not uniform since this is the more realistic assumption. For the case of a round beam of radius r_o in a circular vacuum tube of radius r_w (Fig. 10.4), the fields can be derived by integrating Maxwell's equation $\nabla \times \mathbf{E} = -\frac{[c]}{c}\frac{\partial \mathbf{B}}{\partial t}$ and we get after application of Stoke's law

$$\oint \mathbf{E}\,\mathrm{d}\mathbf{s} = -\frac{[c]}{c}\frac{\partial}{\partial t}\int \mathbf{B}\,\mathrm{d}\mathbf{A}\,, \tag{10.51}$$

where dA is an element of the area enclosed by the integration path s. In Fig. 10.4 the integration path is shown leading to the determination of the electrical field E_{zo} in the center of the beam.

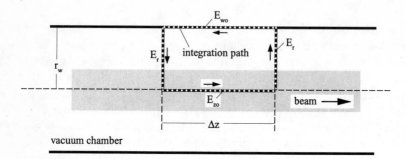

Fig. 10.4. Space-charge fields due to a particle beam travelling inside a circular metallic vacuum chamber

Integrating the l.h.s. of (10.51) along the integration path shown in Fig. 10.4 we get with (10.19,20)

$$E_{zo}\,\Delta z + \int_0^{r_{\rm w}} E_r(z+\Delta z)\mathrm{d}r - E_{zw}\,\Delta z - \int_0^{r_{\rm w}} E_r(z)\mathrm{d}r$$

$$= (E_{zo} - E_{zw})\,\Delta z + \frac{q}{[4\pi\epsilon_0]}\left(1+2\ln\frac{r_{\rm w}}{r_0}\right)\frac{\partial\lambda}{\partial z}\,\Delta z\,,$$

(10.52)

where a Taylor's expansion was applied to the linear particle density $\lambda(z+\Delta z)$ and only linear terms were retained. E_{zw} is the longitudinal electrical field on the vacuum chamber wall.

For the r.h.s. of (10.51) we use the expressions for the magnetic field (10.19,20) and get with $\int B_\varphi \mathrm{d}A = \Delta z \int B_\varphi \mathrm{d}r$

$$-\frac{\beta}{c} q\left(1+2\ln\frac{r_{\rm w}}{r_0}\right)\frac{\partial\lambda}{\partial t}\,\Delta z = \beta^2 q\left(1+2\ln\frac{r_{\rm w}}{r_0}\right)\frac{\partial\lambda}{\partial z}\,\Delta z$$

(10.53)

while using the continuity equation

$$\frac{\partial\lambda}{\partial t} + \beta c\frac{\partial\lambda}{\partial z} = 0\,.$$

(10.54)

The *longitudinal space-charge field* is therefore

$$E_{zo} = E_{zw} - \frac{q}{[4\pi\epsilon_0]}\frac{1}{\gamma^2}\left(1+2\ln\frac{r_{\rm w}}{r_0}\right)\frac{\partial\lambda}{\partial z}$$

(10.55)

and vanishes indeed for a uniform charge distribution because $E_{zw} = 0$ for a dc current. However, variations in the charge distribution cause a longitudinal field which together with the associated ac field in the vacuum chamber wall, acts on individual particles.

The perturbation of a uniform particle distribution in a circular accelerator is periodic with the circumference of the ring and we may set for the longitudinal particle distribution keeping only the nth harmonic for simplicity

$$\lambda = \lambda_{\mathrm{o}} + \lambda_n \, \mathrm{e}^{\mathrm{i}(n\theta - \omega_n t)}, \tag{10.56}$$

where ω_n is the nth harmonic of the perturbation, $\omega_n = n\omega_{\mathrm{o}}$. Of course a real beam may have many modes and we need therefore to sum over all modes n. In case of instability, it is clear that the whole beam is unstable if one mode is unstable; however, in case of stability for one mode it is possible that another mode is unstable because of the frequency dependence of the complex impedance.

With the derivative $\mathrm{d}\lambda/\mathrm{d}z$ and $\theta = z/\bar{R}$ with $\bar{R}$ the average ring radius an integration of (10.55) around the circular accelerator gives the total induced voltage due to space-charge fields

$$V_{zo} = 2\pi \bar{R} \, E_{zw} - \mathrm{i} \, \frac{I_n}{[4\pi\epsilon_{\mathrm{o}}]} \, \frac{2\pi n}{\beta c \, \gamma^2} \left(1 + 2\ln\frac{r_{\mathrm{w}}}{r_{\mathrm{o}}} \right) \mathrm{e}^{\mathrm{i}(n\theta - \omega_n t)}. \tag{10.57}$$

In this expression we have also introduced the nth harmonic of the beam-current perturbation $I_n = \beta c \, q \lambda_n$. Equation (10.57) exhibits a relation of the induced voltage to the beam current. Borrowing from the theory of electrical currents, it is customary to introduce here the concept of an impedance which will become a powerful tool to describe the otherwise complicated coupling between beam current and induced voltage. We will return to this point in Sect. 10.4.

10.3 Beam-Current Spectrum

The importance of variations in the instantaneous beam current for beam stability became apparent in the last section. This is particularly true in circular accelerators where the particle distribution is periodic with the circumference of the ring. On one hand, we have an orbiting particle beam which constitutes a harmonic oscillator with many eigen frequencies and harmonics thereof and on the other hand, there is an environment with a frequency dependent response to electromagnetic excitation. Depending on the coupling of the beam to its environment at a particular frequency, periodic excitations occur which can create perturbations of particle and beam dynamics. This interaction is the subject of this discussion. In this text, we will concentrate on the discussion of basic phenomena of beam-environment interactions or beam instabilities. For a more detailed introduction into the field of beam instabilities, the interested reader is referred to the general references for this chapter.

In this discussion we will follow mainly the theories as formulated by *Chao* [10.25], *Laclare* [10.26], *Sacherer* [10.27] and *Zotter* [10.28]. Since the coupling of the beam to its environment depends greatly on the frequency involved, it seems appropriate to discuss first the frequency spectrum of a circulating particle beam.

Longitudinal beam spectrum: In case of a single circulating particle of charge q in each of n_b equidistant buckets, a pick up electrode located at azimuth φ would produce a signal proportional to the single-particle beam current which is composed of a series of delta function signals

$$i_{\parallel}(t,\varphi) = q \sum_{k=-\infty}^{+\infty} \delta(t - \frac{\varphi}{2\pi} T_o - k \frac{T_o}{n_b} - \tau),$$ (10.58)

where τ is the longitudinal offset of the particle from the reference point, n_b the number of equidistant bunches and T_o the revolution time (Fig. 10.5). Of course, for a single particle n_b would be unity.

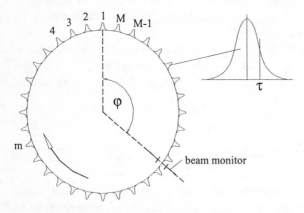

Fig. 10.5. Particle distribution along the circumference of a circular accelerator and definition of parameters

With the revolution frequency $\omega_o = 2\pi/T_o$, we use the mathematical relations $2\pi \sum_{k=-\infty}^{+\infty} \delta(y - 2\pi k) = \sum_{p=-\infty}^{+\infty} e^{ipy}$ and $|c|\,\delta(cy) = \delta(y)$ for

$$\sum_{k=-\infty}^{+\infty} \delta\left(x - \frac{2\pi k}{n_b \omega_o}\right) = \frac{n_b \omega_o}{2\pi} \sum_{p=-\infty}^{+\infty} e^{ipn_b \omega_o x}$$ (10.59)

to replace the delta functions and

$$e^{iy \sin \psi} = \sum_{n=-\infty}^{+\infty} J_n(y) e^{in\psi} \tag{10.60}$$

to replace the term including the synchrotron oscillation $\tau = \hat{\tau} \cos[(m + \nu_s)\omega_o t + \zeta_i]$ where ν_s is the *synchrotron oscillation tune*. The term $m\omega_o t$ reflects the *mode* of the longitudinal particle distribution in all buckets. This distribution is periodic with the periodicity of the circumference and the modes are the harmonics of the distribution in terms of the revolution frequency.

Inserting (10.59) on the r.h.s. of (10.58) and replacing the $e^{-ipn_b\omega_o\tau}$ term with (10.60) one gets

$$i_\parallel(t, \varphi) = \frac{qn_b\omega_o}{2\pi} \sum_{p=-\infty}^{+\infty} \sum_{n=-\infty}^{+\infty} i^{-n} J_n(pn_b\omega_o\hat{\tau})$$

$$\times e^{i[(pn_b + nm + n\nu_s)\omega_o t - pn_b\varphi + n\zeta_i]}. \tag{10.61}$$

To get the single-particle current spectrum, we perform a Fourier transform

$$i_\parallel(\omega, \varphi) = \frac{1}{2\pi} \int_{-\infty}^{+\infty} i(t, \varphi) e^{-i\omega t} \, dt \tag{10.62}$$

and get instead of (10.61) the *longitudinal current spectrum*

$$i_\parallel(\omega, \varphi) = \frac{qn_b\omega_o}{2\pi} \sum_{\substack{p= \\ -\infty}}^{+\infty} \sum_{\substack{n= \\ -\infty}}^{+\infty} i^{-n} J_n(pn_b\omega_o\hat{\tau}) e^{-i(pn_b\varphi - n\zeta_i)} \delta(\Omega), \tag{10.63}$$

where $\Omega = \omega - (pn_b + nm + n\nu_s)\omega_o$ and making use of the identity $\int e^{-i\omega t} \, dt = 2\pi\delta(\omega)$. This spectrum is a line spectrum with harmonics of the revolution frequency separated by $n_b\omega_o$. Each of these main harmonics is accompanied on both sides with satellites separated by $\Omega_s = \nu s\omega_o$. The harmonics $m\omega_o$ finally are generated by a nonuniform particle distribution along the circumference which may be due to synchrotron oscillations of individual bunches or nonuniform bunch intensity or both. Schematically, some of the more important lines of this spectrum are shown in Fig. 10.6 for a single particle.

In the approximation of small synchrotron oscillation amplitudes, one may neglect all terms with $|n| > 1$ and the particle beam includes only the frequencies $\omega = [pn_b \pm (m + \nu_s)]\omega_o$. In Chap. 6 the interaction of this spectrum for $p = h$ with the narrow-band impedance of a resonant cavity was discussed in connection with *Robinson damping*.

A real particle beam consists of many particles which are distributed in initial phase ζ_i as well as in oscillation amplitudes $\hat{\tau}$. Assuming the simple

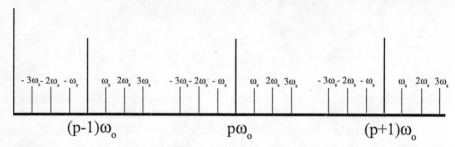

Fig. 10.6. Current spectrum of a single particle orbiting in a circular accelerator and executing synchrotron oscillations

case of equal and equidistant bunches with uniform particle distributions in synchrotron phase ζ_i we may set $n = 0$. The time independent particle distribution is then $\Phi_o(t, \hat{\tau}) = \phi_o(\hat{\tau})$ which is normalized to unity and the total beam-current spectrum is given by

$$I_\|(\omega, \varphi) = I_b \sum_{p=-\infty}^{+\infty} \delta(\omega - \Omega_o)\,e^{-ip\varphi} \int_{-\infty}^{+\infty} J_o(pn_b\omega_o\hat{\tau})\,\phi_o(\hat{\tau})\,d\hat{\tau}, \quad (10.64)$$

where $I_b = q/T_o$ is the bunch current and $\Omega_o = pn_b\omega_o$. All synchrotron satellites vanished because of the uniform distribution of synchrotron phases and lack of coherent bunch oscillations. Observation of synchrotron satellites, therefore, indicates a perturbation from this condition either by coherent oscillations of one or more bunches ($n \neq 0, \hat{\tau} \neq 0$) or coherent density oscillations within a bunch $\Phi_o(t, \hat{\tau}) = f(\zeta_i)$.

The infinite sum over p represents the periodic bunch distribution along the circumference over many revolutions whether it be a single or multiple bunches. The beam-current spectrum is expected to interact with the impedance spectrum of the environment and this interaction may result in a significant alteration of the particle distribution $\Phi(t, \hat{\tau})$. As an example for what could happen the two lowest order modes of bunch oscillations are shown in Fig. 10.7

In lowest order a collection of particles contained in a bunch may perform *dipole mode* oscillations where all particles and the bunch center oscillate coherently (Fig. 10.7 a). In the next higher mode, the bunch center does not move but particles at the head or tail of the bunch oscillate 180° out of phase. This bunch shape oscillation is in its lowest order a *quadrupole mode* oscillation as shown in Fig. 10.7 b. Similarly, higher order mode bunch shape oscillations can be defined.

Transverse beam spectrum: Single particles and a collection of particles in a bunch may also perform transverse betatron oscillations constituting

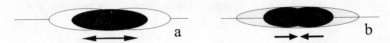

Fig. 10.7. Dipole mode oscillation **a** and quadrupole bunch shape oscillations **b**

a transverse beam current which can interact with its environment. Again, we observe first only a single particle performing betatron oscillations

$$u = \hat{u} \cos \psi(t),$$

$$(10.65)$$

where $u = x$ or $u = y$, $\psi(t)$ is the betatron phase, and the *transverse current* is

$$i_\perp(t, \varphi) = i_\parallel(t, \varphi) \hat{u} \cos \psi(t).$$

$$(10.66)$$

Note that the transverse current has the dimension of a *current moment* represented by the same spectrum as the longitudinal current plus additional spectral lines due to betatron oscillations. The betatron phase is a function of time and depends on the revolution frequency and the chromaticity, which both depend on the momentum of the particle. From the definition of the momentum compaction $\mathrm{d}\omega/\omega_o = \eta_c \delta$, chromaticity $\mathrm{d}\nu = \xi_u \delta$ and relative momentum deviation $\delta = \mathrm{d}p/p_o$, the variation of the betatron phase with time, while keeping only linear terms in δ, is

$$\dot{\psi}(t) = \omega_u = \nu_o \left(1 + \frac{\xi_u}{\nu_o} \delta\right) \omega_o \left(1 + \eta_c \delta\right),$$

$$\approx \nu_o \omega_o + \left(\nu_o + \frac{\xi_u}{\eta_c}\right) \omega_o \dot{\tau},$$

$$(10.67)$$

where we have replaced the relative momentum deviation δ by the synchrotron oscillation amplitude $\dot{\tau} = -\eta_c \delta$. Equation (10.67) can be integrated for

$$\psi(t) = \nu_o \omega_o(t - \tau) - \omega_o \frac{\xi_u}{\eta_c} \tau + \psi_o$$

$$(10.68)$$

and (10.66) becomes with (10.53,60,63)

$$i_\perp(t, \varphi) = i_\parallel(t, \varphi) \hat{u} \cos \psi(t),$$

$$= q \hat{u} \cos \psi(t) \sum_{m=-\infty}^{+\infty} \delta\left(t - \frac{\varphi}{2\pi} T_o - m \frac{T_o}{n_b}\right),$$

$$= q \hat{u} \frac{e^{\mathrm{i}\psi(t)} + e^{-\mathrm{i}\psi(t)}}{2} \frac{n_b}{T_o} \sum_{p=-\infty}^{+\infty} e^{\mathrm{i}[p n_b \omega_o(t - \tau) - p\varphi]}.$$

$$(10.69)$$

Following the derivation for the longitudinal current and performing a Fourier transform we get the *transverse beam spectrum*

$$i_\perp(\omega,\varphi) = \frac{q}{2T_{\mathrm{o}}}\,\hat{u}\,e^{i\psi_{\mathrm{o}}} \sum_{p=-\infty}^{+\infty}\sum_{n=-\infty}^{+\infty} i^{-n} J_n\left\{\left[(p+\nu_{\mathrm{o}})n_{\mathrm{b}}\omega_{\mathrm{o}} - \frac{\xi_u}{\eta_{\mathrm{c}}}\right]\hat{\tau}\right\}$$

$$\times\, e^{-i(p\varphi-n\zeta_i)}\,\delta(\Omega_u)\,, \qquad (10.70)$$

where $\Omega_u = \omega-(p+\nu_{\mathrm{o}})n_{\mathrm{b}}\omega_{\mathrm{o}}+n\Omega_{\mathrm{s}}$ defines the line spectrum of the transverse single-particle current (Fig. 10.8).

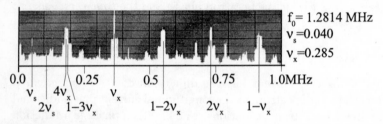

f_0= 1.2814 MHz
v_s=0.040
v_x=0.285

Fig. 10.8. Current spectrum of a single particle orbiting in a circular accelerator and executing betatron and synchrotron oscillations

We note that the betatron harmonics $(p + \nu_{\mathrm{o}})n_{\mathrm{b}}\omega_{\mathrm{o}}$ are surrounded by synchrotron oscillation satellites, however, in such a way that the maximum amplitude is shifted in frequency by $\omega_{\mathrm{o}}\xi_u/\eta_{\mathrm{c}}$. It is interesting to note at this point that the integer part of the tune ν_{o} cannot be distinguished from the integer p of the same value. This is the reason why a spectrum analyzer shows only the *fractional tune* $\Delta\nu\,\omega_{\mathrm{o}}$.

The transverse current spectrum is now just the sum of all contributions from each individual particles. If we assume a uniform distribution $\Phi(t,\hat{\tau},\hat{u})$ in betatron phase, we get no transverse coherent signal because $\langle e^{i\psi_{\mathrm{o}}}\rangle = 0$, although the incoherent space-charge tune shift is effective. Additional coherent signals appear as a result of perturbations of a uniform transverse particle distribution.

10.4 Wake Fields and Impedance

While discussing self fields of a charged particle bunch we noticed a significant effect from nearby metallic surfaces. The dynamics of individual particles as well as collective dynamics of the whole bunch depends greatly on the electromagnetic interaction with the environment. Such interactions must be discussed in more detail to establish stability criteria for particle beams.

The electric field from a charge in its rest frame extends isotropically from the charge into all directions. In the laboratory frame, this field is Lorentz contracted and assumes for a charge in a uniform beam pipe the form shown in Fig. 10.9. The contracted field lines spread out longitudinally only within an angle $\pm 1/\gamma$. This angle is very small for most high energy electron beams and we may describe the single-particle current as well as its image current by a delta function. Some correction must be made to this assumption for lower energy protons and specifically ions for which the angle $1/\gamma$ may still be significant. In the following discussions, however, we will assume that the particle energy is sufficiently large and $\gamma \gg 1$.

The image currents of a charge q travelling along the axis of a uniform and perfectly conducting tube move without losses with the charge and no forces are generated that would act back on the particle. This is not so for a resistive wall where the fields drag a significant distances behind the charge or in case of an obstacle extending into the tube or any other sudden variation of the tube cross section (Fig. 10.9).

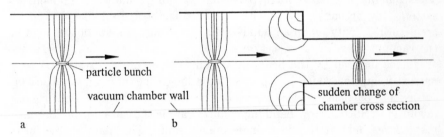

Fig. 10.9. Coupling of a charged particle beam to the environment; uniform chamber cross section **a**, and obstacle on vacuum chamber surface **b**

In any of these cases *wake fields* are created which have the ability to pull or push the charge q or test particles following that charge. Because of causality no such fields exist ahead of a relativistically moving charge.

Energy losses and gains of a single or collection of particles can cause significant modifications in the dynamics of particle motion. Specifically, we are concerned that such forces may lead to particle or beam instability which must be understood in detail to determine limitations or corrective measures in a particular accelerator design. The interaction of a charged particle beam with its environment can be described in *time domain* or *frequency domain* where both have their advantages and disadvantages when it comes to evaluate their effect on particle dynamics.

In *time domain*, the interaction is described by the wake fields which then act on charges. In *frequency domain*, vacuum chamber components can be represented as a frequency dependent impedance. We used this picture before while discussing properties of accelerating cavities. Many vacuum chamber components or sudden changes in cross section behave like cavities and represent therefore frequency dependent impedances. Together with the frequency spectrum of the beam, we find strong coupling to the vacuum chamber if the impedance and particle beam have a significant component at the same frequency. The induced voltage $V(\omega)$ from this interaction is proportional to the collective particle current $I(\omega)$ and the impedance $Z(\omega)$ acting as the proportionality factor, describes the actual coupling from the particle beam via the vacuum chamber environment to the test particle. Mathematically, we set

$$V(\omega) = -Z(\omega) I(\omega) \tag{10.71}$$

indicating by the minus sign that the induced voltage leads to an energy loss for beam particles. The impedance is in general complex and depends for each piece of vacuum chamber including accelerating cavities or accidental cavities, on its shape, material and on the frequency under consideration. The *coupling impedance* for a particular vacuum chamber component or system may be narrow band with a quality factor $Q \gg 1$ like that in an accelerating cavity or broad band with $Q \approx 1$ due to a sudden change in the vacuum chamber cross section.

Fields induced by the beam in a high Q structure are restricted to a narrow frequency width and persist for a long time and can act back on subsequent particle bunches or even on the same particles after one or more revolutions. Such narrow-band impedances can be the cause for *multi-bunch instabilities* but rarely affect single bunch limits. The main source for a narrow-band impedance in a well-designed accelerator comes from accelerating cavities at the fundamental as well as *higher-order mode frequencies*. There is little we can or want do about the impedance at the fundamental frequency which is made large by design for efficiency, but research and development efforts are underway to design accelerating cavities with significantly reduced impedances for *higher-order modes* or *HOM*'s.

The source for broad-band impedances are discontinuities in cross section or material along the vacuum chamber including accelerating cavities, flanges, kicker magnets with ferrite materials, exit chambers electrostatic plates, beam position monitors, etc. Because many higher order modes over a wide frequency range can be excited in such discontinuities by a passing short particle bunch, the fields decohere very fast. Only for a very short time are the fields of all frequencies in phase adding up to a high field intensity but at the time of arrival of the next particle bunch or the same bunch after one or more revolutions these fields have essentially vanished. Broad-band wake fields are therefore mainly responsible for the appearance of single-bunch beam instabilities.

The particle beam covers a wide frequency spectrum from the kHz regime of the order of the revolution frequency up to many GHz limited only by the bunch length. On the other hand, the vacuum chamber environment constitutes an impedance which can become significant in the same frequency regime and efficient coupling can occur leading to collective effects. The most important impedance in an accelerator is that of the accelerating cavity at the cavity fundamental frequency. Since the particle beam is bunched at the same frequency, we observe a very strong coupling which has been extensively discussed in Chap. 6 in connection with beam loading. In this section we will therefore ignore beam loading effects in resonant cavities at the fundamental frequency and concentrate only on higher-order mode losses and interaction with the general vacuum chamber environment.

10.4.1 Definitions of Wake Field and Impedance

A bunched particle beam of high intensity represents a source of electromagnetic fields, called *wake fields* [10.29] in a wide range of wavelengths down to the order of the bunch length. The same is true for a realistic coasting beam where fluctuations in beam current simulate short particle bunches on top of an otherwise uniform beam.

In [Ref. 10.17, Sect. 12.3] very basic ideas of wake fields and higher-order mode losses were introduced distinguishing two groups, the longitudinal and the transverse wake fields. The longitudinal wake fields being in phase with the beam current cause energy losses to the beam particles, while transverse wakes deflect particle trajectories. There is no field ahead of relativistically moving charge due to causality. From the knowledge of such wake fields in a particular environment we may determine the effect on a test charge moving behind a charge q.

The character of local wake fields depends greatly on the actual geometry and material of the vacuum chamber and we may expect a significant complication in the determination of wake field distributions along a vacuum enclosure of an actual accelerator. It is not practical to evaluate these fields in detail along the beam path and fortunately we do not need to. Since the effects of localized fields are small compared to the energy of the particles, we may integrate the wake fields over a full circumference. As we will see, this integral of the field can be experimentally determined.

One may wonder how the existence of an obstacle in the vacuum chamber, like a disk which is relatively far away from the charge q, can influence a test particle following closely behind the charge q. To illustrate this, we consider the situation shown in Fig. 10.10.

Long before the charge reaches the obstruction, fields start to diverge from the charge toward the obstruction to get scattered there. Some of the scattered fields move move again toward the charge and catch up with it due to its slightly faster speed.

Longitudinal wake fields: The details of this catch up process are, however, of little interest compared to the integrated effect of wake fields on the test particle. Each charge at the position z creates a wake field for a particle at location $\tilde{z} < z$ and this wake field persists during the whole travel time along an accelerator segment L assuming that the distance $\zeta = z - \tilde{z}$ does not change appreciably along L. We define now a *longitudinal wake function* by integrating the longitudinal wake fields $\mathbf{E}_\parallel$ along the interaction length L, which might be the length of a vacuum chamber component, a linear accelerator or the circumference of a circular accelerator, and normalize it to a unit charge. By integrating, which is the same as averaging over the length L, we eliminate the need to calculate everywhere the complicated fields along the vacuum chambers. The wake field at the location of a test particle at $\tilde{z}$ from a charge q at location z is then (Fig. 10.10)

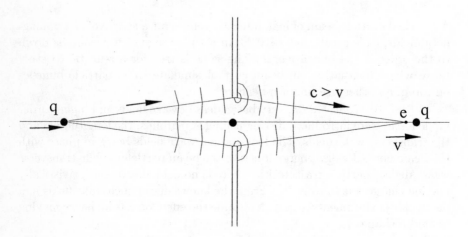

Fig. 10.10. Catch up of wake fields with test particle

$$W_\parallel(\zeta) = \frac{1}{q} \int_L \mathbf{E}_\parallel(s, t - \zeta/\beta c)\, ds \,, \tag{10.72}$$

where $\zeta = z - \tilde{z} > 0$. The wake function is measured in V/Cb using practical units and is independent of the sign of the charge. To get the full wake field for a test particle, one would integrate the wake function over all particles ahead of the test particle.

The longitudinal wake function allows us to calculate the total energy loss of the whole bunch by integrating over all particles. We consider a test particle with charge e at position $\tilde{z}$ and calculate the energy loss of this particle due to wake fields from charges further ahead at $z \geq \tilde{z}$ The total induced voltage on the test charge at $\tilde{z}$ is then determined by the *wake potential*[1]

$$V_{\mathrm{HOM}}(\tilde{z}) = -e \int\limits_{\tilde{z}}^{\infty} \lambda(z)\, W_{\parallel}(z - \tilde{z})\, \mathrm{d}z , \qquad (10.73)$$

where a negative sign was added to indicate that the wake fields are decelerating. Integrating over all slices $\mathrm{d}\tilde{z}$, the total energy loss of the bunch into HOM fields is

$$\Delta U_{\mathrm{HOM}} = - \int\limits_{-\infty}^{\infty} e\lambda(\tilde{z})\, \mathrm{d}\tilde{z} \underbrace{\int\limits_{\tilde{z}}^{\infty} e\lambda(z)\, W_{\parallel}(z - \tilde{z})\, \mathrm{d}z}_{\text{wake potential at } \tilde{z}} , \qquad (10.74)$$

where $\lambda(z)$ is the linear distribution of particles with charge e and normalized to the total number of particles N_{b} in the bunch $\int \lambda(z)\mathrm{d}z = N_{\mathrm{b}}$.

It is interesting to perform the integrations in (10.74) for a very short bunch such that the wake function seen by particles in this bunch is approximately constant and equal to W_{o}. In this case we define the function $w(\tilde{z}) = \int_{\tilde{z}}^{\infty} e\lambda(z)\, \mathrm{d}z$ and the double integral assumes the form $-\int_{-\infty}^{\infty} w\, \mathrm{d}w = \tfrac{1}{2}\,(eN_{\mathrm{b}})^2$ where we have used the normalization $w(-\infty) = eN_{\mathrm{b}}$. Particles in a bunch see therefore only 50% of the wake fields produced by the same bunch consistent with our earlier formulation of the *fundamental theorem of wake fields* discussed in Chap. 6 in connection with wake fields in rf cavities. By the same argument each particle sees only half of its own wake field.

Wake functions describe higher-order mode losses in the time domain. For further discussions, we determine the relationship to the concept of impedance in the frequency domain and replace in (10.73) the charge distribution with the instantaneous current passing by $\tilde{z}$

$$I(\tilde{z}, t) = \hat{I}_{\mathrm{o}}\, \mathrm{e}^{\mathrm{i}(k\tilde{z} - \omega t)} . \qquad (10.75)$$

The beam current generally includes more than one mode k but for simplicity we consider only one in this discussion. Integrating over all parts of the beam which have passed the location $\tilde{z}$ before, the wake potential (10.73) becomes

---

[1] Expression (10.72) is sometimes called the wake potential. We do not follow this nomenclature because the expression (10.72) does not have the dimension of a potential but (10.73) does.

$$V_{\text{HOM}}(\tilde{z}, t) = -\frac{1}{c\beta} \int\limits_{\tilde{z}}^{\infty} I\left(\tilde{z}, t + \frac{z - \tilde{z}}{c\beta}\right) W_{\parallel}(z - \tilde{z}) \, \mathrm{d}z. \tag{10.76}$$

Consistent with a time dependent beam current, the induced voltage depends on location $\tilde{z}$ and time as well. The wake function vanishes due to causality for $z - \tilde{z} < 0$ and the integration can therefore be extended over all values of z. With (10.75), $\zeta = z - \tilde{z}$ and applying a Fourier transform (10.76) becomes

$$V_{\text{HOM}}(t, \omega) = -I(t, \omega) \frac{1}{c\beta} \int\limits_{-\infty}^{\infty} \mathrm{e}^{-\mathrm{i}\omega\zeta/c\beta} W_{\parallel}(\zeta) \, \mathrm{d}\zeta. \tag{10.77}$$

From (10.77) we define the *longitudinal coupling impedance* in the frequency domain

$$Z_{\parallel}(\omega) = \frac{1}{c\beta} \int\limits_{-\infty}^{\infty} \mathrm{e}^{-\mathrm{i}\omega\zeta/c\beta} W_{\parallel}(\zeta) \, \mathrm{d}\zeta \tag{10.78}$$

which has in practical units the dimension Ohm. The impedance of the environment is the Fourier transform of the wake fields left behind by the beam in this environment. Because the wake function has been defined in (10.72) for the length L of the accelerator under consideration, the impedance is an integral parameter of the accelerator section L as well. Conversely, we may express the wake function in terms of the impedance spectrum

$$W_{\parallel}(z) = \frac{1}{2\pi} \int\limits_{-\infty}^{\infty} Z_{\parallel}(\omega) \, \mathrm{e}^{\mathrm{i}\omega z/c\beta} \, \mathrm{d}\omega. \tag{10.79}$$

The interrelations between wake functions and impedances allows us to use the most appropriate quantity for the problem at hand. Generally, it depends on whether one wants to work in the frequency or the time domain. For theoretical discussions, the well defined impedance concept allows quantitative predictions for beam stability or instability to be made. In most practical applications, however, the impedance is not quite convenient to use because it is not well known for complicated shapes of the vacuum chamber. In a linear accelerator, for example, we need to observe the stability of particles in the time domain to determine the head-tail interaction. The most convenient quantity depends greatly on the problem to be solved, theoretically or experimentally.

Loss parameter: In a real accelerator, the beam integrates over many different vacuum chamber pieces with widely varying impedances. We are therefore not able to experimentally determine the impedance or wake function of a particular vacuum chamber element. Only the integrated impedance for the whole accelerator can sometimes be probed at specific frequencies by observing specific instabilities as we will discuss later. The most accurate quantity to measure the total resistive impedance for the whole accelerator integrated over all frequencies is the *loss factor* or *loss parameter* introduced in [Ref. 10.17, Sect. 12.3]. and defined by

$$k_{\text{HOM}} = \frac{\Delta U_{\text{HOM}}}{e^2 \, N_{\text{b}}^2} .$$ (10.80)

This loss factor can be related to the wake function and we get from comparison with (10.74), the relation

$$k_{\parallel \, \text{HOM}} = \frac{1}{N_{\text{b}}^2} \int\limits_{-\infty}^{\infty} \lambda(\tilde{z}) \, \mathrm{d}\tilde{z} \int\limits_{\tilde{z}}^{\infty} \lambda(z) \, W_{\parallel}(z - \tilde{z}) \, \mathrm{d}z .$$ (10.81)

The loss parameter can be defined for the complete circular accelerator or for a specific vacuum chamber component installed in a beam line or accelerator. Knowledge of this loss factor is important to determine possible heating effects which can become significant since the total higher-order mode losses are deposited in the form of heat in the vacuum chamber component. In a circular accelerator, the energy loss rate or heating power of a beam circulating with revolution frequency f_{o} is

$$P_{\text{HOM}} = k_{\text{HOM}} \, \frac{I_{\text{o}}^2}{f_{\text{o}} \, n_{\text{b}}} ,$$ (10.82)

where n_{b} is the number of bunches in the beam and $I_{\text{o}} = n_{\text{b}} \, q N_{\text{b}} \, f_{\text{o}}$ is the average circulating beam current in the accelerator. As an example, we consider a circulating beam of 1mA in one bunch of the LEP storage ring where the revolution frequency is about $f_{\text{o}} = 10$ kHz. The heating losses in a component with loss factor $k_{\text{HOM}} = 0.1$ V/pCb would be 10 Watts. This might not seem much if the component is large and an external cooling fan might be sufficient. On the other hand, if the vacuum component is small and not accessible like a bellows this heating power might be significant and must be prevented by design. The higher-order heating losses scale like the average current, the bunch current and inversely proportional with the revolution frequency. For a given circulating beam current, the losses depend therefore greatly on the number of bunches this beam current is distributed over and the size of the circular accelerator.

In addition, the loss factor depends strongly on the bunch length. In the SPEAR storage ring an overall dependence $k_{\text{HOM}} \propto \sigma^{-1.21}$ has been measured (Fig. 10.11) [10.30].

A drastic reduction in bunch length can therefore significantly increase heating effects. Although we try to apply a careful design to all accelerator components to minimize the impedance it is prudent to be aware of

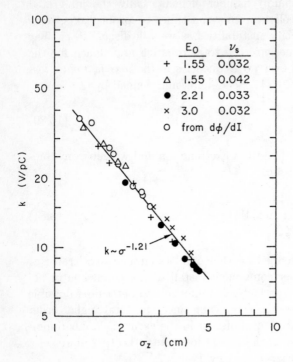

Fig. 10.11. Dependence of the overall loss factor in the storage ring SPEAR on the bunch length [10.30]

this heating effect while developing accelerators that involve significantly reduced bunch length like those in *quasi-isochronous storage rings* or beams accelerated by laser beams.

The loss parameter can be measured by observing the shift in the synchronous phase. A bunch of particles circulating in an accelerator looses energy due to the resistive impedance of the vacuum chamber. This additional energy loss is compensated by an appropriate shift in the synchronous phase which is given by

$$\Delta U_{\mathrm{HOM}} = e N_{\mathrm{b}} V_{\mathrm{rf}} |\sin \phi_{\mathrm{s}} - \phi_{\mathrm{so}}|, \tag{10.83}$$

where ϕ_{so} is the synchronous phase for a very small beam current and V_{rf} the peak rf voltage. The loss factor is then

$$k_{\mathrm{HOM}} = \frac{\Delta U_{\mathrm{HOM}}}{e^2 N_{\mathrm{b}}^2}. \tag{10.84}$$

Performing this measurement as a function of rf voltage one can establish a curve similar to that shown in Fig. 10.11 for the storage ring SPEAR and the dependence of the loss parameter on the bunch length can be used to determine the total resistive impedance of the accelerator as a function of frequency.

To do that we write (10.81) in terms of Fourier transforms

$$k_{\parallel_{\mathrm{HOM}}} = \frac{\pi}{e^2 N_{\mathrm{b}}^2} \int\limits_{-\infty}^{\infty} Z_{\mathrm{res}}(\omega)\,|I(\omega)|^2\,\mathrm{d}\omega \tag{10.85}$$

and recall that the bunch or current distribution is gaussian in a storage ring

$$I(\tau) = \frac{1}{\sqrt{2\pi}\sigma_\tau}\,\mathrm{e}^{-\tau^2/2\sigma_\tau^2}. \tag{10.86}$$

The Fourier transform of a gaussian distribution is

$$I(\omega) = I_{\mathrm{o}}\,\mathrm{e}^{-\frac{1}{2}\omega^2\sigma_\tau^2}, \tag{10.87}$$

where I_{o} is the total bunch current and inserting (10.87) into (10.85) we get

$$k_{\parallel_{\mathrm{HOM}}} = \frac{\pi\,I_{\mathrm{o}}}{e^2 N_{\mathrm{b}}^2} \int\limits_{-\infty}^{\infty} Z_{\mathrm{res}}(\omega)\,\mathrm{e}^{-\omega^2\sigma_\tau^2}\,\mathrm{d}\omega. \tag{10.88}$$

With (10.88) and the measurement $k_{\parallel_{\mathrm{HOM}}}(\sigma_\ell)$, where $\sigma_\ell = c\sigma_\tau$, one may solve for $Z_{\mathrm{res}}(\omega)$ and determine the *resistive-impedance spectrum* of the ring.

Unfortunately, it is not possible to attach a resistance meter to an accelerator to determine its impedance and we will have to apply a variety of wake field effects on the particle beams to determine the complex impedance as a function of frequency. No single effect, however, will allow us to measure the whole frequency spectrum of the impedance.

Transverse wake fields: Similar to the longitudinal case we also observe transverse wake fields with associated impedances. Such fields exert a transverse force on particles generated by either transverse electrical or magnetic wake fields. Generally such fields appear when a charged particle beam passes off center through a nonuniform but cylindrical or through an asymmetric vacuum chamber. Transverse wake fields can be induced only on structures which also exhibit a longitudinal impedance. A beam travelling off center through a round pipe with perfectly conducting walls will not create longitudinal and therefore also no transverse wake fields.

We consider a charge q passing through a vacuum chamber structure with an offset $\Delta u = (\Delta x, \Delta y)$ in the horizontal or vertical plane as shown in Fig. 10.12.

In analogy to the definition of the longitudinal wake function (10.72), we define a *transverse wake function* per unit transverse offset by

$$W_\perp(\zeta,t) = +\frac{\int_L \{\mathbf{E}(t-\zeta/\beta c) + [c]\,[\boldsymbol{\beta}\times\mathbf{B}](t-\zeta/\beta c)\}_\perp\,\mathrm{d}s}{q\,\Delta u} \tag{10.89}$$

which is measured in units of V/Cb/m. Consistent with the definition (10.78) of the longitudinal impedance, the *transverse coupling impedance* is the Fourier transform of the transverse wake functions defined by

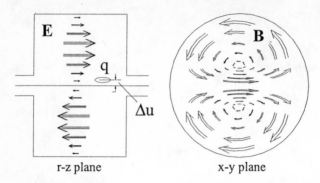

r-z plane   x-y plane

Fig. 10.12. Transverse wake field generated by a charge q with a transverse offset Δu as seen by a test particle q_t

$$Z_\perp(\omega) = -\frac{i}{c\beta} \int\limits_{-\infty}^{\infty} e^{-i\omega\zeta/c\beta}\, W_\perp(\zeta)\, d\zeta \tag{10.90}$$

adding the factor i to indicate that the action of the transverse force is a mere deflection while the particle energy stays constant. This transverse impedance is measured in Ohm/m. The inverse relation is similar to the longitudinal case

$$W_\perp(z) = -\frac{i}{2\pi} \int\limits_{-\infty}^{\infty} Z_\perp(\omega)\, e^{i\omega z/c\beta}\, d\omega\,. \tag{10.91}$$

Panofsky-Wenzel theorem: The general relationship between longitudinal and transverse wake fields is expressed by the *Panofsky-Wenzel theorem* [10.31]. *Panofsky* and *Wenzel* studied the effect of transverse electromagnetic fields on a particle trajectory and applied general relations of electromagnetic theory to derive a relationship between longitudinal and transverse electromagnetic forces. We will derive the same result in the realm of wake fields. The Lorentz force on a test particle at $\tilde{z}$ due to transverse wake fields from charges at location $z > \tilde{z}$ causes a deflection of the particle trajectory and the change in transverse momentum of the test particle is after integration over all charges at locations $z < \tilde{z}$

$$c\mathbf{p}_\perp = \frac{e}{\beta} \int\limits_{-\infty}^{\infty} \left[\mathbf{E} + \frac{[c]}{c}\,(\mathbf{v} \times \mathbf{B})\right]_\perp dz\,. \tag{10.92}$$

Note that the wake fields vanish because of causality for $\zeta < 0$. The fields can be expressed by the vector potential $\mathbf{E}_\perp = -\frac{[c]}{c}\partial \mathbf{A}_\perp/\partial t$ and $\mathbf{B}_\perp = (\nabla \times \mathbf{A})_\perp$.

The particle velocity has only one nonvanishing component $\mathbf{v} = (0, 0, v)$ and (10.92) becomes with $\partial z / \partial t = v$

$$
c\mathbf{p}_\perp = -[c]\,e \underbrace{\int_0^d \overbrace{\left(\frac{\partial}{\partial t} \frac{\partial t}{\partial z} + \frac{\partial}{\partial z} \right)}^{=d/dz} A_\perp \, dz}_{=0} + [c]\,e\nabla_\perp \int_0^d A_\parallel \, dz , \tag{10.93}
$$

where we made use of the vector relation

$$
\mathbf{v} \times (\nabla \times \mathbf{A}) + \underbrace{\mathbf{A} \times (\nabla \times \mathbf{v})}_{=0} = \nabla_\perp (\mathbf{vA}) - (\mathbf{v}\nabla)\,\mathbf{A}_\perp - \underbrace{(\mathbf{A}\nabla)\,\mathbf{v}}_{=0} . \tag{10.94}
$$

noting that the particle velocity is a constant.

The integrand in the first integral of (10.93) is equal to the total derivative $dA_\perp/ds$ and the integral vanishes because the fields vanish for $\zeta = \pm\infty$. After differentiation with respect to the time t (10.93) becomes

$$
\frac{d\mathbf{p}_\perp}{dt} = -e\nabla_\perp \int_{-\infty}^{\infty} E_\parallel \, dz \tag{10.95}
$$

which is in terms of forces

$$
\frac{\partial}{\partial z} \mathbf{F}_\perp = -\nabla_\perp \mathbf{F}_\parallel . \tag{10.96}
$$

The longitudinal gradient of the transverse force or electromagnetic field is proportional to the transverse gradient of the longitudinal force or electromagnetic field and knowledge of one allows us to calculate the other.

10.4.2 Impedances in an Accelerator Environment

The vacuum chamber of an accelerator is too complicated in geometry to allow an analytical expression for its impedance. In principle each section of the chamber must be treated individually. By employing two or three-dimensional numerical codes it may be possible to determine the impedance for a particular component and during a careful design process for a new accelerator, this should be done to avoid later surprises. Yet, every accelerator is somewhat different from another and will have its own particular overall impedance characteristics. For this reason, we focus in these discussions specifically on such instabilities which can be studied experimentally revealing impedance characteristics of the ring. However, depending on the frequency involved, there are a few classes of impedances which are common to all accelerators and may help understand the appearance and strength of certain instabilities. In this section, we will discuss such impedances to be used in later discussions on stability conditions and growth rate of instabilities.

Consistent with (10.72,78) the *longitudinal impedance* for a circular accelerator is defined as the ratio of the induced voltage at frequency ω to the Fourier transform of the beam current at the same frequency

$$
\begin{aligned}
Z_{\|}(\omega) &= -\frac{\int \mathbf{E}_{\|}(\omega)\,\mathrm{ds}}{I(\omega)} \\
&= \left[\frac{1}{4\pi\epsilon_\mathrm{o}}\right] \frac{1}{eN_\mathrm{b}} \int_L \mathbf{E}_{\|}(z, t - \zeta/\beta c)\,\mathrm{e}^{-\mathrm{i}\omega\zeta/\beta c}\,\mathrm{ds}\,.
\end{aligned}
\tag{10.97}
$$

Similarly the *transverse impedance* is from (10.89,90) the ratio of induced transverse voltage to the transverse moment of the beam current

$$
Z_{\perp}(\omega) = -\mathrm{i}\,\frac{\int (\mathbf{E}_{\perp} + [c]\,[\boldsymbol{\beta} \times \mathbf{B}]_{\perp})|_{(z,t-\zeta/\beta c)}\mathrm{e}^{-\mathrm{i}\omega\zeta/\beta c}\,\mathrm{dz}}{I(\omega)\,\Delta u}\,,
\tag{10.98}
$$

where Δu is the horizontal or vertical offset of the beam from the axis.

Space-charge impedance: In (10.57) we found an induced voltage leading to an energy gain or loss due to a collection of charged particles. It is customary to express (10.57) in a form exhibiting the impedance of the vacuum chamber. In case of a perfectly conducting vacuum chamber $E_{zw} = 0$ and (10.57) becomes

$$
V_z = -Z_{\|\mathrm{sc}}\,I_n\,\mathrm{e}^{\mathrm{i}(n\theta - \omega_n t)}\,,
\tag{10.99}
$$

where the longitudinal *space-charge impedance* $Z_{\|\mathrm{sc}}$ is defined by

$$
Z_{\|\mathrm{sc}}(\omega_n) = \frac{\mathrm{i}}{[4\pi\epsilon_\mathrm{o}]}\,\frac{2\pi n}{\beta c\,\gamma^2}\left(1 + 2\ln\frac{r_\mathrm{w}}{r_\mathrm{o}}\right).
\tag{10.100}
$$

This expression is correct for long wavelength below cut off of the vacuum chamber or for $\omega < c/r_\mathrm{w}$. The space-charge impedance is purely reactive and, as we will see, capacitive. It is customary to divide the impedance by the harmonic number n defining a *normalized impedance*

$$
\frac{Z_{\|\mathrm{sc}}(\omega_n)}{n} = \frac{\mathrm{i}}{[4\pi\epsilon_\mathrm{o}]}\,\frac{2\pi}{\beta c\,\gamma^2}\left(1 + 2\ln\frac{r_\mathrm{w}}{r_\mathrm{o}}\right),
\tag{10.101}
$$

where $n = \omega_n/\omega_\mathrm{o}$ and ω_o the nominal revolution frequency.

For a round beam of radius r_o and offset with respect to the axis of a round beam pipe with diameter $2r_\mathrm{w}$ a *transverse space-charge impedance* can be derived which is

$$
Z_{\perp\mathrm{sc}}(\omega_n) = -\frac{\mathrm{i}}{[4\pi\epsilon_\mathrm{o}]}\,\frac{\bar{R}}{\beta^2\,\gamma^2}\left(\frac{1}{r_\mathrm{o}^2} - \frac{1}{r_\mathrm{w}^2}\right).
\tag{10.102}
$$

The transverse space-charge impedance is inversely proportional to β^2 and is therefore especially strong for low energy particle beams.

Resistive-wall impedance: The particle beam induces an image current in the vacuum chamber wall in a thin layer with a depth equal to the skin depth. For less than perfect conductivity of the wall material, we observe resistive losses which exert a pull or decelerating field on the particles. This pull is proportional to the beam current and integrating the fields around a full circumference of the accelerator, we get for the associated *longitudinal resistive wall impedance* in a uniform tube of radius r_{w} at frequency ω_n

$$\left.\frac{Z_{\parallel}(\omega_n)}{n}\right|_{\mathrm{res}} = (1+\mathrm{i})\,\frac{\bar{R}}{cr_{\mathrm{w}}}\,\sqrt{\frac{2\pi\omega_0\mu}{n\,\sigma}} = (1+\mathrm{i})\,\frac{\bar{R}}{nr_{\mathrm{w}}\sigma\,\delta_{\mathrm{skin}}}\,, \qquad (10.103)$$

where the *skin depth* is defined by [10.32]

$$\delta_{\mathrm{skin}}(\omega) = \left[\frac{\sqrt{4\pi}}{c}\right]\frac{c}{\sqrt{2\pi\,\mu\,\omega\,\sigma}}\,. \qquad (10.104)$$

The longitudinal resistive wall impedance decays with increasing frequency and therefore plays an important role only for low frequencies.

Cavity-like structure impedance: The impedance of accelerating cavities or cavity like objects of the vacuum chamber can be described by the equivalent of a parallel resonant circuit for which the impedance is from (6.86)

$$\left.\frac{1}{Z_{\parallel}}\right|_{\mathrm{cy}}(\omega) = \frac{1}{R_{\mathrm{s}}}\left(1+\mathrm{i}Q\,\frac{\omega^2-\omega_{\mathrm{r}}^2}{\omega_{\mathrm{r}}\,\omega}\right)\,, \qquad (10.105)$$

where Q is the quality factor and R_{s} the cavity impedance at the resonance frequency ω_{r} or *cavity shunt impedance*. Taking the inverse, we get for the normalized impedance

$$\left.\frac{Z_{\parallel}}{n}\right|_{\mathrm{cy}}(\omega) = \left.\left|\frac{Z_{\parallel}}{n}\right|\right|_{\mathrm{o}}\frac{1+\mathrm{i}Q\,\frac{\omega^2-\omega_{\mathrm{r}}^2}{\omega_{\mathrm{r}}\,\omega}}{1+Q^2\,\frac{(\omega^2-\omega_{\mathrm{r}}^2)^2}{\omega_{\mathrm{r}}^2\,\omega^2}}\,, \qquad (10.106)$$

where $\left|\frac{Z}{n}\right|_{\mathrm{o}} = R_{\mathrm{s}}$ is purely resistive.

Vacuum chamber impedances occur, for example, due to sudden changes of cross section, flanges, beam position monitors, etc., and are collectively described by a cavity like impedance with a quality factor $Q \approx 1$. This is justified because fields are induced in these impedances at any frequency. From (10.106) the *longitudinal broad-band impedance* is therefore

$$\left.\frac{Z_\parallel}{n}\right|_{bb}(\omega) = \left.\left|\frac{Z_\parallel}{n}\right|\right._o \frac{1 - i\frac{\omega^2 - \omega_r^2}{\omega_r\,\omega}}{1 + \frac{(\omega^2 - \omega_r^2)^2}{\omega_r^2\,\omega^2}} \cdot \qquad (10.107)$$

This broad-band impedance spectrum is shown in Fig. 10.13 and we note that the resistive and reactive part exhibit different spectra.

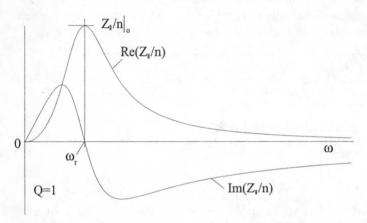

Fig. 10.13. Resistive and reactive broad-band impedance spectrum

The resistive broad-band impedance has a symmetric spectrum and scales like ω^2 for low frequencies decaying again for very high frequencies like $1/\omega^2$. At low frequencies, the broad-band impedance (10.107) is almost purely reactive and inductive scaling linear with frequency

$$\left.\frac{Z_\parallel}{n}\right|_{bb} = i\left.\left|\frac{Z_\parallel}{n}\right|\right._o \frac{\omega}{\omega_r} \qquad \text{for} \quad \omega \ll \omega_r. \qquad (10.108)$$

At high frequencies the impedance becomes capacitive

$$\left.\frac{Z_\parallel}{n}\right|_{bb} = -i\left.\left|\frac{Z_\parallel}{n}\right|\right._o \frac{\omega_r}{\omega} \qquad \text{for} \quad \omega \gg \omega_r \qquad (10.109)$$

and decaying slower with frequency than the resistive impedance. We note, however, that the reactive broad-band impedance spectrum changes sign and beam stability or instability depend therefore greatly on the actual coupling frequency. At resonance, the broad-band impedance is purely resistive as would be expected.

Sometimes it is convenient to have a simple approximate correlation between longitudinal and transverse impedance in a circular accelerator. This has been done for a round vacuum pipe of radius r_w where the *transverse broad-band impedance* can be calculated from [10.33]

$$Z_\perp(\omega_n) \approx \frac{2c}{r_w^2 \omega} Z_\parallel(\omega_n).$$ (10.110)

Although this correlation is valid only for the resistive wall impedance in a round beam pipe, it is often used for approximate estimates utilizing the broad-band impedance.

Overall accelerator impedance: At this point, we have identified all significant types of impedances we generally encounter in an accelerator which are the space charge, resistive wall, narrow-band impedances in high Q cavities, and broad-band impedance. In Fig. 10.14 we show qualitatively these resistive as well as reactive impedance components as a function of frequency.

At low frequency the reactive as well as the resistive component of the resistive wall impedance dominates while the space charge impedance is independent of frequency. The narrow-band cavity spectrum includes the high impedances at the fundamental and higher mode frequencies.

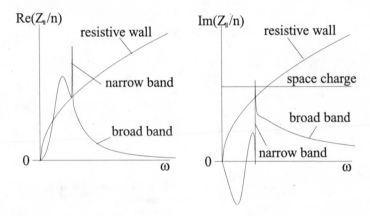

Fig. 10.14. Qualitative spectra of resistive and reactive coupling impedances in a circular accelerator

Generally, it is not possible to use a uniform vacuum chamber in circular accelerators. Deviations from a uniform chamber occur at flanges, bellows, rf cavities, injection/ejection elements, electrostatic plates, etc. It is not convenient to consider the special impedance characteristics of every vacuum chamber piece and we may therefore look for an average impedance as seen by the beam. The broad-band impedance spectrum created by chamber components in a ring reaches a maximum at some frequency and then diminishes again like $1/\omega$. This turn over of the broad-band impedance function depends on the general dimensions of all vacuum chamber components of a circular accelerator and has to do with the cut off frequency for travelling waves in tubes. In Fig. 10.15, the measured impedance spectrum of the SPEAR storage ring [10.34] is reproduced from [Ref. 10.17, Fig. 12.4].

This spectrum is typical for complex storage ring vacuum chambers which are generally composed of similar components exhibiting at low frequencies an *inductive impedance* increasing linearly with frequency and diminishing again at high frequencies. This is also the characteristics of broadband cavity impedance and therefore expressions for broad-band impedance are useful tools in developing theories for beam instabilities and predicting conditions for beam stability. The induced voltage for the total ring circumference scales like $L\,\dot{I}_{\mathrm{w}}$ where L is the wall inductance and $\dot{I}_{\mathrm{w}}$ the time derivative of the image current in the wall. With (10.56) the induced voltage is

$$\Delta V_{zo} = -L\frac{\mathrm{d}I}{\mathrm{d}t} = \mathrm{i}\,\omega\,L(\omega)\,I_n\,\mathrm{e}^{\mathrm{i}(n\theta-\omega_n t)} \tag{10.111}$$

where the *inductive impedance* is defined by

$$Z_{\parallel\mathrm{ind}} = -\mathrm{i}\,\omega\,L(\omega)\,. \tag{10.112}$$

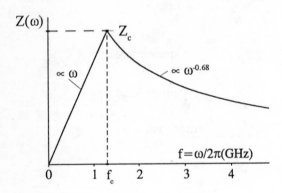

Fig. 10.15. Impedance spectrum of the storage ring SPEAR

The total induced voltage due to space charge, resistive and inductive wall impedance is finally

$$V_{zw} = -Z_{\parallel}\,I_n\,\mathrm{e}^{\mathrm{i}(n\theta-\omega_n t)}\,, \tag{10.113}$$

where the total longitudinal normalized impedance at frequency ω_n is from (10.101,103,112)

$$[4\pi\epsilon_0]\,\frac{Z_{\parallel}(\omega_n)}{n} = \mathrm{i}\,\frac{2\pi}{\beta c\gamma^2}\left(1+2\ln\frac{r_{\mathrm{w}}}{r_{\mathrm{o}}}\right)$$
$$-\mathrm{i}\,\omega_{\mathrm{o}}\,L(\omega_n) - \frac{(\mathrm{i}-1)\bar{R}}{cr_{\mathrm{w}}}\sqrt{\frac{2\pi\omega_{\mathrm{o}}\mu}{n\sigma}}\,. \tag{10.114}$$

From the frequency dependence, we note that space charge and inductive wall impedance becomes more important at high frequencies while the resistive wall impedance is dominant at low frequencies. The inductive wall impedance derives mostly from vacuum chamber discontinuities like sudden change in the vacuum chamber cross section, bellows, electrostatic plates, cavities, etc. In older accelerators, little effort was made to minimize the impedance and total ring impedances of $|Z_\parallel/n| \approx 20$ to 30Ω were not uncommon. Modern vacuum design have been developed to greatly reduce the impedance mostly by avoiding changes of vacuum chamber cross section or by introducing gentle transitions and impedances of the order of $|Z_\parallel/n| \approx 1$ Ω can be achieved whereby most of this remaining impedance comes from accelerating rf cavities.

From (10.114), we note that the space-charge impedance has the opposite sign of the inductive impedance and is therefore capacitive in nature. In general, we encounter in a realistic vacuum chamber resistive as well as reactive impedances causing both real frequency shifts or imaginary shifts manifesting themselves in the form of damping or instability. In subsequent sections, we will discuss a variety of such phenomena and derive stability criteria, beam-current limits or rise times for instability. At this point, it is noteworthy to mention that we have not made any assumption as to the nature of the particles involved and we may therefore apply the results of this discussion to electron as well as proton and ion beams.

Broad-band wake fields in a linear accelerator: The structure of a linear accelerator constitutes a large impedance for a charged particle beam, specifically, since particle bunches are very short compared to the periodicity of the accelerator lattice. Every single cell resembles a big sudden change of the vacuum chamber cross section and we expect therefore a large accumulation of wake fields or impedance along the accelerator. The wake fields can be calculated numerically [10.29] and results for both the longitudinal and transverse wakes from a point charge are shown in Fig. 10.16 as a function of the distance behind this point charge.

Broad-band wake fields for other structures look basically similar to those shown in Fig. 10.16. Specifically, we note the longitudinal wake to be strongest just behind the head of the bunch while the transverse wake builds up over some distance. For an arbitrary particle distribution, one would fold the particle distribution with these wake functions to obtain the wake potential at the location of the test particle.

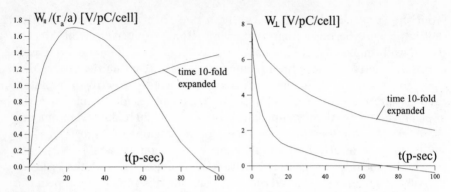

Fig. 10.16. Time dependence of longitudinal (left) and transverse (right) wake fields from a point charge moving through one 3.3 cm long cell of a SLAC type 3GHz linear accelerator structure [10.29]. For other structures the numerical values may be different, but the general time dependence of the wake fields are similar

10.5 Coasting-Beam Instabilities

The space-charge impedance as well as resistive and reactive wall impedances extract energy from a circulating particle beam. As long as the particle distribution is uniform, this energy loss is the same for all particles and requires simple replacement in acceleration cavities. In reality, however, some modulation of the longitudinal particle distribution cannot be avoided and we encounter therefore an uneven energy loss along the *coasting particle beam*. This can have serious consequences on beam stability and we therefore need to discuss stability criteria for coasting beams.

10.5.1 Negative-Mass Instability

Consider a beam in a ring below transition energy. The repulsive electrostatic field from a lump in the charge distribution causes particles ahead of the lump to be accelerated and particles behind the lump to be decelerated. Since accelerated particles will circulate faster and decelerated particles circulate slower, we observe a stabilizing situation and the lumpy particle density becomes smoothed out.

At energies above transition energy the situation changes drastically. Now the acceleration of a particle ahead of a lump leads to a slower revolution frequency and it will actually move closer to the lump with every turn. Similarly a particle behind the lump becomes decelerated and circulates therefore faster, again catching up with the lump. We observe an instability leading to a growing concentration of particles wherever a small perturbation started to occur. We call this instability the *negative-mass instability* [10.35] because acceleration causes particles to circulate slower similar to the acceleration of a negative mass. The same mechanism can lead to stabilization of oscillations if the forces are attractive rather than repulsive. Nature demonstrates this in the stability of *Saturn's rings*.

We will derive conditions of stability for this effect in a more quantitative way. The stability condition depends on the variation of the revolution frequency for particles close to the small perturbation of an otherwise uniform longitudinal particle distribution and we therefore investigate the time derivative of the revolution frequency

$$\frac{d\omega}{dt} = \frac{\partial\omega}{\partial t} + \frac{\partial\omega}{\partial\theta}\frac{\partial\theta}{\partial t}$$

which can also be expressed in the form

$$\frac{d\omega}{dt} = \frac{d\omega}{dE}\frac{dE}{dt} = \frac{\eta_c\omega_o}{\beta^2 E_o}\frac{dE}{dt}, \tag{10.115}$$

where η_c is the *momentum compaction*. The energy change per unit time is for a longitudinal impedance Z_z and nth harmonic of the beam current

$$\frac{dE}{dt} = qV_{zo}\frac{\omega_o}{2\pi} = -qZ_z I_n e^{i(n\theta-\omega t)}\frac{\omega_o}{2\pi}, \tag{10.116}$$

where $q = eZ > 0$ is the electrical charge of the particle and Z the charge multiplicity. Collecting all terms for (10.115) we get with

$$\omega = \omega_o + \omega_n e^{i(n\theta-\Omega t)} \tag{10.117}$$

the relation

$$\omega_n(\Omega - n\omega_o) = -i\frac{q\eta_c\omega_o^2}{2\pi\beta^2}\frac{I_n Z_z}{E_o}. \tag{10.118}$$

This can be further simplified with the continuity equation

$$\frac{\partial\lambda}{\partial t} + \frac{1}{R}\frac{\partial}{\partial\theta}(\beta c\,\lambda) = \frac{\partial\lambda}{\partial t} + \frac{\partial\lambda}{\partial\theta}\omega_o + \frac{\partial\omega}{\partial\theta}\lambda_o = 0$$

and we get with (10.56,117)

$$(\Omega - n\omega_o)I_n = \omega_n n\,I_o. \tag{10.119}$$

Replacing ω_n in (10.118) by the expression in (10.119) we finally get for the perturbation frequency Ω with $I_o = \beta c\lambda_o$

$$\Delta\Omega^2 = (\Omega - n\omega_o)^2 = -i\frac{n\,q\,\eta_c\,\omega_o^2\,I_o}{2\pi\beta^2\,E_o}Z_z. \tag{10.120}$$

Equation (10.120) determines the evolution of the charge or current perturbation λ_n or I_n respectively. With $\Delta\Omega = \Delta\Omega_r + i\Delta\Omega_i$, the current perturbation is

$$I_n \, e^{i(n\theta - n\omega_o t - \Delta\Omega_r t - i\Delta\Omega_i t)} \;=\; I_n \, e^{\Delta\Omega_i t} \, e^{i(n\theta - n\omega_o t - \Delta\Omega_r t)} \tag{10.121}$$

exhibiting an exponential factor which can cause instability or damping since there is a positive as well as negative solution from (10.120) for the frequency shift $\Delta\Omega_i$. The situation in a particular case will depend on initial conditions describing the actual perturbation of the density distribution, however, different initial perturbations must be expected to be present along a real particle distribution including at least one leading to instability.

Beam stability occurs only if the imaginary part of the frequency shift vanishes. This is never the case if the impedance has a resistive component causing the *resistive-wall instability* [10.36] with a growth rate

$$\frac{1}{\tau_{\mathrm{res.wall}}} \;=\; \mathrm{Im}\{\Delta\Omega\} \;=\; \frac{1}{\sqrt{2}} \sqrt{(\sqrt{2}-1)\,\frac{n^2 q \eta_c \omega_o^2 \, I_o \bar{R}}{2\pi \beta^2 \, E_o \, c r_o}} \, \sqrt{\frac{2\pi\omega_o\mu}{n\sigma}} \,. \tag{10.122}$$

This conclusion requires some modification since we know that circular accelerators exist, work, and have metallic vacuum chambers with a resistive surface. The apparent discrepancy is due to the fact that we have assumed a monochromatic beam which indeed is unstable but also unrealistic. In the following sections, we include a finite momentum spread, find a stabilizing mechanism called *Landau damping* and derive new stability criteria.

Below transition energy, $\eta_c > 0$ will assure stability of a coasting beam as long as we consider only a purely capacitive impedance like the space-charge impedance (10.100) in which case $\Delta\Omega_i = 0$. Above transition energy $\eta_c < 0$ and the *negative-mass instability* appears as long as the impedance is capacitive or $Z_i > 0$. For an inductive impedance, the stability conditions are exchanged below and above transition energy. In summary, we have the following *longitudinal coasting beam stability conditions*:

$$\text{if} \quad Z_r \neq 0 \; \rightarrow \; \Delta\omega_i \neq 0 \; \rightarrow \; \begin{cases} \text{always unstable} \\ \text{resistive-wall instability} \end{cases} \tag{10.123}$$

$$\text{if} \quad Z_r = 0 \quad \begin{cases} \begin{cases} Z_i < 0 \\ \text{(inductive)} \end{cases} \rightarrow \begin{cases} \text{stable for } \gamma > \gamma_{\mathrm{tr}}\,,\ \eta_c < 0 \\ \text{unstable for } \gamma \leq \gamma_{\mathrm{tr}}\,,\ \eta_c > 0 \end{cases} \\[2mm] \begin{cases} Z_i > 0 \\ \text{(capacitive)} \end{cases} \rightarrow \begin{cases} \text{stable for } \gamma < \gamma_{\mathrm{tr}}\,,\ \eta_c > 0 \\ \text{unstable for } \gamma \geq \gamma_{\mathrm{tr}}\,,\ \eta_c < 0\,. \end{cases} \end{cases} \tag{10.124}$$

It is customary to plot the stability condition (10.120) in a (Z_r, Z_i)-diagram with $\Delta\Omega_i$ as a parameter. We solve (10.120) for the imaginary impedance Z_i and get

$$Z_i \;=\; \mathrm{sgn}(\eta_c)\, a \left[\left(\frac{Z_r}{2\,\Delta\Omega_i}\right)^2 \mp \left(\frac{\Delta\Omega_i}{a}\right)^2 \right], \tag{10.125}$$

where

$$a \;=\; \frac{n\, q\, |\eta_c|\, \omega_o^2 \, I_o}{2\pi \beta^2 \, E_o} \tag{10.126}$$

and plot the result in Fig. 10.17. Only the case $\eta_c > 0$ is shown in Fig. 10.17 noting that the case $\eta_c < 0$ is obtained by a 180° rotation of Fig. 10.17 about the Z_r-axis. Figure 10.17 demonstrates that beam stability occurs only if $Z_r = 0$ and $Z_i > 0$. Knowing the complex impedance for a particular accelerator, Fig. 10.17 can be used to determine the rise time $1/\tau = \Delta\Omega_i$ of the instability.

The *rise time* or *growth rate* of the negative-mass instability above transition is for a beam circulating within a perfectly conducting vacuum chamber from (10.100) and (10.120)

$$\frac{1}{\tau_{\text{neg.mass}}} = \frac{n\omega_o}{\beta c\gamma}\sqrt{\frac{q\,|\eta_c|\,c\,I_o\left(1 + 2\ln\frac{r_w}{r_o}\right)}{\beta\,E_o}}. \tag{10.127}$$

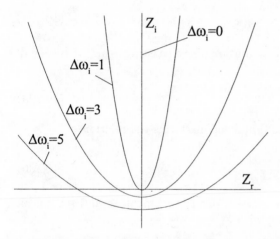

Fig. 10.17. Stability diagram for a coasting monochromatic particle beam

In this section, it was implicitly assumed that all particles have the same momentum and therefore, the same revolution frequency ω_o allowing a change of the revolution frequency only for those particles close to a particle density perturbation. This restriction to a monochromatic beam is not realistic and provides little beam stability for particle beams in a circular accelerator. In the following section, we will discuss the more general case of a beam with a finite momentum spread and review beam stability conditions under more realistic beam parameters.

10.5.2 Dispersion Relation

In the previous section, conditions for beam stability were derived based on a monochromatic particle beam. The rise time of the instability depends critically on the revolution frequency and we may assume that the conditions for beam stability may change if we introduce the more realistic case of a beam with a finite momentum spread and therefore a finite spread of revolution frequencies. In Chap. 8, we discussed the mathematical tool of the *Vlasov equation* to describe collectively the dynamics of a distribution of particles in phase space. We will apply this tool to the *collective interaction* of a particle beam with its environment.

The canonical variables describing longitudinal motion of particles are the azimuth θ and relative momentum error $\delta = \Delta p / p_o$. Neglecting radiation damping, the Vlasov equation is

$$\frac{\partial \Psi}{\partial t} + \dot\theta \, \frac{\partial \Psi}{\partial \theta} + \dot\delta \, \frac{\partial \Psi}{\partial \delta} = 0 \,, \tag{10.128}$$

where $\Psi(\delta, \theta, t)$ is the particle distribution. For a coasting beam with a small perturbation

$$\Psi = \Psi_o + \Psi_n \, e^{i(n\theta - \omega_n t)} \tag{10.129}$$

we get after insertion in (10.128) and sorting terms the relation

$$i \, (\omega_n - n\omega) \, \Psi_n = \frac{\dot\delta}{e^{i(n\theta - \omega_n t)}} \, \frac{\partial \Psi_o}{\partial \delta} \,. \tag{10.130}$$

Making use of the correlation between particle momentum and revolution frequency, we get from (10.130) with $\frac{\partial \Psi_o}{\partial \delta} = \frac{\partial \Psi_o}{\partial \omega} \frac{\partial \omega}{\partial \delta} = \eta_c \omega_o \frac{\partial \Psi_o}{\partial \omega}$

$$\Psi_n = -i \, \frac{\eta_c \omega_o \dot\delta}{e^{i(n\theta - \omega_n t)}} \, \frac{\partial \Psi_o}{\partial \omega} \, \frac{1}{\omega_n - n\omega} \,. \tag{10.131}$$

Integrating the l.h.s. of (10.131) over all momenta, we get for the perturbation current

$$q \, \frac{\beta c}{R} \int\limits_{-\infty}^{\infty} \Psi_n(\delta) \, d\delta = I_n \,.$$

At this point, it is convenient to transform from the variable δ to the frequency ω and obtain the particle distribution in these new variables

$$\Psi(\delta, \theta) = \eta_c \omega_o \, \Phi(\omega, \theta) \,. \tag{10.132}$$

Performing the same integration on the r.h.s. of (10.131), we get with (10.116) and $\dot\delta = (dE/dt)/(\beta^2 E_o)$ the *dispersion relation* [10.37]

$$1 = -\mathrm{i}\,\frac{q^2\,\omega_\mathrm{o}^3\,\eta_\mathrm{c}\,Z_z}{2\pi\,\beta^2\,E_\mathrm{o}} \int \frac{\partial\Phi_\mathrm{o}/\partial\omega}{\omega_n - n\omega}\,\mathrm{d}\omega\,. \tag{10.133}$$

The integration is to be taken over the upper or lower complex plane where we assume that the distribution function Φ vanishes sufficiently fast at infinity.

Trying to establish beam stability for a particular particle distribution, we solve the dispersion relation for the frequency ω_n or frequency shift $\Delta\omega_n = \omega_n - n\omega$ which is in general complex. The real part causes a shift in the frequency while the imaginary part determines the state of stability or instability for the collective motion.

For example, it is interesting to apply this result to the case of a coasting beam of monochromatic particles as discussed in the previous section. Let the particle distribution be uniform in θ and a delta function in energy. In the dispersion relation, we need to take the derivative with respect to the revolution frequency and set therefore

$$\frac{\partial\Phi_\mathrm{o}}{\partial\omega} = \frac{N_\mathrm{p}}{2\pi}\frac{\partial}{\partial\omega}\delta(\omega - \omega_\mathrm{o})\,. \tag{10.134}$$

Insertion into (10.133) and integration by parts results in

$$\int_{-\infty}^{\infty} \frac{\partial\Phi_\mathrm{o}/\partial\omega}{\omega_n - n\omega}\,\mathrm{d}\omega = \frac{N_\mathrm{b}}{2\pi}\frac{n}{(\omega_n - n\omega_\mathrm{o})^2} \tag{10.135}$$

which is identical to the earlier result (10.120) in the previous section. Application of the *Vlasov equation* therefore gives the same result as the direct derivation of the negative-mass instability conditions as it should be.

We may now apply this formalism to a beam with finite momentum spread. In preparation to do that, we note that the integrand in (10.133) has a singularity at $\omega = \omega_n/n$ which we take care of by applying *Cauchy's residue theorem* for

$$\int \frac{\partial\Phi_\mathrm{o}/\partial\omega}{\omega_n - n\omega}\,\mathrm{d}\omega = \int_{n\omega \neq \omega_n} \frac{\partial\Phi_\mathrm{o}/\partial\omega}{\omega_n - n\omega}\,\mathrm{d}\omega - \mathrm{i}\pi\left.\frac{\partial\Phi_\mathrm{o}}{\partial\omega}\right|_{\omega_n/n}\,. \tag{10.136}$$

The dispersion relation (10.133) then assumes the form

$$1 = \mathrm{i}\,\frac{q^2\,\omega_\mathrm{o}^3\,\eta_\mathrm{c}\,Z_z}{2\pi\beta^2 E_\mathrm{o}}\left[\mathrm{i}\,\frac{\pi}{n}\left.\frac{\partial\Phi_\mathrm{o}}{\partial\omega}\right|_{\omega=\frac{\omega_n}{n}} - \mathrm{P.V.}\int \frac{\partial\Phi_\mathrm{o}/\partial\omega}{\omega_n - n\omega}\,\mathrm{d}\omega\right]\,, \tag{10.137}$$

where P.V. indicates that only the principal value of the integral be taken.

The solutions of the dispersion function depend greatly on the particle distribution in momentum or revolution-frequency space. To simplify the expressions, we replace the revolution frequency by its deviation from the reference value [10.38]. With $2S$ being the full-width half maximum of the particular momentum distribution (Fig. 10.18), we define the new variables

$$x = \frac{\omega - \omega_o}{S}, \qquad \text{and}$$
$$x_1 = \frac{\Delta\omega_n}{nS} = \frac{\Omega - n\omega_o}{nS} . \tag{10.138}$$

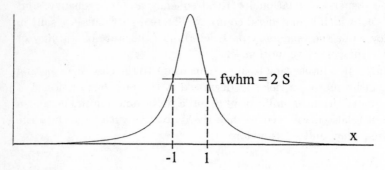

fwhm = 2 S

Fig. 10.18. Particle distribution $f(x)$

In these variables the particle distribution becomes

$$f(x) = \frac{2\pi S}{N_b} \Phi(\omega) \tag{10.139}$$

which is normalized to $f(\pm 1) = \frac{1}{2} f(0)$ and $\int f(x)\mathrm{d}x = 1$. The full momentum spread at half maximum intensity is

$$\frac{\Delta p}{p_o} = \frac{2S}{|\eta_c|\omega_o} \tag{10.140}$$

and (10.137) becomes with (10.140)

$$1 = -\mathrm{i}\,\frac{2q\,Z_z\,I_o}{\pi\beta^2 E_o n\eta_c \left(\frac{\Delta p}{p_o}\right)^2}\left[\mathrm{P.V.}\int\limits_{-\infty}^{\infty} \frac{\partial f_o(x)/\partial x}{x_1 - x}\,\mathrm{d}x - \mathrm{i}\pi\,\frac{\partial f_o}{\partial x}\bigg|_{x_1}\right].\tag{10.141}$$

It is customary to define parameters U, V by

$$V + \mathrm{i}\,U = \frac{2q\,I_o}{\pi\beta^2 E_o \eta_c \left(\frac{\Delta p}{p_o}\right)^2}\,\frac{(Z_r + \mathrm{i}Z_i)_z}{n} \tag{10.142}$$

and the *dispersion relation* becomes finally with this

$$1 = -(V + \mathrm{i}\,U)\,I, \tag{10.143}$$

where the integral

$$I = \left[\text{P.V.} \int\limits_{-\infty}^{\infty} \frac{\partial f_o(x)/\partial x}{x_1 - x} \, \mathrm{d}x - \mathrm{i}\pi \left. \frac{\partial f_o}{\partial x} \right|_{x_1} \right].$$ (10.144)

For a particular accelerator all parameters in (10.143) are known, at least in principle, and we may determine the status of stability or instability for a desired beam current I_o by solving for the generally complex frequency shift $\Delta\omega$. The specific boundary of stability depends on the actual particle distribution in momentum. Unfortunately, equation (10.143) cannot be solved analytically for an arbitrary momentum distribution and we will have to either restrict our analytical discussion to simple solvable distributions or to numerical evaluation.

For reasonable representations of real particle distributions in an accelerator a central region of stability can be identified for small complex impedances and finite spread in momentum. Regions of stability have been determined for a number of particle distributions and the interested reader is referred for more detailed information on such calculations to references [10.39 – 42]. For further discussions only one momentum distribution is used here to elucidate the salient features of the dispersion relation.

As an example, we use a simple particle distribution (Fig. 10.19)

$$f(x) = \frac{1}{\pi} \frac{1}{1 + x^2}$$ (10.145)

and evaluate the dispersion relation (10.143).

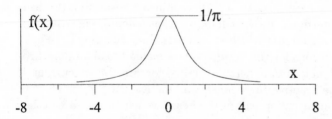

Fig. 10.19. Particle distribution in momentum space

The integral in (10.144) becomes now after integration by parts

$$\text{P.V.} \int\limits_{-\infty}^{\infty} \frac{1}{(1 + x^2)(x_1 - x)^2} \, \mathrm{d}x$$

exhibiting a new singularity at $x = \mathrm{i}$ while the integration path still excludes the other singularity at $x = x_1$. Applying the residue theorem

$$\int \frac{f(z)\mathrm{d}z}{z - z_o} = 2\pi\mathrm{i} \, \text{Res}[f(z), z_o] = 2\pi\mathrm{i} \lim_{z \to z_o} (z - z_o) \, f(z)$$ (10.146)

we get

$$\text{P.V.} \int\limits_{-\infty}^{\infty} \frac{1}{(1 + x^2)\,(x_1 - x)}\,dx \;=\; \frac{1}{(x_1 - i)^2}\,.$$

The second term in (10.144) is

$$-i\pi\,\frac{\partial f_0}{\partial x}\bigg|_{x_1} \;=\; i\,\frac{2x_1}{(1 + x_1^2)^2}$$

and the dispersion relation (10.143) becomes

$$1 \;=\; -i\,(V + iU)\left(\frac{1}{(x_1 - i)^2} + i\,\frac{2x_1}{(1 + x_1^2)^2}\right).\tag{10.147}$$

We solve this for $(x_1 - i)^2$ and get

$$x_1 \;=\; i \pm \sqrt{-i(V + iU)\left(1 + i\,\frac{2x_1}{(i + x_1)^2}\right)}.\tag{10.148}$$

For a small beam current i_0, we get $x_1 \approx i$ and the second term in the square bracket becomes approximately $1/2$. Recalling the definition (10.138) for x_1, we get from (10.148)

$$\Delta\Omega \;=\; inS \pm \sqrt{\tfrac{3}{2}\,n^2\,S^2\,(U - iV)},\tag{10.149}$$

where from (10.140) $S = 1/2|\eta_c|\omega_0\Delta p/p_0$. The significant result in (10.149) is the fact that the first term on the right-hand side has a definite positive sign and provides therefore damping which is called *Landau damping* [10.43].

Recalling the conditions for the *negative-mass instability* of a monochromatic beam, we did not obtain beam stability for any beam current if $Z_r \propto V = 0$ and the reactive impedance was inductive or $Z_i \propto U < 0$. Now with a finite momentum spread in the beam we get in the same case

$$\Delta\Omega_{\text{neg.mass}} \;=\; inS \pm i\sqrt{{}^{3}\!/_{2}\,n^2\,S^2\,|U|},\tag{10.150}$$

where $S^2|U|$ is independent of the momentum spread. We note that it takes a finite beam current, $U \propto I_0$, to overcome Landau damping and cause instability. Of course Landau damping is proportional to the momentum spread S and does not occur for a monochromatic beam. Equation (10.149) serves as a *stability criterion* for longitudinal coasting-beam instabilities and we will try to derive a general expression for it. We write (10.149) in the form

$$\Delta\Omega \;=\; inS \pm \sqrt{a - ib}\tag{10.151}$$

and get after evaluating the square root

$$\Delta\Omega = inS \pm \left(\sqrt{\frac{r+a}{2}} - i\sqrt{\frac{r-a}{2}} \right),$$ (10.152)

where $r = \sqrt{a^2 + b^2}$. Beam stability occurs for $\text{Im}\{\Delta\Omega\} > 0$ or

$$n^2 S^2 = \frac{r-a}{2}$$ (10.153)

which is in more practical quantities recalling the definition (10.140) for S

$$\left(\frac{\Delta p}{p_o} \right)^2 \geq \frac{3}{2\pi} \frac{qI_o}{\beta^2 E_o |\eta_c|} \left(\frac{|Z_z|}{n} - \frac{Z_i}{n} \right).$$ (10.154)

We may solve (10.154) for the impedance and get an equation of the form

$$Z_i = A Z_r^2 - \frac{1}{4A}$$ (10.155)

which is shown in Fig. 10.20.

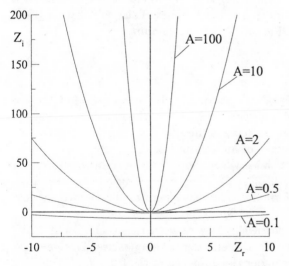

Fig. 10.20. Stability diagram for the particle distribution (10.145)

Any combination of actual resistive and reactive impedances below this curve cause beam instability for the particle distribution (10.145). We note the significant difference to Fig. 10.17 where the impedance had to be purely positive and reactive to obtain beam stability.

Other momentum distributions like $f(x) \propto (1 - x^2)^m$ lead to similar results [10.39] although the stability curves allow less resistive impedance than the distribution (10.145). As a safe stability criterion which is true for many such distributions including a gaussian distribution we define the area of stability by a circle with a radius $R = Z_i|_{Z_r=0} = 1/(4A)$. With this assumption the *stability criterion* for the *longitudinal microwave instability* is

$$\boxed{\frac{|Z_z|}{n} \le F \frac{\beta^2 E_o |\eta_c|}{q I_o} \left(\frac{\Delta p}{p_o}\right)^2 ,}$$

(10.156)

where the form factor $F = \pi/3$ for the distribution (10.145) and is generally of the order of unity for other bell shaped distributions. The criterion (10.156) has been derived by *Keil* and *Schnell* [10.44] and is generally known as the *Keil-Schnell stability criterion*. For a desired beam current and an allowable momentum spread an upper limit for the normalized impedance can be derived.

The impedance seen by the particle beam obviously should be minimized to achieve the highest beam-beam currents. A large value of the momentum compaction is desirable here to increase the mixing of the revolution frequencies as well as a large momentum spread to achieve high beam currents. A finite momentum spread increases beam stability where there was none for a monochromatic coasting beam as discussed earlier. This stabilization effect of a finite momentum spread is called *Landau damping*.

10.5.3 Landau Damping

In previous sections, we repeatedly encountered a damping phenomenon associated with the effect of collective fields on individual particle stability. Common to the situations encountered is the existence of a set of oscillators or particles, under the influence of an external driving force. Particularly, we are interested in the dynamics when the external excitation is caused by the beam itself. *Landau* [10.43] studied this effect bearing his name first, and later *Hereward* [10.45] formulated the theory for application to particle accelerators.

We consider a bunch of particles where each particle oscillates at a different frequency Ω, albeit within a small range of frequencies. The equation of motion for each oscillator under the influence of the force $F e^{-i\omega t}$ is

$$\ddot{u} + \Omega^2 u = F e^{-i\omega t}$$

(10.157)

and the solution

$$u = F \frac{e^{-i\omega t}}{2\omega} \left(\frac{1}{\Omega - \omega} - \frac{1}{\Omega + \omega}\right) .$$

(10.158)

Folding this solution with the distribution function of particles in frequency space

$$\psi(\omega) = \frac{1}{N_{\rm b}} \frac{{\rm d}N_{\rm b}}{{\rm d}\Omega}$$ (10.159)

one obtains the center of mass amplitude of the bunch

$$\bar{u} = F \frac{{\rm e}^{-{\rm i}\omega t}}{2\omega} \int\limits_{-\infty}^{\infty} \left[\frac{\psi(\Omega)}{\Omega - \omega} - \frac{\psi(\Omega)}{\Omega + \omega} \right] {\rm d}\Omega$$

or with $\psi(\Omega) = \psi(-\Omega)$ and $\int_{-\infty}^{\infty} \frac{\psi(\Omega)}{\Omega - \omega} {\rm d}\Omega = -\int_{-\infty}^{\infty} \frac{\psi(\Omega)}{\Omega + \omega} {\rm d}\Omega$

$$\bar{u} = F \frac{{\rm e}^{-{\rm i}\omega t}}{\omega} \int\limits_{-\infty}^{\infty} \frac{\psi(\Omega)}{\Omega - \omega} {\rm d}\Omega .$$ (10.160)

Here we apply again *Cauchy's residue theorem* and get

$$\bar{u} = F \frac{{\rm e}^{-{\rm i}\omega t}}{\omega} \left[+{\rm i}\pi\,\psi(\omega) + {\rm P.V.} \int\limits_{-\infty}^{\infty} \frac{\psi(\Omega)}{\Omega - \omega} {\rm d}\Omega \right] .$$ (10.161)

The derivation up to here appears quite abstract and we pause a moment to reflect on the physics involved here. We know that driving an oscillator at resonance leads to infinitely large amplitudes and that is what the mathematical formulation above expresses. However, we also know that infinite amplitudes take an infinite time to build up and the solutions gained above describe only the state after a long time. The same result can be obtained in a physical more realistic way if we apply the excitation at time $t = 0$ and look for the solution at $t \to \infty$ as has been shown by *Hofmann* [10.46].

As an added result of this time evolution derivation, we obtain the correct sign for the residue which we have tacitly assumed to be negative, but mathematically could be of either sign.

To understand the damping effect, we calculate the velocity $\dot{\bar{u}}$ and get from (10.161)

$$\dot{\bar{u}} = -{\rm i}\omega\,\bar{u}$$

$$= F\,{\rm e}^{-{\rm i}\omega t} \left[+\pi\,\psi(\omega) - {\rm i}\,{\rm P.V.} \int\limits_{-\infty}^{\infty} \frac{\psi(\Omega)}{\Omega - \omega} {\rm d}\Omega \right] .$$ (10.162)

The bunch velocity is in phase with the external force for the residue term allowing extraction of energy from the external force. The principal value term, on the other hand, is out of phase and no work is done. If, for example, the external force is due to a transverse wake field generated by a bunch performing coherent betatron oscillations, the described mechanism would extract energy from the wake field thus damping the coherent betatron oscillation. The question is where does the energy go?

For this, we study the time evolution of the solution for the inhomogeneous differential equation (10.157) in the form

$$u = a \sin \Omega t + \frac{F}{\Omega^2 - \omega^2} \sin \omega t. \tag{10.163}$$

At time $t = 0$ we require that the amplitude and velocity of the bunch motion be zero $u(t = 0) = 0$ and $\dot{u}(t = 0) = 0$. The oscillation amplitude

$$a = -\frac{\omega}{\Omega} \frac{F}{\Omega^2 - \omega^2} \tag{10.164}$$

and the final expression for the solution to (10.157) is for $\Omega \neq \omega$

$$u_{\Omega \neq \omega}(t) = \frac{F}{\Omega^2 - \omega^2} \left(\sin \omega t - \frac{\omega}{\Omega} \sin \Omega t \right). \tag{10.165}$$

Close to or at resonance $\Omega = \omega + \Delta$ and (10.165) becomes

$$u_{\Omega = \omega}(t) = -\frac{F}{2\omega} \left(t \cos \omega t - \frac{\sin \omega t}{\omega} \right). \tag{10.166}$$

The oscillation amplitude of particles at resonance grows continuously with time while the width of the resonance shrinks like $1/t$ thus absorbing energy linear in time.

This Landau damping depends critically on the resistive interaction with the wake fields or external forces and is mathematically expressed by the residue term. This residue term, however, depends on the particle density at the excitation frequency ω and is zero if the particle distribution in frequency space does not overlap with the frequency ω. For effective Landau damping to occur such an overlap is essential.

10.5.4 Transverse Coasting-Beam Instability

Particle beams travelling off center through vacuum chamber sections can induce transverse fields which act back on the beam. We express the strength of this interaction by the *transverse coupling impedance*. In terms of the transverse coupling impedance, the force is

$$F_\perp = i \frac{q Z_\perp I_o u}{2\pi \bar{R}}, \tag{10.167}$$

where I_o is the beam current, u the transverse beam displacement, $Z_\perp / (2\pi \bar{R})$ the average transverse coupling impedance and $2\pi \bar{R}$ the ring circumference. The equation of motion is then

$$\ddot{u} + \nu_o^2 \omega_o^2 u = -i \frac{q Z_\perp I_o}{2\pi \bar{R} m\gamma} (u + \bar{u}) \tag{10.168}$$

with u the betatron oscillation amplitude of an individual particle and $\bar{u}$ the amplitude of the coherent bunch oscillation. Since the perturbation is linear in the amplitudes, we expect tune shifts from the perturbations. The *incoherent tune shift* due to individual particle motion will be incorporated on the l.h.s. as a small tune shift

$$\delta\nu_{\rm o} \;=\; \frac{{\rm i}c\,q\,Z_\perp\,I_{\rm o}}{4\pi\,\nu_{\rm o}\,\omega_{\rm o}\,E_{\rm o}}\,. \tag{10.169}$$

The transverse impedance is generally complex and we get therefore from the real part of the coupling impedance a real tune shift while the imaginary part leads to damping or antidamping depending on the sign of the impedance. The imaginary frequency shift is equal to the *growth rate* of the instability and is given by

$$\boxed{\;\frac{1}{\tau} \;=\; {\rm Im}\{\omega\} \;=\; \frac{q\,{\rm Re}\{Z_\perp\}\,I_{\rm o}}{4\pi\bar{R}\,m\gamma\,\omega_{\beta{\rm o}}}\,.\;} \tag{10.170}$$

For a resistive transverse impedance, we observe therefore always instability known as the *transverse resistive-wall instability*.

Similar to the case of a longitudinal coasting beam, we find instability for any finite beam current just due to the reactive space-charge impedance alone, and again we have to rely on *Landau damping* to obtain beam stability for a finite beam intensity. To derive transverse stability criteria including Landau damping, we consider the *coherent tune shift* caused by the coherent motion of the whole bunch for which the equation of motion is

$$\ddot{u} + \omega_{\beta{\rm o}}^2\,u \;=\; 2\nu_{\rm o}\omega_{\rm o}\left[U + (1+{\rm i})V\right]\bar{u}\,, \tag{10.171}$$

where

$$U + (1+{\rm i})V \;=\; -{\rm i}\,\frac{c\,q\,Z_\perp\,I_{\rm o}}{4\pi\nu_{\rm o}\,E_{\rm o}}\,, \tag{10.172}$$

The coherent beam oscillation must be periodic with the circumference of the ring and is of the form $\bar{u} = \hat{u}\,{\rm e}^{{\rm i}(n\theta-\omega t)}$. As can be verified by back insertion the solution of (10.171) is

$$y \;=\; \left[U + (1+{\rm i})V\right]\frac{2\nu_{\rm o}\omega_{\rm o}}{\nu_1^2\omega_{\rm o}^2 - (n\omega_{\rm o} - \omega)^2}\,\bar{u}\,. \tag{10.173}$$

Now we must fold (10.173) with the distribution in the spread of the betatron oscillation frequency. This spread is mainly related to a momentum spread via the chromaticity and the momentum compaction. The distribution $\psi(\delta)$ where $\delta = \Delta p/p_{\rm o}$, is normalized to unity $\int \psi(\delta){\rm d}\delta = 1$ and the average particle position is $\bar{u} = \int u\psi(\delta){\rm d}\delta$. The *dispersion relation* is then with this from (10.173)

$$1 \;=\; \left[U + (1+{\rm i})V\right]\int\limits_{-\infty}^{\infty}\frac{2\nu_{\rm o}\omega_{\rm o}\psi(\delta){\rm d}\delta}{\nu_1^2\omega_{\rm o}^2 - (n\omega_{\rm o} - \omega)^2}\,, \tag{10.174}$$

or simplified by setting $\nu_1 \approx \nu_o$ and ignoring the fast wave $(n + \nu)\omega_o$ [10.47]

$$1 = [U + (1 + \mathrm{i})V] \int_{-\infty}^{\infty} \frac{\psi(\delta)\mathrm{d}\delta}{\omega - (n - \nu_o)\omega_o}. \tag{10.175}$$

This is the *dispersion relation* for transverse motion and can be evaluated for stability based on a particular particle distribution in momentum. As mentioned before, the momentum spread transforms to a betatron oscillation frequency spread by virtue of the momentum compaction

$$\Delta\nu_\beta = \nu_{\beta o}\, \Delta\omega_o = \nu_{\beta o}\, \eta_c\, \delta\, \omega_o \tag{10.176}$$

and by virtue of the chromaticity

$$\Delta\nu_\beta = \xi_u\, \delta. \tag{10.177}$$

Landau damping provides again beam stability for a finite beam current and finite coupling impedances, and the region of stability depends on the actual particle distribution in momentum.

10.6 Longitudinal Single-Bunch Effects

The dynamics in bunched particle beams is similar to that of a coasting beam with the addition of synchrotron oscillations. The frequency spectrum of the circulating beam current contains now many harmonics of the revolution frequency with sidebands due to betatron and synchrotron oscillations. The bunch length depends greatly on the interaction with the rf field in the accelerating cavities but also with any other field encountered within the ring. It is therefore reasonable to expect that wake fields may have an influence on the bunch length which is know as *potential well distortion*.

10.6.1 Potential-Well Distortion

From the discussions on longitudinal phase space motion in circular accelerators, it is known that the particle distribution or bunch length depends on the variation in time of the rf field interacting with the beam in the accelerating cavities. Analogous, we would expect that wake fields may have an impact on the longitudinal particle distribution [10.48]. For a particular wake field, we have studied this effect in Chap. 6 recognizing that a bunch passing through an rf cavity causes beam loading by exciting fields at the fundamental frequency in the cavity. These fields then cause a modification of the bunch length. In this section, we will expand on this thought and study the effect due to higher-order mode wake fields.

To study this in more detail, we ignore the transverse particle distribution. The rate of change in the particle momentum can be derived from the

integral of all longitudinal forces encountered along the circumference and
we set with $\delta = \mathrm{d}p/p_{\mathrm{o}}$

$$\frac{\mathrm{d}\delta}{\mathrm{d}t} = \frac{qF(\tau)}{\beta^2 E_{\mathrm{o}} T_{\mathrm{o}}}, \tag{10.178}$$

where $qF(\tau)$ is the sum total of all acceleration and energy losses of a
particle at a position $z = \beta c\tau$ from the bunch center or reference point over
the course of one revolution and T_{o} is the revolution time. The change of τ
per unit time depends on the momentum compaction of the lattice and the
momentum deviation

$$\frac{\mathrm{d}\tau}{\mathrm{d}t} = -\eta_{\mathrm{c}}\,\delta. \tag{10.179}$$

Both equations can be derived from the Hamiltonian

$$\mathcal{H} = -\tfrac{1}{2}\,\eta_{\mathrm{c}}\,\delta^2 - \int_{\mathrm{o}}^{\tau} \frac{qF(\bar{\tau})}{\beta^2 E_{\mathrm{o}} T_{\mathrm{o}}}\,\mathrm{d}\bar{\tau}. \tag{10.180}$$

For an electron ring and small oscillation amplitudes, we have

$$qF(\tau) = qV_{\mathrm{rf}}(\tau_{\mathrm{s}} + \tau) - U(E) + qV_{\mathrm{w}}(\tau) = q\frac{\partial V_{\mathrm{rf}}}{\partial \tau}\bigg|_{\tau_{\mathrm{s}}} \tau + qV_{\mathrm{w}}(\tau), \tag{10.181}$$

where we ignored radiation damping and where $V_{\mathrm{w}}(\tau)$ describes the wake
field. In the last form, the equation is also true for protons and ions if we set
the synchronous time $\tau_{\mathrm{s}} = 0$. Inserting (10.181) into (10.180) and using the
definition of the synchrotron oscillation frequency $\Omega_{\mathrm{so}}^2 = \omega_{\mathrm{o}}^2 \frac{h\eta_{\mathrm{c}} q\hat{V}_{\mathrm{rf}} \cos\psi_{\mathrm{so}}}{2\pi\beta^2 E_{\mathrm{o}}}$,
we get the new Hamiltonian

$$\mathcal{H} = -\tfrac{1}{2}\,\eta_{\mathrm{c}}\,\delta^2 - \tfrac{1}{2}\frac{\Omega_{\mathrm{so}}^2}{\eta_{\mathrm{c}}}\,\tau^2 - \int_{\mathrm{o}}^{\tau} \frac{qV_{\mathrm{w}}(\bar{\tau})}{\beta^2 E_{\mathrm{o}} T_{\mathrm{o}}}\,\mathrm{d}\bar{\tau}. \tag{10.182}$$

Synchrotron oscillation tune shift: First we use the Hamiltonian to formulate the equation of motion and determine the effect of wake fields on
the dynamics of the synchrotron motion. The equation of motion is from
(10.182)

$$\ddot{\tau} + \Omega_{\mathrm{so}}^2\,\tau = \mathrm{sign}(\eta_{\mathrm{c}})\frac{2\pi\,\Omega_{\mathrm{so}}^2\,V_{\mathrm{w}}}{\omega_{\mathrm{o}}\,h\,V_{\mathrm{rf}}|\cos\psi_{\mathrm{so}}|}, \tag{10.183}$$

where we have made use of the definition of the unperturbed synchrotron
oscillation frequency Ω_{so}. We express the wake field in terms of impedance
and beam spectrum

$$V_{\rm w}(t) = - \int_{-\infty}^{\infty} Z_{\|}(\omega) \, I(t,\omega) \, e^{i\omega t} \, d\omega \,, \tag{10.184}$$

and use (10.64) for

$$V_{\rm w}(t) = - I_{\rm b} \sum_{p=-\infty}^{\infty} Z_{\|}(p) \, \Psi(p) \, e^{-ip\omega_{\rm o}\tau} \,, \tag{10.185}$$

where $I_{\rm b}$ is the bunch current and

$$\Psi(p) = \int_{-\infty}^{+\infty} J_{\rm o}(p\omega_{\rm o}\hat{\tau}) \, \Phi(t,\hat{\tau}) \, d\hat{\tau} \,.$$

The maximum excursion $\hat{\tau}$ during phase oscillation is much smaller than the revolution time and the exponential factor

$$e^{ip\omega_{\rm o}\tau} \approx 1 + ip\omega_{\rm o}\tau - \tfrac{1}{2}p^2\omega_{\rm o}^2\tau^2 + \mathcal{O}(3) \tag{10.186}$$

can be expanded. After insertion of (10.185,186) into (10.183) the equation of motion is

$$\ddot{\tau} + \Omega_{\rm so}^2 \, \tau \approx \tag{10.187}$$

$$- \, {\rm sign}(\eta_{\rm c}) \, \frac{2\pi \, I_{\rm b} \, \Omega_{\rm so}^2}{\omega_{\rm o} \, h \, V_{\rm rf}| \cos \psi_{\rm so}|} \sum_{p=-\infty}^{\infty} Z_{\|}(p) \, \Psi(p) \, \left(1 - ip\omega_{\rm o}\tau - \tfrac{1}{2}p^2\omega_{\rm o}^2\tau^2\right) \,.$$

The first term is independent of τ and causes a *synchronous phase shift* due to resistive losses

$$\Delta\psi_{\rm s} = {\rm sign}(\eta_{\rm c}) \frac{2\pi \, I_{\rm b}}{V_{\rm rf}| \cos \psi_{\rm so}|} \sum_{p=-\infty}^{\infty} {\rm Re}\{Z_{\|}(p)\} \, \Psi(p) \,. \tag{10.188}$$

For a resistive positive impedance, for example, the phase shift is negative above transition indicating that the beam requires more energy from the rf cavity. By measuring the shift in the synchronous phase of a circulating bunch as a function of bunch current, it is possible to determine the resistive part of the longitudinal impedance of the accelerator. To do this one may fill a small amount of beam in the bucket upstream from the high intensity bunch and use the signal from the small bunch as the time reference against which the big bunch will shift with increasing current.

The second term in (10.187) is proportional to τ and therefore acts like a focusing force shifting the *incoherent synchrotron oscillation frequency* by

$$\Delta\Omega_{\rm s} = - \, {\rm sign}(\eta_{\rm c}) \frac{\pi \, I_{\rm b}\Omega_{\rm so}}{h \, V_{\rm rf}| \cos \psi_{\rm so}|} \sum_{p=-\infty}^{\infty} {\rm Im}\{Z_{\|}(p)\} \, p \, \Psi(p) \,. \tag{10.189}$$

The real part of the impedance is symmetric in p and therefore cancels in the summation over p which leaves only the imaginary part consistent with the expectation that the tune shift be real. At this point, it becomes necessary to introduce a particular particle distribution and an assumption for the impedance spectrum. For long bunches, the frequencies involved are low and one might use for the impedance the space charge and broad-band impedance which both are constant for low frequencies. In this case, the impedance can be extracted from the sum in (10.189) and the remaining arguments in the sum depend only on the particle distribution.

For a parabolic particle distribution, for example, (10.189) reduces to [10.26]

$$\Delta\Omega_s = -\operatorname{sign}(\eta_c)\frac{16\,I_b}{\pi^3\,B^3\,h\,V_{rf}|\cos\psi_o|}\operatorname{Im}\left\{\frac{Z_{\parallel}(p)}{p}\right\}, \qquad (10.190)$$

where B is the bunching factor $B = \ell/(2\pi\bar{R})$ with ℓ the effective bunch length.

A measurement of the *incoherent synchrotron tune shift* as a function of bunch current allows the determination of the reactive impedance of the accelerator for a given particle distribution. This tune shift is derived from a measurement of the unperturbed synchrotron frequency Ω_{so} for a very small beam current combined with the observation of the quadrupole mode frequency Ω_{2s} as a function of bunch current. The incoherent tune shift is then

$$\Delta\Omega_{s,incoh} = \mu\left(\Omega_{2s} - 2\,\Omega_{so}\right), \qquad (10.191)$$

where μ is a distribution dependent form factor of order 2 for a parabolic distribution [10.26, 49].

The third and subsequent terms in (10.187) contribute nonlinear effects making the synchrotron oscillation frequency amplitude dependent similar to the effects of nonlinear fields in transverse beam dynamics.

Bunch lengthening: A synchrotron frequency shift is the consequence of a change in the longitudinal focusing and one might expect therefore also a change in the bunch length. In first approximation, one could derive expressions for the new bunch length by scaling with the synchrotron tune shift. Keeping the phase space area constant in the proton and ion case or keeping only the energy spread constant in the electron case, a rough estimate for bunch lengthening can be obtained for a specific particle distribution. Since the electron bunch length scales inversely proportional to the synchrotron frequency, we have

$$\frac{\sigma_\ell}{\sigma_{\ell o}} = \frac{\Omega_s}{\Omega_{so}} = 1 + \frac{\Delta\Omega_s}{\Omega_{so}}. \qquad (10.192)$$

From (10.192), one can determine for an electron beam the potential-well bunch lengthening or shortening, depending on the sign of the reactive impedance. For a proton or ion beam, the scaling is somewhat different because of the preservation of phase space.

This approach to understanding potential-well bunch lengthening assumes that the particle distribution does not change which is an approximate but not correct assumption. The deformation of the potential well is nonlinear and can create significant variations of the particle distribution specifically, for large amplitudes.

In this discussion, we determine the stationary particle distribution $\psi(\tau, \delta)$ under the influence of wake fields by solving the *Vlasov equation*

$$\frac{\partial \psi}{\partial t} + \dot{\tau} \frac{\partial \psi}{\partial \tau} + \dot{\delta} \frac{\partial \psi}{\partial \delta} = 0. \tag{10.193}$$

For a stationary solution, $\frac{\partial \psi}{\partial t} = 0$ and therefore any function of the Hamiltonian is a solution of the Vlasov equation. Since the Hamiltonian does not exhibit explicitly the time variable, any such function could be the stationary solution which we are looking for and we set therefore $\psi(\tau, \delta) = \psi(\mathcal{H})$. The local particle density is then after integrating over all momenta

$$\lambda(\tau) = N_{\rm b} \int_{-\infty}^{\infty} \psi(\mathcal{H})\, {\rm d}\delta, \tag{10.194}$$

where $N_{\rm b}$ is the number of particles per bunch or with (10.182)

$$\lambda(\tau) = N_{\rm b} \int_{-\infty}^{\infty} \psi \left(-\tfrac{1}{2}\, \eta_{\rm c}\, \delta^2 - \tfrac{1}{2}\, \frac{\Omega_{\rm so}^2}{\eta_{\rm c}}\, \tau^2 - \int_{o}^{\tau} \frac{q V_{\rm w}(\bar{\tau})}{\beta^2\, E_{\rm o}\, T_{\rm o}}\, {\rm d}\bar{\tau} \right) {\rm d}\bar{\delta}. \tag{10.195}$$

Without wake fields, the distribution of an electron beam is gaussian and the introduction of wake fields does not change that for the energy distribution. We make therefore the ansatz [10.50]

$$\psi(\tau, \delta) = A \exp\left(\frac{\mathcal{H}}{\eta_{\rm c}\, \sigma_\delta^2} \right) = A_\delta \exp\left(\tfrac{1}{2} \frac{\delta^2}{\sigma_\delta^2} \right) A_\lambda\, \lambda(\tau), \tag{10.196}$$

where A_δ and A_λ are normalization factors for the respective distributions. Integrating over all momenta, the longitudinal particle distribution is finally

$$\lambda(\tau) = N_{\rm b}\, A_\lambda \exp\left(-\tfrac{1}{2} \frac{\tau^2}{\sigma_\tau^2} - \frac{q}{\eta_{\rm c}\beta^2 E_{\rm o} T_{\rm o} \sigma_\delta^2} \int_{o}^{\tau} V_{\rm w}(\tilde{\tau})\, {\rm d}\tilde{\tau} \right), \tag{10.197}$$

where we used $\sigma_\delta = \Omega_{\rm so}\sigma_\tau/|\eta_{\rm c}|$ from [Ref. 10.17, Eq. (10.44)]. A self-consistent solution of this equation will determine the longitudinal particle distribution under the influence of wake fields. Obviously, this distribution is consistent with our earlier results for an electron beam in a storage ring, in the limit of no wake fields. The nature of the wake fields will then determine the distortion from the gaussian distribution.

As an example, we assume a combination of an inductive (L) and a resistive (R) wake field

$$V_{\mathrm{w}} = -L\frac{\mathrm{d}I}{\mathrm{d}t} - RI_{\mathrm{b}}. \tag{10.198}$$

Such a combination actually resembles rather well the real average impedance in circular accelerators at least at lower frequencies as evidenced in the impedance spectrum of the SPEAR storage ring shown in Fig. 10.15. Inserting (10.198) into (10.197) while setting for a moment the resistance to zero, $R = 0$, we get after integration the transcendental equation

$$\lambda(\tau) = N_{\mathrm{b}} A_{\lambda} \exp\left[-\frac{1}{2}\frac{\tau^2}{\sigma_\tau^2} - \frac{q^2 L N_{\mathrm{b}} \lambda(\tau)}{\eta_{\mathrm{c}} \beta^2 E_{\mathrm{o}} T_{\mathrm{o}} \sigma_\delta^2}\right] \tag{10.199}$$

which must be solved numerically to get the *particle distribution* $\lambda(\tau)$. We note that the inductive wake does not change the symmetry of the particle distribution in τ. For large values of τ, the particle distribution must approach zero to meet the normalization requirement, $\lim_{\tau \to \infty} \lambda(\tau) = 0$, and the particle distribution is always gaussian for large amplitudes. The effect of the inductive wake field is mainly concentrated to the core of the particle bunch.

Evaluating numerically (10.199), we distinguish between an electron beam and a proton or ion beam. The momentum spread σ_δ in case of an electron beam is determined by quantum effects related to the emission of synchrotron radiation and is thereby for this discussion a constant. Not so for proton and ion beams which are subject to Liouville's theorem demanding a strong correlation between bunch length and momentum spread such that the longitudinal phase space of the beam remains constant. Equation (10.199) has the form

$$f(t) = K \exp\left[-\frac{1}{2}t^2 - f(t)\right]$$

or after differentiation with respect to t

$$\frac{\mathrm{d}f(t)}{\mathrm{d}t} = -\frac{t\,f(t)}{1 + f(t)}. \tag{10.200}$$

For strong wake fields $f(t) \gg 1$ and (10.200) can be integrated for

$$f(t) = f_{\mathrm{o}} - \frac{1}{2}t^2. \tag{10.201}$$

The particle distribution in the bunch center assumes more and more the shape of a parabolic distribution as the wake fields increase. Figure 10.21 shows the particle distribution for different strengths of the wake field.

Now we add the resistive wake field component. This field actually extracts energy from the bunch and therefore one expects that the whole bunch is shifted such as to compensate this extra loss by moving to a higher

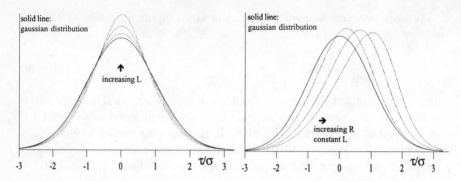

Fig. 10.21. Potential-well distortion of gaussian particle distributions. **a** for an inductive wake field, and **b** for a combination of an inductive and a resistive wake field

field in the accelerating cavities. Inserting the full wake field (10.198) into (10.197) results in the distribution

$$\lambda(\tau) = N_b A_\lambda \exp\left[-\frac{1}{2}\frac{\tau^2}{\sigma_\tau^2} - a L N_b \lambda(\tau) - a R N_b \int\limits_0^\tau \lambda(\bar\tau)\,\mathrm{d}\bar\tau\right], \quad (10.202)$$

where

$$a = \frac{q^2}{\eta_c \beta E_o T_o \sigma_\delta^2}.$$

Looking for a shift of the tip of the particle distribution, we get from $\mathrm{d}\lambda/\mathrm{d}\tau = 0$ the location of the distribution maximum

$$\tau_{\max} \propto N_b \lambda(\tau_{\max}). \qquad (10.203)$$

The maximum of the particle distribution is therefore shifted proportional to the bunch intensity and the general distortion is shown in Fig. 10.21 b for a resistive wake much larger than generally encountered in an accelerator.

The distortion of the particle distribution leads to a deviation from a gaussian distribution and a variation of the bunch length. In the limit of a strong and inductive wake field, for example, the full-width half maximum value of the bunch length increases like

$$\tau_{\text{fwhm}} = \sigma_\tau \sqrt{f_o} = \frac{q\,\sigma_\tau}{\beta\,\sigma_\delta}\sqrt{\frac{\beta L N_b \lambda(\tau)}{\eta_c E_o T_o}}. \qquad (10.204)$$

The bunch length changes as the bunch intensity is increased while the sign and rate of change is dependent on the actual ring impedance spectrum on hand. We have used an induction as an example for the reactive impedance in a ring because it most appropriately represents the real impedance for lower frequencies or longer bunch length. In general, this potential-well bunch lengthening may be used to determine experimentally the nature and quantity of the ring impedance by measuring the bunch length as a function of bunch current.

Turbulent bunch lengthening: At higher bunch currents the bunch lengthening deviates significantly from the scaling of potential well distortion and actually proceeds in the direction of true lengthening. Associated with this lengthening is also an increase in the particle momentum spread. The nature of this instability is similar to the microwave instability for coasting beams.

Considering long bunches, a strong instability with a rise time shorter than the synchrotron oscillation period and high frequencies with wavelength short compared to the bunch length, we expect basically the same dynamics as was discussed earlier for a coasting beam. This was recognized by *Boussard* [10.51] who suggested a modification of the *Keil-Schnell criterion* by replacing the coasting-beam particle density by the bunch density. For a gaussian particle distribution, the peak bunch current is

$$\hat{I} = I_\mathrm{o} \frac{2\pi \bar{R}}{\sqrt{2\pi}\,\sigma_\ell}, \tag{10.205}$$

where I_o is the average circulating beam current per bunch, and the bunch length is related to the energy spread by

$$\sigma_\ell = \frac{\beta c\,|\eta_\mathrm{c}|}{\Omega_\mathrm{so}} \frac{\sigma_\epsilon}{E_\mathrm{o}}. \tag{10.206}$$

With these modifications, the *Boussard criterion* is

$$\boxed{\left|\frac{Z_z}{n}\right| \le F\,\frac{\beta^3 E_\mathrm{o}|\eta_\mathrm{c}|^2}{qI_\mathrm{o}\sqrt{2\pi}\,\nu_\mathrm{so}}\left(\frac{\sigma_\epsilon}{E_\mathrm{o}}\right)^3,} \tag{10.207}$$

where the form factor F is still of the order unity.

As a consequence of this turbulent bunch lengthening we observe an increase of the energy spread as well as an increase of the bunch length. The instability does not necessarily lead to a beam loss but rather to an adjustment of energy spread and bunch length such that the Boussard criterion is met. For very low beam currents the stability criterion is always met up to a threshold where the r.h.s. of (10.207) becomes smaller than the l.h.s. Upon further increase of the beam current beyond the threshold current the energy spread and consequently the bunch length increases to avoid the bunched beam microwave instability.

10.7 Transverse Single-Bunch Instabilities

Transverse wake fields can also greatly modify the stability of a single bunch. Specifically at high frequencies, we note an effect of transverse wake fields generated by the head of a particle bunch on particles in the tail of the same bunch. Such interaction occurs for broad-band impedances where the bunch generates a short wake including a broad spectrum of frequencies. In the first moment all these fields add up being able to act back coherently on particles in the tail but they quickly decoher and vanish before the next bunch arrives. This effect is therefore a true single-bunch effect. In order to affect other bunches passing by later, the fields would have to persist a longer time which implies a higher Q value of the impedance structure which we ignore here.

10.7.1 Beam Break-Up in Linear Accelerators

A simple example of a *transverse microwave instability* is the phenomenon of *beam break-up* in linear accelerators. We noted repeatedly that the impedance of vacuum chambers originates mainly from sudden changes in cross section which therefore must be avoided to minimize impedance and microwave instabilities. This, however, is not possible in accelerating cavities of which there are particularly many in a linear accelerator. Whatever single-pass microwave instabilities exist they should become apparent in a linear accelerator. We have already discussed the effect of longitudinal wake fields whereby the fields from the head of a bunch act back as a decelerating field on particles in the tail. In the absence of corrective measures we therefore expect the particles in the tail to gain less energy than particles in the head of an intense bunch.

Transverse motion of particles is confined to the vicinity of the linac axis by quadrupole focusing in the form of betatron oscillations while travelling along the linear accelerator. However, coherent transverse betatron oscillations can create strong transverse wake fields at high bunch intensities. Such fields may act back on subsequent bunches causing *bunch to bunch instabilities* if the fields persist long enough. Here we are more interested in the effect on the same bunch. For example, the wake fields of the head of a bunch can act back on particles in the tail of the bunch. This interaction is effected by *broad-band impedances* like sudden discontinuities in the vacuum chamber which are abundant in a linear accelerator structure. The interaction between particles in the head of a bunch on particles in the tail of the same bunch can be described by a *two macro particle model* resembling the head and the tail.

Transverse wake fields generated by the head of a bunch of particles are proportional to the transverse oscillation amplitude of the head, and we describe the dynamics of the head and tail of a bunch in a two particle

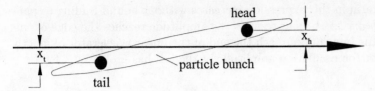

Fig. 10.22. Head-tail dynamics of a particle bunch represented by two macroparticles

model where each particle represents half the charge of the whole bunch as shown in Fig. 10.22.

The head particle with charge $\frac{1}{2}qN_\mathrm{b}$ performs free betatron oscillations while the tail particle responds like a driven oscillator. Since all particles travel practically at the speed of light, the longitudinal distribution of particles remains fixed along the whole length of the linear accelerator. The equations of motion in smooth approximation where $k_\beta = 1/(\nu_\mathrm{o}\,\bar{\beta}_u)$ and $\bar{\beta}_u$ is the average value of the betatron function in the plane u, are for both macroparticles

$$x_\mathrm{h}'' + k_\beta^2\, x_\mathrm{h} = 0\,,$$

$$x_\mathrm{t}'' + k_\beta^2\, x_\mathrm{t} = r_\mathrm{c}\,\frac{x_\mathrm{h}(s,z)}{\gamma(z,s)} \int\limits_{\tilde{z}}^{\infty} \lambda(z)\widetilde{W}_\perp(z - \tilde{z})\,\mathrm{d}z\,,$$

$$= \frac{r_\mathrm{c}\,N_\mathrm{b}\,\widetilde{W}_\perp}{2\gamma}\,x_\mathrm{h}\,, \tag{10.208}$$

where we use the indices $_\mathrm{h}$ and $_\mathrm{t}$ for the head and tail particles respectively and introduce the average wake field per unit length

$$\widetilde{W}_\perp = \frac{W_\perp}{L_\mathrm{acc}}\,. \tag{10.209}$$

For simplicity, it was assumed in (10.208) that the beam is just coasting along the linear accelerator to demonstrate the dynamics of the instability. If the beam is accelerated the adiabatic damping effect through the increase of the energy must be included.

Because of causality only the tail particle is under the influence of a wake field. The transverse wake field $W_\perp(2\sigma_z)$, for example, which is shown in Fig. 10.12, is to be taken at a distance $2\sigma_z$ behind the head particle. Inserting the solution $x_\mathrm{h}(s) = \hat{x}_\mathrm{h}\cos k_\beta s$ into the second equation, we obtain the solution for the betatron oscillation of the tail particle in the form

$$x_\mathrm{t}(s) = \hat{x}_\mathrm{h}\cos k_\beta s + \hat{x}_\mathrm{h}\,\frac{r_\mathrm{c}\,N_\mathrm{b}\,\widetilde{W}_\perp}{4\gamma\,k_\beta}\,s\,\sin k_\beta s\,. \tag{10.210}$$

The second term in this expression increases without bound leading to particle loss or *beam break-up* as soon as the amplitude reaches the edge of the aperture. If the bunch does reach the end of the linear accelerator of length L_{acc}, the betatron oscillation amplitude of the tail has grown by a factor

$$F_{bb} = \frac{\hat{x}_t}{\hat{x}_h} = \frac{r_c N_b \widetilde{W}_\perp L_{acc}}{4 \gamma k_\beta} . \tag{10.211}$$

One consequence of this instability is an apparent increase in beam emittance long before beam loss occurs. A straight bunch with a small cross section becomes bent first like a banana and later like a snake and the transverse distribution of all particles in the bunch occupies a larger cross-sectional area than before. This increase in apparent beam size has a big detrimental effect on the attainable luminosity in *linear colliders* and therefore must be minimized as much as possible. The two particle model adopted here is insufficient to determine a more detailed structure than that of a banana. However, setting up equations similar to (10.208) for more than two macroparticles will start to reveal the oscillatory nature of the transverse bunch perturbation.

One scheme to greatly reduce the beam break-up effect is called *BNS damping* in reference to its inventors *Balakin* et al. [10.52] and has been successful implemented into the *Stanford Linear Collider* [10.53]. The technique utilizes the fact that the betatron oscillation frequency depends by virtue of the chromaticity on the energy of the particles. By accelerating the bunch behind the crest of the accelerating field the tail gains less energy than the head. Therefore the tail is focused more by the quadrupoles than the head. Since the transverse wake field introduces defocusing this additional chromatic focusing can be used for compensation.

Of course this method of damping the beam break-up by accelerating ahead of the crest is counter productive to compensating for the energy loss of tail particles due to longitudinal wake fields. In practice, BNS damping is applied only at lower energies where the instability is strongest and in that regime the energy reducing effect of the longitudinal wake field actually helps to maximize BNS damping. Toward the end of the linear accelerator at high beam energies, the beam break up effect becomes small $\propto 1/\gamma$ and the bunch is now moved ahead of the crest to reduce the energy spread in the beam.

10.7.2 Fast Head-Tail Effect

Transverse bunch perturbations due to broad-band impedances are not restricted to linear accelerators but occur also in circular accelerators. In a circular proton accelerator, for example, the "length" is for all practical purposes infinite, there is no radiation damping and therefore even weak transverse wake fields can in principle lead to transverse bunch blow up and beam loss. This instability is known as the *fast head-tail instability* or *strong* head-tail instability and has been first discussed and analyzed by *Kohaupt* [10.54]. The dynamics in a circular accelerator is, however, different from that in a linear accelerator because particles in the head of a bunch will not stay there but rather oscillate between head and tail in the course of synchrotron oscillations. These synchrotron oscillations disturb the coherence between head and tail and the instability becomes much weaker.

On the other hand particles in circular accelerators and especially in storage rings are expected to circulate for a long time and even a much reduced growth rate of the transverse bunch blow up may still be too strong. The dynamics of interaction is similar to that in a linear accelerator at least during about half a synchrotron oscillation period $1/2 t_s$, but during the next half period the roles are interchanged for individual particles. Particles travelling for one half period in the head of the bunch find themselves in the tail for the next half period only to reach the head again and so forth. To understand the dynamics over many oscillations, we set up equations of motion for two macroparticles resembling the head and tail of a particle bunch similar to (10.208) but we now use the time as the independent variable. The distance ζ between head and tail particle varies between 0 and the maximum distance of the two macro particles 2ℓ during the course of a synchrotron oscillation period and since the transverse wake field increases linearly with ζ, we set $W_\perp(\zeta) = W_\perp(2\sigma_\ell) \sin \Omega_s t$. With this the equations of motion are for $0 \leq t \leq 1/2 t_s$

$$\ddot{x}_1 + \omega_\beta^2 x_1 = 0,$$

$$\ddot{x}_2 + \omega_\beta^2 x_2 = \frac{r_c \beta^2 c^2 N_b \widetilde{W}_\perp(2\sigma_\ell) \sin \Omega_s t}{2\gamma} x_1,$$

(10.212)

where $\widetilde{W}_\perp = W_\perp/(2\pi\bar{R})$ is the wake function per unit length. For the next half period $1/2 t_s \leq t \leq t_s$

$$\ddot{x}_1 + \omega_\beta^2 x_1 = \frac{r_c \beta^2 c^2 N_b \widetilde{W}_\perp(2\ell) \sin \Omega_s t}{2\gamma} x_2,$$

$$\ddot{x}_2 + \omega_\beta^2 x_2 = 0.$$

(10.213)

For further discussions it is convenient to consider solutions to (10.212,213) in the form of phasors defined by

$$\mathbf{x}(t) = \mathbf{x}(0) e^{i\omega_\beta t} = x - i\frac{\dot{x}}{\omega_\beta}.$$

(10.214)

The first equation (10.212) can be solved immediately for

$$\mathbf{x}_1(t) = \mathbf{x}_1(0) e^{i\omega_\beta t}$$

(10.215)

and the second equation (10.212) becomes with (10.215)

$$\ddot{\mathbf{x}}_2 + \omega_\beta^2 \mathbf{x}_2 = A \sin \Omega_s t \, e^{i\omega_\beta t} \mathbf{x}_1(0),$$

(10.216)

where

$$A = \frac{r_c \beta^2 c^2 N_b \widetilde{W}_\perp (2\ell)}{2\gamma}. \tag{10.217}$$

The synchrotron oscillation frequency is generally much smaller than the betatron oscillation frequency, $\Omega_s \ll \omega_\beta$, and the solution of (10.217) becomes with this approximation

$$\mathbf{x}_2(t) = \mathbf{x}_2(0) e^{i\omega_\beta t} + \frac{1}{\omega_\beta} \int_0^t [A \mathbf{x}_1(0) \sin \Omega_s t' e^{i\omega_\beta t'}] \sin \omega_\beta (t - t') \, dt'$$

or after some manipulation

$$\mathbf{x}_2(t) = \mathbf{x}_2(0) e^{i\omega_\beta t} - i \mathbf{x}_1(0) \tfrac{1}{2} a (1 - \cos \Omega_s t) e^{i\omega_\beta t}, \tag{10.218}$$

where $a = A/(\omega_\beta \Omega_s)$. During the second half synchrotron oscillation period, the roles of both macroparticles are exchanged. We may formulate the transformation through half a synchrotron oscillation period in matrix form and get with $1 - \cos \Omega_s \tfrac{1}{2} t_s = 2$ for the first half period

$$\begin{pmatrix} \mathbf{x}_1(\tfrac{1}{2}t_s) \\ \mathbf{x}_2(\tfrac{1}{2}t_s) \end{pmatrix} = e^{i\omega_\beta t_s/2} \begin{pmatrix} 1 & 0 \\ -ia & 1 \end{pmatrix} \begin{pmatrix} \mathbf{x}_1(0) \\ \mathbf{x}_2(0) \end{pmatrix} \tag{10.219}$$

and for the second half period

$$\begin{pmatrix} \mathbf{x}_1(t_s) \\ \mathbf{x}_2(t_s) \end{pmatrix} = e^{i\omega_\beta t_s/2} \begin{pmatrix} 1 & -ia \\ 0 & 1 \end{pmatrix} \begin{pmatrix} \mathbf{x}_1(\tfrac{1}{2}t_s) \\ \mathbf{x}_2(\tfrac{1}{2}t_s) \end{pmatrix}. \tag{10.220}$$

Combining both half periods one gets finally for a full synchrotron oscillation period

$$\begin{pmatrix} \mathbf{x}_1(t_s) \\ \mathbf{x}_2(t_s) \end{pmatrix} = e^{i\omega_\beta t_s} \begin{pmatrix} 1 - a^2 & -ia \\ -ia & 1 \end{pmatrix} \begin{pmatrix} \mathbf{x}_1(0) \\ \mathbf{x}_2(0) \end{pmatrix}. \tag{10.221}$$

The stability of the motion after many periods can be extracted from (10.221) by solving the *eigenvalue equation*

$$\begin{pmatrix} 1 - a^2 & -ia \\ -ia & 1 \end{pmatrix} \begin{pmatrix} \mathbf{x} \\ \mathbf{x} \end{pmatrix} = \lambda \begin{pmatrix} \mathbf{x} \\ \mathbf{x} \end{pmatrix}. \tag{10.222}$$

The *characteristic equation*

$$\lambda^2 - (2 - a^2) \lambda + 1 = 0 \tag{10.223}$$

has the solution

$$\lambda_{1,2} = (1 - \tfrac{1}{2}a^2) \pm \sqrt{(1 - \tfrac{1}{2}a^2)^2 - 1} \tag{10.224}$$

and the eigenvalues can be expressed by

$$\lambda = e^{\pm i\Phi}, \tag{10.225}$$

where $(1 - \frac{1}{2}a^2) = \cos\Phi$ for $|a| \le 2$ or

$$|a| = \frac{r_c\,\beta^2 c^2\,N_b\,\widetilde{W}_\perp(2\ell)}{2\gamma\,\omega_\beta\,\Omega_s} \le 2. \tag{10.226}$$

The motion remains stable since no element of the transformation matrix increases unbounded as the number of periods increases $n \to \infty$. In the form of a *stability criterion*, we find that the single-bunch current $I_b = qN_b f_{rev}$ must not exceed the limit

$$I_b \le \frac{4\,q\,\gamma\,\omega_o^2\,\nu_\beta\,\nu_s}{r_c\,\beta c\,W_\perp(2\ell)}, \tag{10.227}$$

where q is the charge of the particles, (ν_β, ν_s) the betatron and synchrotron tune respectively and $W_\perp = 2\pi\bar{R}\,\widetilde{W}_\perp$. In a storage ring, it is more convenient to use impedance rather than wake fields. Had we set up the equations of motion (10.208,209) expressing the perturbing force in terms of impedance we would get the same results but replacing the wake field by

$$W_\perp(2\ell) = \frac{\omega_o}{\pi}\,\text{Im}\{Z_\perp\} \tag{10.228}$$

and the threshold beam current for the *fast head-tail instability* becomes

$$\boxed{I_b \le \frac{4\pi\,q\,\gamma\,\omega_o\,\nu_\beta\,\nu_s}{r_c\,\beta c\,\text{Im}\{Z_\perp\}}.} \tag{10.229}$$

The bunch current I_b is a threshold current which prevents us from filling more current into a single bunch. Exceeding this limit during the process of filling a bunch in a circular accelerator leads to an almost immediate loss of the excess current. This *microwave instability* is presently the most severe limitation on single-bunch currents in storage rings and special care must be employed during the design to minimize as much as possible the transverse impedance of the vacuum chamber system.

The strength of the instability becomes more evident when we calculate the growth time for a beam current just by an increment ϵ above the threshold. For $|a| > 2$ we have $(1 - \frac{1}{2}a^2) = -\cosh\mu$ and the eigenvalue is $\lambda = e^{\pm\mu}$. The phase $\mu = 0$ at threshold and $\cosh\mu \approx 1 + \frac{1}{2}\mu^2$ for $a = 2 + \epsilon$ and we get

$$\mu = 2\sqrt{\epsilon}. \tag{10.230}$$

In each synchrotron oscillation period the eigenvalues increase by the factor e^μ or at a *growth rate* of

$$\frac{1}{\tau} = \frac{\mu}{t_s} = \frac{2\sqrt{\epsilon}}{t_s}. \tag{10.231}$$

If, for example, the beam current exceeds the threshold by 10%, we have $\epsilon = 0.2$ and the rise time would be $\tau/t_s = 0.89$ or the oscillation amplitudes increase by more than a factor of two during a single synchrotron oscillation period. This is technically very difficult to counteract by a feedback system.

We have assumed that transverse wake fields are evenly distributed around the accelerator circumference. In a well designed accelerator vacuum chamber, however, most of the transverse wake field occur in the accelerating cavities and therefore only the transverse betatron oscillation amplitude in the cavities are relevant. In this case, one recalls the relation $\nu_\beta \approx \bar{R}/\beta_u$ and we replace in (10.227) the average value of the betatron function by the value in the cavities for

$$I_b \le \frac{4\,q\,\gamma\,\Omega_s}{r_c\,\beta_{u,cy}\,W_{\perp,cy}(2\ell)}.\tag{10.232}$$

This result suggest that the betatron function in the plane $u = x$ or $u = y$ at the location of cavities should be kept small and the synchrotron oscillation frequency should be large. The exchange of head and tail during synchrotron oscillation slows down considerably the growth rate of the instability. The result (10.232) is the same as the amplification factor (10.211) if we consider that in a linear accelerator the synchrotron oscillation period is infinite.

As we approach the threshold current, the beam signals the appearance of the head-tail instability on a spectrum analyzer with a satellite below the betatron frequency. The threshold for instability is reached when the satellite frequency reaches a value $\omega_{sat} = \omega_\beta - 1/2\,\Omega_s$. This becomes apparent when replacing the transformation matrix in (10.221) by the eigenvalue

$$\begin{pmatrix} x_1(t_s) \\ x_2(t_s) \end{pmatrix} = e^{i\omega_\beta t_s}\,e^{i\Phi} \begin{pmatrix} x_1(0) \\ x_2(0) \end{pmatrix}.\tag{10.233}$$

The phase reaches a value of $\Phi = \pi$ at the stability limit and (10.233) becomes at this limit

$$\begin{pmatrix} x_1(t_s) \\ x_2(t_s) \end{pmatrix} = e^{i(\omega_\beta - 1/2\Omega_s)\,t_s} \begin{pmatrix} x_1(0) \\ x_2(0) \end{pmatrix}.\tag{10.234}$$

At this point, it should be noted, however, that the shift of the betatron frequency to $1/2\Omega_s$ is a feature of the two macro particle model. In reality there is a distribution of particles along the bunch and while increasing the beam current the betatron frequency decreases and the satellite $\nu_\beta - \nu_s$ moves until both frequencies merge and become imaginary. This is the point of onset for the instability. It is this feature of merging frequencies which is sometimes called *mode mixing* or *mode coupling*.

10.7.3 Head-Tail Instability

Discussing the fast head-tail instability we considered the effect of transverse wake fields generated by the head of a particle bunch on the transverse betatron motion of the tail. We assumed a constant betatron oscillation frequency which is only an approximation since the betatron frequency depends on the particle energy. On the other hand, there is a distinct relationship between particle energy and particle motion within the bunch, and it is therefore likely that the dynamics of the head-tail instability becomes modified by considering the energy dependence of the betatron oscillation frequency.

Like in the previous section, we represent the particle bunch by two macroparticles which perform synchrotron oscillations being 180° apart in phase. The wake fields of the head particle act on the tail particle while the reverse is not true due to causality. However, during each half synchrotron oscillation period the roles become reversed.

In (10.218), we obtained an expression which includes the perturbation term and consider the variation of this term due to chromatic oscillations of the betatron frequency. The perturbation term is proportional to $e^{i\omega_\beta t}$ and we set therefore with $\delta = \Delta p/p_o$

$$\omega_\beta = \omega_\beta(\delta) = \omega_{\beta o} + \frac{\partial \omega_\beta}{\partial \delta} \delta + \mathcal{O}(\delta^2). \qquad (10.235)$$

The *chromaticity* is defined by the betatron tune shift per unit relative momentum deviation

$$\xi_\beta = \frac{\Delta \nu_\beta}{\delta} \qquad (10.236)$$

and (10.235) becomes with $\omega_\beta = \nu_\beta \omega_o$

$$\omega_\beta = \omega_{\beta o} + \xi_\beta \delta \omega_o. \qquad (10.237)$$

The momentum deviation is oscillating at the synchrotron frequency and is correlated with the longitudinal motion by

$$\delta = -\frac{\Omega_s \ell}{\beta c |\eta_c|} \sin \Omega_s t, \qquad (10.238)$$

where $2\hat{\ell}$ is the maximum longitudinal distance between the two macroparticles. Combining (10.237,238) we get

$$\omega_\beta = \omega_{\beta o} - \frac{\Omega_s \ell \xi_\beta}{\nu_\beta \bar{R} |\eta_c|} \sin \Omega_s t, \qquad (10.239)$$

where the second term is much smaller than unity so that we may expand the exponential function of this term to get

$$\mathrm{e}^{\mathrm{i}\omega_\beta t} \approx \mathrm{e}^{\mathrm{i}\omega_{\beta 0} t} \left[1 - \mathrm{i}\, \frac{\Omega_\mathrm{s}\, \ell\, \xi_\beta}{\nu_\beta\, \bar{R}\, |\eta_\mathrm{c}|}\, t\, \sin(\Omega_\mathrm{s} t) \right] . \tag{10.240}$$

The expression in the square bracket is the variation of the scaling factor a in (10.218) and we note specifically, the appearance of the imaginary term which gives rise to an instability. The phase Φ in the eigenvalue equation (10.225) becomes for small beam currents $\Phi \approx a$ and with (10.240)

$$\Phi = a \left[1 - \mathrm{i}\, \frac{\Omega_\mathrm{s}\, \ell\, \xi_\beta}{\pi\nu_\beta\, \bar{R}\, |\eta_\mathrm{c}|}\, t_\mathrm{s} \right] , \tag{10.241}$$

where we have set $t = \frac{1}{2}\, t_\mathrm{s}$ and $\langle \sin \Omega_\mathrm{s} t \rangle \approx 2/\pi$. The first term represents the *fast head-tail instability* with its threshold characteristics discussed in the previous section. The second term is an outright damping or antidamping effect being effective at any beam current. This instability is called the *head-tail effect* discovered and analyzed by *Pellegrini* [10.55] and *Sands* [10.56] at the storage ring ADONE.

Considering only the imaginary term in (10.241), we note an exponential growth of the head-tail instability with a growth rate

$$\frac{1}{\tau} = \frac{\Omega_\mathrm{s}\, a\, \ell\, \xi_\beta}{\pi\nu_\beta\, \bar{R}\, |\eta_\mathrm{c}|} = \frac{\ell\, \xi_\beta\, r_\mathrm{c}\, \beta c\, N_\mathrm{b}\, \widetilde{W}_\perp(2\ell)}{2\pi\, \gamma\, |\eta_\mathrm{c}|\, \nu_\beta^2} . \tag{10.242}$$

Instability may occur either in the vertical or the horizontal plane depending on the magnitude of the transverse wake function in both planes. There are two modes, one stable and one unstable depending on the sign of the chromaticity and momentum compaction. Above transition $\eta_\mathrm{c} < 0$ and the beam is unstable for negative chromaticity. This instability is the main reason for the insertion of sextupole magnets into circular accelerators to compensate for the naturally negative chromaticity. Below transition, the situation is reversed and no correction of chromaticity by sextupoles is required. From (10.242), we would conclude that we need to correct the chromaticity exactly to zero to avoid instability by one or the other mode. In reality, this is not the case because a two particle model overestimates the strength of the negative mode. Following a more detailed discussion including Vlasov's equation [10.25] it becomes apparent that the negative mode is much weaker to the point where, at least in electron accelerators, it can be ignored in the presence of radiation damping.

Observation of the *head-tail damping* for positive chromaticities or measuring the risetime as a function of chromaticity can be used to determine the transverse wake function or impedance of the accelerator [10.57, 58]. Measurements of *head-tail damping rates* have been performed in SPEAR [10.57] as a function of chromaticity and are reproduced in Figs. 10.22,23.

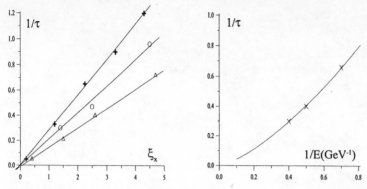

Fig. 10.23. Measurement of the head-tail damping rate in SPEAR as a function of chromaticity **a** and beam energy **b** [10.57]

We clearly note the linear increase of the damping rate with chromaticity. The scaling with energy and beam current is less linear due to a simultaneous change in bunch length. Specifically the bunch length increases with beam intensity causing the wake fields to drop for a smaller damping rate.

10.8 Multi-Bunch Instabilities

Single-bunch dynamics is susceptible to all kinds of impedances or wake fields whether it be narrow or broad-band impedances. This is different for *multi-bunch instabilities* or *coupled-bunch instabilities* [10.49 – 59]. In order for wake fields to generate an effect on more than one bunch it must persist at least until the next bunch comes by the location of the impedance. We expect therefore multi-bunch instabilities only due to high Q or *narrow-band impedances* like those encountered in accelerating cavities. Higher-order modes in such cavities persist some time after excitation and actually reach a finite amplitude in a circular accelerator where the orbiting beam may periodically excite one or the other mode. Because these modes have a high quality factor they are also confined to a narrow frequency spread. The impedance spectrum we need to be concerned with in the study of multi-bunch instabilities is therefore a line spectrum composed of many cavity modes.

To study the effect of these modes on the circulating beam we must fold the beam current spectrum with the mode spectrum and derive from this interaction conditions for beam stability. We will do this for the case of the two lowest order mode oscillations only where all bunches oscillate in synchronism at the same phase or are 90° out of phase from bunch to

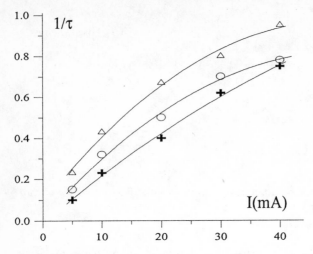

Fig. 10.24. Measurement of the head-tail damping rate in SPEAR as a function of beam current [10.57]

bunch respectively. Of course in a real accelerator higher-order modes can be present too and must be taken into account. Here we must limit ourself, however, to the discussion of the physical effect only and direct the interested reader to more detailed discussions on this subject in references [10.25 – 28].

We consider the dynamics of rigid coupled bunches ignoring the internal motion of particles within a single bunch. The beam spectrum is then from (10.63) with q the bunch charge and observing at $\varphi = 0$ for simplicity

$$I_{\parallel}(\omega, \varphi) = \frac{q n_b \omega_o}{2\pi} \sum_{p=-\infty}^{+\infty} \sum_{n=-\infty}^{+\infty} \mathrm{i}^{-n} J_n(p n_b \omega_o \hat{\tau}) \, \delta(\omega - \Omega_n), \qquad (10.243)$$

where now $\Omega_n = (p n_b + nm + n\nu_s)\omega_o$ and where we have replaced the synchrotron frequency by the synchrotron tune and the phase ζ_i for individual particles by the *mode* of the bunch distribution setting $\zeta_i = m\omega_o t$ with $0 \le m \le n_b$.

A beam of n_b equidistant bunches can oscillate in n_b different modes. Two bunches, for example, can oscillate in phase or 180° out of phase; four bunches can oscillate with a phase difference of 0°, 90°, 180°, and 270° between consecutive bunches. In general the order of the mode m defines the phase difference of consecutive bunches by

$$\Delta\phi = m \frac{360°}{n_b}. \qquad (10.244)$$

To determine the multi-bunch dynamics we calculate first the induced volt-age $V(t)$ by the beam current in the impedance $Z(\omega)$ and then fold the voltage function with the beam function to calculate the energy loss per turn by each particle. Knowing this we will be able to formulate the equation of motion for the synchrotron oscillation as we have done so in [10.17]. Specifically we will be able to formulate frequency shifts and damping or antidamping due to the interaction of the multi-bunch beam with its environment to identify conditions for beam stability.

For simplicity we assume small phase oscillations $\hat{\tau} \ll 1$ and consider only the fundamental beam frequency and the first satellite $n = 0, 1$. With this (10.243) becomes

$$I_{\parallel}(\omega) = \frac{q n_{\mathrm{b}} \omega_{\mathrm{o}}}{2\pi} \sum_{p=-\infty}^{+\infty} J_{\mathrm{o}}(p n_{\mathrm{b}} \omega_{\mathrm{o}} \hat{\tau}) \, \delta(\omega - \Omega_{\mathrm{o}}) - \mathrm{i} J_1(p n_{\mathrm{b}} \omega_{\mathrm{o}} \hat{\tau}) \, \delta(\omega - \Omega_1) \,,$$

$$(10.245)$$

where $\Omega_{\mathrm{o}} = p n_{\mathrm{b}} \omega_{\mathrm{o}}$, $\Omega_1 = (p n_{\mathrm{b}} + m + \nu_{\mathrm{s}}) \omega_{\mathrm{o}}$, and J_i are Bessel's functions. The induced voltage spectrum is $V(\omega) = Z(\omega) I(\omega)$ and its Fourier transform $V(t) = \int V(\omega) \mathrm{e}^{\mathrm{i}\omega t} \, \mathrm{d}\omega$ or

$$V_{\parallel}(t) = \frac{q n_{\mathrm{b}} \omega_{\mathrm{o}}}{2\pi} \sum_{p=-\infty}^{+\infty} \left\{ J_{\mathrm{o}}(\hat{\tau} \Omega_{\mathrm{o}}) \, Z(\Omega_{\mathrm{o}}) \, \mathrm{e}^{\mathrm{i}\Omega_{\mathrm{o}} t} \right.$$

$$\left. - \mathrm{i} J_1(\hat{\tau} \, \Omega_{\mathrm{o}}) \, Z(\Omega_1) \, \mathrm{e}^{\mathrm{i}\Omega_1 t} \right\} \,.$$

$$(10.246)$$

The energy loss per particle is then defined by integrating in time the product of voltage function and single-bunch current function

$$U = \frac{1}{N_{\mathrm{b}}} \int V_{\parallel}(t) \frac{I_{\parallel}(t + \tau)}{n_{\mathrm{b}}} \, \mathrm{d}t \,,$$

$$(10.247)$$

N_{b} is the number of particles per bunch and $T_{\mathrm{b}} = T_{\mathrm{o}}/n_{\mathrm{b}}$ the time between passage of consecutive bunches. The bunch current can be expanded for $\tau \ll 1$

$$I_{\parallel}(t + \tau) \approx I_{\parallel}(t) + \tau \frac{\mathrm{d}}{\mathrm{d}t} I_{\parallel}(t) \,.$$

$$(10.248)$$

The Fourier transforms of both current and its derivative with respect to time are correlated by

$$\frac{\mathrm{d}}{\mathrm{d}t} I_{\parallel}(\omega) = \mathrm{i}\omega I_{\parallel}(\omega)$$

$$(10.249)$$

and (10.248) becomes in frequency domain with (10.245)

$$I_\parallel(t+\tau) = \frac{q n_\mathrm{b} \omega_\mathrm{o}}{(2\pi)^2} \int \sum_{p=-\infty}^{+\infty} (1+\mathrm{i}\omega\tau)\,(J_\mathrm{o}\,\delta_\mathrm{o} - \mathrm{i}J_1\,\delta_1)\,\mathrm{e}^{\mathrm{i}\omega t}\,\mathrm{d}t\,, \qquad (10.250)$$

where we have used some abbreviations which become obvious by comparison with (10.245). Inserting (10.250) and (10.246) into (10.247) we get

$$U = \frac{(q\omega_\mathrm{o})^2 n_\mathrm{b}}{(2\pi)^2\,N_\mathrm{b}} \int_t \int_\omega \sum_p (J_\mathrm{o}\,Z_\mathrm{o}\mathrm{e}^{\mathrm{i}\Omega_\mathrm{o}t} - \mathrm{i}J_1\,Z_1\mathrm{e}^{\mathrm{i}\Omega_1 t})$$
$$\times (1+\mathrm{i}\omega\tau)\sum_r (J_\mathrm{o}\,\delta_{\mathrm{o}r} - \mathrm{i}J_1\,\delta_{1r})\,\mathrm{e}^{\mathrm{i}\omega t}\,\mathrm{d}\omega\,\mathrm{d}t\,. \qquad (10.251)$$

For abbreviation we have set $\delta_i = \delta(\Omega_i)$, $Z_i = Z(\Omega_i)$, $J_\mathrm{o} = J_\mathrm{o}(\Omega_\mathrm{o})$, and $J_1 = J_1(\Omega_\mathrm{o})$. An additional index has been added to indicate whether the quantity is part of the summation over p or r. Before we perform the time integration we reverse the first summation by replacing $p \to -p$ and get terms like $\int \mathrm{e}^{-\mathrm{i}(\Omega_\mathrm{o}-\omega)t}\,\mathrm{d}t = 2\pi\delta_\mathrm{o}$ etc. and (10.251) becomes

$$U = \frac{(q\omega_\mathrm{o})^2 n_\mathrm{b}}{2\pi\,N_\mathrm{b}} \int_\omega \sum_p (J_\mathrm{o}\,Z_\mathrm{o}\delta_{\mathrm{o}r} + \mathrm{i}J_1\,Z_1\delta_{1r})$$
$$\times (1+\mathrm{i}\omega\tau)\sum_r (J_\mathrm{o}\,\delta_\mathrm{o} - \mathrm{i}J_1\,\delta_1)\,\mathrm{d}\omega\,. \qquad (10.252)$$

The integration over ω will eliminate many components. Specifically we not that all cross terms $\delta_\mathrm{o}\delta_1$ vanish after integration. We also note that the terms $\delta_{\mathrm{o}p}\delta_{\mathrm{o}r}$ vanish unless $r = p$. With this in mind we get from (10.252)

$$U = \frac{(q\omega_\mathrm{o})^2 n_\mathrm{b}}{2\pi\,N_\mathrm{b}} \sum_p (J_\mathrm{o}^2\,Z_\mathrm{o} + \mathrm{i}\Omega_\mathrm{o}\tau\,J_\mathrm{o}^2\,Z_\mathrm{o} + J_1^2\,Z_1 + \mathrm{i}\Omega_1\tau J_1^2\,Z_1)\,. \qquad (10.253)$$

Finally the summation over p leads to a number of cancellations considering that the resistive impedance is an even and the reactive impedance an odd function. With $Z_\mathrm{o} = Z_\mathrm{ro} + \mathrm{i}Z_\mathrm{io}$, $Z_\mathrm{ro}(\omega) = Z_\mathrm{ro}(-\omega)$, and $Z_\mathrm{io}(\omega) = -Z_\mathrm{io}(-\omega)$ (10.253) becomes

$$U = \frac{(q\omega_\mathrm{o})^2 n_\mathrm{b}}{2\pi\,N_\mathrm{b}} \sum_{p=-\infty}^{+\infty} [J_\mathrm{o}^2(\hat{\tau}\Omega_\mathrm{o})\,Z_\mathrm{r}(\Omega_\mathrm{o}) + J_1^2(\hat{\tau}\Omega_\mathrm{o})\,Z_\mathrm{r}(\Omega_1) \qquad (10.254)$$
$$+ \mathrm{i}\tau\,\Omega_1 J_1^2(\hat{\tau}\Omega_1)\,Z_\mathrm{r}(\Omega_1) - \tau\,\Omega_1 J_1^2(\hat{\tau}\Omega_1)\,Z_\mathrm{i}(\Omega_1)]\,.$$

The first and second term are the resistive energy losses of the circulating beam and synchrotron oscillations respectively while the third and fourth term are responsible for the stability of the multi bunch beam.

The equation of motion for synchrotron oscillations has been derived in [10.17] and we found that frequency and damping is determined by the accelerating rf field and energy losses. We expect therefore that the energy loss derived for coupled bunch oscillations will also lead to frequency shift and damping or anti damping. Specifically we have for the equation of motion from [Ref. 10.17, Eq. (8.27)]

$$\ddot{\varphi} + \omega_o^2 \frac{h\eta_c}{2\pi\beta c\, cp_o}\, e\, \frac{dV}{d\psi}\bigg|_{\psi_s}\, \varphi - \frac{1}{T_o}\frac{dU}{dE}\bigg|_{E_o}\, \dot{\varphi} = 0, \tag{10.255}$$

where we notice the phase proportional term which determines the unperturbed synchrotron frequency

$$\Omega_{so}^2 = \omega_o^2 \frac{h\eta_c}{2\pi\beta cp_o}\, e\, \frac{dV}{d\psi}\bigg|_{\psi_s} = \omega_o^2 \frac{h\eta_c\, e\widehat{V}_o\,\cos\psi_s}{2\pi\,\beta\,cp_o}\,. \tag{10.256}$$

The term proportional to $\dot{\varphi}$ gave rise to the damping decrement

$$\alpha_{so} = -\frac{1}{2T_o}\frac{dU}{dE}\bigg|_{E_o}. \tag{10.257}$$

The modification of the synchrotron frequency is with $\tau = \varphi/h\omega_o$ from (10.254 – 256) similar to the derivation of the unperturbed frequency

$$\Omega_s^2 = \Omega_{so}^2 + \omega_o^2 \frac{h\eta_c n_b}{\beta cp_o N_b} \sum_{p=-\infty}^{+\infty} \tau\Omega_1\,[qf_o\, J_1(\hat{\tau}\Omega_1)]^2\, Z_i(\Omega_1), \tag{10.258}$$

where $f_o = \omega_o/2\pi$ is the revolution frequency. Note that $\eta_c < 0$ above transition and the additional damping or energy loss due to narrow-band impedances reduces the frequency as one would expect.

Similarly we derive the modification of the damping decrement from the imaginary term in (10.254) noting that the solution of the synchrotron oscillation gives $\dot{\tau} = -i\Omega_s\tau$ with $\varphi = h\omega_o\tau$ and the damping decrement for a multi-bunch beam is

$$\alpha_s = \alpha_{so} - \frac{\omega_o\eta_c n_b}{2cp_o N_b} \sum_{p=-\infty}^{+\infty} \frac{\Omega_1}{\Omega_s}\,[qf_o J_1(\hat{\tau}\Omega_1)]^2\, Z_r(\Omega_1). \tag{10.259}$$

For proton and ion beams we would set $\alpha_{so} = 0$ because there is no radiation damping and the interaction of a multi-bunch beam with narrow-band impedances would provide damping or antidamping depending on the sign of the damping decrement for each term. If, however, only one term is antidamped the beam would be unstable and get lost as was observed first at the storage ring DORIS [10.60]. It is therefore important to avoid the overlap of any line of the beam spectrum with a narrow-band impedance in the ring.

Since this is very difficult to achieve and to control, it is more convenient to minimize higher-order narrow-band impedances in the ring by design as much as possible to increase the rise time of the instabilities. In electron storage rings the situation is similar, but now the instability rise time must exceed the radiation damping time. Even though, modern storage rings are designed for high beam currents and great efforts are being undertaken to reduce the impedance of higher cavity modes by designing *monochromatic cavities* where the higher-order modes are greatly suppressed [10.61 – 65].

We have discussed here only the dipole mode of the longitudinal coupled-bunch instability. Of course there are more modes and a similar set of instabilities in the transverse dimensions. A more detailed discussion of all aspects of multi-bunch instabilities would exceed the scope of this text and the interested reader is referred to the specific literature, specifically to [10.25 – 28].

Problems

Problem 10.1. Specify a damping ring at an energy of 1.5 GeV and an emittance of 10^{-10} m-rad. The rf frequency be 500 MHz and 10^{11} electrons are to be stored into a single bunch at full coupling. Calculate the Touschek lifetime and the coherent and incoherent space-charge tune shift. Would the beam survive if a tune shift of 0.05 were permissible?

Problem 10.2. Derive (10.155) from (10.154) and show that the constant A in (10.155) is given by $A = \frac{3}{4\pi} \frac{q I_o}{\beta^2 E_o |\eta_c| (\Delta p/p_o)^2}$.

Problem 10.3. Explain the stability of saturns rings in the realm of negative-mass instability. Why appears matter so uniformly distributed within the rings?

Problem 10.4. Use the wake field for the SLAC linear accelerator structure (Fig. 10.16) and calculate the energy loss of a particle in the tail of a 1mm long bunch of 10^{11} electrons for the whole SLAC linear accelerator of 3 km length. This energy droop along a bunch is mostly compensated by accelerating the bunch ahead of the crest of the accelerating wave. This way the particles in the head of the bunch gain less energy than the particles in the tail of the bunch. The extra energy gain of the tail particles is then lost again due to wake field losses. How far off the crest must the bunch be accelerated for this compensation?

Problem 10.5. Consider the phenomenon of beam break-up in a linear accelerator and split the bunch into a head, center and tail part with a particle distribution $N_b/4$ to $N_b/2$ to $N_b/4$. Set up the equations of motion for all three particles including wake fields and solve the equations. Show the exponential build up of oscillation amplitudes of the tail particle. Perform the same derivation including BNS damping where each macroparticle has a different betatron oscillation frequency. Determine the condition for optimum BNS damping.

Problem 10.6. Determine the perturbation of a gaussian particle distribution under the influence of a capacitive wake field. In particular derive expressions for the perturbation of the distribution (if any) and the change in the fwhm bunch width as a function of σ_τ in the limit of small wakes. If there is a shift in the distribution what physical effects cause it ? Hint: think of a loss mechanism for a purely reactive but capacitive wake field?

Problem 10.7. During the discussion of the dispersion relation we observed the stabilizing effect of Landau damping and found the stability criterion (10.156) stating that the threshold current can be increased proportional to the square of the momentum spread in the beam. How does this stability criterion in terms of a momentum spread relate to the conclusion in the section on Landau damping that the beam should have a frequency overlap with the excitation frequency? Why is a larger momentum spread better than a smaller spread?

Problem 10.8. Determine stability conditions for the fast head-tail instability in a storage ring of your choice assuming that all transverse wake fields come from accelerating cavities. Use realistic parameters for the rf system and the number of cells appropriate for your ring. What is the maximum permissible transverse impedance for a bunch current of 100 ma? Is this consistent with the transverse impedance of pill box cavities? If not how would you increase the current limit?

Problem 10.9. Calculate the real and imaginary impedance for the first longitudinal and transverse higher-order mode in a pill box cavity and apply these to determine the multi-bunch beam limit for a storage ring of your choice assuming that the beam spectrum includes the HOM frequency. Calculate also the frequency shift at the limit.

11. Insertion Device Radiation

Synchrotron radiation from bending magnets is characterized by a wide spectrum from microwaves up to soft or hard x-rays as determined by the critical photon energy. To optimally meet the needs of basic research with synchrotron radiation, it is desirable to provide specific radiation characteristics that cannot be obtained from ring bending magnet but require special magnets. The field strength of bending magnets and the maximum particle beam energy in circular accelerators like a storage ring is fixed leaving no adjustments to optimize the synchrotron radiation spectrum for particular experiments. To generate specific synchrotron radiation characteristics, radiation is often produced from special *insertion devices* installed along the particle beam path. Such insertion devices introduce no net deflection of the beam and can therefore be incorporated in a beam line without changing its geometry. *Motz* [11.1] proposed first the use of *undulators* or *wiggler magnets* to optimize characteristics of synchrotron radiation. By now such magnets have become the most common insertion devices consisting of a series of alternating magnet poles and deflecting the beam periodically in opposite directions as shown in Fig. 11.1.

During the course of discussions in [11.2] and this volume, we have repeatedly encountered such wiggler magnets and have investigated some of their particular properties. Here we concentrate on more detailed derivations of radiation characteristics from relativistic electrons passing through *undulator* and *wiggler magnets*.

There is no fundamental difference between wiggler and undulator radiation. An undulator is basically a weak wiggler magnet. The transverse deflection in an undulator is weak and the transverse momentum of the particles remains nonrelativistic. The motion is therefore purely sinusoidal in a sinusoidal field, and the emitted radiation is therefore monochromatic at the particle oscillation frequency which is the Lorentz-contracted periodicity of the undulator period. Since the radiation is emitted from a moving source the observer in the laboratory frame of reference then sees a Doppler shifted frequency. We call this monochromatic radiation the *fundamental radiation* or *fundamental line* of the undulator.

As the undulator field is increased the transverse motion becomes stronger and the transverse momentum starts to become relativistic. As a consequence, the so far purely sinusoidal motion becomes periodically distorted causing the appearance of harmonics of the fundamental monochro-

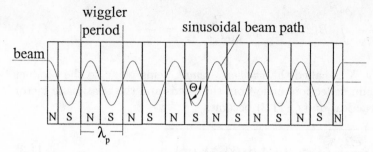

Fig. 11.1. Trajectory of a particle beam in a flat wiggler magnet

matic radiation. These harmonics increase in number and density with further increase of the magnetic field and eventually merge into one broad
spectrum characteristic for wiggler or bending magnet radiation.

An insertion device does not introduce a net deflection of the beam and
we may therefore choose any arbitrary field strength which is technically fea-
sible to adjust the radiation spectrum to experimental needs. The radiation
intensity from a wiggler magnet also can be made much higher compared
to that from a single bending magnet. A wiggler magnet with say ten poles
acts like a string of ten bending magnets or radiation sources aligned in a
straight line along the photon beam direction. The effective photon source
is therefore ten times more intense than the radiation from a single bending
magnet with the same field strength.

Wiggler magnets come in a variety of types with the flat wiggler magnet
being the most common. Here the beam is deflected only in one plane as
shown in Fig. 11.1. To generate circularly or elliptically polarized radiation
a *helical wiggler magnet* [11.3] may be used or a combination of several
flat wiggler magnets deflecting the beam in orthogonal planes which will be
discussed in more detail in Sect. 11.4.

11.1 Particle Dynamics in an Undulator

Particle dynamics and resulting radiation characteristics for an undulator
have been derived first by Motz [11.1] and later in more detail by other
authors [11.4 – 5]. A sinusoidally varying vertical field in the undulator
causes a periodic deflection of particles in the (x, z)-plane shown in Fig.
11.1. We express the wiggler field in the form

$$B(z) = B_\mathrm{o} \cos \frac{2\pi z}{\lambda_\mathrm{p}}, \tag{11.1}$$

where B_o is the maximum amplitude and λ_p the period length. The deflection angle of the particle in the wiggler field for each half pole is (Fig.
11.2)

$$\Theta = [c]\,\frac{e}{cp_{\mathrm{o}}} \int\limits_{\mathrm{o}}^{\lambda_{\mathrm{p}}/4} B(z)\,\mathrm{d}z = [c]\,\frac{e\,B_{\mathrm{o}}\,\lambda_{\mathrm{p}}}{mc^2\,\beta\gamma\,2\pi} = \frac{K}{\gamma} \tag{11.2}$$

and the factor K is called the *wiggler strength* being equal to the product of the maximum deflection angle from the z-axis and the relativistic factor γ. In more practical units, (11.2) becomes

$$\boxed{\; K = [c]\,\frac{eB_{\mathrm{o}}\,\lambda_{\mathrm{p}}}{2\pi\,mc^2\,\beta} = 9.344\,B_{\mathrm{o}}\,(\mathrm{kG})\,\lambda_{\mathrm{p}}(m)\,. \;} \tag{11.3}$$

To describe the particle trajectory, we use the equation of motion

$$\frac{\mathbf{n}}{\rho} = [c]\,\frac{e}{mc\gamma\,v_{\mathrm{o}}^2}\,[\mathbf{v}\times\mathbf{B}]\,, \tag{11.4}$$

where v_{o} is the particle velocity and get with (11.1) the equations of motion in component form

$$\begin{aligned}
\frac{\mathrm{d}^2 x}{\mathrm{d}t^2} &= -\frac{eB_{\mathrm{o}}}{mc\gamma}\,\frac{\mathrm{d}z}{\mathrm{d}t}\,\cos(k_{\mathrm{p}}z)\,, \\[2mm]
\frac{\mathrm{d}^2 z}{\mathrm{d}t^2} &= +\frac{eB_{\mathrm{o}}}{mc\gamma}\,\frac{\mathrm{d}x}{\mathrm{d}t}\,\cos(k_{\mathrm{p}}z)\,,
\end{aligned} \tag{11.5}$$

where we have set $k_{\mathrm{p}} = 2\pi/\lambda_{\mathrm{p}}$ and $\mathrm{d}z = \beta c\,\mathrm{d}t$ with $\beta = v/c$.

Equations (11.5) describe the coupled motion of a particle in the sinusoidal field of a flat wiggler magnet. This coupling is common to the motion in any magnetic field but so far we have always set $\mathrm{d}z/\mathrm{d}t \approx v$ and $\mathrm{d}x/\mathrm{d}t \approx 0$. This approximation is justified in most beam transport applications for relativistic particles, but we will have to be cautious not to neglect effects that might be of relevance on a very short time or small geometric scale comparable to the oscillation period and wave length of synchrotron radiation.

We will refrain from this approximation and find from the first equation (11.5) with $\mathrm{d}z/\mathrm{d}t \approx v$ and after integrating twice that the particle trajectory follows the magnetic field in the sense that the oscillatory motion reaches a maximum where the magnetic field reaches a maximum and crosses the beam axis where the field is zero. We start the time $t = 0$ in the middle of a magnet pole where the transverse velocity $\dot{x}_{\mathrm{o}} = 0$ while the longitudinal velocity $\dot{z}_{\mathrm{o}} = \beta c$ and integrate both equations (11.5) utilizing the integral of the first equation in the second to get

$$\begin{aligned}
\frac{\mathrm{d}x}{\mathrm{d}t} &= -\beta c\,\frac{K}{\gamma}\,\sin(k_{\mathrm{p}}z)\,, \\[2mm]
\frac{\mathrm{d}z}{\mathrm{d}t} &= \beta c\left[1 - \frac{K^2}{2\gamma^2}\,\sin^2(k_{\mathrm{p}}z)\right].
\end{aligned} \tag{11.6}$$

The transverse motion describes the expected oscillatory motion and the longitudinal velocity v exhibits a periodic modulation reflecting the varying projection of the velocity vector to the z-axis. Closer inspection of this velocity modulation shows that its frequency is twice that of the periodic motion. It is convenient to describe the longitudinal particle motion with respect to a cartesian reference frame where the particles move along the z-axis with an average longitudinal velocity $\bar{\beta}c = \langle \dot{z} \rangle$ which can be derived from (11.6)

$$\bar{\beta} = \beta \left(1 - \frac{K^2}{4\gamma^2} \right) . \tag{11.7}$$

In this reference frame the particle follows a figure of eight trajectory composed of the transverse oscillation and a longitudinal oscillation with twice the frequency. We will come back to this point since both oscillations contribute to the radiation spectrum. A second integration of (11.6) results finally in the equation of motion in component representation

$$x(t) = +\frac{K}{\gamma k_{\mathrm{p}}} \cos\left(k_{\mathrm{p}} \bar{\beta} c t\right) ,$$

$$z(t) = \bar{\beta} c t + \frac{K^2}{8 k_{\mathrm{p}} \gamma^2} \sin(2 k_{\mathrm{p}} \bar{\beta} c t), \tag{11.8}$$

where we set $z = \bar{\beta} c t$. The maximum amplitude a of the transverse particle oscillation is finally from (11.8)

$$\boxed{a = \frac{K}{k_{\mathrm{p}}\gamma} = \frac{\lambda_{\mathrm{p}} K}{2\pi \gamma}.} \tag{11.9}$$

This last expression gives another simple relationship between the wiggler strength parameter and the transverse displacement of the beam trajectory.

11.2 Undulator Radiation

The physical process of undulator radiation is not different from the radiation produced from a single bending magnet. However, the radiation received at great distances from the undulator exhibits special features. Basically, we observe an electron performing N oscillations while passing through an undulator where N is the number of undulator periods. The observed radiation spectrum is the Fourier transform of the electron motion and therefore basically monochromatic with a finite line width inversely proportional to the number of oscillations performed or undulator periods. In the reference frame of the electron we have the exact analog of dipole antenna radiation.

Undulator radiation can also be viewed as a superposition of radiation fields from N sources yielding monochromatic radiation as a consequence of interference. To see that we observe the radiation at an angle θ with respect to the path of the electron as shown in Fig. 11.2.

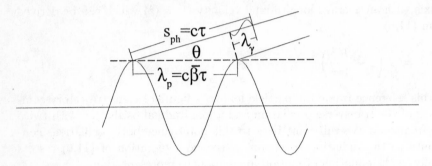

Fig. 11.2. Interference of undulator radiation

The electron moves along its path at an average velocity given by (11.7) and it takes the time

$$\tau = \frac{\lambda_{\mathrm p}}{c\bar\beta} = \frac{\lambda_{\mathrm p}}{c\beta\,[1 - K^2/(4\gamma^2)]} \tag{11.10}$$

to move along one undulator period. During that same time the radiation field proceeds for a distance

$$s_{\mathrm{ph}} = \tau c = \frac{\lambda_{\mathrm p}}{\beta\,[1 - K^2/(4\gamma^2)]} \tag{11.11}$$

moving ahead of the particle. For constructive superposition of radiation from all undulator periods, we require that the difference $s_{\mathrm{ph}} - \lambda_{\mathrm p}\cos\theta$ be equal to a wavelength λ_γ of the radiation spectrum or

$$\lambda_\gamma = \frac{\lambda_{\mathrm p}}{\beta\,[1 - K^2/(4\gamma^2)]} - \lambda_{\mathrm p}\left(1 - \tfrac{1}{2}\theta^2\right), \tag{11.12}$$

where we have assumed a small angle θ of observation. After some manipulations, we get with $K^2/\gamma^2 \ll 1$ for the *fundamental wavelength* of radiation into and angle θ

$$\boxed{\lambda_\gamma = \frac{\lambda_{\mathrm p}}{2\gamma^2}\left(1 + \tfrac{1}{2}K^2 + \gamma^2\theta^2\right).} \tag{11.13}$$

From an infinitely long undulator, the radiation spectrum consists of only one spectral line for a particular direction of observation. In particular we note that the shortest wavelength is emitted into the forward direction while the radiation at a finite angle θ appears red shifted by the Doppler effect. For an undulator with a finite number of periods, the spectral line becomes widened to a line width of about $1/N$ as we will see in the next section.

The radiation power is from (7.44) with $q^2 = r_c\, mc$

$$P = \tfrac{2}{3} r_c\, mc\, |\dot{\boldsymbol{\beta}}^*|^2_{\text{ret}}\,, \tag{11.14}$$

where * indicates that the quantity is to be evaluated in the particle reference system. We may use this expression in the particle system to calculate the total radiated energy from an electron passing through an undulator. The transverse particle acceleration is expressed by $m\dot{\mathbf{v}}^* = \mathrm{d}\mathbf{p}_\perp/\mathrm{d}t^* = \gamma\mathrm{d}\mathbf{p}_\perp/\mathrm{d}t$ where we used $t^* = t/\gamma$ and inserting into (11.14) we get

$$P = \tfrac{2}{3}\frac{r_c\,\gamma^2}{mc}\left(\frac{\mathrm{d}\mathbf{p}_\perp}{\mathrm{d}t}\right)^2. \tag{11.15}$$

The transverse momentum is determined by the particle deflection in the undulator with a period length λ_p and is for a particle of momentum $c\,p_o$

$$p_\perp = \hat{p}\sin\omega_p t\,, \tag{11.16}$$

where $\hat{p} = p_o\Theta$ and $\omega_p = ck_p = 2\pi c/\lambda_p$. The angle $\Theta = K/\gamma$ is the maximum deflection angle defined in (11.2). With these expressions and averaging over one period, we get from (11.15) for the *instantaneous radiation power* from a charge q traveling through an undulator

$$\boxed{P_{\text{inst}} = \tfrac{1}{3}\,c\,r_c\,mc^2\,\gamma^2\,K^2\,k_p^2\,,} \tag{11.17}$$

where r_c is the classical electron radius. The duration of the radiation pulse is equal to the traveling time through an undulator of length L_u and the *total radiated energy* per electron is therefore

$$\Delta E = \tfrac{1}{3}\,r_c\,mc^2\,\gamma^2\,K^2\,k_p^2\,L_u\,. \tag{11.18}$$

In more practical units

$$\Delta E(eV) = C_u\frac{E^2\,K^2}{\lambda_p^2}\,L_u = 725\,\frac{E^2(\text{GeV})\,K^2}{\lambda_p^2(\text{cm})}\,L_u(\text{m}) \tag{11.19}$$

with

$$C_u = \frac{4\pi^2\,r_c}{3\,mc^2} = 7.2567\times10^{-20}\,\frac{\text{m}}{\text{eV}}\,. \tag{11.20}$$

The average total *undulator radiation power* for an electron beam circulating in a storage ring is then just the radiated energy (11.18) multiplied by the number of particles N_b in the beam and the revolution frequency or

$$\langle P \rangle = \tfrac{1}{3} r_c \, c \, m c^2 \, \gamma^2 \, K^2 \, k_p^2 \, N_b \, \frac{L_u}{2\pi \, \bar{R}} \tag{11.21}$$

or

$$\langle P(\mathrm{W}) \rangle = 6.336 \, E^2(\mathrm{GeV}) \, B_o^2(\mathrm{kG}) \, I(\mathrm{A}) \, L_u(\mathrm{m}), \tag{11.22}$$

where I is the circulating electron beam current. The total angle integrated radiation power from an undulator in a storage ring is proportional to the square of the beam energy and maximum undulator field B_o and proportional to the beam current and undulator length.

11.3 Undulator Radiation Distribution

For bending magnet radiation the particle dynamics is relatively simple being determined only by the particle velocity and the bending radius of the magnet. In a wiggler magnet, the magnetic field parameters are different from those in a constant field magnet and we will therefore rederive the synchrotron radiation spectrum for the beam dynamics in a general wiggler magnet. No assumptions on the magnetic field parameters have been made to derive the radiation spectrum in the form of equation (7.87) which we use to calculate the radiation spectrum from a wiggler magnet

$$\frac{\mathrm{d}^2 W}{\mathrm{d}\omega \, \mathrm{d}\Omega} = \frac{r_c \, m c \, \omega^2}{4\pi^2} \left| \int_{-\infty}^{\infty} \mathbf{n} \times [\mathbf{n} \times \boldsymbol{\beta}] \, e^{-\mathrm{i}\omega \left(t_r + \frac{R}{c} \right)} \, \mathrm{d}t_r \right|^2. \tag{11.23}$$

We have now all the information about the particle dynamics to evaluate the integrand in (11.23) noting that all quantities are to be taken at the retarded time t_r. The unit vector from the observer to the radiating particle is from Fig. 11.3

$$\mathbf{n} = -\mathbf{x} \cos\varphi \sin\theta - \mathbf{y} \sin\varphi \sin\theta - \mathbf{z} \cos\theta. \tag{11.24}$$

The exponent in (11.23) includes the term $R/c = \mathbf{n}R/c$. We express again the vector $\mathbf{R}$ from the observer to the particle by the constant vector $\mathbf{r}$ from the origin of the coordinate system to the observer and by the vector $\mathbf{r}_p$ from the coordinate origin to the particle and $\mathbf{R}(t_r) = -\mathbf{r} + \mathbf{r}_p$ as shown in Fig. 11.3.

The $\mathbf{r}$-term gives only a constant phase shift and can therefore be ignored. The location vector $\mathbf{r}_p$ of the particle with respect to the origin of the coordinate system is

$$\mathbf{r}_p(t_r) = x(t_r) \, \mathbf{x} + z(t_r) \, \mathbf{z}$$

and with the solutions (11.8) we have

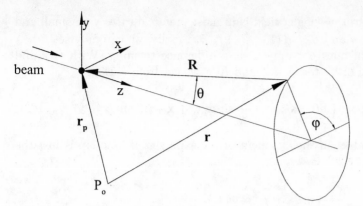

Fig. 11.3. Particle trajectory and radiation geometry in a wiggler field

$$\mathbf{r}_{\mathrm{p}}(t_{\mathrm{r}}) = \frac{K}{k_{\mathrm{p}}\,\gamma}\,\cos(\omega_{\mathrm{p}} t_{\mathrm{r}})\,\mathbf{x} + \left[\bar{\beta}c\,t_{\mathrm{r}} + \frac{K^2}{8\pi\,k_{\mathrm{p}}}\,\sin(2\omega_{\mathrm{p}}\,t_{\mathrm{r}})\right]\mathbf{z}\,, \qquad (11.25)$$

where

$$\omega_{\mathrm{p}} = k_{\mathrm{p}}\,\bar{\beta}c\,. \qquad (11.26)$$

The velocity vector finally is just the time derivative of (11.25)

$$\boldsymbol{\beta}(t_{\mathrm{r}}) = -\frac{K}{\gamma}\,\bar{\beta}\,\sin(\omega_{\mathrm{p}} t_{\mathrm{r}})\,\mathbf{x} + \bar{\beta}\left[1 + \frac{K^2}{4\gamma^2}\,\cos(2\omega_{\mathrm{p}} t_{\mathrm{r}})\right]\mathbf{z} \qquad (11.27)$$

We use these vector relations to evaluate the integrand in (11.23). First, we express the triple vector product $\mathbf{n}\times[\mathbf{n}\times\boldsymbol{\beta}]$ and get with (11.24,27) after some manipulations

$$\begin{aligned}
\mathbf{n}\times[\mathbf{n}\times\boldsymbol{\beta}] = \mathbf{x}&\left[-\frac{K}{\gamma}\,\bar{\beta}\,\sin^2\theta\,\cos^2\varphi\,\cos\omega_{\mathrm{p}} t_{\mathrm{r}} + \frac{K}{\gamma}\,\bar{\beta}\,\sin\omega_{\mathrm{p}} t_{\mathrm{r}}\right.\\
&\left.+\,\bar{\beta}\left(1 + \frac{K^2}{4\gamma^2}\,\cos 2\omega_{\mathrm{p}} t_{\mathrm{r}}\right)\sin\theta\,\cos\theta\,\cos\varphi\right]\\
+\,\mathbf{y}&\left[-\frac{K\bar{\beta}}{\gamma}\,\sin^2\theta\,\sin\varphi\,\cos\varphi\,\sin\omega_{\mathrm{p}} t_{\mathrm{r}}\right. \qquad (11.28)\\
&\left.+\,\bar{\beta}\left(1 + \frac{K^2}{4\gamma^2}\,\cos 2\omega_{\mathrm{p}} t_{\mathrm{r}}\right)\sin\theta\,\cos\theta\,\sin\varphi\right]\\
+\,\mathbf{z}&\left[-\frac{K}{\gamma}\,\bar{\beta}\,\sin\theta\,\cos\theta\,\cos\varphi\,\cos\omega_{\mathrm{p}} t_{\mathrm{r}}\right.\\
&\left.+\,\bar{\beta}\left(1 + \frac{K^2}{4\gamma^2}\,\cos 2\omega_{\mathrm{p}} t_{\mathrm{r}}\right)(\cos^2\theta - 1)\right].
\end{aligned}$$

This expression can be greatly simplified considering that the radiation is emitted into only a very small angle $\theta \ll 1$. Furthermore, we note that the deflection due to the wiggler field is in most practical cases very small and therefore $K \ll \gamma$ and $\bar{\beta} = \beta \left(1 - \frac{K^2}{4\gamma^2}\right) \approx \beta$. Finally, we carefully set $\beta \approx 1$ where this term does not appear as a difference to unity. With this and ignoring second order terms in θ and K/γ we get from (11.28)

$$\mathbf{n} \times [\mathbf{n} \times \boldsymbol{\beta}] = \left(\bar{\beta}\theta \cos\varphi + \bar{\beta}\frac{K}{\gamma} \sin\omega_\mathrm{p} t_\mathrm{r}\right) \mathbf{x} + \bar{\beta}\theta \sin\varphi\, \mathbf{y}. \tag{11.29}$$

The vector product in the exponent of the exponential function is just the product of (11.27,28)

$$\frac{\mathbf{n}\mathbf{r}_\mathrm{p}(t_\mathrm{r})}{c} = -\frac{K\bar{\beta}}{\gamma\omega_\mathrm{p}} \sin\theta \cos\varphi \cos\omega_\mathrm{p} t_\mathrm{r}$$
$$- \left(\bar{\beta}t_\mathrm{r} + \frac{K^2\bar{\beta}}{8\gamma^2\omega_\mathrm{p}} \sin 2\omega_\mathrm{p} t_\mathrm{r}\right) \cos\theta. \tag{11.30}$$

Employing again the approximation $\theta \ll 1$ and keeping only linear terms we get from (11.30)

$$t_\mathrm{r} + \frac{\mathbf{n}\mathbf{r}_\mathrm{p}(t_\mathrm{r})}{c} = t_\mathrm{r}\left(1 - \bar{\beta}\cos\theta\right) - \frac{K\bar{\beta}\theta \cos\varphi}{\gamma\omega_\mathrm{p}} \cos\omega_\mathrm{p} t$$
$$- \frac{K^2\bar{\beta}}{8\gamma^2\omega_\mathrm{p}} \sin 2\omega_\mathrm{p} t. \tag{11.31}$$

With (11.7) and $\cos\theta \approx 1 - \frac{1}{2}\theta^2$, the first term becomes

$$1 - \bar{\beta}\cos\theta = \frac{1}{2\gamma^2}\left(1 + \tfrac{1}{2}K^2 + \gamma^2\theta^2\right) = \frac{\omega_\mathrm{p}}{\omega_1}, \tag{11.32}$$

where we have defined a *fundamental wiggler frequency* ω_1

$$\omega_1^{-1} = \frac{\omega_\mathrm{p}^{-1}}{2\gamma^2}\left(1 + \tfrac{1}{2}K^2 + \gamma^2\theta^2\right) \tag{11.33}$$

or the *fundamental wavelength* of the radiation

$$\lambda_1 = \frac{\lambda_\mathrm{p}}{2\gamma^2}\left(1 + \tfrac{1}{2}K^2 + \gamma^2\theta^2\right). \tag{11.34}$$

At this point, it is interesting to note that the term $\frac{1}{2}K^2$ becomes K^2 for a helical wiggler [11.3]. With the definition (11.32), the complete exponential term in (11.23) can be evaluated to be

$$-\mathrm{i}\omega\left(t_\mathrm{r} + \frac{\mathbf{n}\mathbf{r}_\mathrm{p}(t_\mathrm{r})}{c}\right) =$$
$$-\mathrm{i}\frac{\omega}{\omega_1}\left(\omega_\mathrm{p} t_\mathrm{r} - \frac{K\bar{\beta}\cos\varphi}{\gamma}\frac{\omega_1}{\omega_\mathrm{p}} \cos\omega_\mathrm{p} t_\mathrm{r} - \frac{K^2\bar{\beta}}{8\gamma^2}\frac{\omega_1}{\omega_\mathrm{p}} \sin 2\omega_\mathrm{p} t_\mathrm{r}\right). \tag{11.35}$$

Equation (11.23) can be modified with these expressions into a form suitable for integration by inserting (11.29,35) into (11.23) for

$$\frac{\mathrm{d}^2 W}{\mathrm{d}\omega\,\mathrm{d}\Omega} = \frac{r_{\mathrm{c}}\,mc\,\omega^2}{4\pi^2}\,\bar{\beta}^2 \left| \int_{-\infty}^{\infty} \left[\left(\theta\,\cos\varphi + \frac{K}{\gamma}\,\sin\omega_{\mathrm{p}}t_{\mathrm{r}} \right) \mathbf{x} + \theta\,\sin\varphi\,\mathbf{y} \right] \right.$$

$$\times \exp\left[-\mathrm{i}\,\frac{\omega}{\omega_1}\left(\omega_{\mathrm{p}}t_{\mathrm{r}} - \frac{K\theta\cos\varphi}{\gamma}\,\frac{\omega_1}{\omega_{\mathrm{p}}}\,\cos\omega_{\mathrm{p}}t_{\mathrm{r}} \right.\right. \tag{11.36}$$

$$\left.\left.\left. - \frac{K^2}{8\gamma^2}\,\frac{\omega_1}{\omega_{\mathrm{p}}}\,\sin 2\omega_{\mathrm{p}}t_{\mathrm{r}} \right) \right] \mathrm{d}t_{\mathrm{r}} \right|^2 .$$

We are now ready to perform the integration of (11.36) noticing that the integration over all times can be simplified by separation into an integral along the wiggler magnet alone and an integration over the rest of the time while the particle is traveling in a field free space. We write symbolically

$$\int_{-\infty}^{\infty} = \int_{-\pi N/\omega_{\mathrm{p}}}^{\pi N/\omega_{\mathrm{p}}} (K \neq 0) + \int_{-\infty}^{\infty} (K = 0) - \int_{-\pi N/\omega_{\mathrm{p}}}^{\pi N/\omega_{\mathrm{p}}} (K = 0). \tag{11.37}$$

First we evaluate the second integral for $K = 0$ which is of the form

$$\int_{-\infty}^{\infty} e^{\mathrm{i}\kappa\omega t}\,\mathrm{d}t = \frac{2\pi}{|\kappa|}\,\delta(\omega),$$

where $\delta(\omega)$ is the Dirac delta function. The value of the integral is nonzero only for $\omega = 0$ in which case the factor ω^2 in (11.36) causes the whole expression to vanish.

The third integral has the same form as the second integral, but since the integration is conducted only over the length of the wiggler magnet we get

$$\int_{-\pi N/\omega_{\mathrm{p}}}^{\pi N/\omega_{\mathrm{p}}} e^{-\mathrm{i}\frac{\omega}{2\gamma^2}t_{\mathrm{r}}}\,\mathrm{d}t_{\mathrm{r}} = \frac{2\pi N}{\omega_{\mathrm{p}}}\,\frac{\sin\frac{\pi N}{2\gamma^2}\frac{\omega}{\omega_{\mathrm{p}}}}{\frac{\pi N}{2\gamma^2}\frac{\omega}{\omega_{\mathrm{p}}}}. \tag{11.38}$$

The value of this integral reaches a maximum of $2\pi\frac{N}{\omega_{\mathrm{p}}}$ for $\omega \to 0$. From (11.36) we note the coefficient of this integral to include the angle $\theta \approx 1/\gamma$ and the whole integral is therefore of the order or less than $L_{\mathrm{u}}/(c\gamma)$ where $L_{\mathrm{u}} = N\lambda_{\mathrm{p}}$ is the total length of the wiggler magnet. This value is in general very small compared to the first integral and can therefore be neglected. Actually, this statement is only partially true since the first integral, as we will see, is a fast varying function of the radiation frequency with a distinct line spectrum. Being, however, primarily interested in the peak intensities of the spectrum we may indeed neglect the third integral. Only between the spectral lines does the radiation intensity from the first integral become so small that the third integral would be a relatively significant although absolutely a small contribution.

To evaluate the first integral in (11.37) with $K \neq 0$ we follow Alferov [11.4] and replace the exponential functions by an infinite sum of Bessel's functions

$$e^{i\kappa \sin \psi} = \sum_{p=-\infty}^{p=\infty} J_p(\kappa) e^{ip\psi} . \tag{11.39}$$

The integral in the radiation power spectrum (11.36) has two distinct forms, one where the integrand is just the exponential function multiplied by a time independent factor while the other includes the sine function $\sin \omega_p t_r$ as a factor of the exponential function. We apply the identity (11.39) to the first integral type in (11.36). To simplify the complexity of the expressions, we use the definitions

$$C = \frac{2K\bar{\beta}\gamma\theta \cos\varphi}{1 + \frac{1}{2}K^2 + \gamma^2\theta^2} \qquad S = \frac{K^2\bar{\beta}}{4\left(1 + \frac{1}{2}K^2 + \gamma^2\theta^2\right)} \tag{11.40}$$

and get from (11.36) the exponential functions in the form

$$e^{-i\frac{\omega}{\omega_1}\omega_p t_r} e^{i\frac{\omega}{\omega_1} C \cos \omega_p t_r} e^{i\frac{\omega}{\omega_1} S \sin 2\omega_p t_r} . \tag{11.41}$$

Applying the identity (11.39) to the second and third exponential factors in (11.41), we get with the identity $e^{a\cos x} = e^{a\sin(x+\pi/2)}$ for the product of the exponential functions

$$e^{-i\frac{\omega}{\omega_1}\omega_p t_r} e^{iv \cos \omega_p t_r} e^{iu \sin 2\omega_p t_r}$$

$$= \sum_{m=-\infty}^{\infty} \sum_{n=-\infty}^{\infty} J_m(u) J_n(v) e^{i\frac{\pi}{2}n} e^{-i R_\omega \omega_p t_r} , \tag{11.42}$$

where

$$R_\omega = \frac{\omega}{\omega_1} - n - 2m, \qquad u = \frac{\omega}{\omega_1} S \quad \text{and} \quad v = \frac{\omega}{\omega_1} C. \tag{11.43}$$

The time integration along the length of the wiggler magnet is straight forward for this term since no other time dependent factors are involved and we get

$$\int_{-\pi N/\omega_p}^{\pi N/\omega_p} e^{-i\left(\frac{\omega}{\omega_1} - n - 2m\right)\omega_p t_r} dt_r = \frac{2\pi N}{\omega_p} \frac{\sin \pi N R_\omega}{\pi N R_\omega} . \tag{11.44}$$

The $(\sin x/x)$- term in (11.44) is well known from interference theory and represents the *line spectrum* of the radiation. Specifically, the number N of beamlets, here source points, determines the *spectral purity* of the radiation.

In Fig. 11.4 the $(\sin x/x)$-function is shown for $N = 5$ and $N = 100$. It is clear that the spectral purity improves greatly as the number of undulator periods is increased. The Bessel functions $J_m(u)$ and $J_n(v)$ determine mainly the intensity of the line spectrum. For an undulator, which is a weak wiggler magnet with $K \ll 1$, the higher order Bessel's functions with $m > 0$, $n > 0$ are very small and the radiation spectrum consists therefore only of the fundamental line. For stronger undulators with $K > 1$, the higher order Bessel's functions grow and higher harmonic radiation appears in the line spectrum of the radiation.

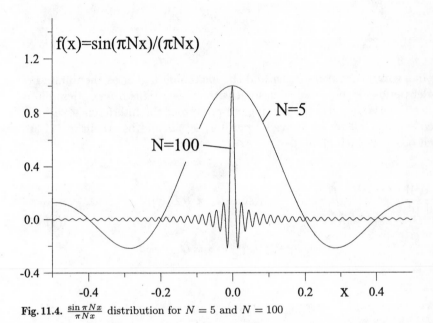

Fig. 11.4. $\frac{\sin \pi N x}{\pi N x}$ distribution for $N = 5$ and $N = 100$

In the second type of the integral we replace the trigonometric factor, $\sin \omega_p t_r$, by exponential functions and get with (11.44) integrals of the form

$$\int_{-\pi N/\omega_p}^{\pi N/\omega_p} \sin \omega_p t_r \, e^{-i R_\omega \omega_p t_r} \, dt_r = -i\frac{1}{2} \int_{-\pi N/\omega_p}^{\pi N/\omega_p} \left(e^{i\omega_p t_r} - e^{-i\omega_p t_r} \right) e^{-iR_\omega \omega_p t_r} \, dt_r$$

$$= i \, \frac{\pi N}{\omega_p} \, \frac{\sin \pi N(R_\omega + 1)}{\pi N(R_\omega + 1)} - i \, \frac{\pi N}{\omega_p} \, \frac{\sin \pi N(R_\omega - 1)}{\pi N(R_\omega - 1)} . \tag{11.45}$$

Both integrals (11.44,45) exhibit the character of a multibeam interference spectrum well known from optical interference theory. The physical interpretation here is that the radiation from the $2N$ wiggler poles consists of $2N$ photon beamlets which have a specific phase relationship such that the intensities are strongly reduced for all frequencies but a few specific frequencies as determined by the $\frac{\sin x}{x}$-factors. By interference all frequencies are eliminated but a few. The line spectrum is the more pronounced the more wiggler periods or beamlets are available for interference. The line width is defined as the width at which the radiation power is reduced to half the maximum value and we solve therefore the equation

$$\sin \pi N \frac{\delta\omega}{\omega_1} = \frac{1}{\sqrt{2}} \pi N \frac{\delta\omega}{\omega_1}$$

for the *line width*

$$\Delta\omega_{1/2} = 2\delta\omega \approx \frac{\omega_1}{N}. \qquad (11.46)$$

The line width becomes smaller and the maximum higher as the number of wiggler periods is increased consistent with expectations from optical interference theory. To get a more complete picture of the interference pattern, we collect now all terms derived separately so far and use them in (11.36) which becomes with (11.41)

$$\frac{d^2W}{d\omega\,d\Omega} = a \left| \int_{-\pi N/\omega_p}^{\pi N/\omega_p} [(A_o + A_1 \sin\omega_p t_r)\,\mathbf{x} + B_o\,\mathbf{y}] \right.$$
$$\left. \times\, e^{-i\frac{\omega}{\omega_1}\omega_p t_r}\, e^{i v \cos\omega_p t_r}\, e^{i u \sin 2\omega_p t_r}\, dt_r \right|^2, \qquad (11.47)$$

where

$$a = \frac{r_c mc\omega^2\bar{\beta}^2}{4\pi^2}; \qquad A_1 = \frac{K}{\gamma};$$
$$A_o = \theta\cos\varphi; \qquad B_o = \theta\sin\varphi. \qquad (11.48)$$

Introducing the identity (11.42), the photon energy spectrum becomes

$$\frac{d^2W}{d\omega\,d\Omega} = a \left| \int_{-\pi N/\omega_p}^{\pi N/\omega_p} [(A_o + A_1 \sin\omega_p t_r)\,\mathbf{x} + B_o\,\mathbf{y}] \right.$$
$$\left. \times \sum_{m=-\infty}^{\infty}\sum_{n=-\infty}^{\infty} J_m(u)\,J_n(v)\, e^{i\frac{\pi}{2}n}\, e^{-i R_\omega \omega_p t_r}\, dt_r \right|^2$$

and after integration with (11.44,45)

$$\frac{\mathrm{d}^2 W}{\mathrm{d}\omega\,\mathrm{d}\Omega} = a\left| \mathbf{x}\,A_\mathrm{o} \sum_{\substack{m=\\-\infty}}^{\infty} \sum_{\substack{n=\\-\infty}}^{\infty} J_m(u)\,J_n(v)\,\mathrm{e}^{\mathrm{i}\pi n/2}\,\frac{2\pi N}{\omega_\mathrm{p}}\,\frac{\sin \pi N\,R_\omega}{\pi N\,R_\omega} \right.$$

$$+\mathbf{x}\,A_1 \sum_{\substack{m=\\-\infty}}^{\infty} \sum_{\substack{n=\\-\infty}}^{\infty} J_m(u)\,J_n(v)\,\mathrm{e}^{\mathrm{i}\pi n/2} \qquad (11.49)$$

$$\times\,\mathrm{i}\,\frac{\pi N}{2\omega_\mathrm{p}}\left[\frac{\sin \pi N(R_\omega+1)}{\pi N(R_\omega+1)} - \frac{\sin \pi N(R_\omega-1)}{\pi N(R_\omega-1)}\right]$$

$$\left. +\mathbf{y}\,B_\mathrm{o} \sum_{\substack{m=\\-\infty}}^{\infty} \sum_{\substack{n=\\-\infty}}^{\infty} J_m(u)\,J_n(v)\,\mathrm{e}^{\mathrm{i}\pi n/2}\,\frac{2\pi N}{\omega_\mathrm{p}}\,\frac{\sin \pi N\,R_\omega}{\pi N\,R_\omega} \right|^2 .$$

All integrals exhibit the resonance character defining the locations of the spectral lines. To determine the frequency and radiation intensity of the line maximas, we simplify the double sum of Bessel's functions by selecting only the most dominant terms. The first and third sums in (11.49) show an intensity maximum for $R_\omega = 0$ at frequencies

$$\omega = (n+2m)\,\omega_1\,, \qquad (11.50)$$

and intensity maxima appear therefore at the frequency ω_1 and harmonics thereof. The transformation of a lower frequency to very high values has two physical components. In the system of relativistic particles, the static magnetic field of the wiggler magnet appears Lorentz contracted by the factor γ, and particles passing through the wiggler magnet oscillate with the frequency $\gamma\omega_\mathrm{p}$ in its own system emitting radiation at that frequency. The observer in the laboratory system receives this radiation from a source moving with relativistic velocity and experiences therefore a Doppler shift by the factor 2γ. The wavelength of the radiation emitted in the forward direction, $\theta = 0$, from a weak wiggler magnet, $K \ll 1$, with the period length λ_p is therefore reduced by the factor $2\gamma^2$. In cases of a stronger wiggler magnet or when observing at a finite angle θ, the wavelength is somewhat longer as one would expect from higher order terms of the Doppler effect.

From (11.49) we determine two more dominant terms originating from the second term for $R_\omega \pm 1 = 0$ at frequencies

$$\omega = (n+2m-1)\,\omega_1\,,$$
$$\omega = (n+2m+1)\,\omega_1\,, \qquad (11.51)$$

respectively. The summation indices n and m are arbitrary integers between $-\infty$ and ∞. Among all possible resonant terms we collect such terms which contribute to the same harmonic k of the fundamental frequency ω_1. To collect these dominant terms for the same harmonic we set $\omega = \omega_k = k\,\omega_1$ where k is the harmonic number of the fundamental and express the index n by k and m to get from (11.50) $\qquad n = k - 2m$,

and from (11.51) $\qquad n = k - 2m + 1$, $\qquad\qquad$ (11.52)

and $\qquad n = k - 2m - 1$.

Introducing these conditions into (11.49) all trigonometric factors assume the form

$$\frac{\sin \pi N \, \Delta\omega_k/\omega_1}{\pi N \, \Delta\omega_k/\omega_1} ,$$

where

$$\frac{\Delta\omega_k}{\omega_1} = \frac{\omega}{\omega_1} - k \qquad\qquad (11.53)$$

and we get the photon energy spectrum of the kth harmonic

$$\frac{\mathrm{d}^2 W_k(\omega)}{\mathrm{d}\omega \, \mathrm{d}\Omega} = \frac{r_{\mathrm{c}} \, mc\,\omega^2 \, \bar\beta^2 \, N^2}{\omega_{\mathrm{p}}^2 \, \gamma^2} \left(\frac{\sin \pi N \, \Delta\omega_k/\omega_1}{\pi N \, \Delta\omega_k/\omega_1} \right)^2$$

$$\times \bigg| + \mathbf{x} \, A_{\mathrm{o}} \sum_{m=-\infty}^{\infty} J_m(v) \, J_{k-2m}(u) \, \mathrm{e}^{\mathrm{i}\pi(k-2m)/2}$$

$$+ \mathbf{y} \, B_{\mathrm{o}} \sum_{m=-\infty}^{\infty} J_m(v) \, J_{k-2m}(u) \, \mathrm{e}^{\mathrm{i}\pi(k-2m)/2} \qquad (11.54)$$

$$+ \mathbf{x} \, \frac{\mathrm{i}}{2} A_1 \sum_{m=-\infty}^{\infty} J_m(u) \, J_{k-2m+1}(v) \, \mathrm{e}^{\mathrm{i}\pi(k-2m+1)/2}$$

$$- \mathbf{x} \, \frac{\mathrm{i}}{2} A_1 \sum_{m=-\infty}^{\infty} J_m(u) \, J_{k-2m-1}(v) \, \mathrm{e}^{\mathrm{i}\pi(k-2m-1)/2} \bigg|^2 .$$

The photon energy spectrum is composed of two distinct components, the σ-mode with polarization in the horizontal plane or normal to the magnetic field and the π-mode with polarization parallel to the magnetic field. Since $B_{\mathrm{o}} \propto \theta$, we conclude that no π-mode radiation is emitted into the forward direction.

Summing over all harmonics of interest one gets the total power spectrum. In the third and fourth terms we use the identities $\mathrm{i}\,\mathrm{e}^{\pm\mathrm{i}\pi/2} = \mp 1$, $J_m(u) \, \mathrm{e}^{\mathrm{i}\pi m} = J_{-m}(u)$ and abbreviate the sums of Bessel's functions by the symbols

$$\Sigma_1 = \sum_{m=-\infty}^{\infty} J_{-m}(u) \, J_{k-2m}(v) ,$$

$$\Sigma_2 = \sum_{m=-\infty}^{\infty} J_{-m}(u) \left[J_{k-2m-1}(v) + J_{k-2m+1}(v) \right] . \qquad (11.55)$$

The total number of photons N_{ph} emitted into a spectral band width $\Delta\omega/\omega$ by a single electron moving through a wiggler magnet is with $N_{\mathrm{ph}}(\omega) = W(\omega)/(\hbar\omega)$

$$\frac{\mathrm{d}N_{\mathrm{ph}}(\omega)}{\mathrm{d}\Omega} = \alpha\,\gamma^2\,\bar{\beta}^2\,N^2\,\frac{\Delta\omega}{\omega}\sum_{k=1}^{\infty}k^2\left(\frac{\sin\pi N\,\Delta\omega_k/\omega_1}{\pi N\,\Delta\omega_k/\omega_1}\right)^2 \qquad (11.56)$$

$$\times\;\frac{(2\gamma\theta\,\Sigma_1\,\cos\varphi - K\,\Sigma_2)^2\,\mathbf{x}^2 + (2\gamma\theta\,\Sigma_1\,\sin\varphi)^2\,\mathbf{y}^2}{(1+\tfrac{1}{2}K^2+\gamma^2\theta^2)^2},$$

where $\alpha = e^2/(\hbar c) = 1/137.036$ is the fine structure constant and we have kept the coordinate unit vectors to keep tack of the polarization modes. The vectors $\mathbf{x}$ and $\mathbf{y}$ are orthogonal unit vectors indicating the directions of the electric field or the polarization of the radiation. Performing the square does not produce therefore cross terms and the two terms in (11.56) including the expressions (11.55) represent the amplitude factors for both polarization directions, the σ-mode and π-mode respectively.

We also made use of (11.53) and the resonance condition

$$\frac{\omega}{\omega_{\mathrm{p}}} = \frac{k\,\omega_1 + \Delta\omega_k}{\omega_{\mathrm{p}}} \approx k\,\frac{\omega_1}{\omega_{\mathrm{p}}} = \frac{2\gamma^2\,k}{1+\frac{K^2}{2}+\gamma^2\theta^2}, \qquad (11.57)$$

realizing that the photon spectrum is determined by the $(\sin x/x)^2$-function which is very small for frequencies away from the resonance condition. This last approximation of frequencies close to the harmonic frequencies seems to be inappropriate. Upon closer inspection of the equation, we note that for a fixed angle θ the photon flux is very much concentrated close to the harmonics k due to the factor $(\sin x/x)^2$ with a relative bandwidth of $1/N$. Radiation into other than harmonic frequencies occurs mainly from a variation in the observation angle θ with lower frequencies emitted into larger angles.

The spatial intensity distribution combines a complex set of different radiation lobes depending on frequency, emission angle and polarization. In Figs. 11.5,6 the radiation intensity distributions described by the last factor in (11.56)

$$I_{\sigma,k} = \frac{(2\gamma\theta\,\Sigma_1\,\cos\varphi - K\,\Sigma_2)^2}{(1+\tfrac{1}{2}K^2+\gamma^2\theta^2)^2} \qquad (11.58)$$

for the σ-mode polarization and

$$I_{\pi,k} = \frac{(2\gamma\theta\,\Sigma_1\,\sin\varphi)^2}{(1+\tfrac{1}{2}K^2+\gamma^2\theta^2)^2} \qquad (11.59)$$

for the π-mode polarization are shown for the lowest order harmonics.

We note clearly the strong forward lobe at the fundamental frequency in σ-mode while there is no emission in π-mode along the path of the particle.

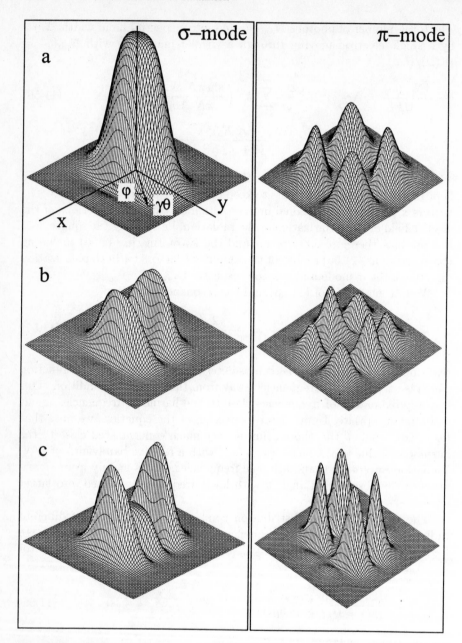

Fig. 11.5. Radiation distribution for the first three harmonics $k = 1, 2, 3$. Distributions on the left are for σ-mode polarization and those on the right for π-mode polarization

The second harmonic radiation vanishes in the forward direction, an observation that is true for all even harmonics. By inspection of (11.56), we note that $v = 0$ for $\theta = 0$ and the square bracket in (11.55) vanishes for all odd indices or for all even harmonics k. There is therefore no forward radiation for even harmonics of the fundamental undulator frequency.

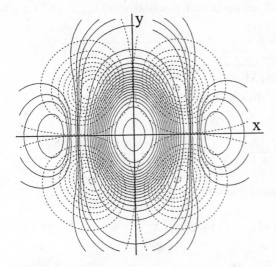

Fig. 11.6. Contour plots for the intensity distribution of the first harmonic σ-mode radiation (solid lines) and π-mode radiation (dashed lines)

To exhibit other important and desirable features of the radiation spectrum (11.56), we ignore the actual frequency distribution in the vicinity of the harmonics and set $\Delta\omega_k = 0$ because the spectral lines are narrow for large numbers of wiggler poles N. Further, we are interested for now only in the forward radiation where $\theta = 0$, keeping in mind that the radiation is mostly emitted into a small angle $\langle\theta\rangle = 1/\gamma$.

There is no radiation for the π-mode in the forward direction and the only contribution to the forward radiation comes from the second term of the σ-mode. From (11.43,46), we get for this case with $\omega/\omega_1 = k$

$$ u_{\mathrm{o}} = k\,\frac{K^2}{4+2K^2} \qquad \text{and} \qquad v_{\mathrm{o}} = 0. \tag{11.60} $$

The sums of Bessel's functions simplify in this case greatly because only the lowest order Bessel's function has a nonvanishing value for $v_{\mathrm{o}} = 0$. In the expression for Σ_2 all summation terms vanish except for the two terms for which the index is zero or for which $k - 2m - 1 = 0$ or $k - 2m + 1 = 0$ and

$$\Sigma_2 = \sum_{m=-\infty}^{\infty} J_{-m}(u_{\rm o}) \left[J_{k-2m-1}(0) + J_{k-2m+1}(0) \right],$$

$$= J_{\frac{-k+1}{2}}(u_{\rm o}) + J_{\frac{-k-1}{2}}(u_{\rm o}). \tag{11.61}$$

Using the identity $J_{-n} = (-1)^n J_n$ and $u_{\rm o} = \frac{k\,K^2}{4+2k^2}$ from (11.60), we get finally with $N_{\rm ph} = W/\hbar\omega$ the *spectral photon intensity* from a highly relativistic particle passing through an undulator per unit solid angle and into a frequency bin $\Delta\omega/\omega$

$$\boxed{\begin{aligned}
\left.\frac{{\rm d}N_{\rm ph}(\omega)}{{\rm d}\Omega}\right|_{\theta=0} &= \alpha\,\gamma^2\,N^2\,\frac{\Delta\omega}{\omega}\,\frac{K^2}{(1+\frac12 K^2)^2} \\
&\quad \times \sum_{k=1}^{\infty} k^2 \left(\frac{\sin \pi N\,\Delta\omega_k/\omega_1}{\pi N\,\Delta\omega_k/\omega_1}\right)^2 \\
&\quad \times \left[J_{\frac{k-1}{2}}\left(\frac{k\,K^2}{4+2K^2}\right) - J_{\frac{k+1}{2}}\left(\frac{k\,K^2}{4+2K^2}\right) \right]^2.
\end{aligned}} \tag{11.62}$$

The amplitude of the harmonics is given by

$$A_k(K) = \sum_{k=1}^{\infty} \frac{k^2\,K^2}{(1+\frac12 K^2)^2} \left[J_{\frac{k-1}{2}}(X_k) - J_{\frac{k+1}{2}}(X_k) \right]^2, \tag{11.63}$$

where the argument $X_k = \frac{k\,K^2}{4+2K^2}$ The strength parameter greatly determines the radiation intensity as shown in Fig. 11.7 for the lowest order harmonics. For the convenience of numerical calculations $A_k(K)$ is also tabulated for odd harmonics only, since $A_k(K) = 0$ for even values of k, in Table 11.1. For weak magnets, $K^2 \ll 1$, the intensity increases with the square of the magnet field or undulator strength parameter K. There is an optimum value for the strength parameter for maximum photon flux depending on the harmonic under consideration. In particular, radiation in the forward direction at the fundamental frequency reaches a maximum photon flux for strength parameters K slightly above unity. The photon flux per unit solid angle increases like the square of the number of wiggler periods N which is a result of the interference effect of many beams which concentrates the radiation more and more into one frequency and its harmonics as the number of interfering beams is increased.

The *opening angle* of the radiation is primarily determined by the $(\sin x/x)^2$-term. We define the rms opening angle by $\theta_1^2 = 2\sigma_{r'}^2$ where θ_1 is the angle at which $\sin x = 0$ for the first time. In this case $x = \pi$ or from (11.56) $N\Delta\omega_k/\omega_1 = 1$. With $\omega_1 = \omega_{\rm p}\frac{2\gamma^2}{1+\frac12 K^2}$, $\omega = k\,\omega_{\rm p}\frac{2\gamma^2}{1+\frac12 K^2+\gamma^2\theta^2}$ and $\frac{\Delta\omega_k}{\omega_1} = \frac{\omega}{\omega_1} - k$ we get $\frac{N\,k\,\gamma^2\,\theta^2}{1+\frac12 K^2+\gamma^2\theta^2} = 1$ or after solving for θ

$$\theta^2 = \frac{1+\frac12 K^2}{\gamma^2\,(1+N\,k)}. \tag{11.64}$$

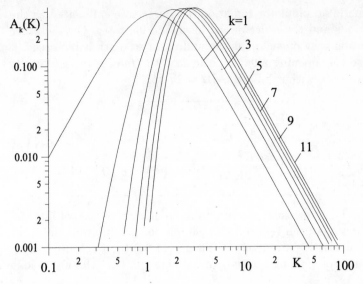

Fig. 11.7. Undulator radiation intensity $A_k(K)$ in the forward direction as a function of the strength parameter K for the five lowest order odd harmonics

Table 11.1. Functions $A_k(K)$ for $k = 1, 3, 5, 7, 9, 11$

K	A_1	A_3	A_5	A_7	A_9	A_{11}
0.1	0.010	0	0	0	0	0
0.2	0.038	0	0	0	0	0
0.4	0.132	0.004	0	0	0	0
0.6	0.238	0.027	0.002	0	0	0
0.8	0.322	0.087	0.015	0.002	0	0
1.0	0.368	0.179	0.055	0.015	0.004	0.001
1.2	0.381	0.276	0.128	0.051	0.019	0.007
1.4	0.371	0.354	0.219	0.118	0.059	0.028
1.8	0.320	0.423	0.371	0.286	0.206	0.142
2.0	0.290	0.423	0.413	0.354	0.285	0.220
5.0	0.071	0.139	0.188	0.228	0.261	0.290
10.0	0.019	0.037	0.051	0.064	0.075	0.085
20.0	0.005	0.010	0.013	0.016	0.019	0.022

Assuming an undulator with many periods we let $kN \gg 1$ and the rms opening angle of undulator radiation is finally

$$\sigma_{r'} \approx \frac{1}{\gamma} \sqrt{\frac{1 + \frac{1}{2}K^2}{2Nk}}. \tag{11.65}$$

The opening angle of undulator radiation becomes more collimated as the number of periods and the order of the harmonic increases. On the other hand the radiation cone opens up as the undulator strength is increased.

We may use this opening angle to calculate the *total photon intensity* into the forward cone with solid angle $d\Omega = 2\pi\,\sigma_{r'}^2$

$$
\begin{aligned}
N_{\text{ph}}(\omega)\Big|_{\theta=0} &= \pi\,\alpha\,N\,\frac{\Delta\omega}{\omega}\,\frac{K^2}{1+\frac{1}{2}K^2} \\
&\times \sum_{k=1}^{\infty} k\left[J_{\frac{k-1}{2}}\left(\frac{k\,K^2}{4+2K^2}\right) - J_{\frac{k+1}{2}}\left(\frac{k\,K^2}{4+2K^2}\right)\right]^2,
\end{aligned}
\tag{11.66}
$$

where $\omega = k\,\omega_1$. The radiation spectrum from an undulator magnet into the forward direction has been reduced to a simple form exhibiting the most important characteristic parameters. Utilizing (11.63) the photon intensity from a single electron passing through an undulator in the kth harmonic is

$$
N_{\text{ph}}(k\omega_1)|_{\theta=0} = \pi\,\alpha\,N\,\frac{\Delta\omega}{\omega}\,\frac{1+\frac{1}{2}K^2}{k}\,A(K).
\tag{11.67}
$$

Equation (11.67) is to be multiplied by the number of particles in the electron beam to get the total photon intensity.

In case of a storage ring articles circulate with a high revolution frequency and we get from (11.67) by multiplication with I/e, where I is the circulating beam current, the photon flux

$$
\frac{dN_{\text{ph}}(k\omega_1)}{dt}\Bigg|_{\theta=0} = \pi\,\alpha\,N\,\frac{I}{e}\,\frac{\Delta\omega}{\omega}\,\frac{1+\frac{1}{2}K^2}{k}\,A_k(K).
\tag{11.68}
$$

The spectrum includes only odd harmonic since all even harmonics are suppressed through the cancellation of Bessel's functions. From a physics point of view this is expected because higher harmonics are caused by relativistic effects in the transverse motion. For weak undulators the transverse momentum is nonrelativistic and the motion is purely sinusoidal generating monochromatic radiation. Only when the undulator strength is increased becomes the transverse momentum relativistic causing a perturbation of the sinusoidal motion. This perturbation of the transverse motion is independent of the direction of motion and therefore introduces only odd harmonics into the motion and radiation spectrum.

In Fig. 11.8 a measured undulator spectrum is shown as a function of the undulator strength K [11.6]. For a strength parameter $K \ll 1$ there is only one line at the fundamental frequency. As the strength parameter increases additional lines appear in addition to being shifted to lower frequencies. The spectral lines from a real synchrotron radiation source are not infinitely narrow as (11.67) would suggest because of the finite opening of a pin hole

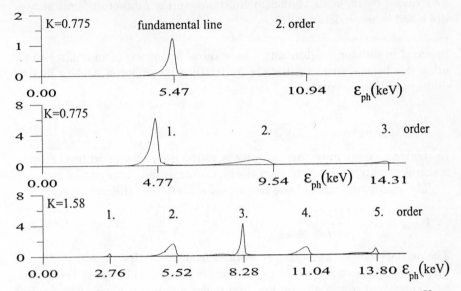

Fig. 11.8. Frequency spectrum from an undulator for different strength parameters K

letting pass some light at small angles with respect to the axis. For the same reason we observe also some signal of the even order harmonic radiation.

Even for an extremely small pin hole we would observe a similar spectrum as shown in Fig. 11.8 because of the finite beam emittance of the electron beam. The electrons follow oscillatory trajectories due not only to the undulator field but also due to betatron oscillations. We observe therefore always some radiation at a finite angle given by the particle trajectory with respect to the undulator axis. Fig. 11.8 also demonstrates the fact that all experimental circumstances must be included to meet theoretical expectations. The amplitudes of the measured low energy spectrum is significantly suppressed compared to theoretical expectations which is due to a Be window being used to extract the radiation from the ultra high vacuum chamber of the accelerator. This material absorbs radiation significantly in the region of a few keV.

Storage rings optimized for very small beam emittance are being used as modern synchrotron radiation sources to reduce the line width of undulator radiation and concentrate all radiation to the frequency desired. The progress in this direction is demonstrated in the spectrum of Fig. 11.8 derived from the first electron storage ring operated at a beam emittance below 10 n-m at 7.1 GeV [11.7]. A finite particle beam emittance contributes additional angles which cause a red shift and broadening of the spectral line width of undulator radiation due to Doppler effect. This line width can be reduced by reducing the beam emittance to the diffraction limit as was discussed in more detail in [11.2].

Spectral undulator brightness: The *spectral brightness* of undulator radiation has been defined in [Ref. 11.2, (9.79)] as the photon density in six-dimensional phase space

$$\mathcal{B} = \frac{\dot{N}_{\text{ph}}}{4\pi^2 \, \sigma_x \sigma_{x'} \sigma_y \sigma_{y'} (d\omega/\omega)} \cdot \tag{11.69}$$

In the laser community this quantity is called the *radiance* while the term spectral brightness is common in the synchrotron radiation community.

The maximum value of the brightness is limited by diffraction to

$$\mathcal{B}_{\text{max}} = \dot{N}_{\text{ph}} \, \frac{(4/\lambda^2)}{d\omega/\omega} \cdot \tag{11.70}$$

The actual photon brightness is reduced from the diffraction limit due to betatron motion of the particles, transverse beam oscillation in the undulator, apparent source size on axis and under an oblique angle. All of these effects tend to increase the source size and reduce brightness.

The particle beam cross section varies in general along the undulator. We assume here for simplicity that the beam size varies symmetrically in the undulator with a waist in its middle. From beam dynamics it is then known that, for example, the horizontal beam size varies like $\sigma_b^2 = \sigma_{\text{bo}}^2 + \sigma'^2_{\text{bo}} s^2$, where σ_{bo} is the beam size at the waist, σ'_{bo} the divergence of the beam at the waist and s the distance from the waist. The average beam size along the undulator length L is then

$$\langle \sigma_b^2 \rangle = \sigma_{\text{bo}}^2 + \frac{\sigma'^2_{\text{bo}} L^2}{12} \cdot$$

Similarly, due to an oblique observation angle θ with respect to the (y, z)-plane or ψ with respect to the (x, z)-plane we get a further additive contribution $\theta L/6$ to the apparent beam size. Finally the apparent source size includes in addition to the particle beam cross section the transverse motion in the undulator field with an amplitude $a = \lambda_{\text{p}} K/(2\pi\gamma)$ from (11.9).

Collecting all contributions and adding in quadrature the total effective beam-size parameters are given by

$$
\begin{aligned}
\sigma_{\text{t},x}^2 &= \tfrac{1}{2}\sigma_r^2 + \sigma_{\text{bo},x}^2 + \left(\frac{\lambda_{\text{p}} K}{2\pi\gamma}\right)^2 + \frac{\sigma_{\text{bo},x'}^2 L^2}{12} + \frac{\theta^2 L^2}{36} \\
\sigma_{\text{t},x'}^2 &= \tfrac{1}{2}\sigma_{r'}^2 + \sigma_{\text{bo},x'}^2 \\
\sigma_{\text{t},y}^2 &= \tfrac{1}{2}\sigma_r^2 + \sigma_{\text{bo},y}^2 + \frac{\sigma_{\text{bo},y'}^2 L^2}{12} + \frac{\psi^2 L^2}{36} \\
\sigma_{\text{t},y'}^2 &= \tfrac{1}{2}\sigma_{r'}^2 + \sigma_{\text{bo},y'}^2,
\end{aligned}
\tag{11.71}
$$

where the particle beam sizes can be expressed by the beam emittance and betatron function as $\sigma^2 = \epsilon\,\beta$, $\sigma'^2 = \epsilon/\beta$, and the diffraction limited beam parameters are $\sigma_r^2 = \lambda/L$, and $\sigma_{r'}^2 = \lambda\,L/(4\pi^2)$.

11.4 Elliptical Polarization

During the discussion of bending magnet radiation in Chap. 7 and insertion radiation in this chapter we noticed the appearance of two orthogonal components of the radiation field which we identified with the orthogonal σ-*mode* and π-*mode* polarization. The π-mode radiation is observable only at a finite angle with the plane including the particle trajectory and the acceleration force vector. In general this plane is the horizontal plane. As we will see both polarization modes can, under certain circumstances, be out of phase giving rise to *elliptical polarization*. In this section we will shortly discuss such conditions.

Elliptical polarization from bending magnet radiation: Upon closer inspection of the radiation field (7.101) from a bending magnet

$$
\mathbf{E}(\omega) \;=\; \frac{1}{[4\pi\epsilon_0]}\frac{\sqrt{3}e}{cR}\frac{\omega}{\omega_c}\,\gamma\,(1+\gamma^2\theta^2)
$$
$$
\times\left[K_{2/3}(\xi)\,\mathbf{u}_\sigma - \mathrm{i}\,\frac{\gamma\theta K_{1/3}(\xi)}{\sqrt{1+\gamma^2\,\theta^2}}\,\mathbf{u}_\pi\right]
\tag{11.72}
$$

we note that both polarization terms are 90° out of phase. As a consequence, the combination of both terms does not just introduce a rotation of the polarization but generates a continuous rotation of the polarization vector which we identify with *circular* or *elliptical polarization*. In this particular case the polarization is elliptical since the π-mode radiation is always weaker than the σ-mode radiation. For large vertical angles, however, the polarization becomes almost circular albeit at low intensity.

We may visualize the polarization property considering that the electrical field and consequently the polarization is proportional to the acceleration vector $\dot{\boldsymbol{\beta}}$. In the deflecting plane which we choose here as the horizontal plane there is only transverse horizontal but no vertical acceleration and we observe therefore only linear polarization of the σ-mode. However, as we observe radiation at an angle with the horizontal plane we note that the acceleration is normal to the trajectory and can be decomposed into two components $\dot{\beta}_x$ and $\dot{\beta}_z$ as shown in Fig. 11.9a.

The longitudinal acceleration component together with a finite observation angle θ gives rise to a vertical acceleration component or vertical electric field

$$
\mathbf{E}_y \propto \dot{\boldsymbol{\beta}}_y = n_y\,\dot{\beta}_z + n_x\,n_y\,\dot{\beta}_x\,.
\tag{11.73}
$$

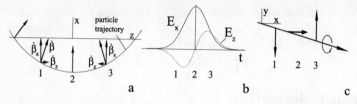

Fig. 11.9. Acceleration along an arc segment of the particle trajectory in a bending magnet **a**, radiation field components as a function of time **b** and resulting polarization as a function of time **c**

An additional component appears if we observe the radiation also at an angle with respect to the (x, y)-plane which we, however, ignore here for this discussion. The components n_x, n_y are components of the observation unit vector from the observer to the source with $n_y = -\sin\theta$. We observe first radiation from an angle $\theta > 0$. The horizontal and vertical radiation field components as a function of time are shown in Fig. 11.9b. Both being proportional to the acceleration (Fig. 11.9a) we observe a symmetric horizontal field E_x and an antisymmetric vertical field E_y. The polarization vector (Fig. 11.9c) therefore rotates with time in a counter clockwise direction giving rise to elliptical polarization with lefthanded helicity. Observing the radiation from below with $\theta < 0$ the antisymmetric field switches sign and and the helicity becomes righthanded. The visual discussion of the origin of elliptical polarization of bending magnet radiation is in agreement with the mathematical result (11.72) displaying the sign dependence of the π-mode component with θ.

The intensities for both polarization modes are shown in Fig. 11.10 as a function of the vertical observation angle θ for different photon energies. Both intensities are normalized to the forward intensity of the σ-mode radiation. From Fig. 11.10 it becomes obvious that circular polarization is approached for large observation angles. At high photon energies both radiation lobes are confined to very small angles but expand to larger angle distributions for photon energies much lower than the critical photon energy.

The elliptical polarization is right or left handed depending on whether we observe the radiation from above or below the horizontal mid plane. Furthermore the *helicity* depends on the direction of deflection in the bending magnet. The sign of the bending radius in (7.91) is significant but got lost in our discussion by introducing the *Larmor frequency*. We must therefore assume here that the Larmor frequency is sign sensitive and as a consequence we find the argument ξ in the modified Bessel's functions $K_{2/3}(\xi)$ and $K_{1/3}(\xi)$ sign sensitive as well. From (7.100) we note that $K_{1/3}(\xi)$ is an even function while $K_{2/3}(\xi)$ is an odd function. By changing the sign of the bending magnet field the helicity of the elliptical polarization can be reversed. This is of no importance for radiation from a bending magnet since we cannot change the field without loss of the particle beam but is of specific

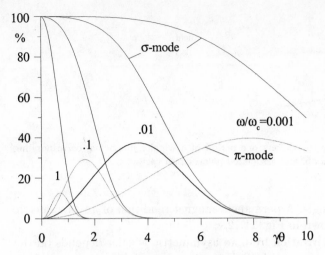

Fig. 11.10. Comparison of σ-mode and π-mode radiation as a function of vertical observation angle θ for different photon energies

importance for the degree of elliptical polarization of wiggler and undulator radiation.

Elliptical polarization from periodic insertion devices: We apply the visual picture for the formation of polarized radiation in a bending magnet to periodic magnetic field of wiggler and undulator magnets. The acceleration vectors are shown in Fig. 11.11a for one period and we do not expect any polarization in the mid plane where $\theta = 0$. Off the mid plane we observe now the radiation from a positive and negative pole. From each pole we get elliptical polarization but the combination of lefthanded polarization from one pole with righthanded polarization from the next pole leads to a cancellation of elliptical polarization from periodic magnets (Fig. 11.11c). In bending magnets this cancellation did not occur for lack of alternating deflection. Since there are generally an equal number of positive and negative poles in a wiggler or undulator magnet the elliptical polarization is compensated. Ordinary wiggler and undulator magnets do not produce elliptically polarized radiation.

The elimination of elliptical polarization in periodic magnets results from a compensation of left and righthanded helicity and we may therefore look for an insertion device in which this symmetry is broken. Such an insertion device is the *asymmetric wiggler magnet* which is designed similar to a wave length shifter with one strong central pole and two weaker poles on either side such that the total integrated field vanishes or $\int B_y \, ds = 0$. A series of such magnets may be aligned to produce an insertion device with many poles to enhance the intensity. The compensation of both helicities does not work anymore since the radiation depends on the magnetic field and not on

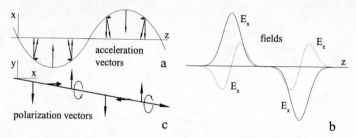

Fig. 11.11. Acceleration vectors along one period of a wiggler magnet **a**, corresponding radiation field components **b**, and associated polarization vectors **c**

the total deflection angle. A permanent magnet rendition of an asymmetric wiggler magnet is shown in Fig. 11.12.

The degree of polarization from an asymmetric wiggler depends on the desired photon energy. The critical photon energy is high for radiation from the high field pole, ϵ_c^+, and lower for radiation from the low field pole, ϵ_c^-. For high photon energies $\epsilon_{\mathrm{ph}} \approx \epsilon_c^+$ the radiation from the low field poles is negligible and the radiation is the same as from a series of bending magnets with its particular polarization characteristics. For lower photon energies $\epsilon_c^- < \epsilon_{\mathrm{ph}} < \epsilon_c^+$ the radiation intensity from high and low field pole become similar and cancellation of the elliptical polarization occurs. At low photon energies $\epsilon_{\mathrm{ph}} < \epsilon_c^-$ the intensity from the low field poles exceeds that from the high field poles and we observe again elliptical polarization although with reversed helicity.

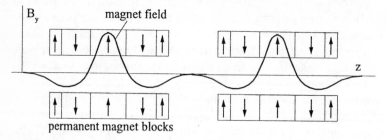

Fig. 11.12. Asymmetric wiggler magnet built from permanent magnet blocks with magnetic field distribution (schematic)

The creation of elliptically and circularly polarized radiation is important for a large class of experiments based on synchrotron radiation and special insertion devices have therefore been developed to meet such needs in an optimal way. Different approaches have been suggested and realized as sources for elliptically polarized radiation [11.8 – 11]. All methods are based on permanent magnet technology sometimes combined with electromagnets

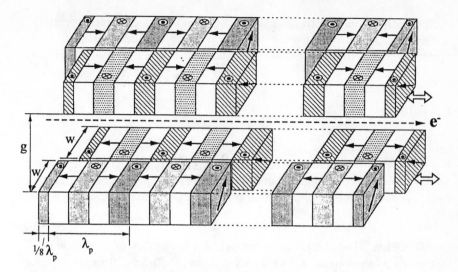

Fig. 11.13. Permanent magnet arrangement for the elliptically polarized radiation source APPLE-1 [11.10]

Fig. 11.14. Actual magnet assembly of APPLE-1 [11.10]

to produce vertical and horizontal fields shifted in phase such that elliptically polarized radiation can be produced. As an example, Fig. 11.13 shows the arrangement of two pairs of permanent magnet arrays [11.10] based on the *phase adjustable undulator principle* developed by *Carr* [11.12]. By shifting the top magnet arrays with respect to the bottom magnets the fundamental frequency of the undulator radiation can be varied. Furthermore, if say the magnet blocks on the left side of the top and bottom arrays are shifted with respect to the blocks on the right side a continuous variation of elliptical polarization from left to linear to right handed helicity can be obtained. The actual magnet with shifted magnets is shown in Fig. 11.14.

Problems

Problem 11.1. Calculate the total energy loss in an undulator by treating it as a series of bending magnets with a sinusoidal field profile. Show that your result for the total energy loss in an undulator is consistent with expression (11.9).

Problem 11.2. Determine quantitatively the trajectory of a particle in the reference system moving with velocity $\bar{\beta}$ from (11.7) for a strength parameter $K = 0.4$ and $K = 4$. Identify in this reference system the two radiation lobes associated with the two orthogonal motions of the particle. What are the fundamental and harmonic frequencies for both oscillations and which are the corresponding radiation lobes in laboratory space?

Problem 11.3. Solve the equations of motion in the moving reference system (11.7) for a strong undulator with strength parameter $K > 1$ by expanding the particle oscillations into a few harmonics. Which harmonics are involved in the perturbation of a pure sinusoidal motion? Can you relate them to the radiation spectrum in the laboratory system ?

Problem 11.4. The radiation at the fundamental frequency is primarily emitted into the forward direction $\theta = 0$. Explain the physical origin of the azimuthal variation of the photon flux within that forward radiation cone.

Problem 11.5. Determine and plot the degree and intensity of elliptically polarized radiation as a fraction of the total forward intensity from a bending magnet as a function of observation angle θ for different photon energies $\epsilon/\epsilon_c = 0.01, 0.1, 1$ and 10.

Problem 11.6. Design an asymmetric wiggler magnet assuming hard edge fields and optimized for the production of elliptical polarized radiation at a photon energy of your choice. Calculate and plot the photon flux of polarized radiation in the vicinity of the optimum photon energy.

Appendix

In this appendix some of the basic assumptions like units and dimensions, conversion from cgs to MKS system, often used relativistic expressions and maxwells equations are reproduced from [.1] for the convenience of the reader. We also include a selection of useful but not exhaustive mathematical formulas.

Conversion from cgs to MKS system: Magnetic fields are quoted either in Gauss or Tesla. Similarly, field gradients and higher derivatives are expressed in Gauss per centimeter or Tesla per meter. Frequently we find the need to perform numerical calculations with parameters given in units of different systems. Some helpful numerical conversions are compiled in Table 1.

Table 1. Numerical conversion factors

quantity:	replace cgs parameter:	by practical units:
potential	1 esu	300 V
electrical field	1 esu	3×10^4 V/m
magnetic induction	1 Gauss	1×10^{-4} Tesla
magnetic field	1 Oersted	$1000/(4\pi)$ A/m
current	1 esu	$10 \cdot c$ A
charge	1 esu	0.3333×10^{-9} C
resistance	1 sec/cm	$9 \times 10^{11} \Omega$
capacitance	1 cm	$1/9 \times 10^{-11}$ Farad
inductance	1 cm	1×10^{-9} Henry
force	1 dyn	10^{-5} N
energy	1 erg	1×10^{-7} J
	1 eV	1.602×10^{-19} J
	1 eV	1.602×10^{-12} erg

Similar conversion factors can be derived for electromagnetic quantities in formulas by comparisons of similar equations in the cgs and MKS system. Table 2. includes some of the most frequently used conversions.

Table 2. Conversion factors for equations

quantity:	replace cgs parameter:	by MKS parameter:
potential	V_{cgs}	$\sqrt{4\pi\epsilon_\text{o}}\, V_{\text{MKS}}$
electric field	$\mathbf{E}_{\text{cgs}}$	$\sqrt{4\pi\epsilon_\text{o}}\, \mathbf{E}_{\text{MKS}}$
current	I_{cgs}	$\frac{1}{\sqrt{4\pi\epsilon_\text{o}}}\, I_{\text{MKS}}$
current density	$\mathbf{j}_{\text{cgs}}$	$\frac{1}{\sqrt{4\pi\epsilon_\text{o}}}\, \mathbf{j}_{\text{MKS}}$
charge	q_{cgs}	$\frac{1}{\sqrt{4\pi\epsilon_\text{o}}}\, q_{\text{MKS}}$
charge density	ρ_{cgs}	$\frac{1}{\sqrt{4\pi\epsilon_\text{o}}}\, \rho_{\text{MKS}}$
conductivity	σ_{cgs}	$\frac{1}{4\pi\epsilon_\text{o}}\, \sigma_{\text{MKS}}$
inductance	L_{cgs}	$4\pi\epsilon_\text{o}\, L_{\text{MKS}}$
capacitance	C_{cgs}	$\frac{1}{4\pi\epsilon_\text{o}}\, C_{\text{MKS}}$
magnetic field	H_{cgs}	$\sqrt{4\pi\mu_\text{o}}\, H_{\text{MKS}}$
magnetic induction	B_{cgs}	$\frac{\sqrt{4\pi}}{\sqrt{\mu_\text{o}}}\, B_{\text{MKS}}$

The absolute dielectric constant is

$$\epsilon_\text{o} = \frac{10^7}{4\pi c^2}\,\frac{\text{C}}{\text{V m}} = 8.854 \times 10^{-12}\,\frac{\text{C}}{\text{V m}}$$

and the absolute magnetic permeability is

$$\mu_\text{o} = 4\pi \times 10^{-7}\,\frac{\text{V s}}{\text{A m}} = 1.2566 \times 10^{-6}\,\frac{\text{V s}}{\text{A m}}\,.$$

Both constants are related by

$$\epsilon_\text{o}\,\mu_\text{o}\,c^2 = 1\,.$$

Basic relativistic formulas: Beam dynamics is expressed in a fixed laboratory system of coordinates but some specific problems are better discussed in the moving coordinate system of a single particle or of the center of charge for a collection of particles. Transformation between the two systems is effected through a *Lorentz transformation*

$$x = x^*, \quad y = y^*, \quad z = \frac{z^* + \beta_z ct^*}{\sqrt{1 - \beta_z^2}}, \quad ct = \frac{ct^* + \beta_z z^*}{\sqrt{1 - \beta_z^2}},$$

where $\beta_z = v_z/c$ and where we have assumed that the particle moves with the velocity v_z along the z-axis with respect to the fixed laboratory system S. The primed quantities are measured in the moving system S^*.

Characteristic for relativistic mechanics is the *Lorentz contraction* and *time dilatation*, both of which become significant in the description of particle dynamics. The Lorentz contraction is the result of a *Lorentz transformation* from one coordinate system to another moving relative to the first system. We consider a rod at rest along the z-coordinate with the length $\Delta z = z_2 - z_1$ in the coordinate system S. In the system S^* moving with the velocity v_z along the z-axis with respect to the system S, this rod appears to have the length $\Delta z^* = z_2^* - z_1^*$ and we get the relationship

$$\Delta z = \gamma (z_2^* + v_z t^*) - \gamma (z_1^* + v_z t^*) = \gamma \Delta z^*$$

or

$$\Delta z = \gamma \Delta z^*,$$

where γ is the total particle energy E in units of the particle rest energy mc^2

$$\gamma = \frac{E}{m c^2} = \frac{1}{\sqrt{1 - \beta_z^2}}.$$

The rod appears shorter in the moving particle system by the factor γ and longest in the system where it is at rest.

Because of the Lorentz contraction the volume of a body at rest in the system S appears also reduced in the moving system S^* and we have for the volume of a body in three dimensional space

$$V = \gamma V^*.$$

Only one dimension of this body is Lorentz contracted and therefore the volume scales only linearly with γ. As a consequence, the charge density ρ of a particle bunch with the volume V is lower in the laboratory system S compared to the density in the system moving with this bunch and becomes

$$\rho = \frac{\rho*}{\gamma}.$$

Similarly, we may derive the *time dilatation* or the elapsed time between two events occurring at the same point in both coordinate systems. From the Lorentz transformations we get with $z_2^* = z_1^*$

$$\Delta t = t_2 - t_1 = \gamma \left(t_2^* + \frac{\beta_z z_2^*}{c} \right) - \gamma \left(t_1^* + \frac{\beta_z z_1^*}{c} \right)$$

or

$$\Delta t = \gamma \Delta t^*.$$

For a particle at rest in the moving system S^* the time t^* varies slower than the time in the laboratory system. This is the mathematical expression for the famous *twin brother paradox* where one of the brothers moving in a space capsule at relativistic speed would age slower than his twin brother staying back. This phenomenon becomes reality for unstable particles. For example, high-energy pion mesons, observed in the laboratory system, have a longer lifetime by the factor γ compared to low-energy pions with $\gamma = 1$. As a consequence we are able to transport high-energy pion beams a longer distance than low energy pions.

The total energy of a particles is given by

$$E = \gamma E_{\rm o} = \gamma mc^2 ,$$

where $E_{\rm o} = mc^2$ is the rest energy of the particle. The kinetic energy is defined as the total energy minus the rest energy

$$E_{\rm kin} = E - E_{\rm o} = (\gamma - 1) mc^2 .$$

The change in kinetic energy during acceleration is equal to the product of the accelerating force and the path length over which the force acts on the particle. Since the force may vary along the path we use the integral

$$\Delta E_{\rm kin} = \int_{L_{\rm acc}} \mathbf{F} \cdot \mathrm{ds}$$

to define the energy increase. The length $L_{\rm acc}$ is the path length through the accelerating field. In discussions of energy gain through acceleration we consider only energy differences and need therefore not to distinguish between total and kinetic energy. The particle *momentum* finally is defined by

$$c^2 p^2 = E^2 - E_{\rm o}^2$$

or

$$cp = \sqrt{E^2 - E_{\rm o}^2} = mc^2 \sqrt{\gamma^2 - 1} = \gamma \beta mc^2 = \beta E ,$$

where $\beta = v/c$. The simultaneous use of the terms energy and momentum might seem sometimes to be misleading as we discussed earlier. In this text, however, we will always use physically correct quantities in mathematical formulations even though we sometimes use the term energy for the quantity cp. In electron accelerators the numerical distinction between energy and momentum is insignificant since we consider in most cases highly relativistic particles. For proton accelerators and even more so for heavy ion accelerators the difference in both quantities becomes, however, significant.

Often we need differential expressions or expressions for relative variations of a quantity in terms of variations of another quantity. Such relations can be derived from the definitions in this section. From the variation of $cp = mc^2 \sqrt{\gamma^2 - 1}$ we get, for example,

$$dcp = \frac{mc^2}{\beta}\, d\gamma = \frac{dE}{\beta} = \frac{dE_{\text{kin}}}{\beta}$$

and

$$\frac{dcp}{cp} = \beta^{-2}\frac{d\gamma}{\gamma}.$$

Varying $cp = \gamma\beta\, mc^2$ and eliminating $d\gamma$ we get

$$dcp = \gamma^3\, mc^2\, d\beta$$

and

$$\frac{dcp}{cp} = \gamma^2\frac{d\beta}{\beta}.$$

In a similar way other relations can be derived.

Electromagnetic fields and the interaction of charged particles with these fields play an important role in accelerator physics. We find it often useful to express the fields in either the laboratory system or the particle system. Transformation of the fields from one to the other system is determined by the Lorentz transformation of electro magnetic fields. We assume again the coordinate system S^* to move with the velocity v_s along the s-axis with respect to a right-handed (x, y, s) reference frame S. The electromagnetic fields in this moving reference frame are identified by primed symbols and can be expressed in terms of the fields in the laboratory frame of reference S:

$$
\begin{aligned}
E_x^* &= \gamma\left(E_x + [c]\,\beta_z\, B_y\right), & [c]\, B_x^* &= \gamma\left([c]\, B_x - \beta_z\, E_y\right), \\
E_y^* &= \gamma\left(E_y - [c]\,\beta_z\, B_x\right), & [c]\, B_y^* &= \gamma\left([c]\, B_y + \beta_z\, E_x\right), \\
E_z^* &= E_z, & B_z^* &= B_z.
\end{aligned}
$$

These transformations exhibit interesting features for accelerator physics, where we often use magnetic or electrical fields, which are pure magnetic or pure electric fields when viewed in the laboratory system. For relativistic particles, however, these pure fields become a combination of electric and magnetic fields.

Maxwell's equations: Accelerator physics is to a large extend the description of charged particle dynamics in the presence of external electromagnetic fields or of fields generated by other charged particles. We use *Maxwell's equations* in a vacuum environment or in a material with well behaved values of the permittivity and permeability to describe these fields and the *Lorentz force* to formulate the particle dynamics under the influence of these electromagnetic fields

$$\nabla \left(\epsilon \mathbf{E} \right) = \frac{4\pi}{[4\pi\epsilon_o]}\,\rho\,, \qquad\qquad \nabla \times \mathbf{E} = -\frac{[c]}{c}\frac{\partial}{\partial t}\,\mathbf{B}\,,$$

$$\nabla\,\mathbf{B} = 0\,, \qquad\qquad \frac{c[c]}{\mu}\,\nabla \times \mathbf{B} = \frac{4\pi}{[4\pi\epsilon_o]}\,\rho\mathbf{v} + \frac{\partial}{\partial t}\,\mathbf{E}\,.$$

Here ρ is the charge density and $\mathbf{v}$ the velocity of the charged particles. In general, we are interested in particle dynamics in a material free environment and set therefore $\epsilon = 1$ and $\mu = 1$. For specific discussions we do, however, need to calculate fields in a material filled environment in which case we come back to the general form of Maxwell's equations.

Whatever the interaction of charged particles with electromagnetic fields and whatever the reference system may be, we depend in accelerator physics on the invariance of the *Lorentz force equation* under coordinate transformations. All acceleration and beam guidance in accelerator physics will be derived from the Lorentz force which is defined in the case of a particle with charge q in the presence of an electric $\mathbf{E}$ and magnetic field $\mathbf{B}$ by

$$\mathbf{F} = q\,\mathbf{E} + q\,\frac{[c]}{c}\left(\mathbf{v}\times\mathbf{B} \right).$$

For simplicity we use throughout this text particles with one unit of electrical charge e like electrons and protons unless otherwise noted. In case of multiply charged ions the single charge e must be replaced by eZ where Z is the *charge multiplicity* of the ion. Both components of the Lorentz force are used in accelerator physics where the force due to the electrical field is mostly used to actually increase the particle energy while magnetic fields are used mostly to guide the particle beams along desired beam transport lines. This separation of functions, however, is not exclusive as the example of the betatron accelerator shows where particles are accelerated by time dependent magnetic fields. Similarly electrical fields are used in specific cases to guide or separate particle beams.

If we integrate the Lorentz force over the time a particle interacts with the field we get a change in the particle momentum,

$$\Delta\mathbf{p} = \int \mathbf{F}\,\mathrm{d}t\,.$$

On the other hand, if the Lorentz force is integrated with respect to the path length we get the change in *kinetic energy* E_{kin} of the particle

$$\Delta E_{\mathrm{kin}} = \int \mathbf{F}\,\mathrm{d}s\,.$$

Comparing the last two equations we find with $\mathrm{d}s = \mathbf{v}\,\mathrm{d}t$ the relation between the momentum and kinetic energy differentials

$$c\boldsymbol{\beta}\,\mathrm{d}\mathbf{p} = \mathrm{d}E_{\mathrm{kin}}\,.$$

With the Lorentz and $\mathrm{d}s = \mathbf{v}\,\mathrm{d}t$ in the second integral we get

$$\Delta E_{\text{kin}} = q \int \mathbf{E} \, d\mathbf{s} + \frac{q}{c} \int (\mathbf{v} \times \mathbf{B}) \, \mathbf{v} \, dt.$$

It becomes obvious that the kinetic energy of the particle increases whenever a finite accelerating electric field $\mathbf{E}$ exists and the acceleration occurs in the direction of the electric field. This acceleration is independent of the particle velocity and acts even on a particle at rest $\mathbf{v} = 0$. The second component of the Lorentz force in contrast depends on the particle velocity and is directed normal to the direction of propagation and normal to the magnetic field direction. We find therefore the well-known result that the kinetic energy is not changed by the presence of magnetic fields since the scalar product $(\mathbf{v} \times \mathbf{B}) \, \mathbf{v} = 0$ vanishes. The magnetic field causes only a deflection of the particle trajectory. The Lorentz force is used to derive the *equation of motion* of charged particles in the presence of electromagnetic fields

$$\frac{d}{dt} \mathbf{p} = \frac{d}{dt} (m \gamma \mathbf{v}) = eZ\mathbf{E} + e \frac{[c]}{c} Z (\mathbf{v} \times \mathbf{B}),$$

where Z is the charge multiplicity of the charged particle. For simplicity we drop from here on the factor Z since the charge multiplicity is different from unity only for ion beams. For ion accelerators we note therefore that the particle charge e must be replaced by eZ.

The fields can be derived from electrical and magnetic potentials in the well-known way

electric field: $\quad \mathbf{E} = -\dfrac{[c]}{c} \dfrac{\partial \mathbf{A}}{\partial t} - \nabla \Phi,$

magnetic field: $\quad \mathbf{B} = \nabla \times \mathbf{A},$

where Φ is the electric scalar potential and $\mathbf{A}$ the magnetic vector potential. The particle momentum is $\mathbf{p} = \gamma m \mathbf{v}$ and it's time derivative

$$\frac{d\mathbf{p}}{dt} = m \gamma \frac{d\mathbf{v}}{dt} + m \mathbf{v} \frac{d\gamma}{dt}.$$

With

$$\frac{d\gamma}{dt} = \frac{d}{d\beta} \frac{1}{\sqrt{1 - \beta^2}} \frac{d\beta}{dt} = \gamma^3 \frac{\beta}{c} \frac{dv}{dt}$$

we get the equation of motion

$$\mathbf{F} = \frac{d\mathbf{p}}{dt} = m \left(\gamma \frac{d\mathbf{v}}{dt} + \gamma^3 \frac{\beta}{c} \frac{dv}{dt} \mathbf{v} \right).$$

For a force parallel to the particle propagation $\mathbf{v}$ we have $\dot{v} \mathbf{v} = \dot{\mathbf{v}} v$ and

$$\frac{d\mathbf{p}_\parallel}{dt} = m \gamma \left(1 + \gamma^2 \beta \frac{v}{c} \right) \frac{d\mathbf{v}_\parallel}{dt} = m \gamma^3 \frac{d\mathbf{v}_\parallel}{dt}.$$

On the other hand, if the force is directed normal to the particle propagation we have $dv/dt = 0$ and the last equation reduces to

$$\frac{d\mathbf{p}_\perp}{dt} = m\gamma\frac{d\mathbf{v}_\perp}{dt}.$$

It is obvious from these results how differently the dynamics of particle motion is affected by the direction of the Lorentz force. Specifically the dynamics of highly relativistic particles under the influence of electromagnetic fields depends greatly on the direction of the force with respect to the direction of particle propagation. The difference between parallel and perpendicular acceleration will have a great impact on the design of electron accelerators. As we will see later the acceleration of electrons is limited due to the emission of synchrotron radiation. This limitation, however, is much more severe for electrons in circular accelerators where the magnetic forces act perpendicular to the propagation compared to the acceleration in linear accelerators where the accelerating fields are parallel to the particle propagation. This argument is also true for protons or for that matter, any charged particle, but because of the much larger particle mass the amount of synchrotron radiation is generally negligible small.

Vector relations:

$$\mathbf{a}(\mathbf{b} \times \mathbf{c}) = \mathbf{b}(\mathbf{c} \times \mathbf{a}) = \mathbf{c}(\mathbf{a} \times \mathbf{b})$$

$$\mathbf{a} \times (\mathbf{b} \times \mathbf{c}) = \mathbf{b}(\mathbf{ac}) - \mathbf{c}(\mathbf{ab})$$

$$(\mathbf{a} \times \mathbf{b})(\mathbf{c} \times \mathbf{d}) = (\mathbf{ac})(\mathbf{bd}) - (\mathbf{bc})(\mathbf{ad})$$

$$\nabla(\mathbf{a}\varphi) = \varphi\nabla\mathbf{a} + \mathbf{a}\nabla\varphi$$

$$\nabla \times (\mathbf{a}\varphi) = \varphi(\nabla \times \mathbf{a}) - (\mathbf{a} \times \nabla\varphi)$$

$$\nabla(\mathbf{a} \times \mathbf{b}) = \mathbf{b}(\nabla \times \mathbf{a}) - \mathbf{a}(\nabla \times \mathbf{b})$$

$$\nabla \times (\mathbf{a} \times \mathbf{b}) = (\mathbf{b}\nabla)\mathbf{a} - (\mathbf{a}\nabla)\mathbf{b} + \mathbf{a}(\nabla\mathbf{b}) - \mathbf{b}(\nabla\mathbf{a})$$

$$\nabla \times (\nabla\varphi) = 0$$

$$\nabla(\nabla \times \mathbf{a}) = 0$$

Some useful constants:

velocity of light	c	$2.9979246 \cdot 10^8$	m/s
Planck constant, reduced	$\hbar$	$1.0545887 \cdot 10^{-27}$	erg s
		$6.582173 \cdot 10^{-16}$	eV s
	$\hbar c$	$1.9732858 \cdot 10^{-7}$	eV m
Avogadro's number	N_A	$6.022045 \cdot 10^{23}$	1/mol
Boltzmann constant	k	$1.380662 \cdot 10^{-23}$	J/°K
Stefan-Boltzmann constant	σ	$5.67032 \cdot 10^{-12}$	J/sec/cm^2/°K^{-4}
molar volume	V_M	22.4138	lt/mol
electron charge	e	$4.803242 \cdot 10^{-10}$	esu
		$1.6021892 \cdot 10^{-19}$	Cb
electron mass	m_e	$9.109534 \cdot 10^{-28}$	g
		0.511003	MeV/c^2
classical electron radius	$r_e = \frac{e^2}{m_e c^2}$	$2.817938 \cdot 10^{-15}$	m
proton mass	m_p	928.2796	MeV/c^2
fine structure constant	$\alpha = \frac{e^2}{\hbar c}$	$\frac{1}{137.036}$	
Thomson cross section	$\sigma_T = \frac{8\pi}{3} r_e^2$	$6.652448 \cdot 10^{-29}$	m^2
energy conversions	1 cal $\equiv$	4.184	J
	1 eV $\equiv$	$1.602189 \cdot 10^{-19}$	J
	1 eV $\equiv$	11,604.8	°K
	1 eV $\equiv$	1.239854	μm

References

Chapter 1

1.1   H. Goldstein: *Classical Mechanics* (Addison-Wesley, Reading 1950)
1.2   L.D. Landau, E.M. Lifshitz: *Mechanics*, 3rd edn. (Pergamon, Oxford 1976)
1.3   L.D. Landau, E.M. Lifshitz: *The Classical Theory of Fields*, 4th edn. (Pergamon, Oxford 1975)
1.4   L. Lur'é *Mécanique Analytique* (Librairie Universitaire, Louvain 1968)
1.5   H. Wiedemann: *Particle Accelerator Physics I* (Springer, Berlin, Heidelberg 1993)
1.6   A. Schoch: Theory of linear and non linear perturbations of betatron oscillations in alternating gradient synchrotrons. CERN Rpt. CERN 57-23 (1958)
1.7   G. Guignard: The general theory of all sum and difference resonances in a three dimensional magnetic field in a synchrotron. CERN Rpt. CERN 76-06 (1976)
1.8   S. Ohnuma: Quarter integer resonance by sextupoles. Int. Note, Fermilab TM-448 040 (1973)
1.9   E.D. Courant, M.S. Livingston, H.S. Snyder: Phys. Rev. **88**, 1190 (1952)
1.10 J. Safranek: SPEAR lattice for high brightness synchrotron radiation, Ph.D Thesis (Stanford University, Stanford 1992)
1.11 B. Chirikov: A universal instability of many-dimensional oscillator systems. Phys. Repts. **52**, 263 (1979)
1.12 *Nonlinear Dynamics Aspects in Particle Accelerators*. Lect. Notes Phys., Vol.247 (Springer, Berlin, Heidelberg 1986)
1.13 *Physics of Particle Accelerators*, AIP Conf. Proc., Vol.184 (American Institute of Physics, New York 1989)
1.14 J.M. Greene: A method for determining a stochastic transition. J. Math. Phys. **20**, 1183 (1979)
1.15 *Nonlinear Dynamics and the Beam-Beam Interaction*, ed. by M. Month, J.C. Herrera. AIP Conf. Proc., Vol.57 (American Institute of Physics, New York 1979)
1.16 *Phase Space Dynamics*. Lect. Notes Phys., Vol.296 (Springer, Berlin, Heidelberg 1988)
1.17 E.D. Courant, H.S. Snyder: Theory of the alternating-gradient synchrotron. Annals Phys. **3**, 1 (1958)
1.18 H. Poincaré: *Nouvelle Méthods de la Mécanique Celeste* (Gauthier-Villars, Paris 1892) Vol.1, p.193

Chapter 2

2.1   H. Wiedemann: *Particle Accelerator Physics I* (Springer, Berlin, Heidelberg 1993)
2.2   K.L. Brown: Adv. Particle Phys. **1**, 71 (1967)
2.3   G. Leleux: Int. Note SOC/ALIS 18 (1969), Saclay, Département du Synchrotron Saturne

2.4   K.W. Robinson, G.A. Voss: Int'l Symp. Electron and Positron Storage Rings (Saclay) (Presses Universitaire de France, Paris 1966) Proc. p.III-4

2.5   J.M. Paterson, J.R. Rees, H. Wiedemann: Stanford Linear Accelerator Center, Int. Note PEP-124 (1975)

2.6   H. Wiedemann: Nucl. Instrum. Methods A **266**, 24 (1988)

2.7   H. Motz: J. Appl. Phys. **22**, 527 (1951)

2.8   W.R. Smythe: *Static and Dynamic Electricity* (McGraw-Hill, New York 1950)

2.9   L.R. Elias, W.M. Fairbank, J.M.J. Madey, H.A. Schwettmann, T.J. Smith: Phys. Rev. Lett. **36**, 717 (1976)

2.10 B. Kincaid: J. Appl. Phys. **48**, 2684 (1977)

2.11 L. Smith: Effects of wigglers and undulators on beam dynamics. Laboratory ESG Tech. Note 24, Lawrence Berkeley Laboratory, Berkeley, CA (1986)

2.12 M.N. Wilson: *Superconducting Magnets* (Clarendon, Oxford 1983)

2.13 E. Willen, P. Dahl, J. Herrera: Superconducting magnets. *AIP Conf. Proc.* **153/2**, 1120 (American Institute of Physics, New York 1987)

2.14 Proc. CERN Accelerator School at Haus Rissen, Hamburg (1988), CERN 89-04

2.15 I.I. Rabi: Rev. Sci. Instrum. **5**, 78 (1934)

2.16 R.A. Beth: Proc. 6th Int'l Conf. on High Energy Accelerators at CEA, Cambridge (1967) p.387

2.17 R.A. Beth: J. Appl. Phys. **37**, 2568 (1966)

2.18 E. Weber: *Electromagnetic Fields* (Wiley, New York 1950)

Chapter 3

3.1   G. Ripken: DESY Internal Rept. R1-70/04 (1970)

Chapter 4

4.1   H. Wiedemann: *Particle Accelerator Physics I* (Springer, Berlin, Heidelberg 1993)

4.2   SLAC-LINEAR-COLLIDER. Conceptual Design Report, SLAC- 229 (1981)

4.3   G.E. Fischer: *Ground Motion and its Effects in Accelerator Design*, ed. by M. Month, M. Dienes, AIP Conf. Proc., Vol.153 (American Institute of Physics, New York 1987)

4.4   N. Vogt-Nielsen: Expansions of the characteristic exponents and the Floquet solutions for the linear homogeneous second order differential equation with periodic coefficients. Int. Note, MURA/NVN/3 (1956)

4.5   H. Wiedemann: Chromaticity correction in large storage rings. Int. Note PEP-220, SLAC (1976)

4.6   P.L. Morton: Derivation of nonlinear chromaticity by higher-order "smooth approximation". Int. Note PEP-221, SLAC (1976)

Chapter 5

5.1   H. Wiedemann: *Particle Accelerator Physics I* (Springer, Berlin, Heidelberg 1993)

5.2   K.L. Brown, R. Belbeoch, P. Bounin: Rev. Sci. Instrum. **35**, 481 (1964)

5.3   K.L. Brown: Proc. 5th Int'l Conf. on High Energy Accelerators, Frascati, Italy (1965) p.507

5.4   K.L. Brown: Adv. Particle Phys. **1**, 71 (1967)

5.5   K.L. Brown, D.C. Carey, Ch. Iselin, F. Rothacker: TRANSPORT - a computer program for designing charged particle beam transport systems, SLAC-75 (1972), CERN 73-16 (1973), and revisions in SLAC-91 (1977), CERN 80-4 (1980)

5.6   S. Kheifets, T. Fieguth, K.L. Brown, A.W. Chao, J.J. Murray, R.V. Servranckx, H. Wiedemann: 13th Int'l Conf. on High Energy Accelerators, Novosibirsk, USSR (1986)

5.7   PEP Conceptual Design Report. Stanford Linear Accelerator Center, Stanford Rept. SLAC-189 and LBL-4288 (1976)

5.8   H. Wiedemann: Chromaticity correction in large storage rings. SLAC Int. Note PEP-220 (1976)

5.9   K.L. Brown, R.V. Servranckx: *Proc. 11th Int'l Conf. on High Energy Accelerators* (Birkhäuser, Basel 1980) p.656

5.10 J.J. Murray, K.L. Brown, T. Fieguth: 1987 IEEE Particle Accelerator Conf., Washington, DC. IEEE Catalog No. 87CH2387-9, 1331 (1987)

5.11 L. Emery: A wiggler-based ultra-low-emittance damping ring lattice and its chromatic correction. Ph.D. Thesis, Stanford University (1990)

5.12 R.V. Servranckx, K.L. Brown: IEEE Trans. NS-**26**, 3598 (1979)

5.13 B. Autin: Nonlinear betatron oscillations. *AIP Conf. Proc.* **153**, 288 (American Institute of Physics, New York 1987)

5.14 B. Autin: The CERN anti proton collector. CERN 84-15, 525 (1984)

5.15 M.H.R. Donald: Chromaticity correction in circular accelerators and storage rings, Pt.I, a users guide to the HARMON program. SLAC Note PEP-311 (1979)

5.16 K.L. Brown, D.C. Carey, Ch. Iselin: DECAY TURTLE - a computer program for simulating charged particle beam transport systems, including decay calculations. CERN-74-2 (1974)

5.17 D.R. Douglas, A. Dragt: IEEE Trans. NS-**28**, 2522 (1981)

5.18 A. Wrulich: Proc. Workshop on accelerator orbit and particle tracking programs. Brookhaven National Laboratory, Rept. BNL-31761, 26 (1982)

5.19 PATPET is a combination of the programs PATRICIA and PETROS. The program PETROS allows to study the effect of errors and has been developed by Kewish and Steffen [5.20]. The combination of both programs was performed by Emery, Safranek and H. Wiedemann: Int. Note SSRL ACD-36, Stanford (1988)

5.20 J. Kewisch, K.G. Steffen: Int. Rept. DESY PET 76/09 (1976)

5.21 F.T. Cole: Nonlinear transformations in action-angle variables. Int. Note, Fermilab TM-179, 2040 (1969)

Chapter 6

6.1   F.T. Cole: Longitudinal motion in circular accelerators. *AIP Conf. Proc.* **153**, 44 (American Institute of Physics, New York 1987)

6.2   W.T. Weng: Fundamentals - longitudinal motion. *AIP Conf. Proc.* **184**, 4243 (American Institute of Physics, New York 1989)

6.3   H. Wiedemann: *Particle Accelerator Physics I* (Springer, Berlin, Heidelberg 1993)

6.4   J.D. Jackson: *Classical Electrodynamics* (Wiley, New York, 1975)

6.5   W.H.K. Panofsky, M. Phillips: *Classical Electricity and Magnetism* (Addison-Wesley, Reading, MA 1962)

6.6   A.B. Baden Fuller: *Microwaves* (Pergamon, Oxford 1969)

6.7   *Waveguide Handbook* ed. by N. Marcuvitz (McGraw-Hill, New York, 1951)

6.8   S. Ramo, J.R. Whinnery, T. van Duzer: *Fields and Waves in Communication Electronics* (Wiley, New York, 1984)

6.9   M. Abramowitz, I. Stegun: *Handbook of Mathematical Functions* (Dover, New York 1972)

6.10   I.S. Gradshteyn, I.M. Ryzhik: *Table of Integrals, Series, and Products* (Academic, New York 1965)

6.11   K. Halbach, F. Holsinger: Particle Accelerators **7**, 213 (1976)

6.12   T. Weiland: Nucl. Instrum. Methods **212**, 13 (1983)

6.13   M.A. Allen, L.G. Karvonen, J.L. Pellegrin, P.B. Wilson: IEEE Trans. NS-**24**, 1780 (1977)

6.14   A. Piwinski: *Proc. 9th Int'l Conf. on High Energy Accelerators*, Stanford (1974) p.405

6.15   J.D. Bjorken, S.K. Mtingwa: Particle Accelerators **13**, 115 (1983)

6.16   M. Borland, M.C. Green, R.H. Miller, L.V. Nelson, E. Tanabe, J.N. Weaver, H. Wiedemann: Proc. Linear Accelerator Conf., Albuquerque, NM (1990); SLAC-PUB-5333 (1990)

6.17   P.B. Wilson: Transient beam loading in electron positron storage rings. SLAC Int. PEP Note 276 (1978); or CERN Int. Note ISR-TH/78-23 (1978)

6.18   P.L. Morton, V.K. Neil: The interaction of a ring of charge passing through a cylindrical rf cavity. Proc. Symp. on Electron Ring Accelerators, LBL Rept. UCRL-18103, 365 (1968)

6.19   E. Keil, W. Schnell, B. Zotter: CERN Rept. CERN-ISR-LTD/76-22 (1976)

6.20   K.W. Robinson: Stability of beam in radiofrequency system, Cambridge Electron Accelerator Int. Reports CEA-11 (1956) and CEAL-1010 (1964)

6.21   A.W. Chao, J. Gareyte: SLAC Int. Note SPEAR-197 (1976)

6.22   A. Hofmann: Frontiers of particle beams. *Lect. Notes Phys.* **296**, 99 (Springer, Berlin, Heidelberg 1986)

6.23   R. Helm: Discussion of focusing requirements for the Stanford two-mile accelerator. Int. Note SLAC-2 (August 1962)

6.24   H. Wiedemann: Strong focusing in linear accelerators. DESY Rept. 68/5 (1968)

Chapter 7

7.1   H. Wiedemann: *Particle Accelerator Physics I* (Springer, Berlin, Heidelberg 1993)

7.2   J. Larmor: Philos. Mag. **44**, 503 (1897)

7.3   J.S. Schwinger: Proc. Nat. Acad. of Sci. USA **40**, 132 (1954)

7.4   M. Sands: In *Physics with Intersecting Storage Rings*, ed. by B. Touschek (Academic, New York 1971)

7.5   R. Coisson: Opt. Commun. **22**, 135 (1977)

7.6   R. Coisson: Nucl. Instrum. Methods **143**, 241 (1977)

7.7   R. Bossart, J. Bosser, L. Burnod, R. Coisson, E. D'Amico, A. Hofmann, J. Mann: Nucl. Instrum. Methods **164**, 275 (1979)

7.8   R. Bossart, J. Bosser, L. Burnod, E. D'Amico, G. Ferioli, J. Mann, F. Meot: Proton beam profile measurements with synchrotron light. Nucl. Instrum. Methods **184**, 349 (1981)

7.9   J.D. Jackson: *Classical Electrodynamics* (Wiley, New York 1975)

7.10   A.A. Sokolov, I.M. Ternov: *Synchrotron Radiation* (Pergamon, Oxford 1968)

7.11   L.I. Schiff: Rev. Sci. Instrum. **17**, 6 (1946)

7.12   J.S. Schwinger: On the classical radiation of accelerated electrons. Phys. Rev. **75**, 1912 (1949)

7.13   G.N. Watson: *Bessel Functions* (Macmillan, New York 1945) p.188

7.14 V.O. Kostroun: Nucl. Instrum. Methods **172**, 371 (1980)
7.15 G.A. Schott: Ann. Physik **24**, 635 (1907)
7.16 G.A. Schott: Phil. Mag.[6] **13**, 194 (1907)
7.17 G.A. Schott: *Electromagnetic Radiation* (Cambridge Univ. Press, Cambridge 1912)
7.18 I.S. Gradshteyn, I.M. Ryzhik: *Table of Integrals, Series, and Products* (Academic, New York 1965)
7.19 V.N. Baier: Radiative polarization of electrons in storage rings, in *Physics with Intersecting Storage Rings*, ed. by B. Touschek (Academic, New York 1971)
7.20 M. Abramowitz, I. Stegun: *Handbook of Mathematical Functions* (Dover, New York 1972)
7.21 A. Hofmann: *Theory of Synchrotron Radiation* (Cambridge Univ. Press, Cambridge 1995)
7.22 D. Ivanenko, A.A. Sokolov: DAN (USSR) **59**, 1551 (1972)
7.23 The large hadron collider in the LEP tunnel, ed. by G. Brianti, K. Hübner, CERN 87-05 (1985)

Chapter 8

8.1 H. Wiedemann: *Particle Accelerator Physics I* (Springer, Berlin, Heidelberg 1993)
8.2 K. W. Robinson: Phys. Rev. **111**, 373 (1958)
8.3 H. Risken: *The Fokker-Planck Equation* (Springer, Berlin, Heidelberg 1989)
8.4 I.S. Gradshteyn, I.M. Ryzhik: *Tables of Integrals, Series and Products* (Academic, New York 1965)
8.5 A.W. Chao: In *Frontiers of Particle Beams*, ed. by M. Month, S. Turner, Lect. Notes Phys., Vol.296 (Springer, Berlin, Heidelberg 1988) p.51
8.6 Y.H. Chin: Quantum lifetime. DESY Rept. DESY 87-062 (1987)

Chapter 9

9.1 H. Wiedemann: Basic lattice for the SLC. SLAC Int. Note AATF/79/7 (September 1979)
9.2 M. Sands: In *Physics with Intersecting Storage Rings*, ed. by B. Touschek (Academic, New York 1971)
9.3 H. Wiedemann: *Particle Accelerator Physics I* (Springer, Berlin, Heidelberg 1993)
9.4 H. Weyl: *The Classical Groups* (Princeton Univ. Press, Princeton, NJ 1946)
9.5 S.P. Kapitza, V.M. Melekhin: *The Microtron* (Harwood, London 1978)
9.6 H. Wiedemann, P. Kung, H.C. Lihn: Nucl. Instrum. Methods A **319**, 1 (1992)
9.7 E. Tanabe, M. Borland, M.C. Green, R.H. Miller, L.V. Nelson, J.N. Weaver, H. Wiedemann: 14th Meeting on Linear Accelerators, Nara, Japan (1989)
9.8 M. Borland, M.C. Green, R.H. Miller, L.V. Nelson, E. Tanabe, J.N. Weaver, H. Wiedemann: Linear Accelerator Conf., Albuquerque, NM (1990)
9.9 H.A. Enge: Rev. Sci. Instrum. **34**, 385 (1963)
9.10 M. Borland: High brightness thermionic microwave electron gun. Ph.D. Thesis, Stanford University (1991) and SLAC Rept. 402 (1991)
9.11 P. Kung, H.C. Lihn, H. Wiedemann, D. Bocek: Phys. Rev. Lett. **73**, 967 (1994)
9.12 D.C. Schultz, J. Clendenin, J. Frisch, E. Hoyt, L. Klaisner, M. Woods, D. Wright, M. Zolotorev: *Proc. 3rd Europ. Particle Accelerator Conf.*, ed. by H. Henke, H. Homeyer, Ch. Petit-Jean-Genaz (Editions Frontiere, Gif-sur Yvette 1992) p.1029

9.13 Ya. S. Derbenev, A. M. Kondratenko: Sov. Phys. - JETP **37**, 968 (1973)
9.14 A. Sokolov, I.M. Ternov: *Synchrotron Radiation* (Pergamon, Oxford 1966)
9.15 V.N. Baier: *Proc. Int'l School of Physics*, ed. by B. Touschek (Academic, New York, 1971) Course XLVI
9.16 A.W. Chao: *Physics of High Energy Particle Accelerators*, ed. by R.A. Carrigan, F.R. Huson, M. Month, AIP Conf. Proc. No.87 (American Institute of Physics, New York 1982) p.395

Chapter 10

10.1 W. Schottky: Ann. Physik **57**, 541 (1918)
10.2 J. Borer, P. Bramham, H.G. Hereward, K. Hübner, W. Schnell, L. Thorndahl: 9th Int'l Conf. on High Energy Accel., Stanford (1974) p.53
10.3 H.G. Hereward: CERN Rept. CERN 77-13, 281 (1977)
10.4 W. Schnell: CERN Rept. CERN 77-13, 290 (1977)
10.5 J. Gareyte: *AIP Conf. Proc.* **184**, 343 (American Institute of Physics, New York 1989)
10.6 D. Boussard: CERN Rept. CERN 87-03, Vol II, 416 (1987)
10.7 S. Van der Meer: CERN Rept. CERN-ISR-PO/72-31 (1972)
10.8 CAS CERN accelerator school "Antiprotons for Colliding Beam Facilities". CERN Rept. CERN-84-15 (1984)
10.9 C. Bernardini G.F. Corazza, G. DiGiugno, G. Ghigo, J. Haissinski, P. Marin, R. Querzoli, B. Touschek: Phys. Rev. Lett. **10**, 407 (1963)
10.10 B. Touschek: 1963 summer study on storage rings, accelerators and experiments at super high energies (Brookhaven). BNL Rept. 7534, 171 (1963)
10.11 H. Bruck: *Accélérateurs Circulaires de Particules* (Presses Universitaires de France, Paris 1966)
10.12 U. Völkel: Particle loss by Touschek effect in a storage ring. DESY Rept. DESY 67-5 (1967)
10.13 J. LeDuff: CERN Rept. CERN 89-01, 114 (1989)
10.14 A. Piwinski: Proc. 9th Int'l Conf. on High Energy Accelerators, Stanford (1974) p.405
10.15 A. Piwinski: CERN Accelerator School, Gif-sur-Yvette, Paris (1984). CERN 85-19 (1985)
10.16 J.D. Bjorken, S.K. Mtingwa: Particle Accelerator **13**, 115 (1983)
10.17 H. Wiedemann: *Particle Accelerator Physics I* (Springer, Berlin, Heidelberg 1993)
10.18 L.J. Laslett: On intensity limitations imposed by transverse space-charge effects. Rept. BNL-7534, Brookhaven National Laboratory, Upton (1963)
10.19 D.W. Kerst: Phys. Rev. **60**, 47 (1941)
10.20 J.P. Blewett: Phys.Rev. **69**, 87 (1946)
10.21 M. Reiser: *Theory and Design of Charged Particle Beams* (Wiley, New York 1994)
10.22 W.T. Weng: Space charge effects - tune shifts and resonances, in *Physics of Particle Accelerators*, ed by M. Month, AIP Conf. Proc., Vol.154 (American Institute of Physics, New York 1987)
10.23 J. P. Delahaye G. Gelato, L. Magnani, G. Nassibian, F. Pedersen, K.H. Reich, K. Schindl, H. Schönauer: In *Proc. 11th Int'l Conf. on High Energy Accelerators*, ed. by W.S. Neumann (Birkhäuser, Basel 1980) p.299
10.24 E. Raka, L. Ahrens, W. Frey, E. Gill, J.W. Glenn, R. Sanders, W.T. Weng: IEEE Trans. NS-**32**, 3110 (1985)
10.25 A.W. Chao: *Physics of Collective Beam Instabilities in High Energy Accelerators* (Wiley, New York 1993)

10.26 J.L. Laclare: CERN 87-03, 264 (1987)

10.27 F. Sacherer: Proc. 9th Int'l Conf. on High Energy Accelerators (Stanford). SLAC (1974) p.347

10.28 B. Zotter, F. Sacherer: CERN Rept. 77-13, 175 (1977)

10.29 P.B. Wilson: Introduction to wake fields and wake functions. SLAC PUB-4547 (January 1989)

10.30 P.B. Wilson, R.V. Servranckx, A.P. Sabersky, J. Gareyte, G.E. Fischer, A.W. Chao: IEEE Trans. NS-24, 1211 (1977)

10.31 W.K.H. Panofsky, W.A. Wenzel: Rev. Sci. Instrum. 27, 967 (1956)

10.32 J.D. Jackson: Classical Electrodynamics (Wiley, New York 1975)

10.33 W. Schnell: CERN Rept. ISR-RF/70-7 (1970)

10.34 A.W. Chao, J. Gareyte: SLAC Int. Note SPEAR-197 (1976)

10.35 C.E. Nielsen, A.M. Sessler, K.R. Symon: Proc. Int'l Conf. on High Energy Accelerators (Geneva) (1959) p.239

10.36 V.K. Neil, A.M. Sessler: Rev. Sci. Instrum. 36, 429 (1965)

10.37 L.J. Laslett, V.K. Neil, A.M. Sessler: Rev. Sci. Instrum. 36, 436 (1965)

10.38 K. Hübner: CERN Rept. ISR-TH/70-44 (1970)

10.39 A.G. Ruggiero, V.G.Vacarro: CERN Rept. CERN ISR-TH/68-33 (1968)

10.40 K. Hübner, A.G. Ruggiero, V.G. Vacarro: 8th Int'l Conf. on High Energy Accelerators, Yerevan (1969)

10.41 K. Hübner, A.G. Ruggiero, V.G. Vaccaro: CERN Rept. CERN ISR-TH-RF/69-23 (1969)

K. Hübner, P. Strolin, V.G. Vaccaro, B. Zotter: CERN Rept. CERN ISR-RF-TH/70-2 (1970)

K. Hübner, V.G. Vaccaro: CERN Rept. CERN ISR-TH/70-44 (1970)

10.42 B. Zotter, F. Sacherer: Rept. CERN 77-13 (1977)

10.43 J. Landau: J. Phys. USSR 10, 25 (1946)

10.44 E. Keil, W. Schnell: CERN Rept. ISR-TH-RF/69-48 (1969)

10.45 H.G. Hereward: CERN Rept. CERN-65-20 (1965)

10.46 A. Hofmann: Lect. Notes Phys. 296, 112 (Springer, Berlin, Heidelberg 1988)

10.47 L.R. Evans: AIP Conf. Proc. (ed. by M. Month, F. Dahl, M. Dienes) 127, 243 (American Institute of Physics, New York 1985)

10.48 C. Pellegrini, A.M. Sessler: Nuovo Cimento A 3, 116 (1971)

10.49 F. Sacherer: CERN Rept. SI-BR/72-5 (1972)

10.50 J. Haissinski: Nuovo Cimento B 18, 72 (1973)

10.51 D. Boussard: Int. CERN Rept. Lab II/RF/Int 75-2 (1975)

10.52 V. Balakin, A. Novokhatsky, V. Smirnov: Proc. 12th Int'l Conf. on High Energy Accelerators, FNAL (1983) p.119

10.53 J.T. Seemann: Frontiers of particle beams: Intensity limitations. Lect. Notes Phys. 400, 255 (Springer, Berlin, Heidelberg 1992)

10.54 D. Kohaupt: IEEE Trans. NS-26, 3480 (1979)

10.55 C. Pellegrini: Nuovo Cimento A 64, 447 (1969)

10.56 M. Sands: SLAC Int. Repts. TN-69-8 and TN-69-10 (1969)

10.57 J.M. Paterson, B. Richter, A.P. Sabersky, H. Wiedemann, P.B. Wilson, M.A. Allen, J.-E. Augustin, G.E. Fischer, R.H. Helm, M.J. Lee, M. Matera, P. L. Morton: Proc. 9th Int'l Conf. on High Energy Accelerators, SLAC (1974) p.338

10.58 J. Gareyte, F. Sacherer: Proc. 9th Int'l Conf. on High Energy Accelerators, SLAC (1974) p.341

10.59 R.D. Kohaupt: Excitation of a transverse instability by parasitic cavity modes. Int. Note DESY H1-74/2 (May 1972)

10.60 J. Le Duff, J. Maidment, E. Däskowski, D. Degèle, H.D. Dehne, H. Gerke, D. Heins, K. Hoffmann, K. Holm, E. Jandt, R.D. Kohaupt, J. Kouptsidis, F. Krafft, N. Lehnart, G. Mülhaupt, H. Nesemann, S. Pätzold, H. Pingel, A. Piwinski, R. Rossmanith, H.D. Schulz, K.G. Steffen, H. Wiedemann, K. Wille, A. Wrulich: Proc. 9th Int'l Conf. on High Energy Accelerators, SLAC (1974) p.43

10.61 A. Massarotti, M. Svandrlik: Particle Accelerators 35, 167 (1991)

10.62 F. Voelker, G. Lambertson, R. Rimmer: Conf on Particle Accelerators, San Francisco, IEEE Cat. No. 91CH3038-7 (1991) p.687

10.63 R. Rimmer, F. Voelker, G. Lambertson, M. Allen, J. Hodgeson, K. Ko, R. Pendleton, H. Schwartz: Conf. on Particle Accelerators, San Francisco, IEEE Cat.No. 91CH3038-7 (1991) p.801

10.64 S. Bartalucci, R. Boni, A. Gallo, L. Palumbo, R. Parodi, M. Serio, B. Spataro, G. Vignola: *Proc. 3rd Europ. Conf on Part. Accel.*, Berlin, ed. by H. Henke, H. Homeyer, Ch. Petit-Jean-Genaz (Edition Frontiere, Gif-sur-Yvette 1992)

10.65 M. Svandrlik, G. D'Auria, A. Fabris, A. Massarotti, C. Pasotti, C. Rossi: A pillbox resonator with very strong suppression of the H.O.M. spectrum. Sincrotrone Trieste Int. Rept. ST/M-92/14 (1992)

Chapter 11

11.1 H. Motz: J. Appl. Phys. 22, 527 (1951)

11.2 H. Wiedemann: *Particle Accelerator Physics I* (Springer, Berlin, Heidelberg 1993)

11.3 B.M. Kincaid: J. Appl. Phys. 48, 2684 (1977)

11.4 D.F. Alferov, Y.A. Bashmakov, E.G. Bessonov: Sov. Phys. - Tech. Phys. 18, 1336 (1974)

11.5 S. Krinsky: Nucl. Instrum. Methods 172, 73 (1979) or IEEE Trans. NS-30, 3078 (1983)

11.6 W.M. Lavender: Observation and analysis of x-ray undulator radiation from PEP. Ph.D. thesis, Stanford University (1988)

11.7 A. Bienenstock, G. Brown, H. Wiedemann, H. Winick: Rev. Sci. Instrum. 60, 1393 (1989)

11.8 P. Elleaume: Nucl. Instrum. Methods A 291, 371 (1990)

11.9 B. Diviacco, R.P. Walker: Nucl. Instrum. Methods A 292, 517 (1990)

11.10 S. Sasaki, K. Kakuno, T. Takada, T. Shimada, K. Yanagida, Y. Miyahara: Nucl. Instrum. Methods A 331, 763 (1993)

11.11 R. Carr, S. Lidia: The adjustable phase planar helical undulator. SPIE Proc. 2013 (SPIE, Bellingham, WA 1993)

11.12 R. Carr: Nucl. Instrum. Meth. A 306, 391 (1991)

Appendix

A.1 H. Wiedemann: *Particle Accelerator Physics I* (Springer, Berlin, Heidelberg 1993)

Author Index

Subject Index

$C_\perp$ 321
$C_\parallel$ 322
C_γ 239
C_ω 244
$C_\perp$ 321
C_{ph} 255
$-\mathcal{I}$-transformation 140
$\lambda/4$-lens 77, 85
π-mode 250, 422, 429
σ-mode 250, 422, 429
ϑ-parameter 283

aberration 136
– in quadrupoles 56
– chromatic 127, 147
– geometric 127, 137
accelerating cavities 167
– voltage 191
acceleration and FODO focusing 220
ACOL 146
action 1
action-angle variables 10, 13
– for oscillator 10
AdA (Anello di Accumulatione) 330
adiabatic damping 226, 272, 281, 282, 305
ADONE 398
Airy's functions 260
alignment error, statistical 96
alpha magnet 316
Ampere's law 175
amplitude dependent tune shift 160
angular distribution 256
aperture 43, 66
aperture, beam lifetime 298
AS(x.y.z), see footnote 256
astigmatism 55
asymmetric wiggler magnet 431

beam break up 390, 392
beam current spectrum 200, 345
beam dynamics, with acceleration 222
beam emittance, horizontal 309
– vertical 310

beam lifetime, quantum 298
beam loading 181, 187, 192
beam matrix 91
beam path correction 95
beam size, perturbation 122
beam-beam effect 24, 39
– interaction 342
– tune shift 342
beat factor 18, 19
Bessel's differential equation 223
Bessel's function, modified 260
beta beat 147
betatron oscillation amplitude 129
betatron oscillations, frequency spectrum 151
Biot Savart fields 235
BNS damping 392
Boussard criterion 389
brightness, spectral 428
broadband impedances 390
Brookhaven National Laboratory, BNL 12
bunch compression 314
bunch length 298, 311
bunch lengthening, potential well distortion 207
bunch - bunch instabilities 390
bunching factor, Laslett 335

canonical coordinates 5
– momentum 4
– variables 5, 8
– transformation 6, 8, 152
– –, generating functions 7
– –, verify 9
capacitive detuning 186, 200
Cauchy Riemann equations 68
Cauchy's residue theorem 68, 373, 379
cavity admittance 185
– damping time 183
– filling time 179
– impedance 204
– losses 175, 176, 177
– shunt impedance 184, 363